Practical Aspects of Waterflooding

Practical Aspects of Waterflooding

Steven W. Poston
Marcelo Laprea-Bigott

Society of Petroleum Engineers

Printed in the United States of America.

Disclaimer

This book was prepared by members of the Society of Petroleum Engineers and their well-qualified colleagues from material published in the recognized technical literature and from their own individual experience and expertise. While the material presented is believed to be based on sound technical knowledge, neither the Society of Petroleum Engineers nor any of the authors or editors herein provide a warranty either expressed or implied in its application. Correspondingly, the discussion of materials, methods, or techniques that may be covered by patents implies no freedom to use such materials, methods, or techniques without permission through appropriate licensing. Nothing described within this book should be construed to lessen the need to apply sound engineering judgment nor to carefully apply accepted engineering practices in the design, implementation, or application of the techniques described herein.

ISBN 978-1-959025-00-9

First Printing 2023

Society of Petroleum Engineers
http://www.spe.org
service@spe.org
1.972.952.9393

Dedication

Marcelo and I would like to dedicate this book to our parents, Richard and Lauretta Poston and Isabel Teresa Bigott de Laprea and Marcelo Laprea Finamore, who made this work possible by imparting and fostering their intellectual activity and energy in their progeny.

In particular, I wish to thank my parents for allowing me to stumble along in life to finally find my goals.

—Steve Poston

I would like to add to Steve's words our deep appreciation for the support given by my wife, Mariela, and our children, Adrianna, Marcelo Eduardo, and Daniel during the many hours of work on this book. Special memories of Chloe Poston, who although unseen and unheard has always been near Steve.

—Marcelo Laprea-Bigott

Introduction

The depletion of an oil reservoir should be designed around four considerations:

- Maintain reservoir pressure as high as possible for dissolved gas to remain in solution. The evolution of solution gas increases oil viscosity, which decreases the relative permeability to oil.
- Updip high gas/oil ratio wells should not be allowed to continue producing excess gas from an oil reservoir. An excessive loss of free gas depletes reservoir energy and unnaturally accelerates the loss of reservoir pressure.
- The evolution of solution gas increases the density of the residual oil, and as a result, more stock-tank barrels of oil are lost at abandonment.
- Maintaining reservoir pressure reduces the potential need for installing artificial-lift equipment.

Oil recovery is highest when the oil is displaced by water or gas invasion. Gas reservoirs, as a result of their high compressibilities, realize the highest recovery factors when pressure depletions exist. Gas expansion is the primary driving force in this case.

Oil and gas reservoirs should be depleted as soon as is practical, to maximize investment and take advantage of the time value of money.

Water is often injected into an oil reservoir to physically displace an otherwise immovable oil toward a producing well to prevent further pressure loss. Water injection increases oil recovery by means of gravity segregation, physical displacement, or a combination of the two.

Density differences between gas, oil, and water often cause a rearrangement of reservoir fluids during depletion. The effectiveness of the time-dependent gravity segregation process is a function of the sand permeability and thickness, oil viscosity, and formation dip. Gravity segregation can provide an effective recovery mechanism in homogeneous, good-quality reservoirs.

Waterflooding is the primary method to enhance oil recovery over and above normal depletion operations. Drilling development and outpost wells define the primary emphasis in the original flow system. Field depletion is oriented to oil migrating toward producing wells, with reservoir pressure providing an incentive for oil movement. However, injecting water into a reservoir reorients the fluid flow normally induced by reservoir pressure decline into a movement caused by higher-pressure sources installed at the water-injection wells. The depletion system changes from global pressure loss → producer into one of local injection → producer.

Waterflooding a reservoir is an organic process. The production response typically changes with maturity. Each facet of the project must function effectively for the successful completion of the waterflood operation. Startup costs approach that of the initial field development but without the resulting significant production increase during primary depletion. Initial project planning must not be based solely on geological and economic fundamentals; potential unanticipated factors that can influence the introduction of foreign waters into the reservoir, altering the recovery process, must also be considered. A waterflood is not a set piece but requires continual updates through the observation of changes in the dynamic production and pressure records.

Waterflooding increases the ultimate recovery but also increases the cost structure and complexity of the recovery process. Production now depends on economic variables such as well connectivity, using the proper well locations, treating and surface equipment, reactions to the introduction of foreign fluids, water source, and responses to unforeseen events.

The great majority of published engineering studies concern the study of static information. These studies often manipulate a series of mathematical relationships and show the results from a known database. Unfortunately, microscopic geological variations often define the flow relationships within a reservoir. The effectiveness and special relationships of these properties are essentially impossible to quantify in most instances.

Reservoir simulation accounts for changing rock and fluid properties as a function of the intrafield location. Viability is ensured by a history match of individual-well production after initialization to reflect changes in the reservoir flow system. History-matching techniques are numerous and often produce questionable results because of unforeseen geological relationships. Averaging techniques allow matching to run more efficiently but do not necessarily replicate actual subsurface conditions. The sample size is miniscule compared to field or reservoir volume.

An efficient waterflood is typically the result of a thorough and practical initial study as well as active monitoring of well performance after the installation and initiation of the waterflood. Compensating for unforeseen problems is a must. This book applies well-known geologic and engineering techniques to study a candidate reservoir. Replicating the results derived from different investigative techniques allows one to draw conclusions about the "goodness" of the estimate.

This book explains the screening and analysis fundamentals for applying the waterflood displacement process to increase oil reserves. The information used in Part 3, Field Study, is typically available for normal field sites (i.e., logs, core analysis, pressure production histories, and geological interpretation). All calculations can be performed on a calculator or using standard computer software.

The data for study are typically available at the operations level and include

- Structure and isopach maps and type sections
- Some rock properties and core and fluid analysis
- Current well-test information and well and reservoir performance histories
- Well pressure histories

As shown in **Fig. 1,** the book is divided into four parts.

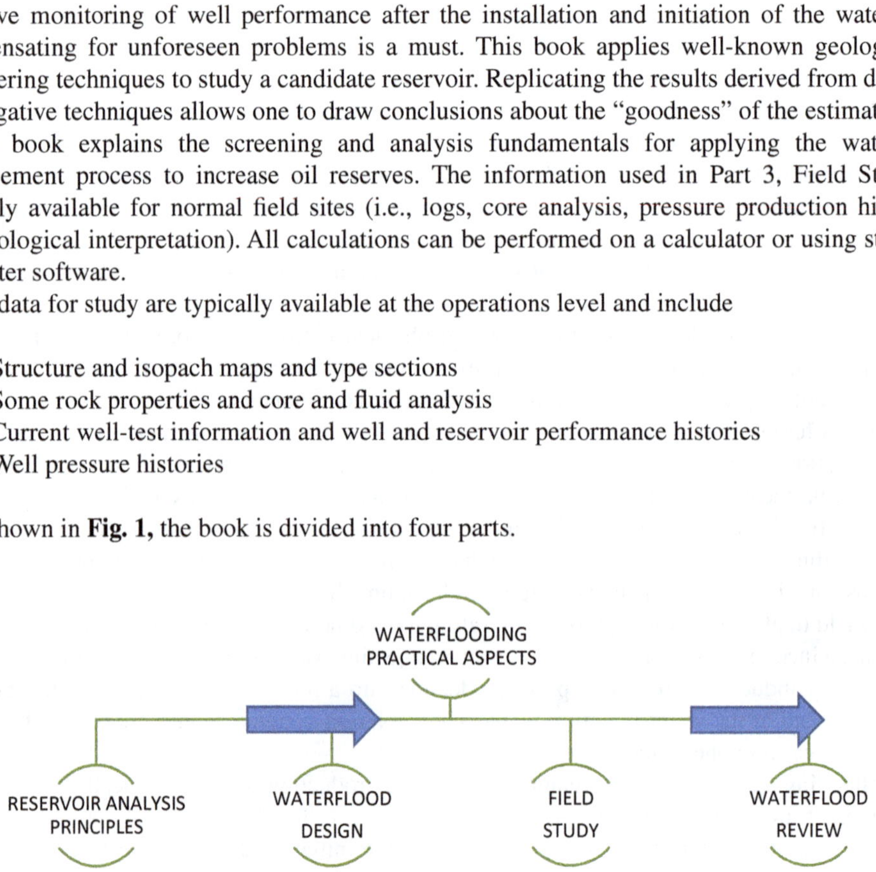

Fig. 1—Flow chart for *Practical Aspects of Waterflooding.*

The first part, Reservoir Analysis Principles, provides a general introduction to and a review of basic reservoir analysis techniques involved in a field review. Data gathering and reservoir analysis methods are explained and illustrated with examples and work problems. All calculation procedures are basic and designed for application at the operational level.

The second part, Waterflood Design, reviews the geological and well producing conditions amenable to the waterflood process. The principles of physical-plant needs and possible changes in well patterns are discussed.

The third part, Field Study, discusses an actual case history of a sandstone oil reservoir. Commonly available surface and subsurface information, typically stored in the field files, should be analyzed by the reader to evaluate the remaining potential of this essentially exhausted reservoir. An accompanying study guide consisting of 23 assigned tasks guides the reader along the analysis process, with proposals for maximizing recovery included. These field study proposals include projections for installing a waterflood and the expected water-injection pressures and rates, forecast oil and gas production rates, and new drills and workovers.

Along the way, decision points are included to consider the feasibility of continuing to a conclusion or abandoning the effort. By the end of the discussion, the reader should understand the basics of applying a waterflood process to a field situation.

In the fourth part, Waterflood Review, the importance of monitoring waterflood performance is emphasized. Because subsurface flow patterns often change as a function of operating time, constant well-performance monitoring is of primary interest to ensure that a waterflood project operates efficiently.

The ultimate goal of *Practical Aspects of Waterflooding* is to acquaint the reader with the waterflood process and provide some knowledge regarding potential problems and the need for an active monitoring program.

Contents

Part 1: Introduction to Reservoir Analysis Principles

Part 1, Reservoir Analysis Principles, provides an overview of the geological and reservoir engineering elements discussed in the second, third, and fourth parts of this book.

Chapter 1, Reservoir Rocks and Framework, reviews the various geological considerations involved when analyzing an ongoing project or when contemplating a new waterflood. Interwell log correlations typically provide the starting point for determining the reservoir architecture and flow system. These studies can result in a high-quality model (providing a recognizable flow system) to a nondescript model (that is so heterogeneous that well correlation is qualitative at best). Optimizing the injector/producer density and subsurface flow units is always a primary consideration.

Chapters 2 and 3 discuss rock and fluid properties. Pore volume is usually known, but pore-size distribution can be very uncertain as a result of a paucity of data. Waterflood reservoirs typically contain a moderate-gravity oil. Problems arise with a high-viscosity oil because relative permeability effects typically result in high water/oil-ratio wells. Depleting a gassy oil causes excessive water injection to create "fill-up" of the gas-filled pore spaces. Nevertheless, fluid properties are typically included in all rate and reserves estimates in the calculation process.

Chapter 4, Well and Reservoir Models, reviews the Darcy and the van Everdingen and Hurst reservoir flow models. Rock, fluid, and reservoir character parameters allow theoretical calculations to be applied to field conditions. The Darcy equation includes rock and fluid characteristics to solve for rate vs. pressure drop for steady-state flow conditions. van Everdingen and Hurst (1949), on the other hand, solve the diffusivity equation for constant inner pressure and constant rate solutions and for various boundary conditions.

Simplified qualitative estimates calculating oil-in-place and reserves are listed and discussed in Chapter 5, Oil Reservoirs. The material-balance equation is a particularly useful tool when studying simple reservoir models. All calculations can be set up on a calculator or spreadsheet.

Chapter 6, Inflow Performance, reviews prediction measures to study the effects of choke size, changing reservoir pressure, the onset of two- and three-phase flow, formation damage, and fracturing on the production rate. Wellhead and downhole chokes control well rates. Flow rates are subject not only to mechanical restrictions but also to changes in well integrity and pressure differentials. Methods for estimating flow rates as a function of changes in the gas/oil ratio, estimating the effects of wellbore damage, and estimating future performance are developed.

Water influx is discussed in Chapter 7. Aquifer-water encroachment caused by reservoir pressure drawdown commonly occurs when producing an oil reservoir. The degree of water influx affecting recovery is often uncertain, and it can vary from being negligible to providing a major reservoir energy source. Applying a producing history to study the water-influx effect provides an additional tool to study reservoir behavior. The van Everdingen and Hurst (1949) solution technique is probably the most widely applied method, but attenuated and much-easier techniques exist and are discussed in this chapter.

Fractional flow, frontal advance, and Stiles idealized reservoir models are discussed in Chapter 8, Displacement Concepts.

The majority of these principles are applied during the course of the reading. However, knowledge of these physical occurrences is important, and they should always be considered when studying subsurface fluid flow although the discussions here represent idealized conditions.

Chapter 1

Reservoir Rocks and Framework

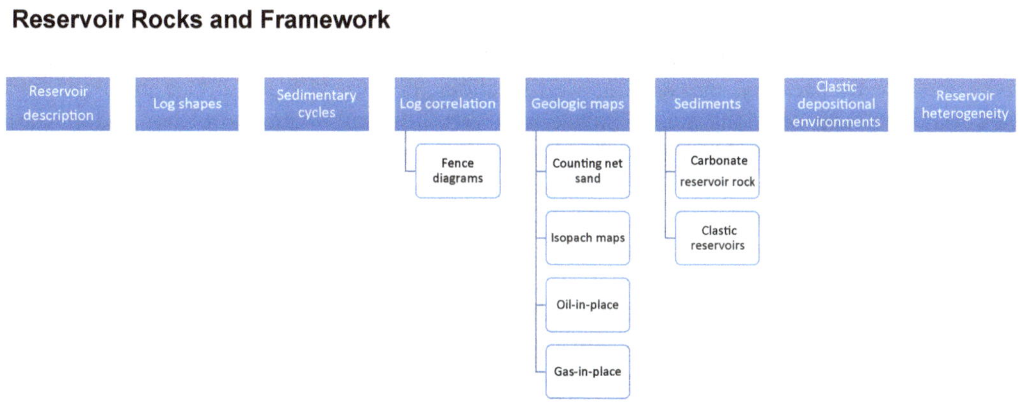

1.1 Reservoir Description

Analyzing the feasibility of installing a waterflood requires a comprehensive vision. Significant segments of interest are reservoir description and analysis, well locations in the context of fluid flow, potential well mechanical issues, a general layout of project development, and the economics describing project viability. These studies require input and analysis from a number of different disciplines, each of which is equally important.

The first chapter concerns the application of both quantitative and qualitative aspects of reservoir architecture. Eventually, these studies are integrated with dynamic data (pressure/volume/temperature, pressure, and production histories) to predict recovery efficiencies as well as optimize reservoir performance. The management of such an enterprise requires a foundation in all of these aspects to understand how a particular facet fits into the picture.

Understanding the elements that confirm the reservoir requires a study of the characteristics and evolution of the Earth and the different changes that have occurred in geologic time. This includes studying the development of petroleum accumulations by investigating rocks and their histories, origins, and diagenesis and determining how to characterize the changes in the properties of those substances occupying the porous media.

Seismic elevated subsurface temperatures and pressures convert abundant organic matter trapped in sedimentary source rock to lighter-than-water hydrocarbon micelles. These long-chained molecular colloids migrate updip through the water-saturated subsurface reservoir rock until they encounter an impermeable barrier rock (such as mudstone or anhydrite), which is the seal rock, through which oil and gas cannot move effectively.

Hydrocarbon accumulation occurs if the overlying surface of a closed, sealed structure acts as a trap of structural and stratigraphic configuration, concentrating oil and gas into an accumulation. The continued accumulation of these hydrocarbons gradually forces the underlying water downward and beyond the spill point, which causes it to migrate to another location. Oil and gas deposits are formed by this oil, gas, and water gravity segregation process.

Thus, several conditions must be satisfied for an economic hydrocarbon accumulation to exist. First, there must be sedimentary rocks possessing good source-rock characteristics and that have reached thermal maturity. Second, the hydrocarbons must have migrated from the source rock to a potential reservoir, which must have adequate porosity (the space between rock grains in which oil and gas accumulate) and permeability (the ease with which oil and gas can move through the pore space between the grains). Finally, there must be a trap to arrest the hydrocarbon migration and to hold sufficient quantities of hydrocarbons to make the recovery prospect economic. Hydrocarbon traps usually consist of an impervious layer (seal), such as shale, above the reservoir and a barrier such as a fault or facies pinch that terminates the reservoir **(Fig. 1.1).**

From Fig. 1.1, it is observed that the nature of the reservoir and the geologic data are unique. Obtaining the data is expensive, and the available data represent only a minimum sample because these data are fragmentary, measured at well locations, or derived from surface manifestations. However, important decisions must be made with very little information available, and all these different types of data must be integrated and cannot be manufactured and are nonrepeatable.

Fig. 1.2 shows a cross section of a petroleum system with hydrocarbon accumulations. Subsurface traps are the result of structural deformations, such as domes or fault traps, or of stratigraphic discontinuities, such as a permeability pinchout that forms the overlying seal. Closed geologic structures can be developed under a variety of conditions and are widespread in the geologic record but usually cover only a small to moderate surface area. On the other hand, stratigraphic traps can be regional in nature. For this reason, the majority of oil and gas fields are found in structural traps, but the largest fields are usually formed by a stratigraphic discontinuity.

Oil and gas reservoirs are usually found in either clastic or carbonate rocks. Different sets of log types identify productivity in carbonate reservoirs. However, carbonates are susceptible to secondary alterations, which can significantly modify the originally deposited sediments. Little work has been performed to relate the present-day formation of carbonate sediments to relic hydrocarbon accumulations. However, knowledge of the order and arrangement of clastic material present in a particular reservoir can furnish clues to the reservoir architecture.

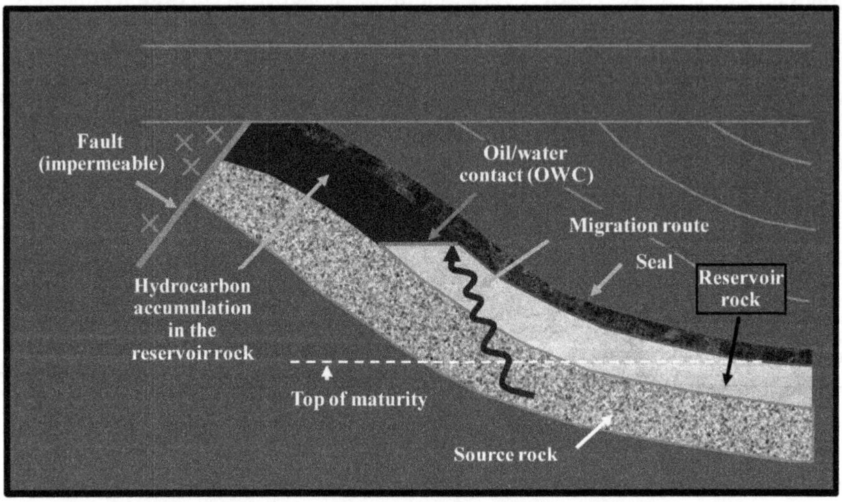

Fig. 1.1—The generation, migration, and trapping of hydrocarbons (after Schlumberger NExT 2007).

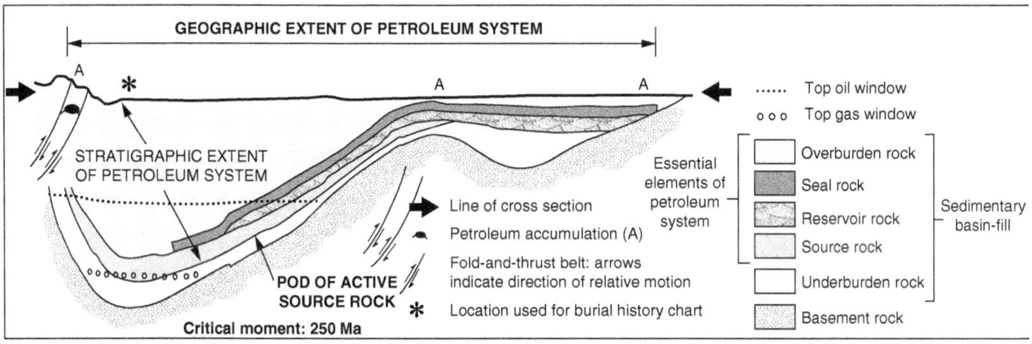

(a)

(b)

Fig. 1.2—Hydrocarbon accumulations—(a) cross section of a petroleum system (after Magoon and Dow 1994) and (b) hydrocarbon migration path (after Schlumberger NExT 2007).

In summary, we can say that carbonate reservoirs often display highly variable porosity and permeability distribution trends resulting from the chemical solution/dissolution activities of groundwater and the presence of silt and shale particles, and this secondary porosity can also be associated with fractures. Carbonate rocks are of two primary types: limestone (calcite is the primary mineral component) and dolomite (dolomite is the primary mineral component). Carbonate rocks are mineralogically much simpler than sandstones, but the mineral components of carbonate rocks are quite reactive chemically. On the other hand, clastic reservoirs are usually laid down in an ordered depositional environment, which can result in discernable and correlative porosity and permeability trends.

Clastic material is classified according to grain size. Listed in decreasing order, it includes boulders, cobbles, pebbles, granules, sandstones (very coarse, coarse, medium, fine, and very fine), siltstones (coarse, medium, fine, and very fine), and shales **(Table 1.1).**

Name	Size (mm)	Size (μm)
Boulder	4096	
Cobble	256	
Pebble	64	
Granule	4	4000
Sandstone—very coarse	2	2000
Sandstone—coarse	1	1000
Sandstone—medium	0.5	500
Sandstone—fine	0.25	250
Sandstone—very fine	0.125	125
Siltstone—coarse	0.063	63
Siltstone—medium	0.031	31
Siltstone—fine	0.016	16
Siltstone—very fine	0.008	8
Shale (clay)	0.004	4

Table 1.1—Grain-size classification for clastic sediments (after LeBlanc 1972).

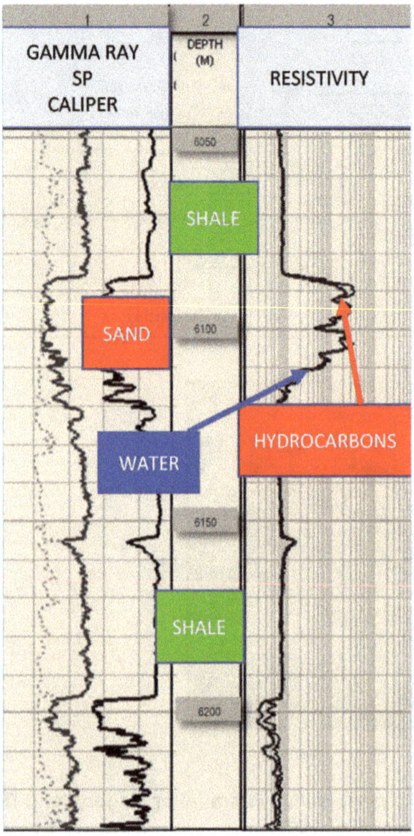

Fig. 1.3—Log profile of a typical sand/shale section (after Schlumberger NExT 2005). Sand sections are evident, with the sand at 6,086–6,120 ft appearing resistive and possibly containing hydrocarbons and increasing water saturations below 6,110–6,120 ft.

Both rock types consist of a pore space containing oil, gas, or water; a rock matrix volume representing the solid structural support; and cementing material gluing together the grain material. The pore-space volume within a reservoir rock can range from < 1 to 35% and can vary according to the original depositional environment and secondary alteration.

The interpretation of reservoir size and its limits are primarily derived from well logs. Spontaneous potential (SP) logs measure the electrical potential across a permeable/impermeable interface while gamma ray (GR) logs measure the radioactivity of the rock material. Because shale is generally more radioactive than sand, log deflections measure the relative fractions of sand- and shale-sized particles located within a particular interval. Large-grained sands are the product of a higher-energy depositional environment than those that form smaller-grained sand bodies. Therefore, the constancy and level of energy present during the deposition of the sedimentary material are reflected in the SP and GR log readings.

1.2 Log Shapes

Log interpretation is the primary source for determining sand continuity, identifying flow channels, and estimating reserves. For clastic sediments, logs basically measure pore space on the left side and fluid type on the right side. **Fig. 1.3** shows a typical log profile through a sand/shale section. Important characteristics for identifying potential pay zones in sand/shale sequences and resistivity in the generally saltwater environment are

- SP and GR logs in the left track
- Resistivity logs in the right track
- The depth in the middle track

Different logging tools relate the shale/sand content, porosity, resistivity, and fluid content as a function of depth. Digitized logs have the ability to collect this information into a single page displaying different properties discussed in the chapter. Other log sets are applied when interpreting carbonate sediments.

Fig. 1.4 shows a computed log section retrieved from a Brunei well. Note that the color-coded GR curve locates zones of differing porosity/permeability because increasing shale volumes imply smaller pore spaces that in turn reduce permeability.

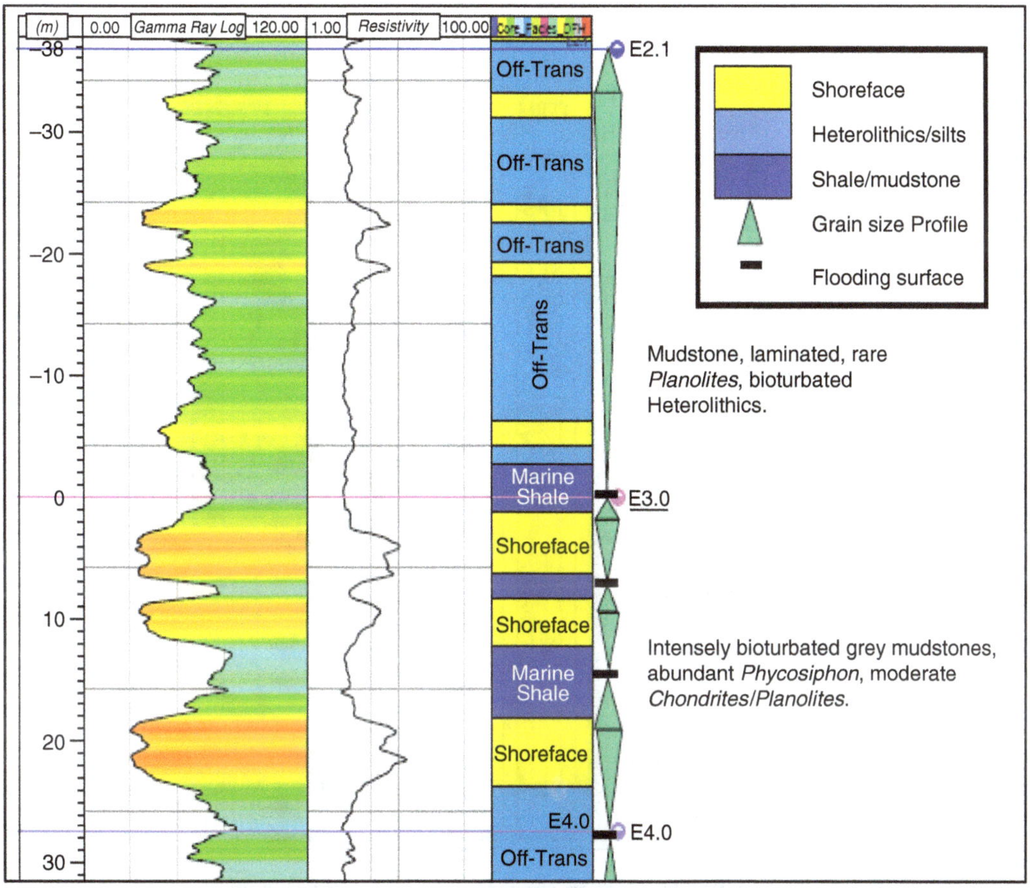

Fig. 1.4—Computed log section retrieved from a Brunei well (after Hadley et al. 2006). Note the triangular diagram shown in the upper right.

The resistivity curve in Fig. 1.4 indicates all sand bodies are loaded with oil or gas. Log and core analysis permitted geologists to relate the depositional environment to the GR curve deflections. A vertical triangle initiates at a low-depositional-energy level (shale), and the base widens to maximum as the sand size increases (increasing energy level). **Fig. 1.5** shows digitized logged sections of wells penetrating a sand/shale section. Digitizing the log information and converting it to a more-legible form greatly enhances readability for the nonexpert. Including the results of geological studies provides the origin of the sediments.

1.3 Sedimentary Cycles

Each change in the sand/shale sequence represents a change in the energy level at the time of deposition, which in reality is a function of water depth. Water transgressing landward becomes deeper with time. On the other hand, a regressive sequence confirms a retreating shoreline. Interpreting this information from well logs coupled with core analysis aids in determining the original depositional environment.

As previously stated, the leftward movement of the GR or SP curve identifies the presence of increasing sand accumulation. The shape of the curve provides an indication of the energy level and

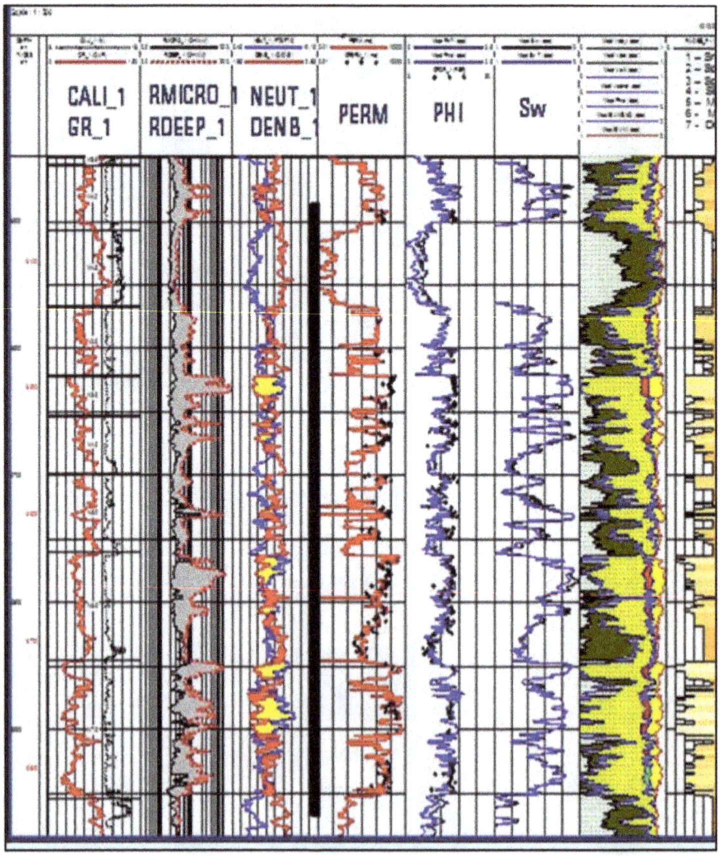

Fig. 1.5—A computed log section showing zones of differing porosity (PHI)/permeability (PERM) with different strata of shale and sand volumes and calculated values of water saturation (Sw) (after Schlumberger NExT 2006).

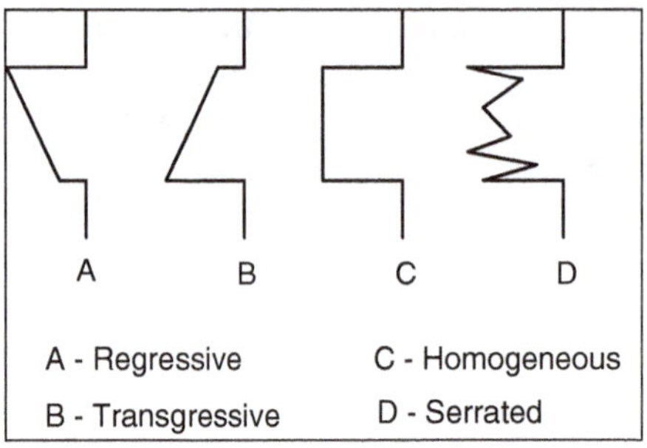

Fig. 1.6—The four major log profiles. A, B, and D reflect changes in energy levels during sand deposition,while a constant energy level is maintained during the formation of profile C. In a log, the rightmost GR or SP deflection is defined as the shale line and the leftmost as the sand line.

the constancy of that level during deposition. Therefore, interpreting the shape of the curve can furnish clues regarding the original depositional conditions. Krueger (1968) identified four basic sand-body types that can be interpreted from electric logs. **Fig. 1.6** reviews the four common shapes.

Directional diagrams representing either regressive, upward-pointing triangles (grain size increasing upward) or transgressive, downward-pointing triangles (grain size decreasing upward) can be applied in the study of sedimentary cycles and correlations. This method is particularly useful when interpreting sedimentary units laid down in a fluvial environment. In **Fig. 1.7,** the base of the triangle denotes the beginning of a cycle and the apex denotes the end of that cycle. The initiation of a cycle usually begins at a major shale break.

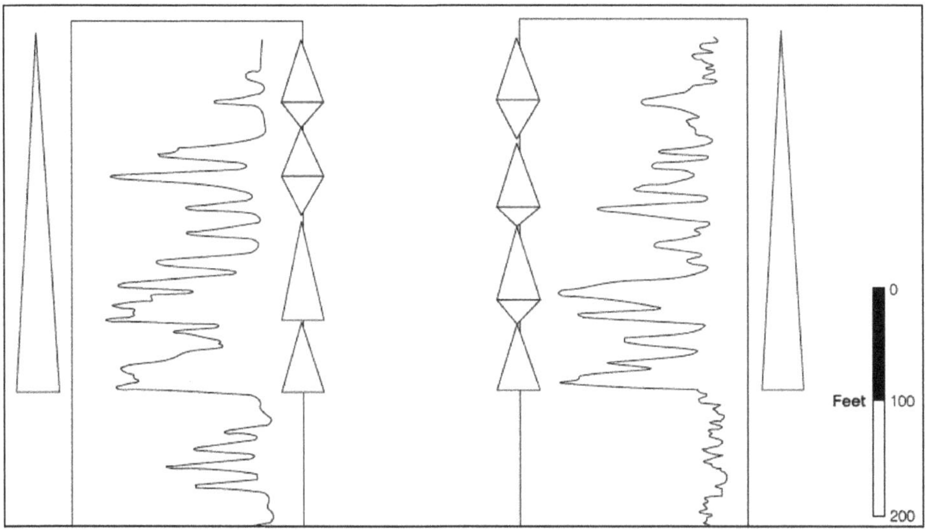

Fig. 1.7—Illustrating the application of the major and minor triangle correlation method to show the overall sedimentary profile of two adjacent Barrancas wells (after Ramón and Cross 1997). Subset sedimentary episodes make up portions of the entire interval. In which well was the depositional energy greatest? How can you tell?

1.4 Log Correlation

Lang (1986) states, "The companion shapes of the resistivity and SP and gamma ray are the principal tools applied when correlating subsurface strata." Tearpock and Bischke (1991) provide an excellent review on log correlation. A priori knowledge of the expected extent of individual flow channels is an important aspect of waterflood design.

A modified version of an example from Tearpock and Bischke (1991) is shown in **Fig. 1.8.** Interwell correlation involves matching common shapes between wells.

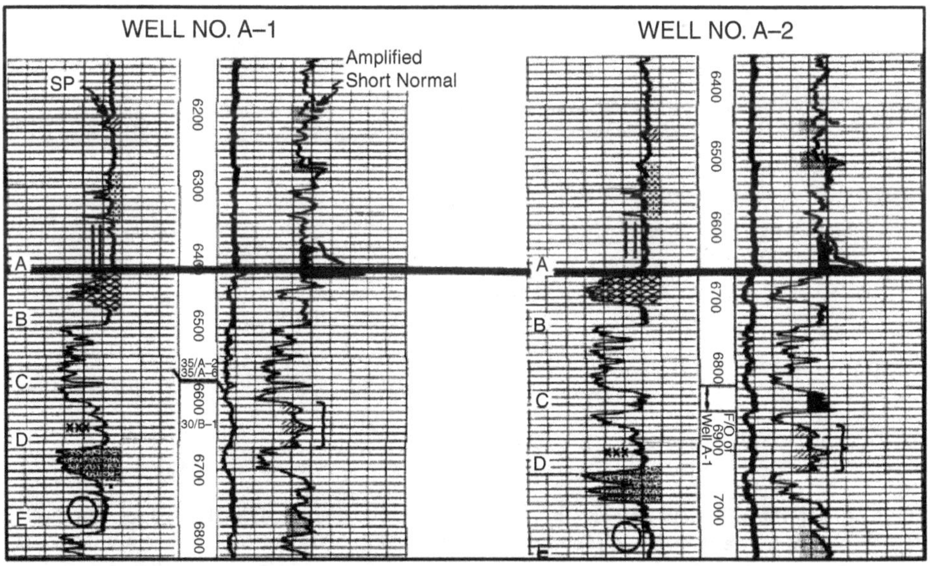

Fig. 1.8—Applying the SP and short normal resistivity curves to correlate between two adjacent wells (after Tearpock and Bischke 1991). The SP and the resistivity tracks display distinctive characteristics that aid in the correlation. Well correlations are not always obvious, and sometimes uncertain, when faults and dramatic changes in stratigraphy occur. Note that marker C is partially faulted out in Well No. A-1.

There is a good correlation between the sandy shale members down to sand member A. The bell-shaped curve in Well No. A-2 indicates a rather well-developed sand. The serrated A sand in Well No. A-1 reflects an overbank deposit with a dramatic reduction in net pay.

Fig. 1.9 presents a well-log correlation, with a stratigraphic cross section. Fig. 1.10 on the left shows a stratigraphic section and a structural cross section on the right. The goal of the structural modeling is to be "anchored" at the well and to guide the large-scale structure and continuity from seismic data. We can then generate a structural cross section to visualize and perform a quality check as shown in **Fig. 1.10**.

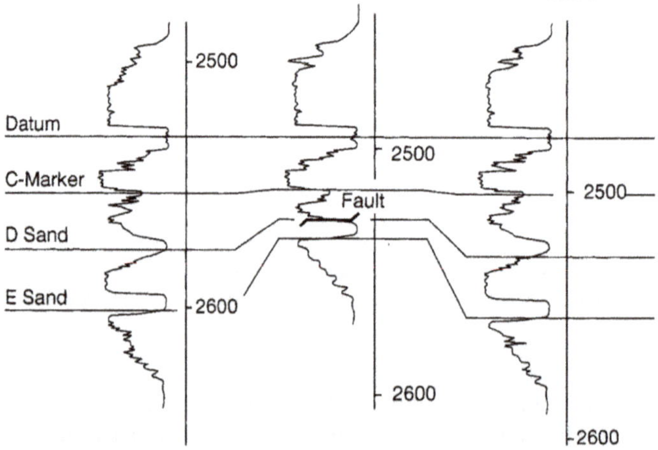

Fig. 1.9—Applying the GR curve to correlate between three adjacent wells (Jahn et al. 1998). The figure shows a stratigraphic cross section.

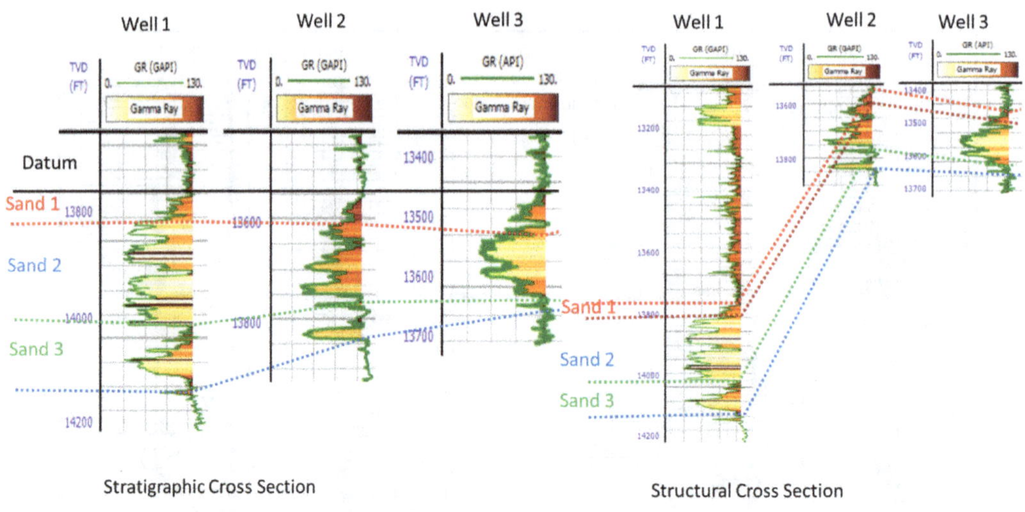

Fig 1.10—Applying the GR curve to correlate between three adjacent wells (Horn Mountain Field data). The left side shows a stratigraphic cross section, and the right side shows a structural cross section. There is an obvious correlation between the Sand 2 bodies. Sand 3 has nearly "shaled out" in Well 3, where it is probably not productive. Very erratic sand development can be seen below Sand 2. Figure provided courtesy of Pahala Sinurat of Texas A&M University.

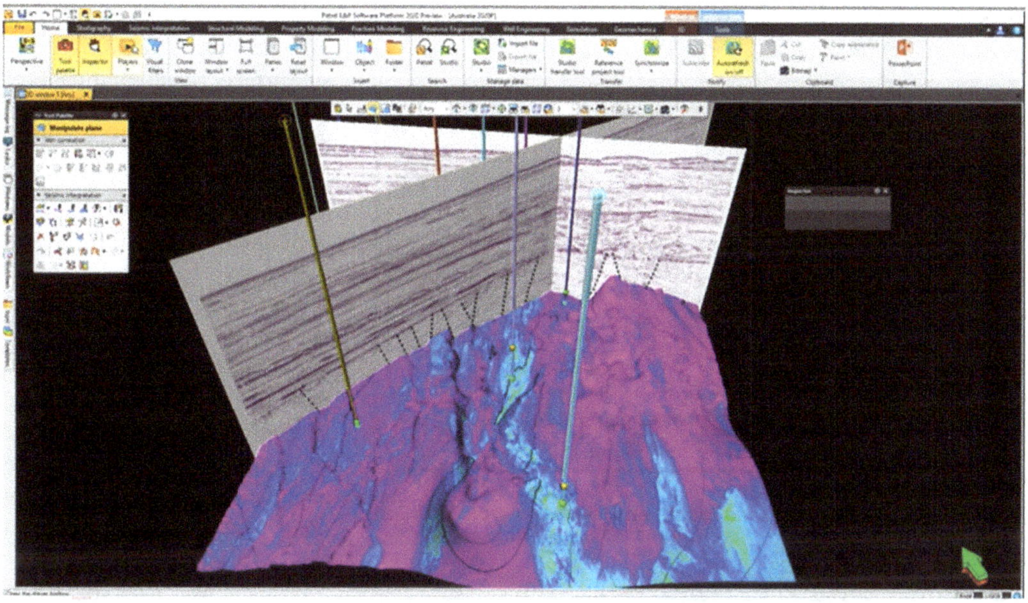

Fig. 1.11—3D visualization (after Schlumberger 2020). Where is the best sand quality located? Where would you inject water? Will injected water spread uniformly throughout the unit?

1.4.1 Fence Diagrams. These diagrams are graphical visualizations of 3D data and interpretations in a 2D-perspective view. Geological and petrophysical data such as log correlations, cross sections, and stratigraphic changes shown in a fence diagram are useful for displaying subsurface structures and formation properties such as gross thickness, net thickness, porosity, water saturation, and other required petrophysical characteristics for reservoir characterization. Stratigraphic changes can also be clearly displayed. **Fig. 1.11** shows an example of an interwell fence diagram. Several companies offer specialized software for correlating and building these types of visualization tools.

1.5 Geologic Maps

Subsurface structure and isopach maps represent the underground topography and volume of oil and gas reservoirs. Well locations and completions are typically included to show well drainage effects. Both the vertical and the horizontal reservoir volumes of the irregularly shaped container are represented by isopach maps, which allow one to calculate reservoir volumes. The type of formation material and the presence of flow discontinuities define the relative volume of hydrocarbons expected to be recovered from a reservoir. These two variables profoundly affect the economics of the oil and gas recovery process.

Oil- and gas-producing sands are commonly laid down in some type of depositional environment. The resulting reservoir morphology assumes characteristics that were formed during deposition and secondary deformation. These characteristics can be studied through the drawing of structure maps, isopach maps, and cross sections. **Fig. 1.12** shows a cross section drawn by correlating GR and resistivity logs.

A structure map relates the subsurface topography and areal configuration of a hydrocarbon accumulation. The structure map plays a critical role in the net-pay determination, which is one of the key economic criteria used to evaluate the commercial viability of a project. **Fig. 1.13** relates a structure map to a cross-sectional slice. Note the elevations shown on the right side of the cross section. The oil-column thickness is approximately 45 ft.

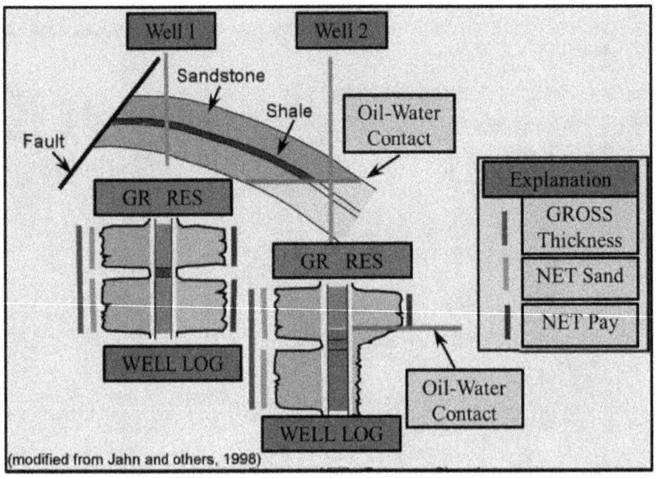

Fig. 1.12—Gross thickness, net sand, and net pay—cross section (after Jahn et al. 1998).

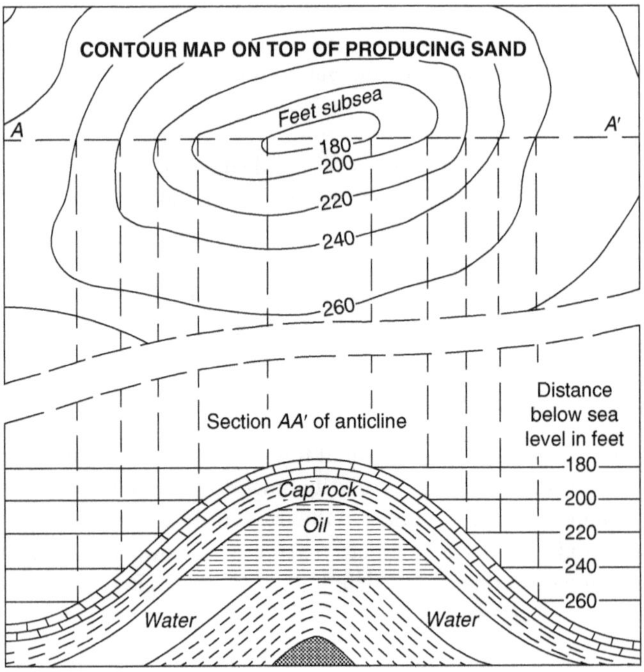

Fig. 1.13—Structure map and cross section of the Austin gas storage field (after Katz et al. 1959). What is missing on the structure map?

Maps and cross sections provide 2D views describing the subsurface configurations of oil and gas reservoirs. Isopach maps provide a 3D view of an irregular shape. **Fig. 1.14** shows the resulting isopach of the oil accumulation in the Austin Field.

The reservoir characteristics can be studied through the drawing of structure maps, isopach maps, and cross sections. In many cases, these reservoir configurations become quite complicated given the highly faulted structure, as seen in **Fig. 1.15.**

Calculating reservoir volumes provides the basic methods for determining original oil-in-place (OOIP) or original gas-in-place (OGIP). The majority of hydrocarbon reservoirs are heterogeneous and composed of productive intervals interspersed with impermeable shales or hard streaks.

1.5.1 Counting Net Sand. Expanded-scale induction electrical survey (IES) logs are commonly used to interpret the net thickness of a pay zone containing both sand and shale sections. **Fig. 1.16** provides an example for calculating net pay from an actual completion log. Note that the completion log contains perforated and production test intervals in addition to logged depth and corrected subsea depth.

1.5.2 Isopach Maps. These maps use equal-value-thickness lines to illustrate a solid figure in two dimensions. Dividing a volume map into a series of equal-thickness layers provides a means for calculating the volume of a 3D figure. The construction of these maps is mechanical, and they do not necessarily infer a producing trend, as in the case of a structure map.

***1.5.2.1 Gross Sand Map* (h).** This map type shows the total thickness of a producing interval. The thicknesses of both the nonproductive and productive intervals of a geologic section are included.

Reservoir volume (acre-ft) = gross sand volume (acre-ft). . . . . (1.1)

***1.5.2.2 Net Sand Map* (h).** This map type shows a sand section with sufficient porosity and permeability to transmit fluids in commercial quantities. The sand interval might or might not contain hydrocarbons.

Reservoir volume (acre-ft) = net sand volume (acre-ft). . . . . (1.2)

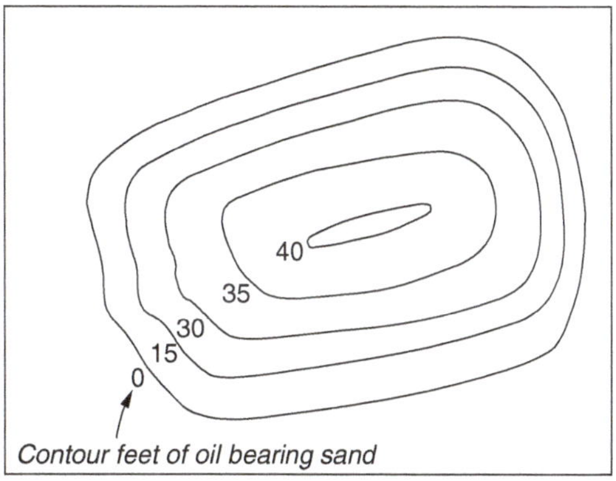

Fig. 1.14—Isopach map of the Austin gas storage field (after Katz et al. 1959). One is able to calculate the volume of the mapped oil shown using the isopach drawing.

***1.5.2.3 Net Oil or Gas Sand Map* (h_o *or* h_g).** Oil and gas accumulations are defined in terms of bulk volume. Average porosity and water saturation values should be included to calculate in-place oil or gas volumes.

$$OOIP = 7{,}758\ \text{bbl/acre-ft}\ \frac{[\text{isopach thickness volume (acre-ft)}](\phi)(1-S_{wi})}{B_{oi}\ (\text{RB/STB})},\ \text{STBO} \quad (1.3)$$

***1.5.2.4 Pore Volume Map* (ϕh).** Contours are drawn through the product of the average porosity and thickness values (ϕh, ft) calculated at each well location.

$$OOIP = 7{,}758\ \text{bbl/acre-ft}\ \frac{[\text{isopach } \varphi h \text{ volume (acre-ft)}](1-S_{wi})}{B_{oi}\ (\text{RB/STB})},\ \text{STBO} \quad (1.4)$$

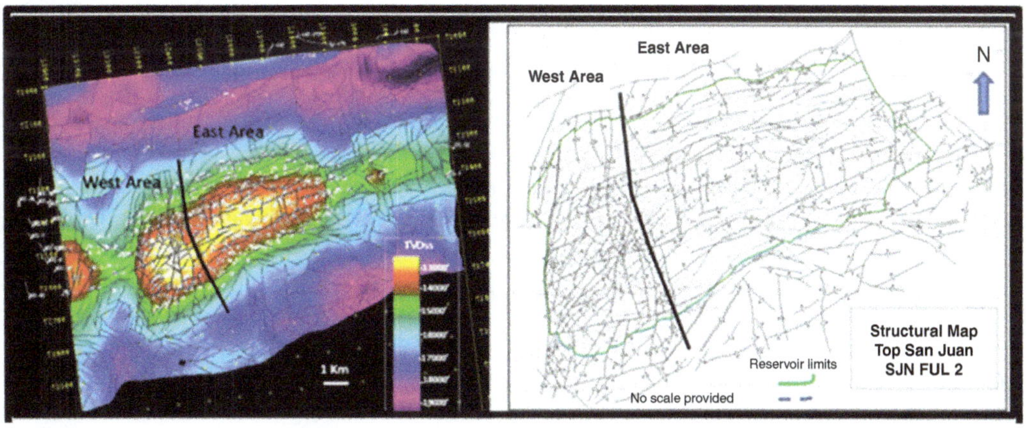

Fig. 1.15—El Furrial Field, east Venezuela (after Pérez Amaro et al. 2020). Structure map of SJN FUL 2 Reservoir, where a waterflood project has been implemented since 1992. TVDss = true vertical depth subsea.

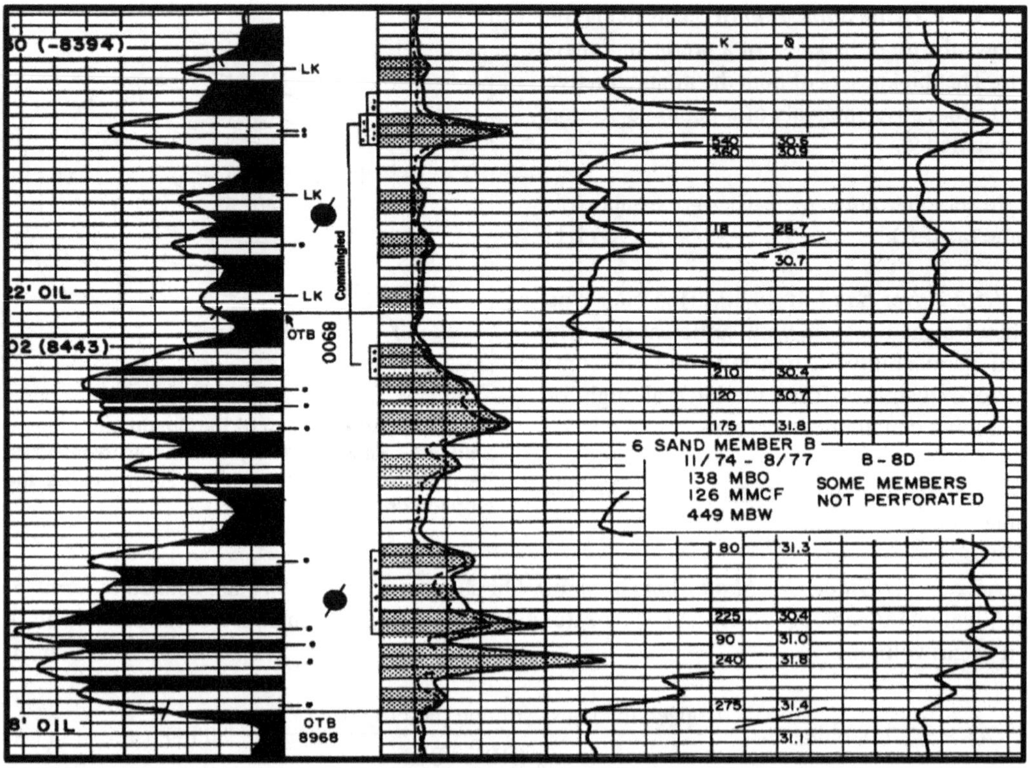

Fig. 1.16—Counting net and gross sand on an IES completion log (after Tearpock and Bischke 1991). Dark sections are counted as nonproductive. Would you perforate additional intervals?

1.5.2.5 Hydrocarbon Pore Volume (HCPV) Map. Estimating the true average porosity and initial water saturation values for heterogeneous reservoirs is often uncertain because of variation in a vertical as well as an areal sense within the reservoir boundaries. The hydrocarbon map provides a more-appropriate model to represent variable-character reservoirs. Contours representing net hydrocarbon values can be calculated as follows:

$$\frac{[\phi h(1-S_{wi})]}{B_{oi}}, \text{ ft-STB/RB.} \quad (1.5)$$

The hydrocarbon isopach map is particularly useful because the mapped volume is transformed to a weighted average area of the porosity and water saturation values. Average porosity and water saturation values are plotted for each well. The resulting contours represent an oil accumulation expressed in acre-ft. A similar representation can be developed to evaluate gas reservoirs.

$$\text{Well hydrocarbon net pay} = \phi h(1-S_{wi}), \text{ ft,} \quad (1.6)$$

$$OOIP = 7{,}758 \text{ bbl/acre-ft} \frac{[\phi(1-S_{wi})]}{B_{oi}}, \text{ STBO/acre-ft,} \quad (1.7a)$$

$$OOIP = 7{,}758 \text{ bbl/acre-ft} \frac{[\phi(1-S_{wi})(HCPV)]}{B_{oi}}, \text{ STBO.} \quad (1.7b)$$

Irregularly shaped land areas outlined on a map can be measured digitally, with a planimeter, or estimated from a transparent overlay of known area. An isopach map divides the reservoir volume into slices representing a series of equal-thickness trapezoids. The average surface area of a trapezoid is calculated using Eq. 1.8; when averaging the top (A_1) and the bottom (A_2) areas, the trapezoid changes into a rectangle. Eq. 1.9 shows the product of the thickness (h) and the average area, which results in a volume estimate of the irregular shape.

$$V = \frac{h(A_1 + A_2)}{2} \quad (1.8)$$

Summing successive trapezoids results in

$$V = \frac{h}{2}\,[A_0 + 2A_1 + 2A_2 + 2A_{(n-1)} \ldots A_n], \quad (1.9)$$

where h is the constant thickness contour interval, A_0 is the area enclosed by the 0-ft thickness contour, $A_{(n-1)}$ represents all areas lying between the first (A_0) and last (A_n) layers, and A_n is the area enclosed by the last contour interval.

Generally, there is a thickness less than h remaining at the top of the structure. This irregularity is usually ignored because it is typically of minor consequence. If deemed necessary, the value of h should be reduced.

Standard US thickness units are feet, and area is measured in acres. Therefore, in these maps, isopach volumes are usually expressed in acre-ft units.

Example 1.1—Calculating Volume. Calculate the volume of the pinchout reservoir when

$A_0 = A_{owc} = 100$ acres, $A_{10} = 75$ acres, $A_{20} = 50$ acres, and $A_{30} = 25$ acres.

Solution. Apply the trapezoid rule to calculate reservoir volume:

$$V = \frac{h}{2}\,(A_0 + 2A_{10} + 2A_{20} + A_{30}) = \frac{10}{2}\,(100 + 2 \times 75 + 2 \times 50 + 25) = 1{,}875 \text{ acre-ft}.$$

The area of each layer or slice is measured and included in the rule of successive trapezoids to calculate the mapped volume.

1.5.3 Oil-in-Place. Reservoir oil volumes can be expressed in unit volume or total reservoir volume terms. Correlations, reserves estimates, and reservoir-quality comparisons often include the unit volume concept shown in Eq. 1.7a.

The unit volume should be multiplied by the reservoir bulk volume to calculate a total reservoir volume, as shown in Eq. 1.7b. Answers are presented at surface conditions.

1.5.4 Gas-in-Place. The general gas law relates reservoir volume to the volume measured at standard conditions. Gas volume equations can be expressed in terms of either the z-factor or the gas formation volume factor (B_g). A shorthand MM term signifying 1 million reduces the final answer to a more-manageable size. Shorthand notations for larger volumes are B for billions and T for trillions.

In gas reservoirs, the gas-in-place is calculated using Eqs. 1.10a, 1.10b, and 1.10c:

$$OGIP = \frac{43{,}560 \text{ ft}^3/\text{acre-ft}(\phi)(1 - S_{wi})\, p_{res}\, T_{sc}}{z_{res}\, p_{sc}\, T_{res} 10^6}, \text{ MMscf/acre-ft}, \quad (1.10a)$$

$$OGIP = \text{Reservoir volume (acre-ft)} \times \text{MMscf/acre-ft, MMscf.} \quad (1.10b)$$

In oil reservoirs at initial conditions at or above the bubblepoint pressure, all available gas is in solution and the volume of dissolved gas in the reservoir oil is calculated using Eq. 1.10c. Under these conditions, these reservoirs are called undersaturated.

$$OGIP = OOIP \text{ (STBO)} \times R_{si} \text{ (scf/STBO).} \quad (1.10c)$$

The presence of an initial gas cap in a reservoir typically indicates saturated conditions because it is evident that the maximum-possible gas volume is in solution.

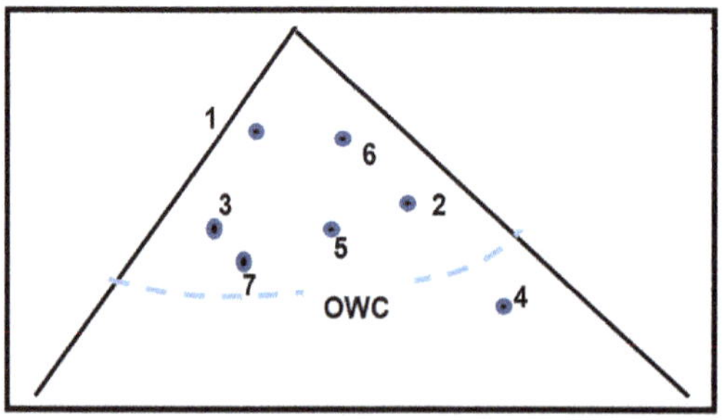

Fig. 1.17—Location map of example field.

Example 1.2—Calculating Original Hydrocarbons-in-Place. Fig. 1.17 shows the Guinea 2A reservoir unit defined by two intersecting faults and an underlying oil/water contact (OWC). Six of the seven wells penetrate the oil portion of the reservoir while the No. 4 well encountered only the downdip water leg. Measured rock properties for each well are shown in **Table 1.2,** and B_{oi} = 1.31 RB/STB and R_{si} = 790 scf/STB (R_{si} = gas/oil ratio at initial conditions).

Calculate the original hydrocarbons-in-place (OOIP and OGIP) by using the average porosity and water saturation and the weighted average porosity and water saturation methods. Which of the two methods should result in the more accurate answer?

	Rock Properties			Calculations				
	1	2	3	4	5	6	7	8
Well No.	Pay Thickness (ft)	Porosity (%)	S_{wi} (%)	Oil-in-Place per Well (ft)	Net Oil—Isopach		Oil PV—Isopach	
				Column 1 × Column 2 × (1−S_{wi})	Interval (ft)	Volume (acre-ft)	Interval (ft)	Volume (acre-ft)
1	19	28	26	3.9	0	233	0	233
2	18	24	29	3.1	5	216	0.5	228
3	14	28	27	2.9	10	200	1	223
4	0	30	100	0.0	15	173	1.5	217
5	18	19	33	2.3	20	49	2	210
6	20	18	38	2.2			2.5	190
7	18	26	28	3.4			3	130
Average Values		23.8	30.2	3.0				
Total (acre-ft)						3,650		624.75

Table 1.2—Average sand properties and results of calculations. PV = pore volume.

Solution. Column 0 = well number. Columns 1–3 = rock properties from a particular well.

Column 4 = net oil thickness calculated on a per-well basis. Using Well No. 1 as an example, the in-place oil is calculated using Eq. 1.6:

$$\phi h(1-S_{wi}) = 0.28 \times 19 \text{ ft}(1-0.26) = 3.9 \text{ ft}.$$

Columns 5 and 6 = OOIP volume calculation by applying the trapezoid rule; the 5-ft slice values and selected contour intervals are based on maximum thickness, which equals 20 ft. Area is obtained using a planimeter. Field area = 233 acres. Column 5 shows the selected slice interval (5 ft); Column 6 shows the computed area covered by the slice interval (5 ft) of the isopach map. Apply the successive trapezoids rule to calculate the isopach reservoir volume:

$$V = \frac{h}{2}(A_0 + 2A_5 + 2A_{10} + 2A_{15} + A_{20}) = \frac{5}{2}(233 + 2\times 216 + 2\times 200 + 2\times 173 + 49) = 3{,}650 \text{ acre-ft}.$$

Calculate the unit volume using Eq.1.7a and the average reservoir properties values shown in Table 1.2:

$$OOIP = 7{,}758 \text{ bbl/acre-ft}\frac{[\phi(1-S_{wi})]}{B_{oi}} = 7{,}758 \text{ bbl/acre-ft}\frac{[0.238\times(1-0.302)]}{1.31 \text{ RB/STB}} 985.7 \text{ STBO/acre-ft}.$$

Multiply the calculated unit volume (STBO/acre-ft) by reservoir volume V to calculate the reservoir OOIP (STBO) using Eq. 1.7b:

$$OOIP = 7{,}758 \text{ bbl/acre-ft}\frac{[\phi(1-S_{wi})\,(HCPV)]}{B_{oi}}, \text{ STBO}.$$

Total OOIP = N = 985.7 STBO/acre-ft × 3,650 acre-ft = 3.598 × 10^6 STBO.

Columns 7 and 8 show the OOIP volume calculation when the 0.5-ft slice values and selected contour intervals are based on the average oil-in-place value (Column 4 = 3 ft). Area is obtained using a planimeter. Field area = 233 acres.

- Column 7 shows the selected slice interval (0.5 ft).
- Column 8 shows the computed area covered by the slice interval (0.5 ft) of the oil isopach map.

Applying the trapezoid rule, 0.5-ft slice values, and weighted average reservoir porosity and water saturation values, calculate the oil pore volume:

$$V = \frac{0.5}{2}(233 + 2\times 228 + 2\times 223 + 2\times 217 + 2\times 210 + 2\times 190 + 130) = 624.75 \text{ acre-ft},$$

$$OOIP = 7{,}758 \text{ bbl/acre-ft}\frac{624.75 \text{ acre-ft}}{1.31 \text{ RB/STB}} = 3.70\times 10^6 \text{ STBO}.$$

The results of the net oil and oil pore-volume calculations are slightly different because the weighted average volume accounts for changes in rock properties within the reservoir and results in a higher-quality answer.

Calculate the original gas volume using Eq. 1.10c:

$$OGIP = OOIP\ (\text{STBO}) \times R_{si}\ (\text{scf/STBO}) = (3.70 \times 10^6\ \text{STBO}) \times (790\ \text{scf/STBO}) = 2.92 \times 10^9\ \text{scf}.$$

1.6 Sediments

Understanding depositional and diagenetic controls on porosity and permeability distribution is crucial for predicting and modeling reservoir quality, determining hydraulic flow units, and maximizing hydrocarbon recovery.

Sedimentary rocks are exposed to physical and chemical changes that occur from the time of deposition through burial and can be broadly classed as carbonate or clastic. Each type possesses generally differing characteristics. Clastics (sandstone, conglomerate, siltstone, and shale) are composed mainly of silicate minerals and are classified on the basis of grain size and mineral composition; carbonates (limestone and dolomite) consist mainly of the carbonate minerals calcite (limestone) or dolomite (dolostone). Shales (clays) are a special form of fine-grained rock material because they can be formed in place or deposited from an outside source. Clastic rocks represent the end process of the erosion, transportation, and redeposition of previously consolidated rock fragments.

As specified in Table 1.1, rock fragments are generally classed according to grain size (i.e., shale, silt, sand, and larger). The erosion and transformation of these materials are subject to a number of identifiable criteria that can be related to the study of present-day depositional environments (LeBlanc 1972). In other words, knowledge of current depositional environments allows one to identify and interpret porosity/permeability and flow channel distribution in petroleum reservoirs by identifying the origin of the rock.

The nature of the pore system initially is controlled by mineral composition and the depositional environment in which the sediment was deposited. The pore system goes through an evolution (changes in porosity), which depends on texture, fluid/rock interactions, and time/temperature/pressure relations after deposition burial to reservoir depths. However, this is not necessarily the case for interpreting the subsurface makeup of carbonate oil and gas fields.

1.6.1 Carbonate Reservoir Rock. These accumulations are usually the product of the deposition of mainly carbonate materials over wide geographical areas. The majority of carbonate rocks are collectively composed of limestone ($CaCO_3$) and dolomite [$CaMg(CO_3)_2$] minerals. The resulting reservoir rock can be composed of very-low-porosity material resulting from the chemical precipitation of calcite; the origin of high-porosity reservoirs is the buildup of marine skeletal material. However, calcite is highly susceptible to secondary alteration by means of mineral dissolution and redeposition, and carbonate reservoirs tend to fracture easily compared to other rock types and/or clastic material (i.e., shale, silt, and sand grains can be transported and included in the sedimentary section during the depositional process). These processes often result in increasing secondary porosity and permeability.

The resulting effect of these processes on the architecture of carbonate reservoirs usually induces a wide range of porosity types, shapes, and porosity distributions within the accumulation. It is difficult to predict trends in the reservoir character because these secondary effects often occur in a random manner. Inferences made by knowing the original depositional environment are not applicable as in the case of sandstone reservoirs.

Fig. 1.18 shows the amalgamation of the different material that can be present in a clastic limestone, such as sand grains and carbonate material. Note how primary porosity can be filled with calcite and dolomite secondary material. Greater and lesser amounts of each component in a sample can profoundly affect the porosity and permeability properties.

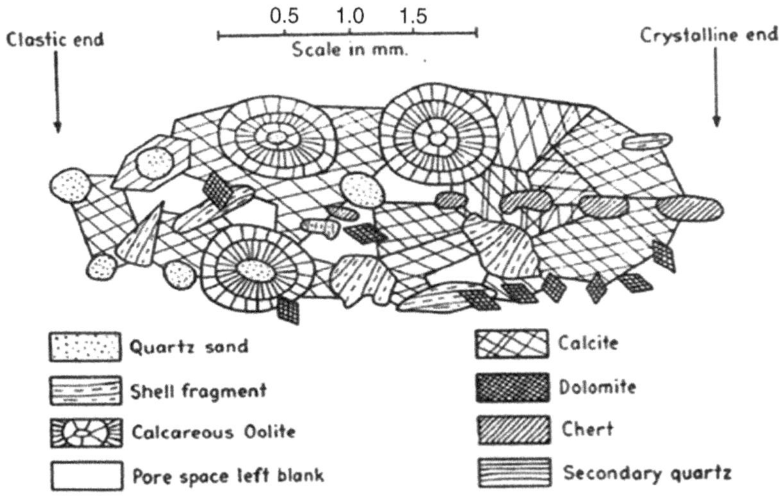

Fig. 1.18—The contents of a carbonate rock can include skeletal material, sand grains, and calcite (after Pirson 1963). An increase of skeletal remains and the sand content increases the in-situ porosity and permeability. On the other hand, the solution/dissolution properties of groundwater movement can alter the pore volume to form a secondary relationship.

Pirson (1963) divided carbonate rocks into four general classes. These classes are

- Accretion (reef) limestone: The vertical buildup of the skeletal remains of organic matter (i.e., shells, sponges, corals) resulting from rising water levels. These remains generally produce good-quality oil and gas reservoirs. The original primary porosity can be disguised by the alteration and rearrangement of the rock pore-volume relationship by means of secondary processes.
- Clastic limestone: Formed by the erosion, transportation, and redeposition of carbonate sediments.
- Chemical limestone: Formed by the direct precipitation of calcium carbonate in a marine environment. It is commonly formed over large areas and possesses minimal primary porosity. The intracrystalline porosity between grains is generally quite small because of the volume of cement in the matrix. Production is a function of the formation of secondary porosity, the inclusion of silt and sand rocks, and/or natural fractures. Solution cavities (vugs) can be formed as a result of leaching from the movement of groundwater. These void spaces may or may not contribute to production end voids such as vugs or other solution cavities and can result in a considerable difference between the measured total and effective interconnected pore volumes.
- Dolomite: The percolation of magnesium-rich groundwater through the formation can replace calcium with the magnesium molecule to form dolomite. Dolomite is denser than the limestone, and as a result, the process increases pore space in the matrix. Dolomitization forms porosity because of the smaller unit volume of dolomite relative to calcite. Dolomite can form contemporaneously with or shortly after deposition or upon burial. Dolomitization can obliterate pre-existing fabric, and it typically forms intercrystalline porosity.

Fig. 1.19 shows an example of idealized carbonate porosity types on the left side and isolated, partly connected and connected vug spaces on the right. Predicting the locations is difficult because of the discontinuous nature of their distribution. Pore types in carbonate rocks are more diverse than those in sandstones as a result of the nature of the deposition and lithification of

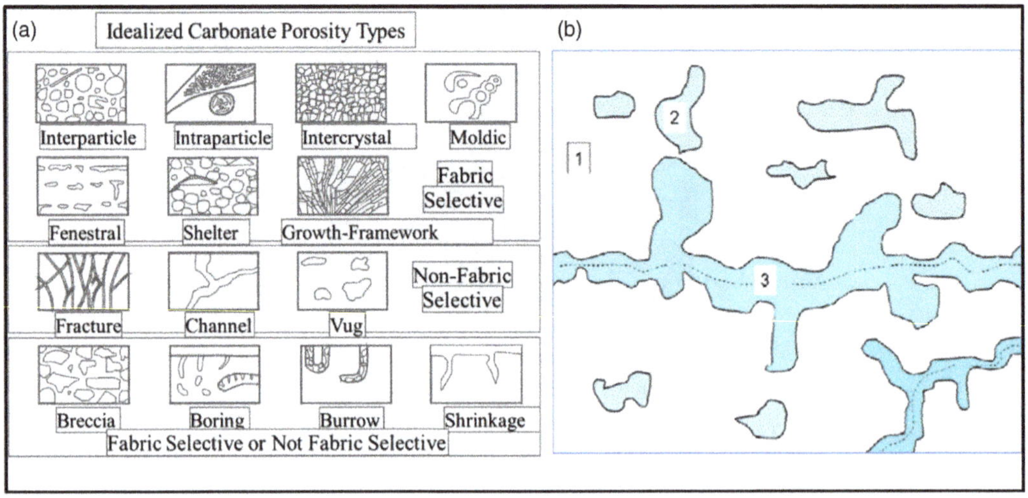

Fig. 1.19—(a) Illustrating idealized carbonate porosity types (after Choquette and Pray 1970); (b) isolated, partly connected, and connected vug spaces, with the possible spatial relation of high-porosity vugs to the much-finer-grained matrix (Vik et al. 2007). Vugs can be (1) not present, (2) isolated with no contribution to production, or (3) have a horizontal extension over a considerable area with a significant contribution to production.

carbonate rocks. Pores in carbonate rocks are classified in terms of whether they are related (fabric selective) or unrelated (nonfabric selective) to depositional texture. In contrast to sandstones, much carbonate diagenesis occurs shortly after deposition, before significant burial. Thus, pore-system characteristics in carbonate rocks become established early in the burial history. However, late-stage diagenesis that affects the pore system can also occur. Note that many porosity types in carbonates result from the presence of organisms (e.g., from the burrowing or dissolution of shell material).

Extended natural fractures are commonly found in relatively brittle carbonate rocks. The presence of natural fractures in a low-matrix-permeability field creates a dual-porosity system. Production forecasting is complicated by the fact that the high initial production rate might only reflect unloading of the high-permeability natural fractures. A very high production-rate loss occurs when the matrix hydrocarbons become the predominant source fluid. Natural fractures act as conduits for oil and gas migrating from the low-permeability matrix at this point.

Fig. 1.20 shows the results on the production history of an Austin Chalk well. Extrapolation of the early decline to a minimum rate indicates the well would last for only 14 months. The second line indicates the well should produce as long as the rate remains above the economic limit and mechanical problems do not arise.

Clastic rocks generally exhibit similar total and effective pore volumes because grain sorting might have occurred at the time of deposition.

In many cases, fine-grained carbonate oil and gas reservoirs containing measurable amounts of sand and silt-sized particles are encountered. These accumulations are sourced from a time of increased erosional depositional forces, which have resulted in areas of higher-than-normal productivity and potential economic opportunities, such as in the Permian in west Texas.

1.6.2 Clastic Reservoirs. These reservoirs are the product of sediment carried in from an outside source. The terms sand, silt, and clay denote the grain or particle size in sedimentary rocks. The sediment size—sandstone (sand: 63 µm–2 mm), siltstone (62–4 µm), or shale (less than 4 µm)—is derived from some external source. The constancy and energy level of the transporting medium and

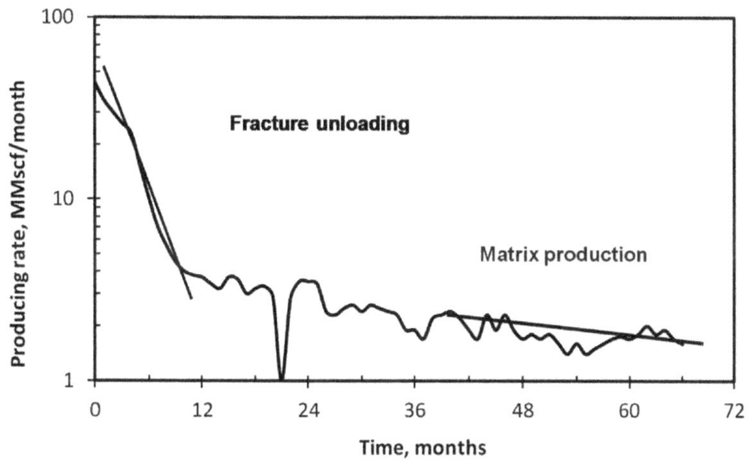

Fig. 1.20—The fracture-unloading decline segment underestimates reserves while extrapolation of the low-permeability-matrix production segment usually overestimates ultimate production because income approaches operating costs. Any additional expense for well work exceeds the economic limit, and the well is shut in. After Poston et al. (2019).

the proximity of the deposition site to the source usually define the size and type of sediment. Both carbonate and clastic sediments can be altered by post-depositional processes, but carbonate sediments are more affected.

Fig. 1.21 shows illustrations of a sandstone rock with shale inclusions. These shales can be brought in and deposited or formed in place by chemical reactions. Fig. 1.21a represents a well-sorted sandstone with equally sized sand particles. The pore system is open, without many restrictions preventing flow between adjacent pores. Thus, this sandstone has an effective pore system and "high" permeability. The material coating the quartz and feldspar sand grains is the clay mineral chlorite. Fig. 1.21b shows that pore throats in sandstone can be lined with a variety of cement minerals that affect petrophysical properties; this illustration shows a pore surrounded by secondary growths of fibrous illite.

Fig. 1.22 shows additional illustrations of a sandstone rock with shale inclusions. Note the accordion-shaped clay grains in Fig. 1.22a. Kaolinite platelets have large diameters relative to pore

Fig. 1.21—Illustrations of sandstones relating sand and shale grains and the pore spaces in between (after Schlumberger NExT 2005, based on scanning electron micrographs by R. L. Kugler). The clay particles could have been brought in during the original deposition or formed as secondary growths.

Fig. 1.22—Illustrations of sandstones with clay minerals in pore spaces (after Schlumberger NExT 2005, based on secondary electron micrographs by R. L. Kugler).

throats, and as a result, under certain reservoir production conditions, kaolinite can migrate to and block pore throats, thus reducing permeability. The introduction of fresh water would cause swelling and the reduction of pore volume. Kaolinite can fill pores or replace detrital grains, such as feldspar. Fig. 1.22b shows in greater magnification a pore space, the surrounding calcite crystals, and a fibrous illite shale particle.

1.7 Clastic Depositional Environments

Depositional systems exert significant control over reservoir properties, and they transport sediments and deposit them in predictable, laterally adjacent environments, forming intertonguing sedimentary facies. Clastic depositional systems are those in which the facies are composed primarily of transported rock and mineral fragments. Reservoir quality, extent, and internal heterogeneity vary among different sedimentary facies; stratigraphic petroleum traps can occur where reservoir facies pinch out. Hydrocarbon reservoirs often vary in both vertical and areal extent and in lithology. Reservoir sediments are commonly deposited in a distinctive manner that reflects energy levels and source material. Relating known characteristics of present-day deposition sites allows one to infer the reservoir character that will be encountered in the subsurface.

LeBlanc (1972) classified reservoir characteristics as a function of the depositional environment. Generally, each reservoir type exhibits a characteristic log signature and architecture. The study of present-day depositional environments to develop an understanding of the framework of petroleum reservoirs became a branch of petroleum geology in the late 1960s and 1970s. The seminal work by LeBlanc (1972) arranged depositional environments into an orderly set of frameworks that can be used as an aid to describing reservoir geometry. Each environment is a function of the type of transporting medium (wind, water, or gravity), distance from the erosional source, eroded rock type, energy level, and constancy of the energy level during the time of deposition.

Reservoir fabric is defined as the general grain-size distribution within the depositional unit, and reservoir framework is defined as the general shape of the reservoir flow units. Facies are defined in a geological sense as the type of sediment accumulation present within a depositional unit.

Fig. 1.23 provides a map showing the relative locations of the recognized depositional environments within continental, transitional, and marine regions. Continental accumulations are generally comprised of cut-and-fill deposits reflecting periodic high-energy environments. Transitional sediments, on the other hand, are usually the result of the moderating influence of decreasing surface slope and near-shore winnowing, with grain size determined by near-shore currents and wave size. Marine sediments range from deepwater dump deposits (turbidites) to rather homogeneous sands formed by deepwater currents.

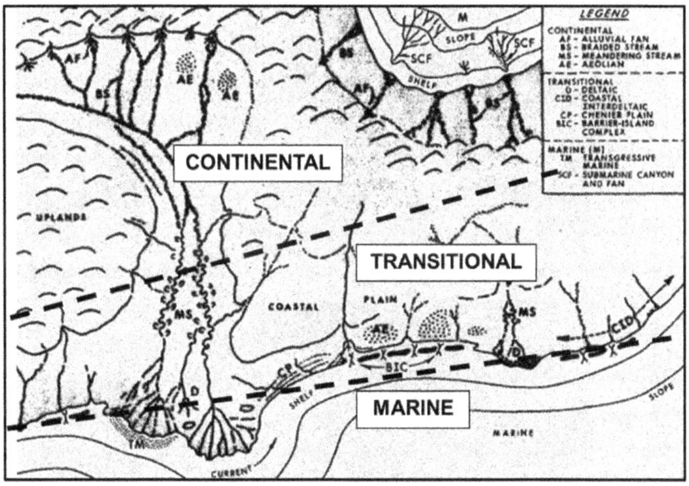

Fig. 1.23—Environments of deposition. Note the division of these environments into continental, transitional, and marine (after LeBlanc 1972).

As shown in **Fig. 1.24,** the log signature of a delta front is bell shaped. Fine-grained sediments drift farther from the river mouth. The outgoing current tends to increase as the delta builds up.

Also shown in Fig. 1.24, a long shore-bar log signature tends to be blocky and elongate parallel to the ocean currents. The entire accumulation is typically encased in shale.

1.8 Reservoir Heterogeneity

The nature of reservoirs is that they are highly complex, largely unexplored, largely unsampled or undersampled, and never truly random. Lateral and vertical continuity in reservoir studies can lead to inaccurate models. Even where such continuity at the megascopic level is apparent, heterogeneities at other levels of study can lead to poor reservoir description. The areal sweep efficiency (E_A) and the vertical sweep efficiency (E_V) are highly dependent on rock property variations and heterogeneities in both directions.

Reservoir heterogeneity implies a variation in properties as a function of space, and it will depend on their measurements at different scales: microscopic, at the pore level (pore and grain size, throat opening, lithology, and mineralogy of rock); macroscopic, at the core level (porosity, permeability, saturation, and special core analysis); megascopic, at the simulator grid level [log data and pressure transient data (repeat formation test, single-well tracer)]; and gigascopic, at the interwell level (permeability from well test, surface seismic, major fault locations). **Fig. 1.25** shows heterogeneities from scanning electron microscope (SEM) images of shales at the microscopic/pore level.

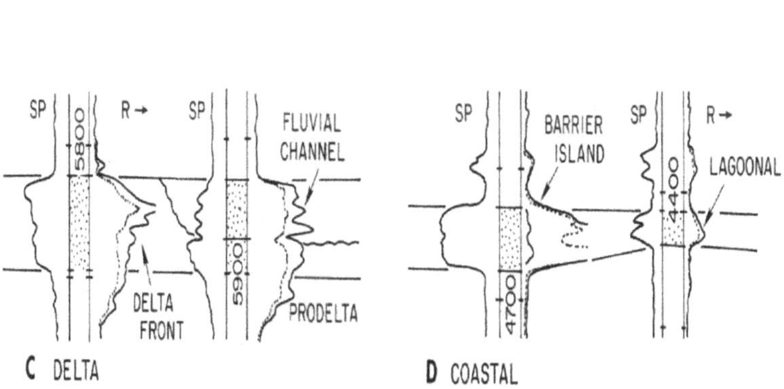

Fig. 1.24—Comparing SP and resistivity log signatures. Note that the SP log expresses the sand/shale content while the resistivity side assumes a gas or oil content. The lateral extent of most of these reservoir types is typically rather limited (after Berg 1986).

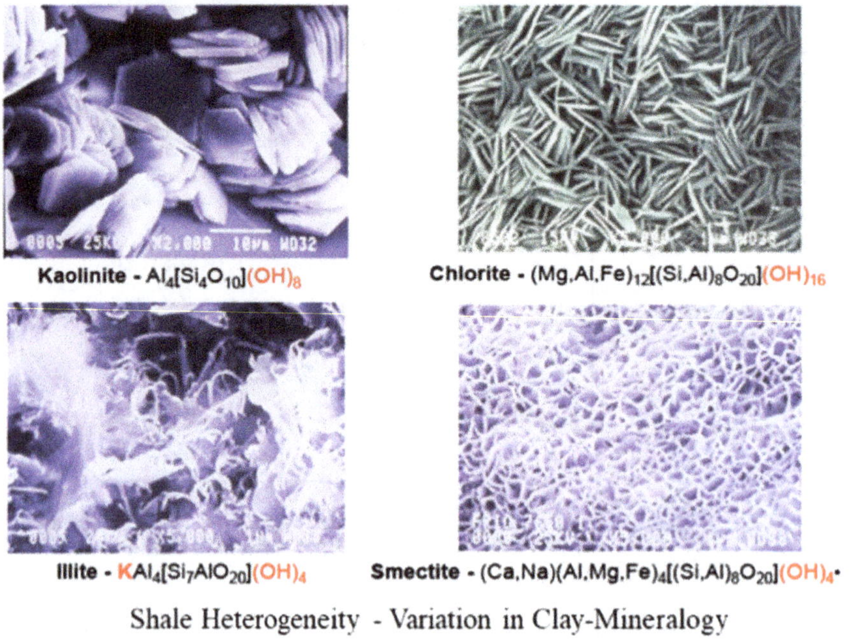

Fig. 1.25—SEM from clastic rocks. Different crystallographic structures of clay minerals are characterized by 2D sheets (courtesy of Mamdouh A. Shebl).

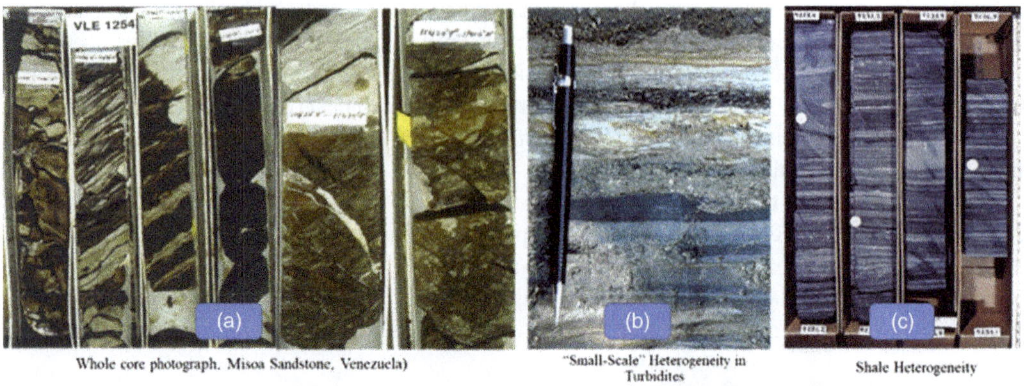

Fig. 1.26—Core photographs from clastic rocks. (a) Eocene sandstone (courtesy of Schlumberger NExT); (b) turbidity deposits showing sediment gravity flow responsible for distributing vast amounts of different clastics (courtesy of Mamdouh A. Shebl); (c) fine-grained, clastic sedimentary rock (courtesy of Mamdouh A. Shebl).

Fig. 1.26 shows different cores of clastic rocks, allowing one to observe at the macroscopic level different depositional and diagenetic features that increase the tortuosity of fluid flow. Faults and fractures can be open (conduits) or closed (barriers) to fluid flow.

Variation of a given measured property decreases as the scale of measurement increases. At a smaller scale (micro), there is greater heterogeneity, but at a higher scale (giga), the property might appear more homogeneous.

Outcrop studies of the shale layer extent as a function of the depositional environment were conducted and core samples were collected by Zeito (1965) and Visher (1984). **Fig. 1.27** shows the extent of shale layers as a function of the depositional environment.

Fig. 1.28 shows the high heterogeneity at a gigascopic/interwell level and different features of shale outcrops in the US. There are several things to note in these very productive source rocks. The Marcellus Shale shows a high degree of fracturing. The Eagle Ford Shale shows an extreme amount of vertical heterogeneity, with layers ranging from fractions of an inch to a few inches thick. Changes in total organic content, mineralogy, and mechanical properties occur continuously over the net height of the reservoir. Horizontally, the reservoir is much more consistent, with changes occurring over several hundred feet. At the base of this outcrop, one can see a gray section that has a higher organic content.

Fig. 1.29 compares pay continuity in a cartoon of a waterflood project as a function of well spacing and completion intervals. In heterogeneous reservoirs, changes in facies trap oil or shield the oil from the sweeping action of injection water. Reducing the interwell distance through infill drilling reduces the effect of these lateral pay discontinuities, thus improving areal and vertical sweep efficiencies and increasing recovery. Successful waterflooding of the entire pay is contingent on having all of the pay open and stimulated. The producing well on the left would probably be converted to injection. As can be seen, if a reservoir is planned to be waterflooded, the areal sweep efficiency will be highly dependent on areal heterogeneities.

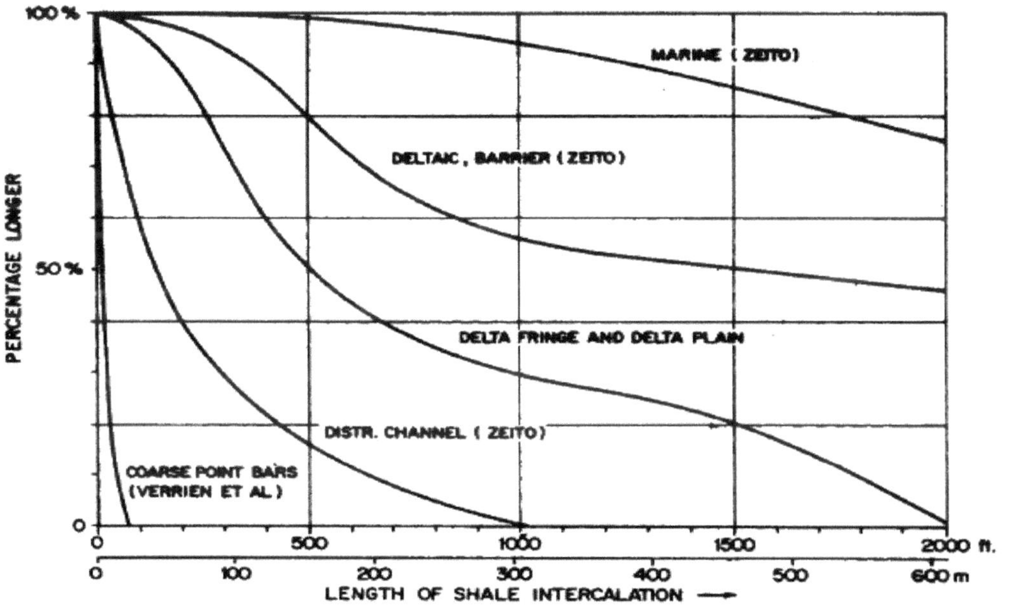

Fig. 1.27—Although this figure from Visher (1984) is somewhat subjective, knowledge of the environment of deposition is helpful when drilling outpost and infill wells. Correlating logs between wells would be more difficult for a distributary channel deposit than for a marine sand.

Fig. 1.28—Outcrop photographs from shale rocks in the US (courtesy of Mamdouh A. Shebl).

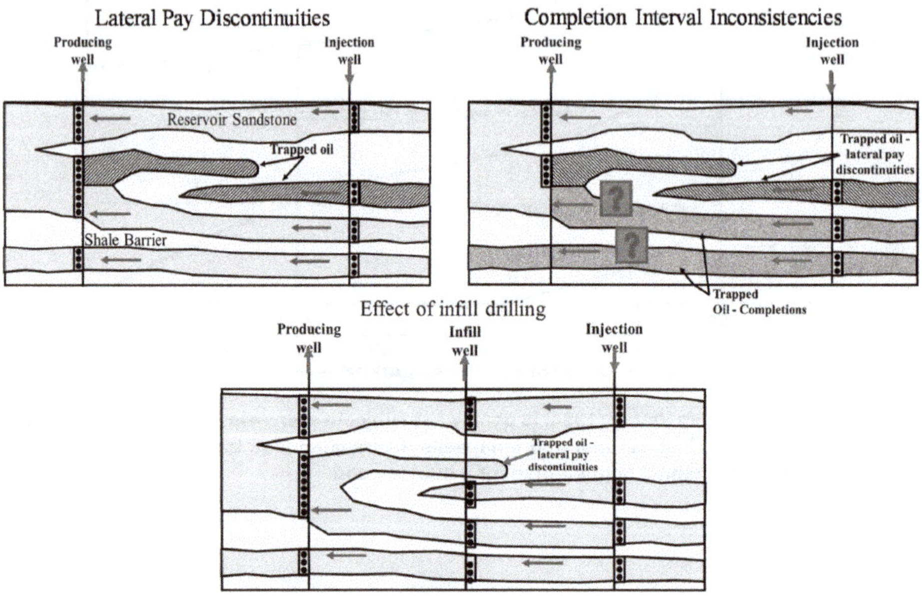

Fig. 1.29—Comparing intrawell reservoir continuity. Infill drilling should possibly be contemplated in the very heterogeneous case (after Schlumberger NExT 2005).

1.9 Summary

Organic material trapped in sedimentary rocks converts to long-chained, lighter-than-water molecular colloids that eventually migrate updip. Hydrocarbon accumulation occurs when the overlying surface of a closed, sealed structure acts as a trap. Subsurface traps are the result of structural deformation or stratigraphic discontinuities.

Oil and gas are generally found in clastic or carbonate sedimentary rock types. Clastic sediments are typically characterized by the physical processes that occurred during deposition. Knowledge of this information is of considerable help when estimating secondary reserves. Carbonate reservoirs, on the other hand, tend to have been altered from their original characters by dissolution and/or natural fracturing processes. These secondary alterations can change the original character of a carbonate reservoir in an erratic and significant manner.

The similarity of log deflection shapes between contiguous wells allows inferences to be made regarding the integrity of the flow system. Well log and computed log examples have been presented to illustrate these principles because the shape of a log deflection can provide a hint regarding the architecture of the reservoir. Cross sections and fence diagrams can provide a practical view of the available flow system.

Knowledge of the pore space available for storage of oil and gas is particularly important when contemplating a waterflood. Basic equations for estimating this value have been developed. Clean sands typically display a commonality of rock properties, while dirty sands contain a combination of sand and clay-sized particles, which can result in less ultimate recovery. The reaction of clay minerals to the introduction of injected water varies according to the composition of both the in-situ and injected water and the mineralogy. A priori knowledge of the mineralogy involved can be very important when designing a waterflood.

Oil recovery from the waterflood process is founded on the interplay between oil production and water-injection wells. The displacement efficiency between the two well systems should be viewed in the vertical and in the horizontal sense. The vertical displacement efficiency results from perforating flow channels common to adjacent wells. The horizontal displacement efficiency is the relation of well density to the degree of reservoir heterogeneity.

1.10 Questions

- How can the analysis of log profiles of clastic rocks help to predict reservoir behavior in a waterflood project? (Answer: The use of log profiles provides knowledge regarding the depositional environment and helps to characterize the possible extent of the interwell flow channels. Well connectivity is the primary influence on achieving an efficient recovery mechanism in a waterflood.)
- Why are carbonate reservoirs difficult to characterize? (Answer: The reservoir architecture of carbonate reservoirs is primarily influenced by migrating groundwater alternately dissolving and redepositing calcium carbonate in an erratic manner throughout the reservoir. The result is a reservoir displaying an unpredictable permeability distribution. The effect establishes zones of high and low permeability in a highly irregular manner. Additionally, carbonates are susceptible to fracturing, which establishes zones of high permeability in a low-permeability matrix.)
- Why is it much easier to waterflood clean sand than dirty sands? (Answer: Reactive minerals are typically absent in clean sands. Reactive clays present in dirty sands can expand when contacted by a foreign water. The expansion of a clay particle reduces the in-situ reservoir permeability, thereby reducing well productivity.)
- Why is it much easier to waterflood sandstones than carbonate reservoirs? (Answer: Sandstones are typically laid down in an orderly manner, which typically results in a reasonable interwell sand/shale correlation over a considerable distance.)

- Why is it important to know the clay type when designing a waterflood? (Answer: Certain clay minerals are extremely reactive when contacted by fresh water. Montmorillonite is an example. Other types, such as kaolinite, are less reactive. Big pumps or extensive water treatment should be included in the project design if this problem is anticipated.)
- Why is it important to recognize the difference between total and effective porosity in reservoir evaluation? (Answer: Effective porosity defines the volume of interconnected pore spaces. A pore space separated from those adjacent to it remains undisturbed because it is impossible to induce a pressure drop from adjacent production.)

1.11 Nomenclature

A_n = area enclosed by last contour interval, acres
$A_{(n-1)}$ = all areas lying between first (A_0) and last (A_n) layers, acres
A_{owc} = oil/water contact area, acres
A_0 = area enclosed by the 0-ft thickness contour, acres
B_g = gas formation volume factor, Rcf/scf, RB/scf, or RB/Mscf
B_{oi} = initial oil formation volume factor, RB/STB
E_A = areal sweep efficiency, percent
E_V = vertical sweep efficiency, percent
h = thickness, ft
h_g = gas thickness, ft
h_o = oil thickness, ft
N = total original oil-in-place, STBO
p_{res} = reservoir pressure, psi
p_{sc} = pressure at standard conditions (sometimes called at base conditions) (in this book, p_{sc} = 14.65 psia), psia
R_{si} = solution gas/oil ratio at initial conditions, scf/STBO
S_{wi} = initial water saturation, percent
$\bar{S}_{wi}$ = average initial water saturation, percent
T_{res} = reservoir temperature, °F
T_{sc} = temperature at standard conditions (sometimes called at base conditions) (in this book, T_{sc} = 60°F or 520°R), °F or °R
V = volume, acre-ft
z_{res} = compressibility factor (also called the gas deviation factor, supercompressibility, or z-factor at reservoir conditions)
ϕ = porosity, percent
$\bar{\phi}$ = average porosity, percent

1.12 Problems

Exercise 1.3—Correlation Using Logs. Correlate the sand bodies using the logs shown in **Fig. 1.30.**

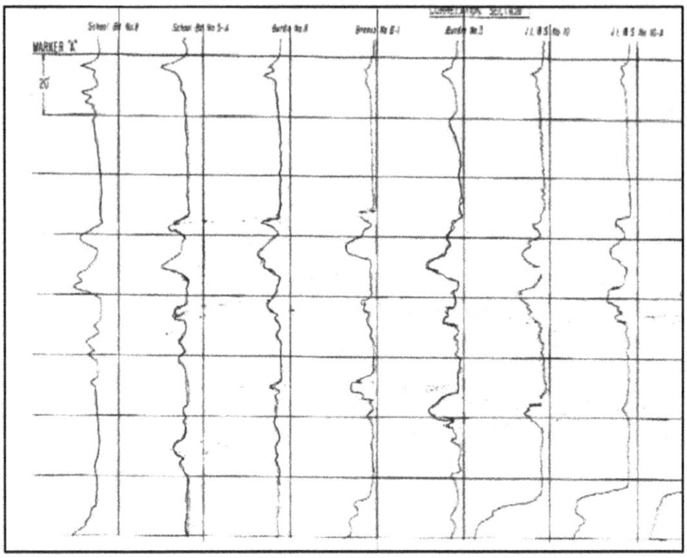

Fig. 1.30—GR logs.

Solution. Tracing an imaginary vertical baseline (north to south) through the farthest GR or SP reading (shale baseline) allows one to visually correlate and trace lines horizontally (west to east), and they are drawn by identifying similar deviations from the vertical baseline. See **Fig. 1.31.**

Exercise 1.4—Oil Isopach Problem. Apply the net oil and net hydrocarbon methods to calculate OOIP and OGIP for the 6600 RA Reservoir shown in **Fig. 1.32** by using **Table 1.3.** The reservoir extends through the subsurface of the two properties, Franks and Cotton, that are shown in Fig. 1.32.

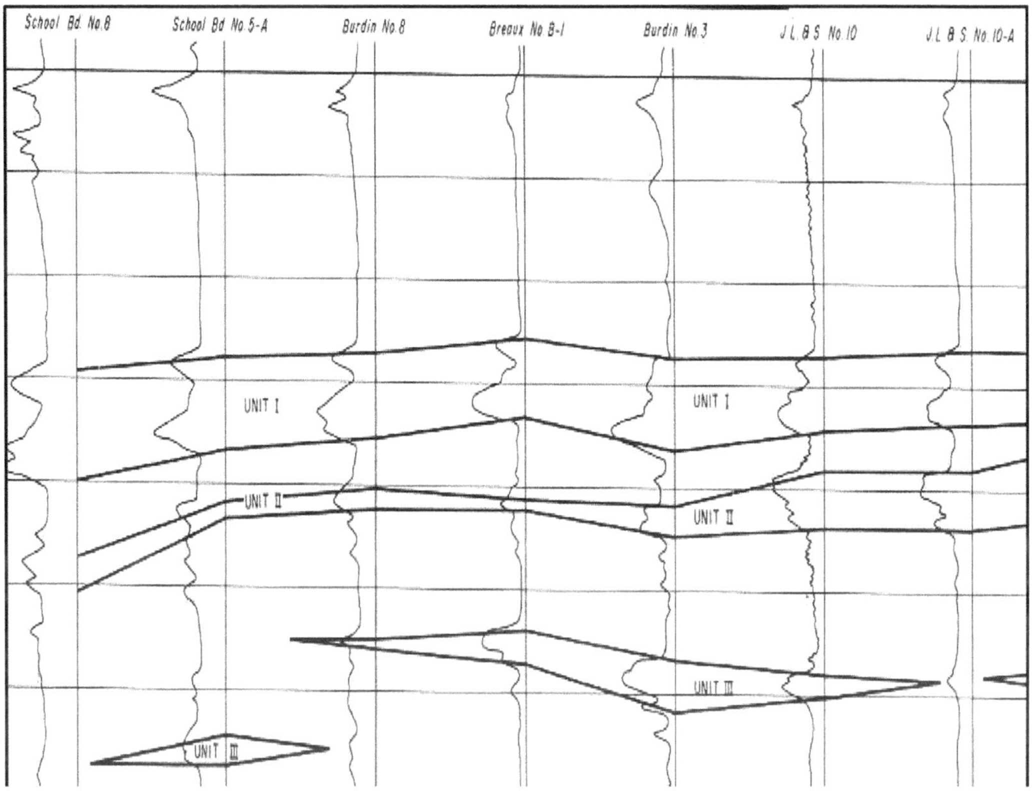

Fig. 1.31—Answer: correlation using GR logs.

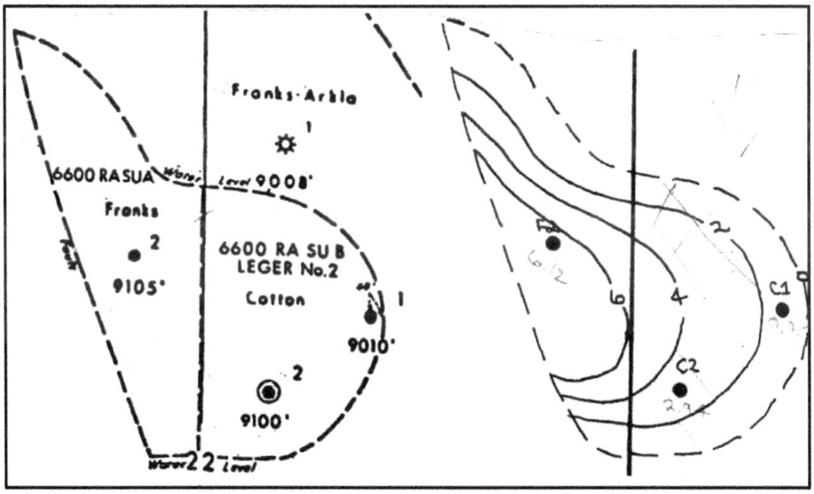

Fig. 1.32—Oil isopach maps—Franks and Cotton leases. Map scale: 1 in. = 1,000 ft; 1 in.^{2} = 23 acres. The left side shows well locations, while the right side shows the 2-ft isopach intervals.

	Franks No. 2	Cotton No. 2	Cotton No. 1	Total
h (ft)	29	21	7	57
ϕ (%)	33	25	20	
S_{wi} (%)	20	30	32	
$\phi h(1-S_{wi})/1.25$ (STB-ft/RB)	6.12	2.94	0.95	
Contour (STB-ft/RB)	0	2	4	
Measured areas (in.^{2})	3.13	1.87	0.50	5.5
Measured isopach areas				
Average pore volume—oil isopach (acre-ft, 2-ft interval)	72.0	67.6	22.1	
Weighted average pore volume—net oil (STB/acre-ft, 2-ft interval)	18.0	10.7	2.9	

Table 1.3—Fluid and rock properties.

The following average rock properties are used in the proposed calculations: $\bar{\phi} = 33\%$, $\bar{S}_{wi} = 25\%$, $B_{oi} = 1.25$ RB/STB, and $R_{si} = 690$ scf/STB.

Perform the following tasks:

1. Calculate field area (acres).
2. Using the average rock properties, calculate OOIP on a unit basis (Eq. 1.7a) and fill in the resulting values in tabular form. Does the Franks or the Cotton lease contain the better-quality reservoir?
3. Use the average rock properties for the field to calculate the OOIP and gas-in-place. Assume there is no primary gas cap.
4. Use the weighted average rock properties to calculate the rock isopach volume and OOIP in STB for the results found in Step 3.
5. Which method results in the better-quality answer?
6. Which method would you favor when the Franks and Cotton properties are unitized if you owned the Leger lease?

Solution.

1. Field area (acres):

   Field area = (5.5 in.^{2})(23 acre/in.^{2}) = 126.5 acres.

2. Calculate OOIP on a unit basis (Eq. 1.7a) for each well using the average rock properties. It can be seen from **Table 1.4** that the Franks lease contains the better-quality reservoir.

	Franks No. 2	Cotton No. 2	Cotton No. 1
h (ft)	29	21	7
ϕ (%)	33	25	20
S_{wi} (%)	20	30	32
OOIP (STB/acre-ft)	1,638.49	1,086.12	844.07

Table 1.4—Resulting OOIP values on a unit basis using the average rock properties.

3. OOIP using the specified average rock property isopach:

$$OOIP = 7{,}758 \text{ bbl/acre-ft} \frac{[\bar{\phi}(1-\bar{S}_{wi})]}{B_{oi}} = 7{,}758 \text{ bbl/acre-ft} \frac{0.33 \times (1-0.25)}{1.25 \text{ RB/STB}} = 1{,}536 \text{ STB/acre-ft.}$$

Net oil isopach:

$$V = \frac{2 \text{ ft}}{2}(72.0 \text{ acres} + 2 \times 67.6 \text{ acres} + 22.1 \text{ acres}) = 229.3 \text{ acre-ft,}$$

OOIP and OGIP are calculated using Eqs. 1.10a, 1.10b and 1.10c, respectively:

$$OOIP = 229.3 \text{ acre-ft} \times 1{,}536 \text{ STB/acre-ft} = 352{,}205 \text{ STB.}$$

$$OGIP = 352{,}205 \text{ STB} \times 690 \text{ scf/STB} = 243 \text{ MMscf.}$$

Table 1.5 provides a summary of these calculated values.

Net Oil Isopach Volume Calculations	
Average volume	229.3 acre-ft
OOIP	352,045 STBO

Table 1.5—Summary of calculated values.

4. Weighted average volume:

$$V = \frac{2[18.0 + (2 \times 10.7) + 2\ .9]}{2} = 42.3 \text{ acre-ft.}$$

Weighted average volume:

$$\text{OOIP} = (7{,}758 \text{ bbl/acre-ft})(42.3 \text{ acre-ft}) = 328{,}163.4 \text{ STB.}$$

Table 1.6 provides a summary of these calculated values.

Weighted average volume	42.3 acre-ft
OOIP	328,163.4 STB

Table 1.6—Summary of calculated values.

5. The results of the net oil and oil pore-volume calculations are slightly different because the weighted average volume accounts for changes in rock properties within the reservoir and results in a better-quality answer.
6. Using the average porosity and water saturation method would be a better choice for the owner of the Leger lease to shift as much of the pore volume to the Leger's side of the unit.

1.13 References

- Berg, R. R. 1986. *Reservoir Sandstones*, 479. Englewood Cliffs, New Jersey: Prentice-Hall, Inc.
- Choquette, P. W. and Pray, L. C. Geological Nomenclature and Classification of Porosity in Sedimentary Carbonates. *AAPG Bulletin* **54** (2): 207–250.
- Hadley, D. F., Arochukwu, E., Nishi, K. et al. 2006. Depositional Modelling of Champion Field, Brunei: Assessing the Impact of Reservoir Architecture on Secondary Recovery. Paper presented at the SPE Asia Pacific Oil & Gas Conference and Exhibition, Adelaide, Australia, 11–13 September. SPE-101033-MS. https://doi.org/10.2118/101033-MS.
- Jahn, F., Cook, M., and Graham, M. 1998. *Hydrocarbon Exploration and Production*, first edition. Amsterdam: Elsevier Science.
- Katz, D. L. et al. 1959. *Handbook of Natural Gas Engineering*. New York, New York: McGraw-Hill Book Company, Inc.
- Lang, W. H. 1986. Correlation with Multiple Logs. *The Log Analyst* **27** (1). SPWLA-1986-vXXVIIn1a5.
- LeBlanc, R. J. 1977. Distribution and Continuity of Sandstone Reservoirs - Part 1, Journal of Petroleum Technology, July, 1977
- Magoon, L. B. and Dow, W. G. 1994. The Petroleum System. In *The Petroleum System—From Source to Trap*, ed. L. B. Magoon and W. G. Dow, 3–24. Vol. 60, AAPG Memoir, American Association of Petroleum Geologists.
- Perez Amaro, I., Chelhod, N. C.-Y., and Laprea-Bigott, M. 2020. Workflow for Optimization of Waterflood Performance: Field Example from El Furrial Field – East Venezuela Basin. Paper presented at the SPE Latin American and Caribbean Petroleum Engineering Conference, Bogota, Colombia, 27–31 July. SPE-199121-MS. https://doi.org/10.2118/199121-MS.
- Pirson, S. J. 1963. Handbook of Well Log Analysis: For Oil and Gas Formation Evaluation, 332. Englewood Cliffs, New Jersey: Prentice-Hall, Inc.
- Poston, S. W., Laprea-Bigott, M., and Poe, B. D., Jr. 2019. Analysis of Oil and Gas Production Performance. Richardson, Texas: Society of Petroleum Engineers. ISBN: 978-1-61399-665-2
- Ramón, J. C. and Cross, T. 1997. Characterization and Prediction of Reservoir Architecture and Petrophysical Properties in Fluvial Channel Sandstones, Middle Magdalena Basin, Colombia. *CT&F – Ciencia, Tecnología y Futuro* **1** (3): 19–46.
- Schlumberger. 2020. Petrel 2020 Features, https://www.software.slb.com/products/petrel/features#projectexplorer (accessed 12 October 2020).
- Tearpock, D. J. and Bischke, R. E. 1991. *Applied Subsurface Geological Mapping*, 41. Englewood Cliffs, New Jersey: Prentice-Hall, Inc.
- Vik, B., Djurhuus, K., Spildo, K. et al. 2007. Characterisation of Vuggy Carbonates. Paper presented at the SPE/EAGE Reservoir Characterization and Simulation Conference, Abu Dhabi, UAE, 28–31 October. SPE-111434-MS. https://doi.org/10.2118/111434-MS.
- Zeito, G. A. 1965. Interbedding of Shale Breaks and Reservoir Heterogeneities. *Jour. Pet. Tech.* **17** (10): 1223–1228. https://doi.org/10.2118/1128-PA

Chapter 2

Reservoir Rock Properties

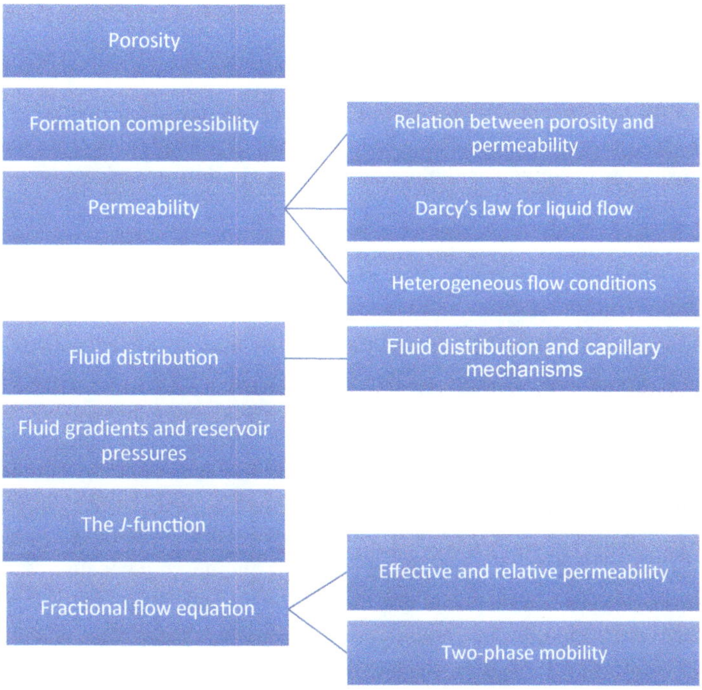

Designing and monitoring oil displacement by means of water encroachment requires a priori knowledge of the basic physical properties of the waterflood system. In this chapter, fundamental rock and rock/fluid interaction properties are reviewed. The aim is to understand the distribution and movement of immiscible fluids in porous media.

A reservoir rock must display two basic properties to possess commercial potential. These properties are

- A capacity to store fluids
- A capability to transmit these fluids to the wellbore

The porosity of a reservoir rock defines the storage capacity while absolute permeability describes the ability of the rock to transmit fluids. These properties will not necessarily remain constant over the producing life of the reservoir because of potential compressibility effects.

Additionally, the ability of a particular fluid to flow in a reservoir in a two- or three-fluid system is controlled by the interaction between the fluids. These relative permeability effects are also a function of the saturation history of the reservoir.

Significant rock properties such as porosity (ϕ) and permeability (k), in addition to subsidiary but no less important fluid saturations (S_w, S_o, S_g) and their distributions, are discussed in this chapter.

2.1 Porosity

Reservoir rock comprises pore spaces surrounded by a rock framework. Rock grains and cementing material compose the framework or matrix. The pore space or porosity can be calculated from well logs or cores. Porosity (ϕ) is defined as the pore volume divided by a total unit volume:

$$\phi = \frac{V_p}{V_t} = \frac{V_t - V_{\text{grain}}}{V_t}. \quad (2.1)$$

The resulting porosity value is typically multiplied by 100 to express the answer as a percent. The unit volume is the sum of the pore spaces and the matrix volume: Total rock volume = V_t = pore volume + matrix volume.

Well logs measure total porosity, which is the sum of the isolated and connected pore volumes located within the rock sample. Laboratory core analysis, on the other hand, involves injecting a measured fluid volume into the rock sample to determine the pore volume. The pore spaces must be connected to allow the intrusion of the foreign fluid into the center of the sample. The resulting volume—defined as the effective porosity—is a measure of the volume of the interconnected pore spaces. Effective porosity is of greater interest in reservoir engineering because reserves are a function of how much oil and gas can be produced, not just how much is present. Trapped fluids in the dead-end pore spaces are nonproducible and therefore of little economic significance. Also, porosity can be greatly reduced by compaction if sediments are unconsolidated.

2.2 Formation Compressibility

Reservoir rocks are compressible; they expand as pore pressure decreases during production and provide a source of expulsive energy. The downward force exerted by the weight of the overburden on a reservoir rock must be offset by a compensating upward vector. The overburden pressure causing the downward force is defined as lithostatic pressure. Resistance to the overburden pressure vector is the sum of the fluid pressure and the supporting force supplied by the rock framework. Eq. 2.2 expresses this force diagram:

$$p_{ob} = p_{\text{fluid}} + p_{\text{matrix}}. \quad (2.2)$$

For most cases, the lithostatic gradient is 1.0 psi/ft, while the salt- and freshwater gradients are 0.465 and 0.433 psi/ft, respectively. The rock matrix strength is a function of the grain type and the degree of sorting and cementation. Cementation of the sand grains increases the matrix strength, while dissolution occurring during post-deposition groundwater movement can weaken the framework.

Well-sorted sand grains tend to bridge at point-to-point contacts. The compression of this type of uncemented rock material is primarily a function of the mineral compressibility of the grains. Heterogeneously sized material will not bridge to as great a degree because the small grains tend to slide into the larger pore spaces.

Immature sediments are usually unconsolidated and possess high water saturations and relatively high permeabilities. Continued burial by the accumulation of sediments increases the overburden pressure, which in turn expels more water from the rock. There is a compensating loss of porosity as the rock grains compact, which increases the cross-sectional area of the rock framework and, hence, its strength. Flat-lying, tabular-shaped clay particles generally exhibit high porosity at initial conditions. Clays and shales are easily compacted at low to moderate overburden pressures if pathways exist for the in-situ water to escape.

Eq. 2.2 indicates that the incremental loss of fluid pressure caused by oil and gas production must be transferred to the matrix. Compression of the rock increases the cross-sectional area of the matrix. The grain compressibility (c_{gr}) equals approximately 16×10^{-8} psi^{-1}. On the other hand, grain crushing and microfracturing occur when the stress exerted on the matrix becomes greater than the strength of the rock framework. In either case, there will be a resultant loss of pore volume. Connate water will migrate to other geologic horizons unless an impermeable shield bounds the volume. An overpressure condition is created in this instance.

Reservoir engineering calculations are often concerned with evaluating the reduction of pore volume brought about by the declining reservoir pressure. A rewritten form of the compressibility definition, shown in Eq. 2.3, accounts for the effect of compressibility on reservoir pore volume. The c_f term is defined as the formation compressibility even though it is a measure of the loss of pore volume as a function of decreasing fluid pressure. Compressibility on a pore-volume basis is defined as

$$c_f = -\frac{1}{V_p}\left(\frac{dV_p}{dp}\right) = \frac{1}{\phi_i}\left(\frac{\phi_i - \phi_2}{p_i - p_2}\right), \text{ psi}^{-1} \quad (2.3)$$

The rock compressibility for most competent rocks = 2 to 20×10^{-6} psi^{-1}. Rock-sample formation compressibility vs. decreasing reservoir pressure relationships are generally measured in the laboratory at room-temperature conditions. These values are difficult to relate to a field study because a small rock sample cut from a core is relatively homogeneous, but producing horizons typically are a collection of a number of rock types of different strengths and compositions. Therefore, it is difficult in the laboratory to measure the true average compressibility value for a particular reservoir.

Laboratory and reservoir condition relationships should be developed, particularly when studying high-temperature and -pressure reservoirs. The formation compressibility term is nonlinear to changes in reservoir pressure. Therefore, its value can change as a function of pressure even though it is often considered as a single-valued expression.

Rock-compressibility correlations are available in the literature. Those of Hall (1953) and van der Knaap (1959) are presented in **Fig. 2.1,** which shows that rock-compressibility correlations can be inaccurate in particular for friable and unconsolidated sandstones. Unfortunately, therefore, the use of these correlations is restricted to well-consolidated formations. The Hall plot reflects c_f = 3 to 4×10^{-6} psi^{-1} for most petroleum reservoir rocks. Newman (1973) indicated that formation compressibility for friable and unconsolidated rocks can vary in an unpredictable manner and can be as high as $c_f = 40 \times 10^{-6}$ psi^{-1}. Assuming that the formation compressibility is a single-valued term is not necessarily accurate.

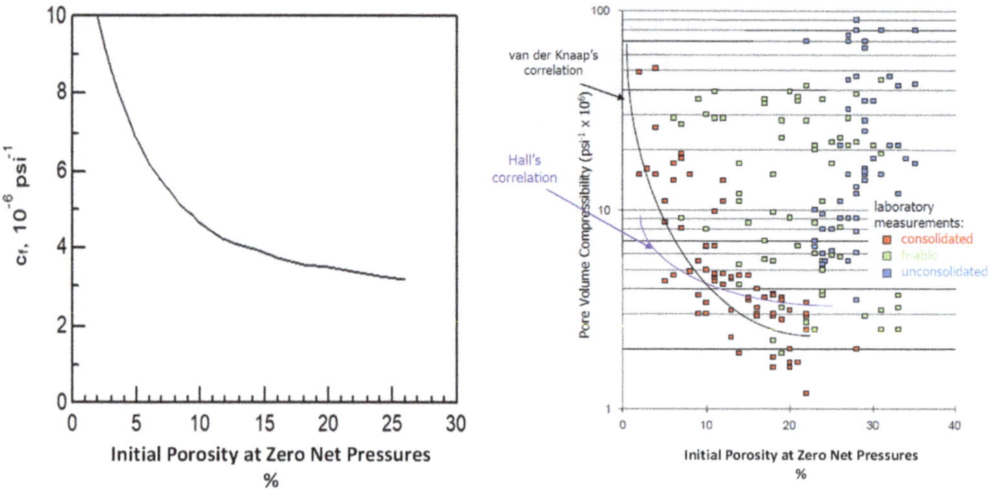

Fig. 2.1—(left) The Hall compressibility correlation for consolidated sandstones (after Bass 1987). The Hall correlation indicates formation compressibility should average 3 to 4 × 10^{-6} psi^{-1} for the majority of petroleum reservoirs located in well-consolidated rocks. (right) van der Knaap's correlation indicates higher values of compressibility at porosity values below 10% (after Bass 1987).

Example 2.1—Determining Porosity Loss. The pressure in an oil reservoir with a datum depth of 10,000 ft is 4,650 psia. Apply the Hall correlation to determine the expected porosity loss if production draws the reservoir pressure down to 1,500 psia when $\phi_i = 25\%$.

Solution. Use a rewritten form of Eq. 2.3 to calculate the expected loss in porosity caused by the decline in the reservoir pressure. Interpreting from Fig. 2.1 indicates that $c_f = 3.2 \times 10^{-6}$ psi^{-1}.

$$\phi_2 = \phi_i - \phi_i c_f \left(p_i - p_2\right) = 25.0 - 25.0\left(3.2 \times 10^{-6}\ \text{psi}^{-1}\right)\left(4,650 - 1,500\right) = 24.75\%.$$

Assume that the sand is poorly consolidated and the value developed from the Hall correlation could be incorrect by a factor of 10. For this case, $c_f = 32 \times 10^{-6}$ psi^{-1}.

$$\phi_2 = \phi_i - \phi_i c_f \left(p_i - p_2\right) = 25.0 - 25.0\left(32 \times 10^{-6}\ \text{psi}^{-1}\right)\left(4,650 - 1,500\right) = 22.48\%.$$

An incorrect estimate of the formation compressibility changes the loss-of-porosity effect that results from compaction by 9%. In certain instances, this apparently small error can profoundly influence the outcome of material-balance reservoir studies.

Example 2.2—Calculating Subsidence. A shallow, 600-ft-thick unconsolidated oil sand is under consideration for development. Its initial reservoir pressure is estimated at 1,800 psi. Artificial lift must be installed because of the lack of a solution-gas-drive effect in the low-gravity reservoir. Pressure depletion is expected and is probably the primary producing mechanism. The field is located in a populated area, and the potential for subsidence must be considered when reservoir pressure declines to 600 psi. Estimated rock properties: $\phi = 40\%$, $c_f = 80 \times 10^{-6}$ psi^{-1}. Calculate the possible subsidence using the ratio of original porosity to predicted porosity as a function of the estimated thickness.

Solution. Rearrange the compressibility equation (Eq. 2.3) to calculate the potential for subsidence:

$$\phi_2 = \phi_i - \phi_i c_f \left(p_i - p_2\right) = 40.0 - 40.0\left(80 \times 10^{-6}\right)\left(1,800 - 600\right) = 36.16\%.$$

Shrinkage = $h(\phi_i / \phi_2) = 600$ ft (36.16/40) = 542 ft.

Subsidence = h – shrinkage. Thus, the subsidence could be approximately 58 ft.

2.3 Permeability

The ability of interconnected pore spaces to transmit fluid is a function of the absolute permeability. Darcy studied the flow of water through a sandpack (Craft and Hawkins 1959). The outcome of this effort was the development of an equation describing the laminar flow of a fluid saturating 100% of the pore space through a porous medium. The generalized Darcy equation, shown in Eq. 2.4a, is expressed in terms of flux velocity v and includes a viscosity expression to account for the resistance to flow exhibited by different types of fluids. All fluids are assumed to be nonreactive to the rock and cementing material.

$$v = \frac{q}{A} = \frac{k}{\mu}\frac{d\psi}{ds}, \qquad (2.4a)$$

where q is the rate of flow, μ is the fluid viscosity, A is the cross-sectional area perpendicular to flow, and $d\psi$ is the potential energy gradient applied across the differential of the length ds of the flow system.

Hydrologists use the flux velocity term to describe the flow rate of water passing through or across some known cross-sectional area, such as a dam where the width and water depth are easily measured. Therefore, flux velocity is equal to q/A, whose units are L/t. The $d\Psi/ds$ potential function of the pressure drop across the system includes the addition or subtraction of any gravity effects.

The coefficient k, defined as the permeability, is a measure of the cross-sectional area available to flow, with dimensions of length squared (L^2). Permeability is typically defined in millidarcy units, or md. The majority of the simplified petroleum reservoir fluid-flow equations discount gravity effects; however, in certain instances, gravity can profoundly influence reservoir performance. Absolute or specific permeability refers to permeability with one fluid present at 100% saturation; air and water permeability are examples, and the units are darcies or millidarcies. When a second or third phase is introduced, the resulting permeability to each phase is called the effective permeability, and the units are darcies or millidarcies.

In oilfield units, the form of Eq. 2.4a for calculating porous medium permeability is

$$k = \frac{0.001127 q B_o \mu L}{A \Delta p}, \qquad (2.4b)$$

where q is the rate of flow (STB/D), B_o is the formation volume factor (res bbl/STB), μ is the fluid viscosity (cp), L is the length of the flow system (ft), A is the cross-sectional area perpendicular to flow (ft^2), and Δp is the pressure (psi) applied across L.

Two corollaries to the Darcy equation are of interest in petroleum reservoir calculations. They are the Klinkenberg effect and the Forchheimer equation.

The Darcy equation can be used to calculate the permeability of a core sample when the pressure differential exerted across the sample and the gas flow rate are measured. However, the mean free path of the gas molecules at low pressures can approach the size of the capillaries of very-low-permeability rock. This induced slippage of the molecules moving through the capillary distorts the calculated permeability value because Darcy's law assumes laminar flow conditions when liquid velocity is zero at the capillary walls. Klinkenberg (1941) introduced Eq. 2.5 to define the influence of pore size and the mean free path of the gas molecule. Notice that the influence of the b coefficient on the equation is reduced as permeability increases.

$$k_g = k_L\left(1+\frac{b}{p_m}\right) \quad (2.5)$$

As shown in **Fig. 2.2,** a straight line results when the calculated gas permeability (k_g) is plotted against the reciprocal of the mean ($1/p_m$), or average pressure, of the test. The intersection of the line on the *y*-axis is the true permeability (k_L). The Klinkenberg effect should be considered when studying low-permeability core samples because core analysis studies generally are conducted at moderate to low pressures using gas. Determining permeability using a gas is similar to the method using a liquid. However, extra terms are required to account for the changes in gas properties. The Klinkenberg correction could be significant in lower-permeability rocks and is dependent on the pore-network tortuosity and the size of the gas molecules used in the flow test, as observed in Fig. 2.2a.

Because gases have very low viscosities relative to liquids, the use of gas is often preferable because it helps speed up the core testing. The transient time reduction can be very significant for low-permeability rocks.

An additional pressure drop is induced into the system in the presence of turbulent flow. A quadratic equation suggested by Forchheimer (1901) adds this effect to express Darcy's law in terms of laminar and turbulent flow conditions. The right-hand term accounts for the effect of turbulence. Eq. 2.6 reverts to the Darcy equation for laminar flow conditions.

$$-\frac{\mathrm{d}p}{\mathrm{d}L} = \frac{\mu v}{k} + \beta\rho v^2, \quad (2.6)$$

where $\mathrm{d}p/\mathrm{d}L$ is the pressure drop across the length of the system, ρ is the fluid density, μ is the fluid viscosity, v is the flux velocity, k is the permeability, and β is the turbulence factor.

An isotropic reservoir displays the same permeability in all directions [vertical (k_v) and horizontal (k_h)]. Anisotropic reservoirs can display permeability in a particular geographical direction. In some reservoirs, permeability can vary significantly with direction. To maximize recovery efficiency, the

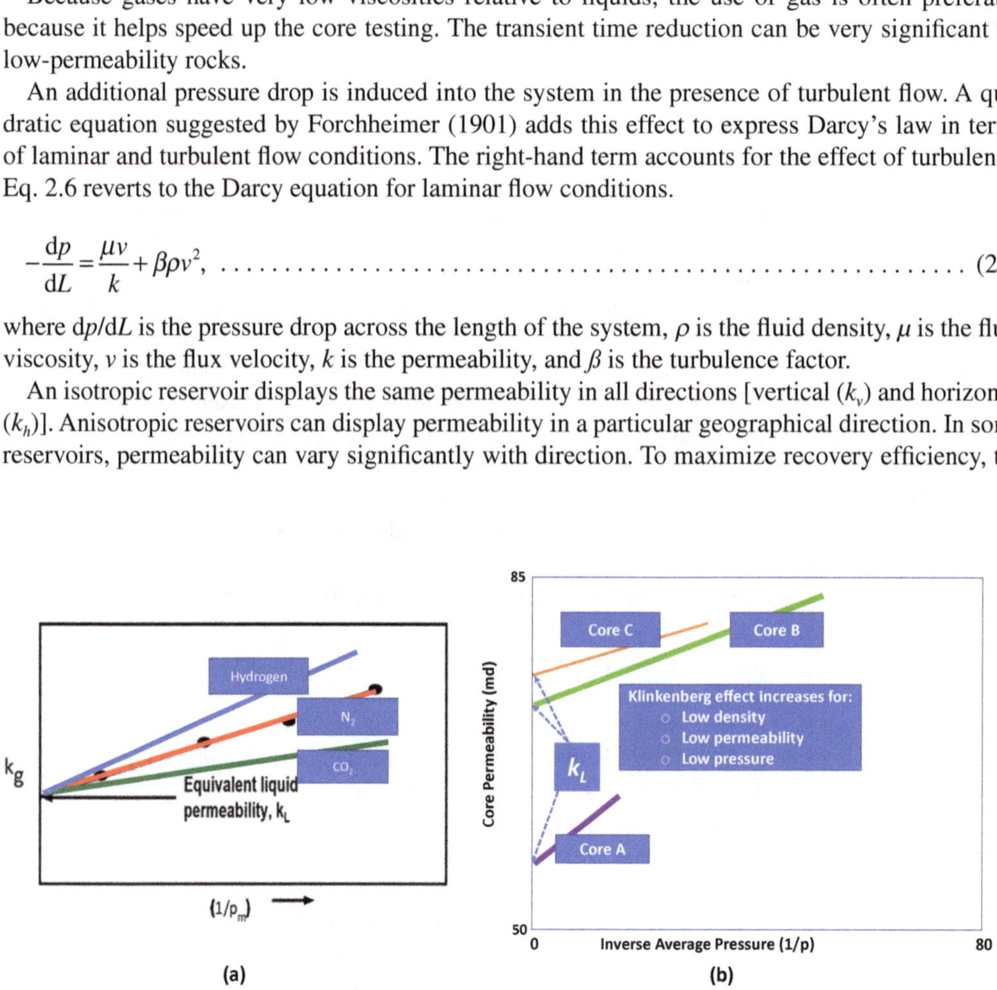

Fig. 2.2—Klinkenberg effect. (a) As the pressure used in the laboratory tests increases, the measured value of permeability increases. The data obtained with lowest-molecular-weight gas yield the straight line with the greatest slope, which is indicative of a greater slippage effect. (b) High gas velocities in the cores can lead to erroneously high values of permeability. This is most pronounced for measurements at low pressures and in low-permeability rocks, as seen in Core A. Extrapolated values of the straight line to the reciprocal of the mean ($1/p_m$), or average pressure, of the tests provide an estimate of k_L.

path of water movement from an injector to a producer should be in the direction of least permeability in a waterflood.

2.3.1 Relation between Porosity and Permeability. The most-common practice is to measure porosity and permeability in the laboratory. Upon arrival, the core is cut into two pieces (one-third and two-thirds pieces). The one-third section is slabbed, often photographed using regular and ultraviolet light, described by geologists, and eventually archived for future study or review. The two-thirds section is the part from which core plugs are cut for routine and special core analysis. Core plugs for routine analysis might be selected on a one per 1-ft or one per 2-ft basis. Core plugs for special core analysis might not be selected until the core is sufficiently described.

If the reservoir is very heterogeneous, core testing will be conducted using the entire core or full-diameter pieces from the entire core. **Fig. 2.3** shows values of permeability vs. porosity for different clastic and carbonate formation samples.

2.3.2 Darcy's Law for Liquid Flow—Closed Outer Boundary Conditions—Oilfield Units. Assumptions:

- Porous medium: isotropic and homogeneous
- Slightly compressible fluid
- Laminar/radial flow
- Steady-state conditions

$$q=\frac{0.00708kh\left(\bar{p}-p_{wf}\right)}{\mu B\left[\ln\left(r_e/r_w-0.75\right)\right]},\ \text{STB/D}, \qquad (2.7)$$

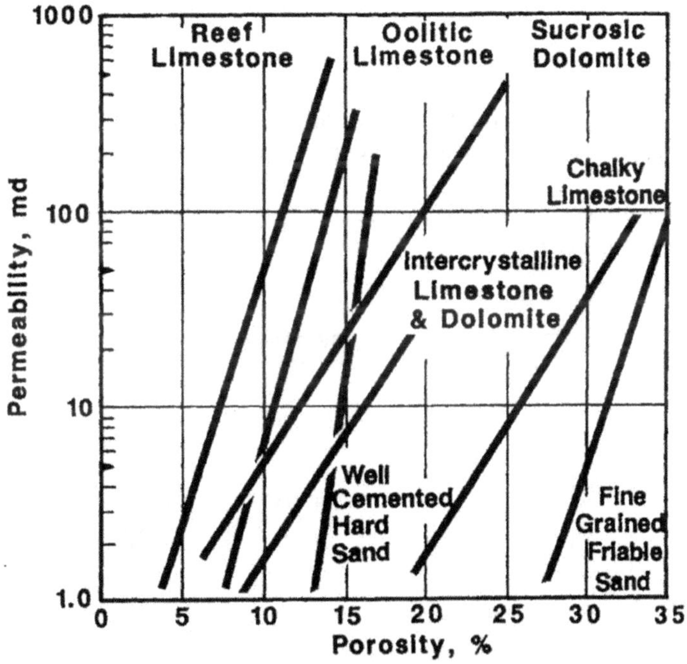

Fig. 2.3—Permeability vs. porosity laboratory measurements, showing a logarithmic relationship (after Tiab and Donaldson 1996).

where k is the permeability (md), B is the formation volume factor (res volume/STB volume), r_e is the radius to the edge of the reservoir (ft), h is the net sand thickness (ft), r_w is the radius to the edge of the borehole (ft), μ is the fluid viscosity (cp), p_{wf} is the bottomhole flowing pressure (psia), and $\overline{p}$ is the average static pressure (psia).

2.3.3 Heterogeneous Flow Conditions. The method used to average permeability depends on how the permeability is distributed in the reservoir. To describe a reservoir thoroughly, one needs to know the vertical and lateral permeability distributions. Lateral distribution can be estimated from individual-well and multiwell pressure transient tests. Vertical permeability distributions can be estimated from core analysis in individual wells.

The geometric average is used when permeability varies randomly. The arithmetic average is applicable to flow parallel to depositional bedding boundaries and when layer permeabilities are similar over a relatively large areal extent. The harmonic average is applicable to flow perpendicular to depositional bedding boundaries and when layer permeabilities are similar over a relatively large areal extent.

Electrical circuits can be used as an analogy when we consider averaging permeability. Permeability is a measurement of "conductance," which is the reciprocal of "resistance." In an electrical circuit, the average resistance of two resistors connected in parallel is the harmonic average.

In this analogy, the average permeability of two parallel beds is the arithmetic average, and the opposite is true for beds connected in series (arithmetic mean for resistance but harmonic mean for permeability). For more-complex configurations (random flow), the geometric mean is the most appropriate to use. **Fig. 2.4** shows the layered or parallel flow, series flow, and random flow systems.

Eqs. 2.8a and 2.8b show the expressions for calculating rate and average permeability for the parallel flow system:

$$q_t = q_1 + q_2 + q_3 + \ldots + q_n, \qquad (2.8a)$$

$$\overline{k}h_t = k_1 h_1 + k_2 h_2 + \ldots + k_n h_n. \qquad (2.8b)$$

Fig. 2.4b shows the concept of series flow, which applies in the case of near-wellbore permeability alteration. This is a function of the change in the reservoir permeability and can act in a positive or negative manner.

Eq. 2.9 shows an expression for calculating the average permeability displayed by an affected well (either damaged or stimulated):

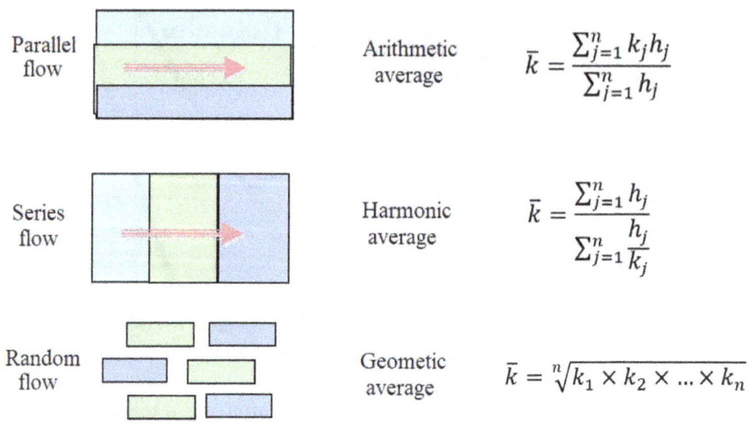

Fig. 2.4—Parallel flow system; series flow system; and random flow system.

$$\bar{k} = \frac{\ln(r_e / r_w)}{\dfrac{\ln(r_d / r_w)}{k_d} + \dfrac{\ln(r_e / r_d)}{k_{\text{res}}}}. \quad (2.9)$$

Random flow will occur when the porous medium shows a statistically random distribution of permeability, such as that seen in Fig. 2.4c.

2.4 Fluid Distribution

Petroleum reservoirs are generally assumed to contain only water at the time of formation because sediments are typically deposited in an aqueous environment. Eventually, oil and gas molecules are formed in-situ or migrate in from an outside source to redistribute the original reservoir water saturation to some combination of the water, oil, and gas phases. Saturation values assume pore spaces are 100% fluid-filled. The total fluid saturation must equal unity when the different fluid saturation values are expressed as fractions.

Dividing pore water volume by total pore volume results in the basic saturation equation. Comparable equations for calculating the oil and gas saturations can be written when the appropriate subscripts are inserted into the numerator of Eq. 2.10.

$$S_w = \frac{V_w}{V_p}, \text{ fraction}. \quad (2.10)$$

The sum of the fluid saturations in the reservoir is equal to unity when water, oil, and gas saturation values are expressed as fractions in Eq. 2.11:

$$S_w + S_o + S_g = 1. \quad (2.11)$$

2.4.1 Fluid Distribution and Capillary Mechanisms. Fundamental physical properties govern the distribution of reservoir fluids. These properties are (1) adhesion tension, which is the molecular attraction between a fluid and the rock material; (2) interfacial tension, which acts as a membrane-like interface between two immiscible fluids; and (3) the density difference between fluids.

The preference of a particular fluid to adhere to a rock surface in a two- or three-phase system is defined as wettability, which is also defined as the tendency of one fluid to spread on or adhere to a solid surface in the presence of other immiscible fluids. Wettability is the measure of the preference the solid has for a particular fluid over others. In relation to capillary pressure, wettability determines which pores retain a given fluid.

Wettability is a function of the type of fluid and the solid surface. Oil contains chemicals that are capable at times of changing the initial wettability, and drilling and injected fluids often contain surfactants that affect wettability and the contact angles between oil and water.

The pressure difference existing across the water/oil interface is defined as the capillary pressure. In a more general sense, capillary pressure is the pressure difference existing across the interface separating two immiscible fluids. The difference between the pressure of the nonwetting phase and that of the wetting phase is known as capillary pressure. The basic capillary pressure equation for an oil/water system is shown in Eq. 2.12. This energy differential is the result of interfacial tension and the adhesion effects discussed previously.

$$p_c = p_{nwt} - p_{wt}, \quad (2.12)$$

where p_{nwt} is the pressure of the nonwetting phase and p_{wt} is the pressure of the wetting phase.

An example of a water/oil system in contact with a quartz plate is illustrated in **Fig. 2.5** for a water-wet condition.

The system is deemed to be water-wet because the water drop is able to displace the oil phase and preferentially adhere to the quartz plate.

Fig. 2.6 provides another example (image on the left) of a water-wet system. The contact angle of the water drop in the presence of oil is less than 90°. This implies water wettability. The water phase coats or wets the rock grains and occupies the very small pore spaces. The oil phase is located in the larger pore spaces and is prevented from touching the rock grains by the water phase. The other drop (image on the right in Fig. 2.6) has a contact angle of water resting on a solid surface in the presence of oil. The adhesion tension between the oil and the quartz plate is less than the water/quartz plate adhesion tension. The capillary pressure of the oil/water system shown on the left in Fig. 2.6 is represented by Eq. 2.13. The smaller pores retain the water phase at higher capillary pressures in water-wet rocks.

$$p_c = p_o - p_w = \frac{2\sigma_{ow}\cos\theta}{r}, \text{ psi}, \quad (2.13)$$

where σ_{ow} is the interfacial tension between the two immiscible fluids, which in this case are oil and water; r is the average radius of the capillary; and $\cos\theta$ is the angle of the interface at the solid/oil/water contact. This angle is measured through the denser phase to the contact wall; see **Fig. 2.7,** in addition to Figs 2.5 and 2.6. In oil/brine systems, it is the angle within the water phase between the solid surface and the oil/brine interface.

The normal order of wetting preference with regard to reservoir rock is water, oil, and gas. As expected, the wettability effects between gas and a sand grain are minimal. The marine origin of most reservoir rocks usually ensures the presence of a water-wet system at initial conditions.

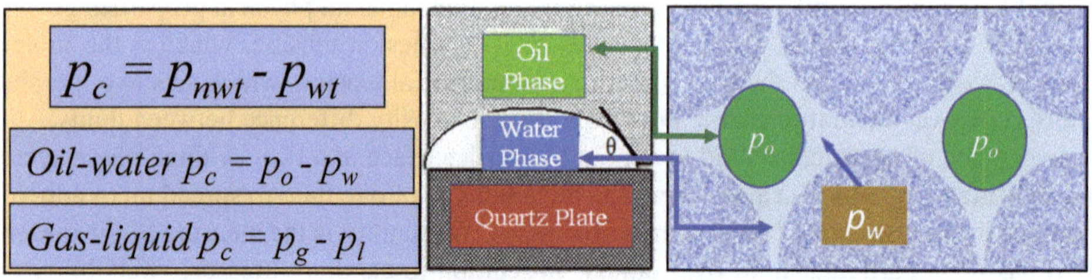

Fig. 2.5—Oil and water locations in a water-wet system representing results from a laboratory experiment. The figure shows oil and water phases and the contact angle between the wetting phase (water) and the surface (quartz plate).

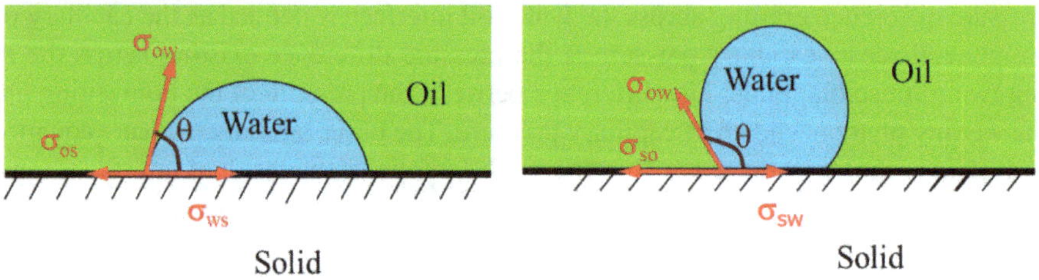

Fig. 2.6—Oil and water locations in a (left) water-wet system and an (right) oil-wet system. Courtesy of Schlumberger NExT.

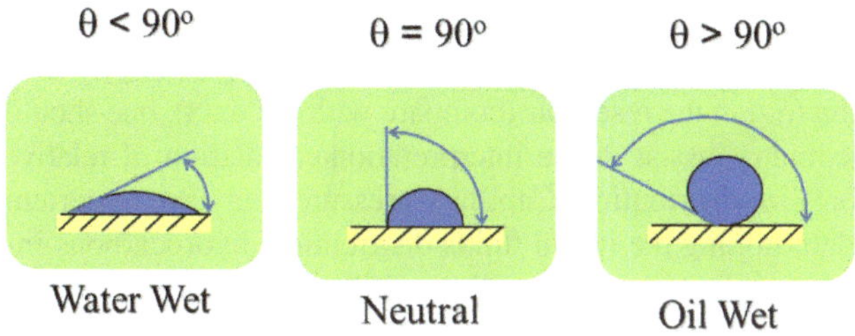

Fig. 2.7—Solid, oil, and water locations in a water-wet system on the left; a neutral-wet system in the center; and an oil-wet system on the right. The contact angle is measured through the denser phase. Courtesy of Schlumberger NExT.

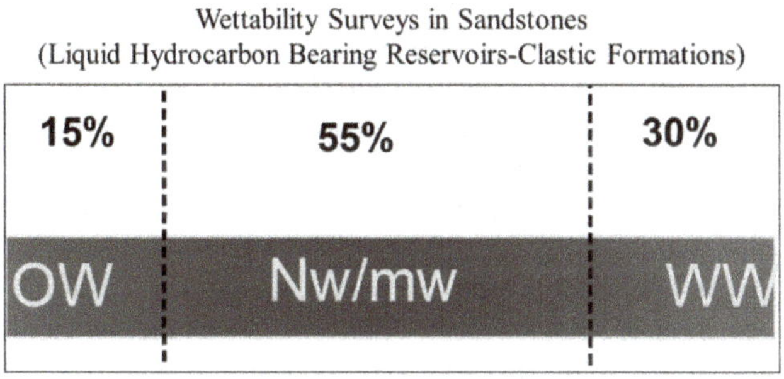

Fig. 2.8—Worldwide reservoir wettability surveys in liquid hydrocarbon sandstone reservoirs (clastic formations) (after Bennion 2010). OW = oil-wet; Nw/mw = neutral-wet/mixed-wet; and WW = water-wet.

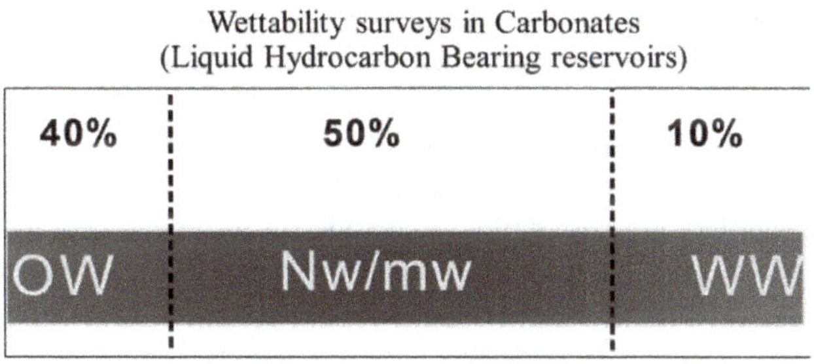

Fig. 2.9—Worldwide reservoir wettability surveys in liquid hydrocarbon carbonate reservoirs (after Bennion 2010). OW = oil-wet; Nw/mw = neutral-wet/mixed-wet; and WW = water-wet.

Sandstones (in general) tend to be water-wet; however, organic acids sometimes convert reservoirs to an oil-wet condition, but the majority of oil fields are intermediate- to water-wet. The silica crystals can react with the brine to form a weak acid, and this repels the acidic crude components. Exceptions depend on rock mineralogy and crude composition. **Fig. 2.8** shows the results of worldwide reservoir wettability surveys in sandstones (Bennion 2010). Thirty percent of the reservoirs were water-wet.

Carbonate reservoirs exhibit a higher incidence of the mixed to oil-wet condition than clastic reservoirs do. Carbonate grain surfaces are basic in nature, and crude oil often contains polar compounds that are acidic in nature. Hence, carbonate grains are attracted to the crude oil and are intermediately to strongly oil-wet. The polar compounds can be chemisorbed on the grain surfaces. There are exceptions because not all crudes have acidic components, and the presence of clay can render the carbonate water-wet. **Fig. 2.9** shows the results of worldwide reservoir wettability surveys in carbonates (Bennion 2010). Only 10% of the reservoirs were water-wet.

Wettability and interfacial tension effects typically cause water to inhabit the smallest pore spaces while gas occupies the largest pores in a three-phase system. Oil remains trapped in the regions between the water and gas accumulations. Changing either the water or the gas saturation values during reservoir depletion can rapidly reduce the average cross section available to oil, which in turn reduces the ability of oil to flow through the reservoir. For instance, establishing a gas saturation in a pressure-depleting oil reservoir can result in a significant increase in the gas/oil ratio because liberated gas tends to reside in the larger-sized pore spaces. Gas flow is increased in this instance while the oil flow rate is reduced. Water adheres to the grain faces while gas resides in the largest pore spaces and oil is trapped between the two phases.

2.4.1.1 Displacement Processes. Capillary pressure data have important applications in reservoir development when planning, designing, and monitoring a waterflood project. In particular, when predicting the water saturation distribution (S_w) in the reservoir (compare with log data), one should predict the free-water level and pore geometry, assist in the interpretation/calculation of relative permeability, and calculate the pore-throat size distribution. Capillary pressure data have important uses in reservoir management, such as determining the initial fluids distribution, hydrocarbons-in-place, reserves, and residual oil; establishing the criteria for well placement and well completion design; and optimizing fluid displacement efficiency.

Differential fluid movement within a system can be defined in the context of two different displacement processes: imbibition displacement and drainage displacement (**Fig. 2.10**). Understanding these displacement relations is particularly important when studying relative permeability curves and reservoir-depletion concepts.

Imbibition Displacement. This term is defined as the spontaneous encroachment of the wetting fluid into a two-phase system. Imbibition displacement implies the wetting fluid saturation increases at the expense of the nonwetting fluid saturation. The mobility of the wetting phase increases as the wetting phase saturation increases. A water-displacing-oil imbibition-displacement process occurs when natural water encroachment or waterflooding processes displace reservoir oil. For this particular case, a pressure differential also helps in the oil-displacement process. For example, waterflooding a fractured oil reservoir in which the reservoir rock is preferentially water-wet will increase liquid hydrocarbon recovery.

Drainage Displacement. The wetting phase saturation is reduced by introducing additional nonwetting phase fluid into the system. An external pressure must be applied to overcome the capillary force of the wetting fluid wanting to remain in the smaller-sized pores. The mobility of the nonwetting fluid phase increases as the nonwetting phase saturation increases. A gas cap expanding downward into the oil zone is an example of such a process. Other examples include the injection of gas into a naturally fractured reservoir where the nonwetting phase replaces the wetting phase (gravity drainage), a hydrocarbon (oil or gas) filling the pore space and displacing the original water of deposition in water-wet rock (migration from the source rock), waterflooding an oil reservoir where the reservoir rock is preferentially oil-wet, and pressure maintenance or gas cycling by means of gas injection into a retrograde condensate reservoir. Drainage displacement implies that a nonwetting fluid displaces some portion of the wetting fluid.

Fig. 2.11 compares the shapes of the two different capillary pressure curves. At a given water saturation, the drainage curve requires a higher capillary pressure. The difference in behavior between the two curves is termed hysteresis. The capillary pressure drainage curve is always higher than the capillary pressure imbibition curve.

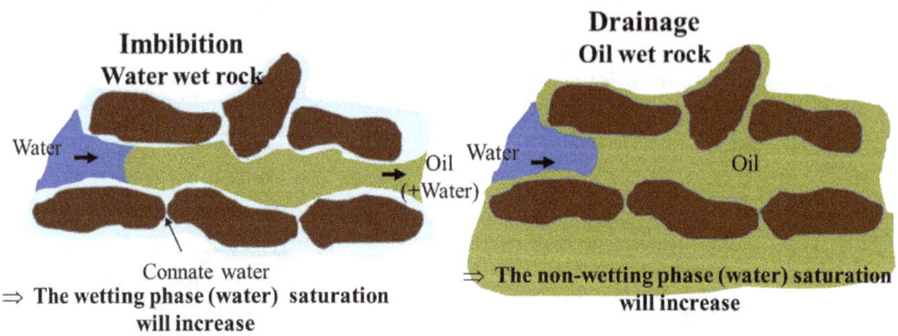

Fig. 2.10—Imbibition and drainage displacement (courtesy of Schlumberger NExT). Imbibition: The water saturation will increase when the nonwetting oil (big pores) is displaced by the wetting-phase water. Drainage: When the wetting oil (small and big pores when S_o = 1) is displaced by the nonwetting water, the water saturation will increase.

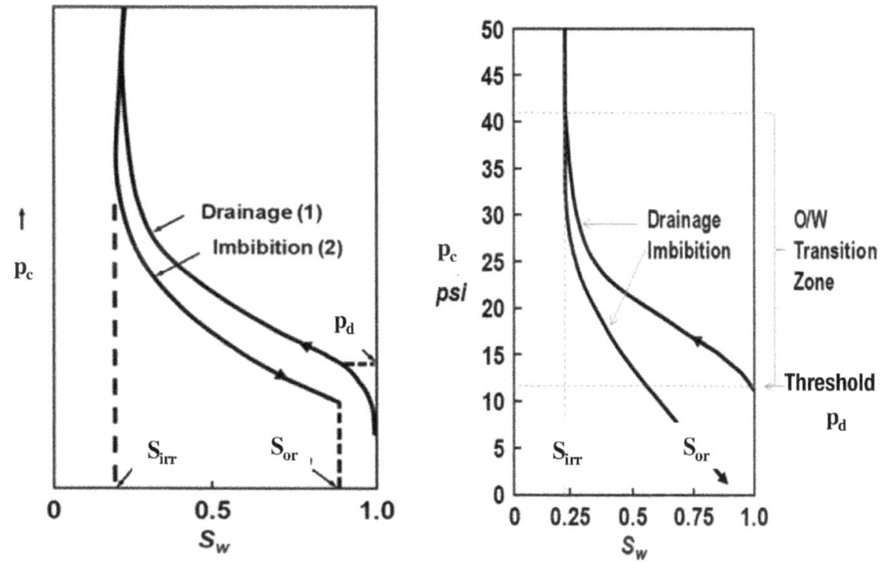

Fig. 2.11—Typical drainage and imbibition capillary pressure curves and the effect of saturation history on the shape of the capillary pressure curve. The imbibition process ends at the residual oil saturation value while the drainage process ends at the irreducible water saturation (S_{irr}) value.

In Fig. 2.11, corresponding to the drainage curve, p_d is the displacement or threshold pressure required to force nonwetting fluid into a sample initially saturated with the wetting phase. The displacement or threshold pressure is required to drain the largest pore. The entry (displacement) pressure is defined as the pressure required to force nonwetting fluid through an initially wetting-phase-saturated sample, and this corresponds to the drainage curve.

Spontaneous imbibition (S_i) does not require displacement pressure to push the wetting phase out of the core and the nonwetting phase into the core, as shown in the left-hand side of Fig. 2.11. During the drainage process, an irreducible water saturation S_{irr} will be obtained, as shown in the right-hand side of Fig. 2.11, and the imbibition process shows the residual oil saturation S_{or}. The difference between the two endpoints S_{irr} and S_{or} describes the rock quality.

The initial fluid saturation in a reservoir is generally governed by the drainage saturation history because the pores are initially 100% water filled. Oil and gas migrate into the structural trap and displace a portion of the connate water later in geologic history. Gravity effects supply the necessary force during this particular displacement process.

2.4.1.2 Laboratory Studies. Laboratory capillary pressure studies are commonly performed using the centrifuge method (**Fig. 2.12**) because of the speed and simplicity of the analysis procedure. Initially, the core sample is completely saturated with brine. A laboratory capillary pressure curve is developed by placing the saturated sample in a centrifuge and measuring the RPM/water saturation relation at different speeds. Similar studies should be conducted on a variety of rock samples with different porosity and permeability values representing the entire producing interval. Unfortunately, the capillary pressure relationships established by the centrifuge method are measured in an air/brine environment. These results must be converted to reservoir conditions.

In converting laboratory data to reservoir conditions, the assumption is that the average radius of the capillaries of the laboratory sample and the field rock is the same. Applying this concept results in Eq. 2.14, relating laboratory conditions to those of the reservoir:

$$\frac{2\sigma_{aw}\cos\theta_l}{p_{cl}} = \frac{2\sigma_{ow}\cos\theta_r}{p_{cr}} \quad (2.14)$$

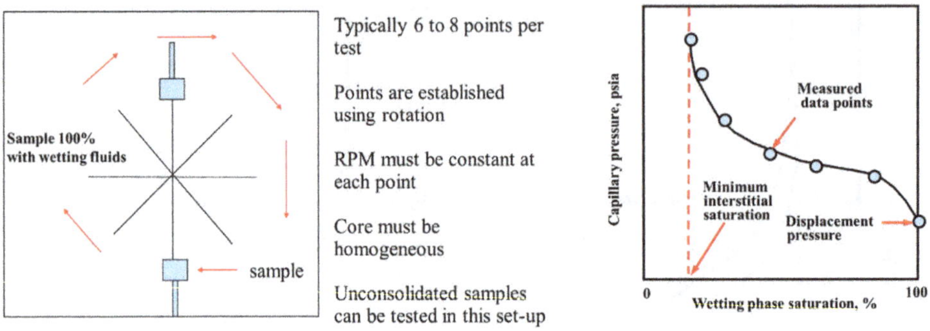

Fig. 2.12—Typical capillary pressure using the centrifuge method. Courtesy of Schlumberger NExT.

Solving for the equivalent capillary pressure at reservoir conditions,

$$p_{cr} = \left(\frac{\sigma_{ow} \cos\theta_r}{\sigma_{aw} \cos\theta_l} \right) p_{cl} \quad \text{(2.15)}$$

Eqs. 2.14 and 2.15 indicate the laboratory and reservoir capillary pressure curves are related when the contact angle and the interfacial tension at each condition are known. Capillary sizes are assumed to be the same.

Other widely used laboratory methods for measuring capillary pressure are mercury injection (destructive test) and the porous plate method (nondestructive, very slow test).

Example 2.3—Calculating Capillary Pressure. Laboratory studies indicate p_{cl} = 8.3 psia when S_w = 50%. What is the equivalent capillary pressure at reservoir conditions? Rock and fluid properties measured in the laboratory are σ_{ow} = 24 dynes/cm, σ_{aw} = 72 dynes/cm, cos $\theta_r = 20°$, and cos $\theta_l = 0°$.

Solution. Applying Eq. 2.15 results in

$$p_{cr} = \left(\frac{\sigma_{ow} \cos\theta_r}{\sigma_{aw} \cos\theta_l} \right) p_{cl} = \left[\frac{(24)\cos 20°}{(72)\cos 0°} \right] 8.3 \text{ psia} = 2.6 \text{ psia.}$$

2.4.1.3 Curve Shapes. Capillary pressure curves for three different porosity sands are compared in **Fig. 2.13** to illustrate the relation between water saturation, capillary pressure, and the saturation height above the free-water level.

The free-water level occurs when only water is present and $p_c = 0$. The significant points of interest defining a drainage-process capillary pressure curve are

- Entry pressure is typically defined at S_w = 100%. This is the pressure required to initially force the oil phase into the core sample. The introduction of the oil phase into the core sample reduces the saturation of the wetting phase. As a rule of thumb, entry-level capillary pressure increases as a function of decreasing pore size. Cossé (1993) states that the entry pressure can be observed at S_w < 100% values when extremely large openings such as vugs are present in the rock sample.

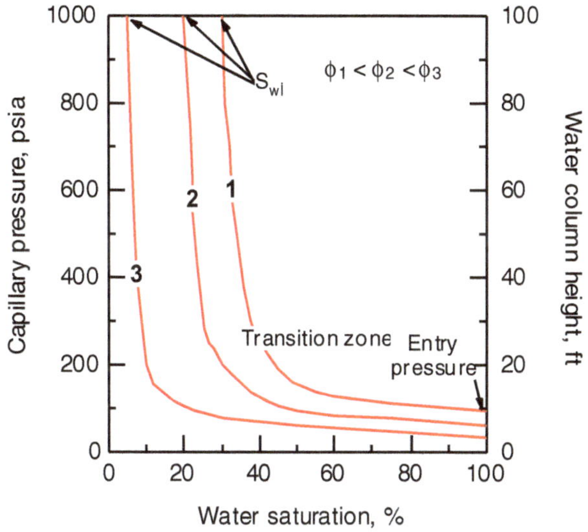

Fig. 2.13—The effect of pore-space size on capillary pressure. The lower-porosity smaller-pore-size samples exhibit the highest capillary pressure.

- The average value of the vertical portion of the capillary pressure curve represents the irreducible water saturation S_{irr} value. The value indicates the state in the core when the addition of more oil into the system will not materially reduce the water saturation. For most cases, the irreducible and initial water saturation values can be assumed to be the same.
- The transition zone reflects the pore-size distribution of the rock sample. Poorly sorted sands exhibit an extended transition zone.

Porosity values for the three curves shown in Fig. 2.13 are as follows: $\phi_1 < \phi_2 < \phi_3$. Higher capillary pressure entry levels and irreducible water saturation values characterize the lower-porosity ϕ_1 curve. The sharp curvature of all three lines at the point of transition reflects samples that are composed of rather uniform sets of pore sizes but varying porosity. The right-hand vertical scale represents the calculated height of the saturation value above the oil/water contact (OWC), which corresponds to values of capillary pressure greater than the $p_c = 0$ level. A comparison of the three curves indicates the water saturation increases as a function of decreasing porosity sand at any given level above the free-water zone.

Fig. 2.14 illustrates the effect of pore-size variation on the shape of the capillary pressure curve. Pore-size distribution is typically a function of grain-size variation for clastic rocks. A homogeneous system of approximately the same-sized pores restricts the transition zone represented in Curve 3 as the curve trends from essentially a horizontal to a vertical direction. Core Sample 1, which is composed of a wide variety of pore sizes, extends the transition zone over the entire capillary pressure range. The capillary pressure curve gradually inclines upward because the entry pressure must continually increase to force the nonwetting fluid into the next set of smaller-sized pore spaces. Forcing water into Core Sample 3 occurs at essentially a single pressure level because the majority of the pores in the sample are entered simultaneously when the minimum entry-level pressure has been reached. Curve 2 in Fig. 2.14 is drawn to show that a particular core sample can exhibit any degree of sorting ranging between the two boundary curves. An estimation of the irreducible water saturation for Sample 1 is much more difficult than for Sample 3.

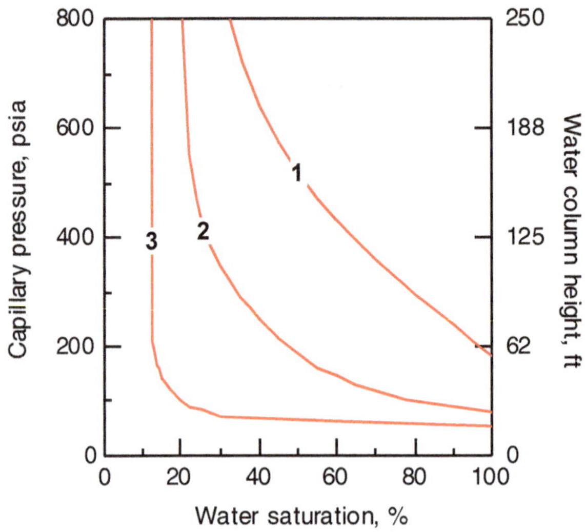

Fig. 2.14—The effect of particle sorting on capillary pressure. The curvature becomes less pronounced as the particle-size distribution increases. Interpreting the oil/water contact from a well log would be much easier for Sample 3 than for Sample 1.

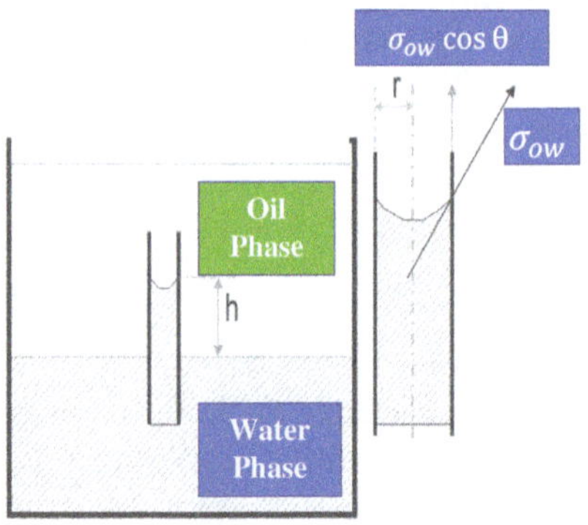

Fig. 2.15—Water/oil system showing the rise of water in the capillary tube.

Resistivity log deflections through the productive interval should mirror the shape of the capillary pressure curves shown in Fig. 2.14. For instance, the OWC displayed by a log run through a productive interval comprised of Core Sample 3 rock material should be obvious while that for Core Sample 1 would not be as much.

2.4.1.4 Height of a Water Column. Eq. 2.15 indicates capillary pressure is an inverse function of capillary size and is directly related to interfacial tension effects. This equation can be reformed to define the height of the rise of the wetting fluid above the free-water level, which is defined as a point of zero capillary pressure in a two-phase system. There is no pressure difference across the interface between the two phases when capillary pressure effects are zero because the water phase predominates. The following discussion concerns an oil/water system, but other combinations can be chosen as long as the two fluids are immiscible.

Fig. 2.15 shows a capillary tube immersed in water/oil at equilibrium. It can be observed that water rises in the capillary tube, which is preferentially water-wet, and a forces balance applies where interfacial tension forces = hydrostatic forces.

Pressure resulting from gravity forces is expressed by Eq. 2.16a, and constructing a forces balance between the interfacial tension attraction and gravity forces results in Eq. 2.16b:

Force up = force down,

$$p=\rho gh. \qquad (2.16a)$$

$$2\pi r\sigma_{nwt-wt}\cos\theta=\pi r^2 h\left(\rho_{wt}-\rho_{nwt}\right). \qquad (2.16b)$$

Eq. 2.16b can be rearranged to solve for the height to which the water will rise in the capillary tube:

$$h = \frac{2\sigma_{nwt-wt} \cos\theta}{r(\rho_{wt} - \rho_{nwt})} \quad (2.17)$$

The height h (and, hence, p_c) in the tube experiment depends on the adhesion tension between the oil and water, the radius of the tube, and the density difference between the immiscible fluids.

Eq. 2.12 can be rearranged to define the height of the rise of the wetting fluid above the oil/water interface as a function of capillary pressure and the difference in density between the two fluids:

$$p_c = p_{nwt} - p_{wt} = \frac{2\sigma_{nwt-wt}}{r} \cos\theta. \quad (2.18)$$

The capillary pressure between oil (nonwetting phase) and water (wetting phase) is defined in Eq. 2.19a:

$$p_{cow} = p_o - p_w. \quad (2.19a)$$

Substituting Eq. 2.19a for Eq. 2.17, an expression for the height h above the free-water level is given as

$$h = \frac{(144 \text{ in.}^2/\text{ft}^2)\, p_c}{\rho_w - \rho_o}, \quad (2.19b)$$

where h is the height above the zero capillary pressure level (free-water level) in ft, p_c is the capillary pressure in psi, and $\rho_w - \rho_o$ is the appropriate fluid density in lbm/ft^3.

Fig. 2.16 illustrates the physical meaning of Eqs. 2.17 and 2.19b in the context of a water-wet water/oil system occupying three capillary tubes of differing cross-sectional areas when $r_s < r_m < r_l$. Capillary effects cause water to rise in opposition to gravity and displace oil as a function of the capillary size. This effect, as would be expected, is more pronounced in the smaller-sized capillaries because the radius is located in the denominator of Eq. 2.18. Fig. 2.16 shows why water saturations are generally higher in fine-grained, low-porosity sands when compared to higher-porosity sands encountered at essentially the same depth in the well. This pore-size/column height/water saturation relationship was also illustrated in Fig. 2.15.

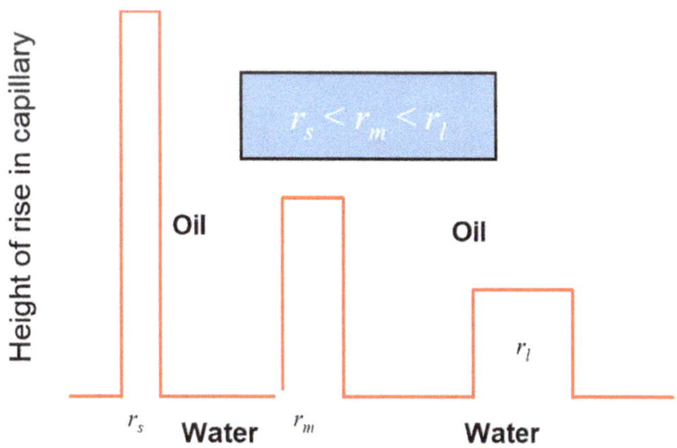

Fig. 2.16—The height of the rise of water in a capillary is a function of the capillary radius of the pores containing the oil/water system. r_s = small radius, r_m = medium radius, and r_l = large radius.

The discussion shows that very-low-porosity porous media can act as a barrier to the passage of hydrocarbons. Berg and Avery (1995) discussed how the grinding and pulverizing action in growth fault shear zones can dramatically reduce the original reservoir porosity. The formation of the fine-grained shear zones in a poorly consolidated sedimentary section is the result of the dragging of ductile shale along the fault zone. Sand grains are ground down to a smaller particle size during the deformation process. The resulting zone of very-small-sized pores exhibits an extremely high water saturation, which can act as a trapping seal to migrating hydrocarbons.

Fig. 2.17 presents the results of one of the Berg and Avery (1995) case studies. Compare the k = 0.09 md of the shear-zone material to the k = 158 md of the reservoir sand. Capillary effects raise the height of the apparent OWC from approximately 3 ft in the unaffected reservoir rock to nearly 320 ft in the shear zone.

Example 2.4—Calculating Equivalent Height. Example 2.3 indicated that p_{cr} = 2.6 psia when S_w = 50%. What is the equivalent height above the zero capillary pressure line in the reservoir when ρ_w = 67.0 lbm/ft^3 and ρ_o = 52.4 lbm/ft^3?

Solution. Apply Eq. 2.19b to relate the capillary pressure to the height of the rise in the capillary:

$$h = \frac{\left(144 \text{ in.}^2/\text{ft}^2\right) p_c}{\rho_w - \rho_o} = \frac{\left(144 \text{ in.}^2/\text{ft}^2\right) 2.6 \text{ psia}}{67.0 - 52.4} = 25.6 \text{ ft.}$$

The calculation indicates a 50% water saturation value can be found in an oil sand at 25.6 ft above the free-water interface. The pore size is assumed to remain constant for this case.

In summary, the capillary pressure equation used to determine the height above the free-water level is expressed as

$$p_c = \frac{h(\rho_w - \rho_o)}{144} = p_o - p_w. \quad \text{(2.20)}$$

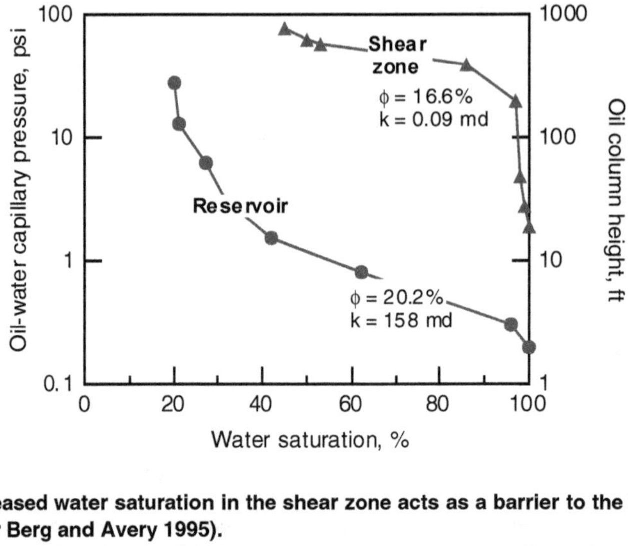

Fig. 2.17—The increased water saturation in the shear zone acts as a barrier to the upward migration of hydrocarbons (after Berg and Avery 1995).

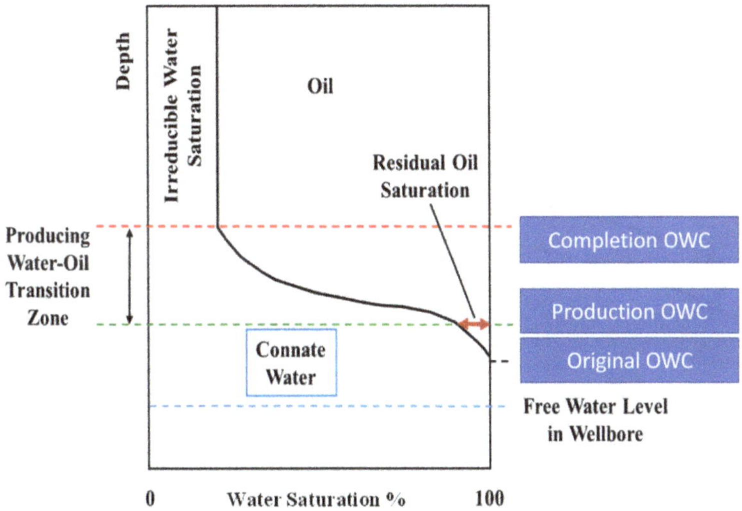

Fig. 2.18—Distribution of fluids in a uniform sand. Courtesy of Schlumberger NExT.

Eq. 2.20 relates capillary pressure and height in the reservoir, where p_c is the capillary pressure (psia) at height h (ft), p_o is the pressure (psia) in the oil at height h (ft), p_w is the pressure (psia) in the water at height h (ft), ρ_o is the density of oil (lbm/ft^3), ρ_w is the density of water (lbm/ft^3), and h is the height above the free-water level (ft).

The distribution of fluids in a uniform sand is shown in **Fig. 2.18.**

The following definitions summarize this section's discussions:

- Free-water level: The imaginary level where capillary pressure is equal to zero.
- OWC: The depth in the reservoir below which essentially no oil is found.
- Threshold capillary pressure ($p_{ct} = p_d$): The pressure required to overcome the capillary pressure of the largest pores and force the first small drop of nonwetting fluid into the rock.
- Water/oil transition zone: The vertical zone of variable water saturation between the OWC and the point above this contact where the water saturation is irreducible.

2.5 Fluid Gradients and Reservoir Pressures

Fig. 2.19 shows graphically the concepts and definitions that correspond to reservoir static conditions, fluid distribution, and capillarity. In the figure, water exists at all levels below 2, and both water and oil exist at all levels above 2. Level 2 is known as "FWL," or the free-water level. Oil and water hydrostatic pressure gradients are different because their densities are different. At Level 2, pressure in the water and oil phases is the same. At any level above 2, such as Level 3, water and oil pressures are different. This difference in pressure between the phases is the capillary pressure.

Capillary pressure effects can cause measured reservoir pressures in gas sands to be greater than the pressure calculated from the hydrostatic gradient. The easiest way to illustrate these phenomena is with an example problem.

Example 2.5—Calculating Reservoir Pressure. The following water, oil, and gas gradients are known:

- $g_w = 0.465$ psi/ft.
- $g_o = 0.37$ psi/ft.
- $g_g = 0.056$ psi/ft.

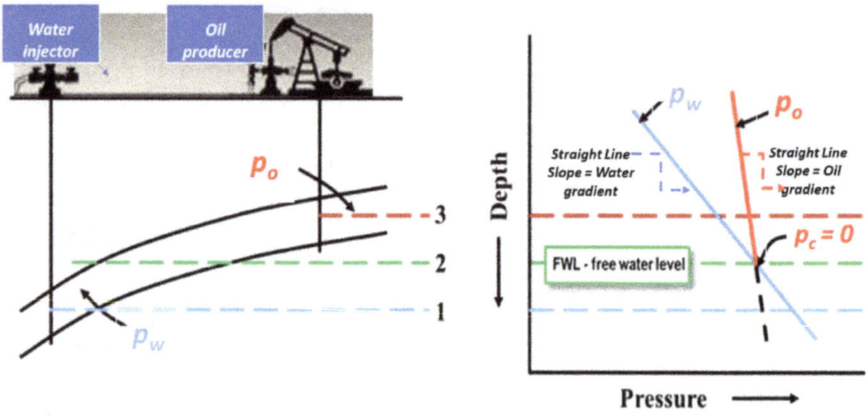

Fig. 2.19—Distribution of fluids and capillarity. Courtesy of Schlumberger NExT.

The free OWC is at 7,000 ft, and the gas/oil contact (GOC) is at 6,800 ft. What is the expected reservoir pressure in the gas cap at the 6,700-ft level when all three fluids are assumed to be in the same reservoir?

Solution.

Hydrostatic pressure at 7,000 ft at free OWC: 7,000 ft (0.465 psi/ft) = 3,255 psig.
Hydrostatic pressure at 6,700 ft at gas cap: 6,700 ft (0.465 psi/ft) = 3,116 psig.

The solid blue line in Fig. 2.19 illustrates the hydrostatic gradient and the pressure calculated at the gas/oil 6,800-ft contact and the free oil/water 7,000-ft contact.

The next step involves calculating the pressure in the oil phase at the 6,800-ft GOC and in the gas phase at the 6,700-ft gas-cap level.

The pressure at the GOC at 6,800 ft is the pressure at the free OWC minus the pressure exerted by the oil phase measured over the intervening 200-ft interval: 3,255 – 200 ft (0.37 psi/ft) = 3,181 psig. Similarly, the pressure at 6,700 ft is 3,181 – 100 ft (0.056 psi/ft) = 3,175 psig.

The red line in **Fig. 2.20** compares these calculated pressures and gradients to the hydrostatic pressure gradient (blue line). The difference between these two lines is the system capillary pressure at that particular depth. Compare the 3,175-psig reservoir pressure in the gas cap to the expected 3,116-psig hydrostatic pressure at the same level. Encountering unexpectedly high reservoir pressures caused by this phenomenon when drilling a well with a balanced mud program could cause a blowout if control is lost. The pressure difference between the phases at some elevation *h* above this level is given by Eq. 2.20. This pressure difference, which is equal to the capillary pressure, obviously increases with height above the hydrocarbon/water contact, and a gradient in the capillary pressure implies a gradient in the water saturation ("water/oil transition zone").

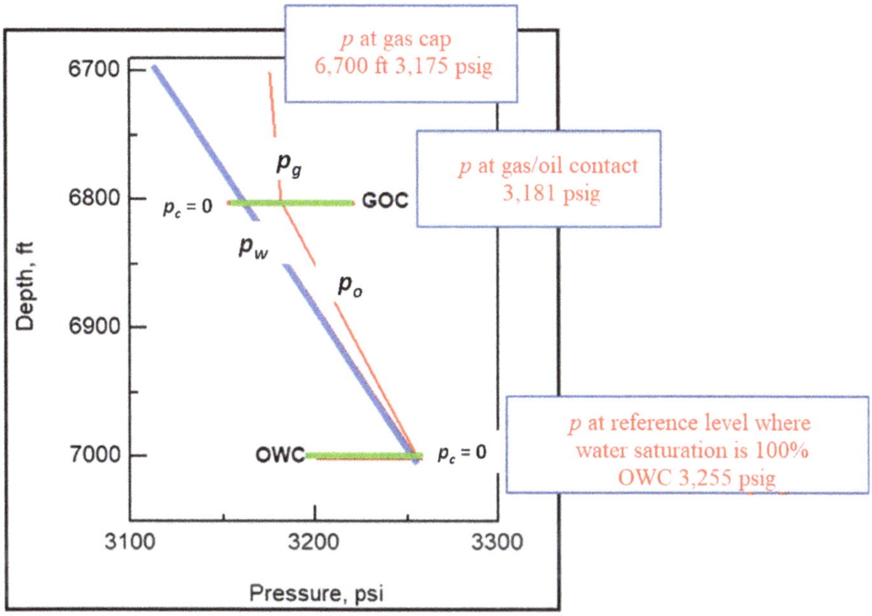

Fig. 2.20—Relating capillary effects to subsurface pressures. The difference between the red line and the blue line is the capillary pressure at that particular depth.

2.6 The *J*-Function

Leverett and Lewis (1940) developed a dimensionless correlation variable defined as the *J*-function to convert capillary pressure data to a single universal curve. The correlation, shown in Eq. 2.21, is effective as long as the pore geometry throughout the system does not change. The k/ϕ dimensions are L^2 because porosity is dimensionless. Therefore, capillary pressure and the *J*-function dimensions are the same.

$$J(S_w) = \frac{p_c}{\sigma \cos\theta}\left(\frac{k}{\phi}\right)^{0.5}, \quad (2.21)$$

where k is in md, p_c is in psi, and ϕ is a fraction.

A universal curve encompassing the characteristics of a wide producing area is probably impossible to achieve because of the variety of rock types commonly encountered. However, the concept is widely applied to develop a rock-property correlation for a reservoir or field. The concept is particularly useful because the correlation mathematically relates log-derived porosity and water saturation values to a permeability estimate.

The method is a function of the number of capillary pressure curves developed in the laboratory and the constancy of the pore sizes and pore-size distribution. The data spread could be minimal for rather homogeneous reservoirs while there might be no obvious correlation for layered or heterogeneous reservoirs.

Example 2.6—Using the *J*-Function. Capillary pressure curves from two core samples from the Clausen Reservoir were developed. The data sets are listed in **Table 2.1** and plotted in **Fig. 2.21.** Apply the *J*-function concept to develop a rock property relationship.

The interfacial tension between the oil and the water is 70 dynes/cm while the contact angle is 0°. The average rock properties of the rather homogeneous producing interval for each well are also listed in Table 2.1.

ϕ_1: 20.8%, k_1: 100 md, ϕ_2: 27.5%, k_2: 500 md					
S_{w1} (%)	p_{c1} (psia)	J	S_{w1} (%)	p_{c2} (psia)	J
20.0	10.00	3.13	15.0	10.00	6.09
22.0	8.00	2.10	15.5	7.00	4.26
25.0	5.00	1.57	16.0	6.00	3.65
30.0	3.30	1.03	16.5	5.00	3.05
35.0	2.50	0.78	20.0	3.20	2.05
40.0	1.86	0.58	25.0	2.40	1.46
45.0	1.45	0.45	30.0	1.80	1.10
50.0	1.36	0.43	40.0	1.16	0.71
55.0	1.26	0.39	50.0	0.80	0.49
60.0	1.15	0.36	60.0	0.68	0.41
70.0	1.02	0.32	70.0	0.63	0.38
80.0	0.99	0.31	80.0	0.59	0.36
90.0	0.95	0.30	90.0	0.58	0.35
100.0	0.90	0.28	100.0	0.56	0.34

Table 2.1—*J*-function example problem. (*continued*)

Solution. Apply the J-function equation (Eq. 2.21) to normalize the two curves. Include the interfacial tension and contact angle particular to the field to develop a general solution.

$$J(S_w)=\frac{p_c}{\sigma\cos\theta}\left(\frac{k}{\phi}\right)^{0.5}=\frac{p_c}{(70)(1.0)}\left(\frac{k}{\phi}\right)^{0.5}=0.0143p_c\left(\frac{k}{\phi}\right)^{0.5}.$$

This calculation is further refined by including the characteristic permeability and porosity defining a particular capillary pressure curve. This equation is reduced to the following form for Core Sample 1:

$$J(S_w)=0.0143p_c\left(\frac{k}{\phi}\right)^{0.5}=0.0143p_c\left(\frac{100}{0.208}\right)^{0.5}=0.314p_c.$$

Capillary pressure values equivalent to a given water saturation obtained from Core Sample 1 are included in the previous equation to generate a J-function vs. S_w plot. The results of the J-function calculations for Core Sample 1 and Core Sample 2 are shown in Fig. 2.21. The shapes of the two capillary pressure curves in red are similar to an approximate value of $S_w > 30\%$. However, differences in the S_{irr} values over the 15 and 21% range indicate that the smaller pore sizes of the two rock samples were different.

This example problem shows the need to obtain a large number of data points from additional core samples to develop a true average reservoir property curve.

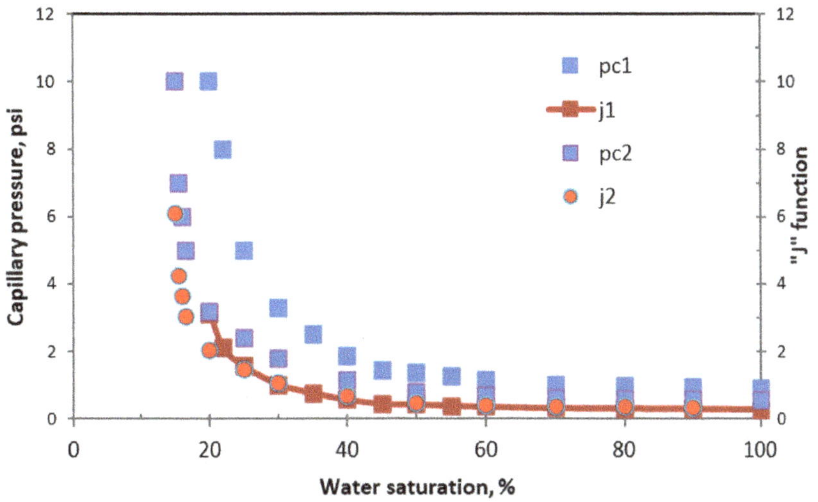

Fig. 2.21—Normalizing a set of capillary pressure curves to the *J*-function curve. The red line represents the *J*-function for the data set. Note the discontinuity at 15% water saturation.

2.7 Fractional Flow Equation

2.7.1 Effective and Relative Permeability. Reservoir rocks are typically saturated with two or more fluids, such as water, oil, and gas. Therefore, it is necessary to introduce the concept of effective permeability to describe the simultaneous flow of multiphase fluid in a porous medium and define it as a direct measure of the ability of the porous system to conduct fluids when there is more than one fluid present and those fluids are immiscible.

Oil and gas reservoirs typically produce some combined volume of oil, gas, and water fluids. Saturation changes, the history of the saturation process, the pore-size distribution, and wettability affect the flow capabilities or permeability of each fluid present in the system. The permeability of a fluid flowing through a porous medium at any particular time is defined as the effective permeability of that fluid, and it is a relative measure of the conductance of the porous medium for one fluid when the medium is saturated with more than one fluid. Effective permeability is clearly a function of the prevailing fluid saturation, the rock wetting characteristics, and the geometry of the rock's pores.

It is very interesting to note that

$$k_o + k_w + k_g < k, \qquad (2.22)$$

where k_o is the effective permeability to oil, k_w is the effective permeability to water, and k_g is the effective permeability to gas.

The generalized Darcy definition expressed in terms of the flow rate of a fluid moving across some cross-sectional area can be expressed using Eq. 2.4a.

The appropriate subscript should be used to express the flow rate in terms of the effective permeability of a particular fluid. Subscripts define the effective term representing oil, gas, or water. In this case, Eq. 2.23 is for the oil effective permeability:

$$q_o = \frac{k_o A}{\mu_o}\left(\frac{\mathrm{d}\psi}{\mathrm{d}s}\right). \qquad (2.23)$$

Effective permeability curves for an oil/water system and a gas/oil system are shown in **Fig. 2.22.**

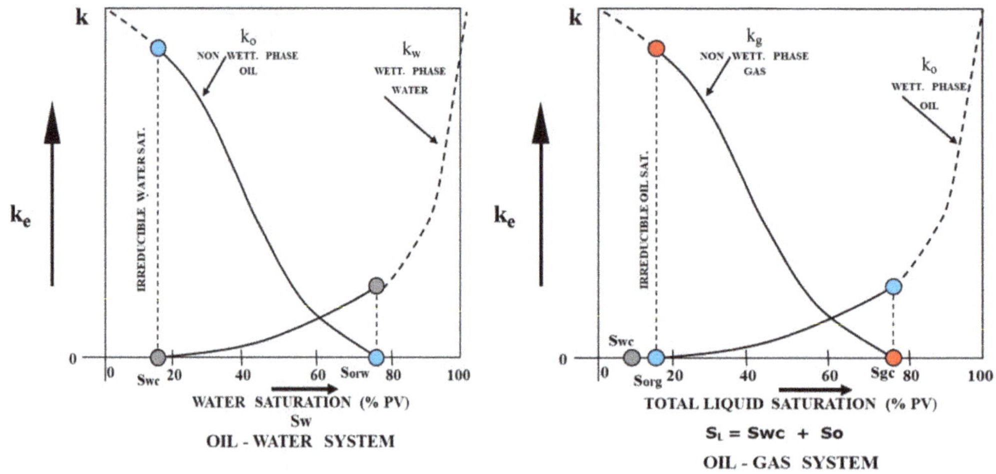

Fig. 2.22—Effective permeability curves for (left) an oil/water system and (right) an oil/gas system.

The concept of relative permeability has been defined as the effective permeability divided by a base value, which is usually the absolute permeability to normalize effective permeability measures. Recall, the absolute permeability value is the permeability when the rock is 100% saturated with the fluid. Eq. 2.24 represents the expression for the relative permeability to oil. Similar normalizing expressions can be developed for gas and water.

$$k_{ro} = \frac{k_o}{k} \rightarrow k_r = \frac{k_e}{k} \quad \text{(2.24)}$$

The nonwetting fluid in a two-phase system occupies the majority of the large-sized pore spaces if the saturation is sufficiently high. Permeability to water is at a maximum when oil or gas saturations are at minimum values. These characteristics can be displayed by a set of wetting- and nonwetting-phase effective or relative permeability curves.

Typically, relative permeability curves are plotted as a function of water saturation. The shapes and locations of the curves reflect the type of displacement process. Recall, water enters increasingly larger pores during imbibition displacement to continually increase the water saturation until the nonwetting phase is at residual conditions. A nonwetting phase initially enters the largest pores and progressively invades smaller pore spaces during the drainage displacement process to cause a continual decrease in the water saturation.

Fig. 2.23 illustrates the case for imbibition displacement in a water/oil system (image on the left). A gas/oil system is shown in the right-hand image of Fig. 2.23. Gas/water systems can be displayed in a similar manner. Endpoints and the type of displacement process define the shapes and locations of the curves during the desaturation process. Irreducible water saturation and residual oil saturation values define the saturation boundaries of the relative permeability curve. Water flows for all saturations greater than the irreducible water saturation while oil flows for all saturations greater than the residual oil saturation. These two points are noted in Fig. 2.23.

The relative permeability curves are never equal to unity because of the dual fluid saturation nature of the core at initial conditions. Interference always causes the sum of the effective permeabilities to be less than the total permeability. The direction of the changing water saturation defines the type of displacement process.

The laboratory procedure to develop a set of relative permeability curves typically requires some type of flow experiments. The process begins with rock that is completely saturated with the wetting phase (i.e., S_{wp} = 100%). The wetting phase is displaced by the nonwetting phase (i.e., the drainage process)

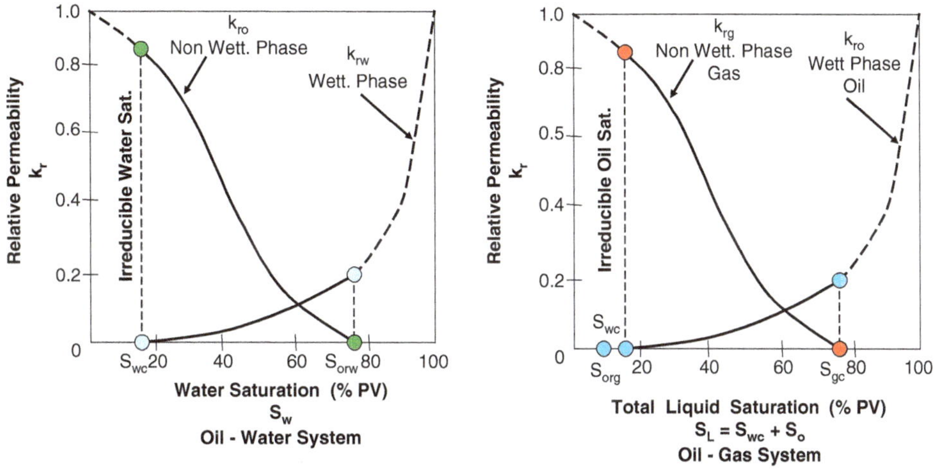

Fig. 2.23—Relative permeability curves. (left) For an oil/water system, a typical imbibition water/oil relative permeability curve. The residual oil saturation nd irreducible water saturation values define the two-phase flow region. (right) An oil/gas system.

until the wetting phase ceases to flow. At this point, the wetting phase saturation equals the minimum interstitial wetting phase saturation. In a water-wet system, the core sample must be completely saturated with brine to begin the imbibition displacement process. Oil is then forced through the core to reduce the water saturation to the irreducible water saturation. The core sample now contains an irreducible water saturation, with the remaining pore spaces filled with oil.

Water is then introduced to the core sample through the input face and is allowed to be naturally imbibed and drawn into the rock. Expelled oil volumes are measured as a function of time. This information is periodically used to calculate the apparent permeability. The arrows in the lines shown in **Fig. 2.24** illustrate the process and compare the imbibition and drainage curves. The disparity between the two curves is caused by the water always residing in the smaller pore spaces during the imbibition displacement process. The term hysteresis describes the process in which the relative permeabilities are different when measurements are made in different directions.

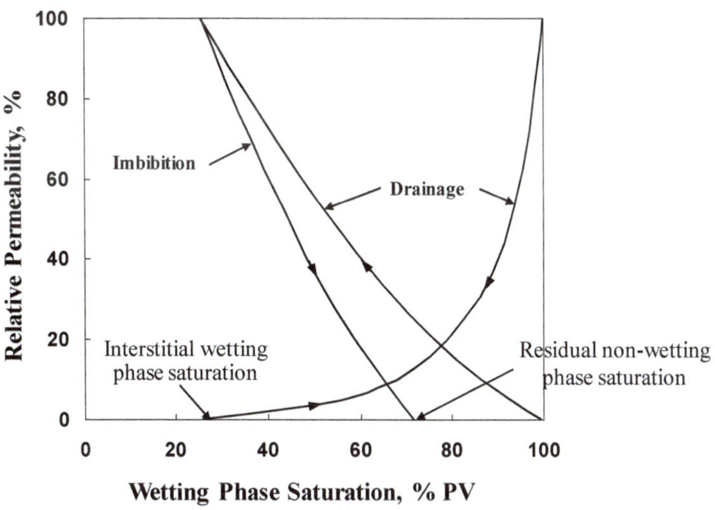

Fig. 2.24—Relative permeability ratio curves illustrating the effect of saturation history (imbibition and drainage processes).

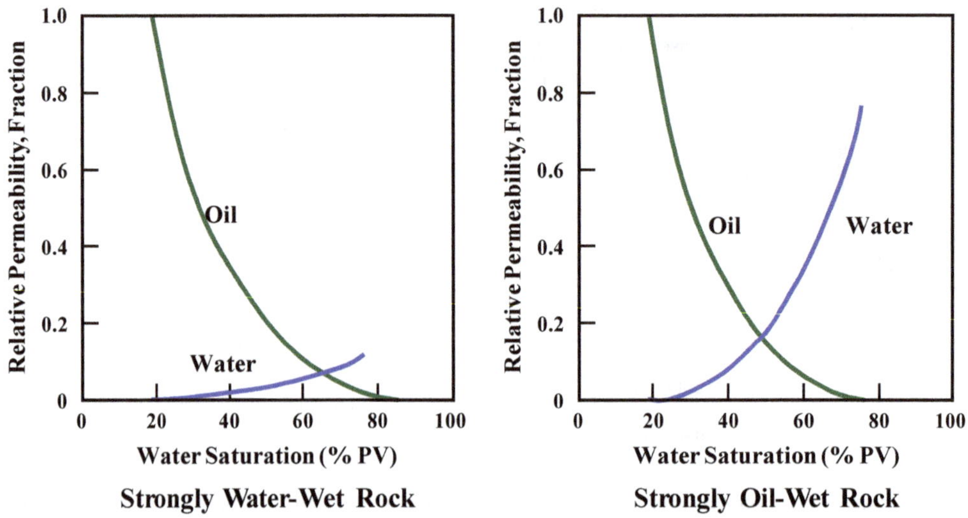

Fig. 2.25—Relative permeability ratio curves illustrating the effect of wettability.

From Fig. 2.24, it can be observed that

- In the drainage curve, the wetting phase is displaced by the nonwetting phase (i.e., the wetting phase saturation is decreasing).
- In the imbibition curve, the nonwetting phase is displaced by the wetting phase (i.e., the wetting phase saturation is increasing).

When simulating the waterflood of a water-wet reservoir rock, imbibition relative permeability curves should be used. When modeling gas injection into an oil reservoir, drainage relative permeability curves should be used.

The effect of wettability is shown in **Fig. 2.25.** Water is expected to flow easier than in a strongly water-wet system. In addition, one generally would expect that the residual oil saturation will be higher.

Two endpoints are shown in the left-hand plot. The 20% irreducible water saturation corresponds to an oil relative permeability of 1.0 because the absolute or base permeability was measured at the irreducible water saturation. The water relative permeability is zero at the irreducible water saturation, signifying water is immobile at this point.

The other endpoint is the residual oil saturation, at approximately 75%. The relative permeability to oil is zero at the residual oil saturation, meaning oil no longer flows. The relative permeability to water is less than 10%, meaning the effective permeability to water is very low. The shape of the blue curve is characteristic of water-wet relative permeability curves. Because water does not flow easily, the relative permeability curve is lower, and crossover between the oil and water relative permeability curves is to the right of 50%, which is a qualitative indicator of water-wet behavior. A high relative permeability to water is an indicator of oil-wet rock and high water production. A low relative permeability to water indicates a water-wet rock and low water production.

Fig. 2.26 shows two sets of water/oil relative permeability data. The data set for Rock Type 1 is drawn with red triangles, and the data set for Rock Type 2 is drawn with blue squares. The oil relative permeability k_{ro} is almost the same for the two sets; however, the water relative permeability k_{rw} is different. In the right-hand side of Fig. 2.26, the y-axis shows oil recovery as a percentage of the

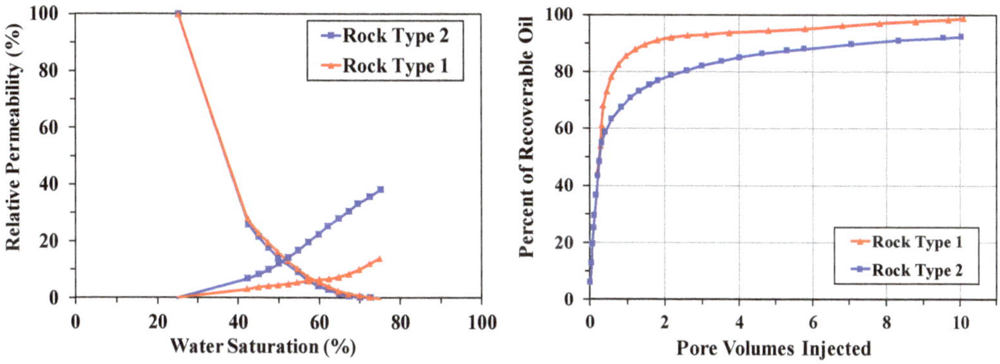

Fig. 2.26—Relative permeability curves illustrating the effects of different rock types. Courtesy of Schlumberger NExT.

movable oil in the reservoir. Fig. 2.26 clearly shows that waterflooding in Rock Type 1 is more efficient than in Rock Type 2. In Rock Type 1, water breakthrough (when water reaches the producer) occurs later in time than in Rock Type 2 because of the shape of the relative permeability data. Oil recovery for a process of water displacing oil can be 25 to 65% of the original oil-in-place. Recovery is a percentage of the recoverable oil, meaning that the irreducible oil saturation has been excluded from the calculation.

2.7.2 Two-Phase Mobility. Viscosity and capillary pressure influence the flow of fluids through a porous medium. These two terms are combined to form a mobility term used in the study of fluid displacement processes. Eq. 2.25 defines mobility λ as the relative or effective permeability exhibited by a porous rock to a particular fluid divided by the fluid viscosity. The mobility term continuously changes as a function of the effective permeability because for most cases viscosity can be assumed to remain constant. As would be expected, the mobility of a high-viscosity oil in a waterflood would be quite low when compared to waterflooding a low-viscosity, high API gravity oil.

$$\lambda = \frac{k_e}{\mu} = \frac{k_r}{\mu} \quad (2.25)$$

The mobility ratio compares the ability of one fluid to displace another fluid in a two-phase system. The mobility ratio M is the mobility of the displacing fluid divided by the mobility of the displaced fluid:

$$M = \frac{\text{mobility of the displacing phase}}{\text{mobility of the displaced phase}} \quad (2.26a)$$

Eq. 2.26b illustrates the mobility ratio in terms of water displacing an oil system when the effective permeabilities of both fluids are evaluated at their maximum saturation levels.

$$M = \frac{\dfrac{k_{rw}}{\mu_w} \text{ at } S_{or} \text{ (water-phase mobility behind the water front)}}{\dfrac{k_{ro}}{\mu_o} \text{ at } S_{wi} \text{ (oil-phase mobility ahead of the water front)}} \quad (2.26b)$$

Notice that the effective permeability value of a fluid is measured at the maximum saturation level when the second fluid is immobile at its residual saturation. Oil and water flow equally well when $M = 1$. An inefficient displacement process occurs when $M > 1$ because water flows more easily than the displaced oil. High mobility ratios cause small perturbations (fingering effects) to develop across the interface of the displacing/displaced phases. These discontinuities gradually become exaggerated and result in an uneven distribution of the fluids at the interface. Fingering causes early breakthrough of the displacing phase before the displaced phase is completely swept ahead of the interface. The desired waterflood displacement process occurs when water flows less easily than oil (i.e., when $M < 1$).

The efficiency of any fluid displacement process is a function of the ability of the displaced oil to flow more easily than the displacing fluid. Enhanced oil recovery projects are designed to alter the mobility ratio relationship in a reservoir. The primary alteration categories are:

- Decrease a very uneven permeability profile by injecting high-viscosity polymers into the more-permeable layers. The long-stringed molecules tend to accumulate in the larger pore throats to even out the permeability profile across the zone. This effect permits the displacing phase to present a more-uniform flood front. The apparent mobility of the displacing phase increases.
- Reduce the oil viscosity with carbon dioxide (CO_2), steam injection, or in-situ combustion to decrease the mobility of the displaced phase and also the mobility ratio.
- Introducing surfactants or CO_2 causes the oil volume to swell and reduces the residual oil saturation. Creating a zone of miscibility reduces interfacial tension effects between the displacing and displaced phases. The net effect allows oil to flow at higher water saturation values and reduces the mobility ratio.

Oil displacement in an oil/water system could be achieved by the dynamic imbibition process when water naturally displaces oil, or the drainage process occurs when an artificially induced, higher-pressured water phase displaces the oil phase. Two-phase flow effects are a function of the mobility ratio for either case.

The Buckley-Leverett frontal advance concept permits the study of a two-phase system when the water-displacing-oil process is a combination of imbibition and external pressure forces. Buckley and Leverett (1942) presented an analytical model for waterflood performance prediction that is based on two fundamental equations: the fractional flow equation and the frontal advance equation. The fractional flow equation gives the fractional flow of water (i.e., the water cut) as a function of oil and water relative permeabilities, which in turn are functions of saturation distribution. This means that if the saturation distribution is known at a given point in the reservoir (e.g., at the producing well), the fractional flow of water can be estimated. The frontal advance equation describes the advance of a given saturation over time in the reservoir. The velocity of a fixed saturation is proportional to the total flow rate (i.e., the water-injection rate) and the derivative of its fractional flow with respect to the saturation. The fractional flow equation provides the basis for the development of the frontal advance study. This method is concerned primarily with the mechanism of oil displacement by means of water in a linear system when predicting waterflood performance. This is a classic prediction method and does not require the use of a numerical simulator providing a first estimate of the performance of a single pattern element. The fractional flow concept and its practical applications will be discussed in detail in the Displacement Concepts chapter.

2.8 Summary

A petroleum reservoir is a porous and permeable rock located in a container underlying an impermeable seal. Porosity (ϕ) defines the pore-space volume, while permeability (k) defines the ability of the pore spaces to transmit fluids to the wellbore. Log and core analysis is applied to determine reservoir pore volume. Flow rate tests measure relations between rate, bottomhole flowing pressure, and reservoir pressure and can be thought of as quantitative measurements to provide permeability estimates.

Normally, the radial flow condition is applied to study subsurface flow toward a wellbore. The mathematical solution indicates a logarithmic pressure drop between the inner and outer boundary. This effect results in the greatest pressure drop occurring close to the wellbore. Rock damage that occurs during completion can create a secondary skin of reduced permeability, thereby reducing the well flow rate. On the other hand, hydraulic fracturing increases near-wellbore permeability, thus increasing the ability of a well to flow at a higher rate.

Studies have shown that reservoir rock parameters can be a function of the rock types encountered in a heterogeneous reservoir. The study and knowledge of these clustered values can provide a useful tool for estimating the architecture of and recoveries in reservoirs with diverse rock properties.

Fluid distribution within a reservoir rock is a function of wettability and capillary pressure. Knowledge of capillary pressure is particularly important because rocks are generally water-wet. This phenomenon causes unwanted water to reside in the smaller pore spaces while oil and gas reside in the larger pore spaces. Laboratory-derived capillary pressure curves provide information regarding average pore size, pore-size distribution, and wettability.

The relation between porosity (whose value is easily obtained) and permeability (which can be highly variable) pertaining to the field and to the rock type should be determined. However, the precision between the two data sets is typically so poor that porosity/permeability correlations should be viewed in a qualitative sense.

2.9 Questions

2.9.1 General Questions.

- Why is effective porosity more important than total porosity when evaluating an oil or gas reservoir? (Answer: Because the interconnected pore spaces are able to transmit fluids in a variety of directions. Total porosity, on the other hand, includes sealed pore spaces that are unable to communicate with those adjacent to them.)
- Why are production tests typically used in place of the Darcy equation to determine the permeability of a producing interval? (Answer: Because the majority of the reservoir parameters expressed in the Darcy equation are unknown or difficult to interpret. A flow equation expresses these values as a single composite value determined from an analysis of the results of the flow test.)
- Fresh water was used during the completion of a newly drilled well. The completion zone was comprised of a shaly sand. Could this cause production problems? (Answer: Yes. Shale particles are often reactive to the introduction of fresh water, and they swell and reduce rock permeability when contacted by fresh water.)
- Why can formation compressibility reach high values when sands are undercompacted? (Answer: Because the undercompacted sand grains still have an ability to collapse onto each other. Insufficient time has elapsed for circulating groundwater to deposit cementing material to increase matrix rock strength. Additionally, continuing burial is interrupted, thereby discontinuing compaction.)

2.9.2 Capillary Pressure and Relative Permeability Questions.

- In which size pore spaces do the oil, gas, and water phases tend to reside in an oil-wet undeveloped saturated reservoir? (Answer: Gas is in the large pores, water is in the medium-sized pores, and oil is in the smallest pores.)
- Relate the capillary pressure curve example for poorly sorted sands and for well-sorted sands to the expected change in the water/oil ratio when the reservoir is undergoing waterflooding. Exclude the effects of gravity. (Answer: There is a gradual increase in the water/oil ratio in the poorly sorted sands because there is a variety of pore sizes present in the sample. There should be a sharp rise in water cut for the well-sorted case because the producing interval would be comprised of similar-sized pore spaces.)

- Why should core studies be conducted with reservoir water and not fresh water? (Answer: Because some shale particles react to the introduction of fresh water by expanding the grain rock volume. Rock expansion causes a reduction of both porosity and permeability.)
- Would effective permeability be expected to be close to the absolute permeability for high-porosity well-sorted sands? (Answer: Yes.)

2.9.3 Relative Permeability Questions.

- Would you use the imbibition or drainage type relative permeability curve to model the behavior of the depletion of a water-wet oil zone in a solution-gas-drive oil reservoir? (Answer: Drainage because gas expands into the oil reservoir.)
- Would you use the imbibition or drainage type relative permeability curve to model the behavior of the depletion of a water-wet oil zone undergoing water influx? (Answer: Imbibition because water is encroaching.)
- Would you use the imbibition or drainage type relative permeability curve to model the behavior of a secondary gas cap expanding into an oil zone? (Answer: Drainage because gas expands into the oil reservoir.)
- The F2B Reservoir is known to be oil-wet. Will the relative permeability to oil increase or decrease if the saturation changes seen in **Table 2.2** occur? (Answer: The relative permeability to oil will decrease because the rock is oil-wet, causing the oil saturation to reside in smaller pores.)

Fluid	Initial Saturation (%)	Final Saturation (%)
Water	40	10
Oil	50	50
Gas	10	40

Table 2.2—Example saturation changes.

2.10 Nomenclature

A = cross-sectional area perpendicular to flow, ft^2
B = formation volume factor, res volume/STB volume
B_o = oil formation volume factor, RB/STB
B_w = water formation volume factor, RB/STB
c_f = formation compressibility, psi^{-1}
c_{gr} = grain compressibility, psi^{-1}
g = gravity
g_g = gas gradient, psi/ft
g_o = oil gradient, psi/ft
g_w = water gradient, psi/ft
h = height or thickness, ft
h_j = height or thickness of layer j, ft
h_n = height or thickness of layer n, ft
h_t = total height or thickness, ft
j = layer numbering
J (function) = capillary pressure correlation variable, dimensionless
k = permeability, L^2, md
$\overline{k}$ = average permeability, md
k_d = permeability of damaged portion, md
k_e = effective permeability, md
k_g = gas permeability, md
k_h = horizontal permeability, md
k_j = permeability of layer j, md
k_L = liquid permeability or true permeability, md
k_n = permeability of layer n, md
k_o = effective permeability to oil, md
k_r = relative permeability, md
k_{res} = reservoir permeability, md
k_{rg} = relative permeability to gas, md
k_{ro} = relative permeability to oil, md
k_{rw} = relative permeability to water, md
k_v = vertical permeability, md
k_w = effective permeability to water, md
k_∞ = extrapolated value to $1/p_m = 0$ of measured average permeability in core tests

when using gas (Klinkenberg effect), md
L = length, ft
M = mobility ratio
n = layer or reservoir strata numbering
p = pressure, psia
$\overline{p}$ = average pressure, psia
p_c = capillary pressure, psia
p_{cl} = capillary pressure at laboratory conditions, psia
p_{cow} = capillary pressure between oil and water, psia
p_{cr} = capillary pressure at reservoir conditions, psia
p_{ct} = threshold capillary pressure, psia
p_d = displacement or threshold pressure, psia
p_{fluid} = fluid pressure, psia
p_g = gas pressure, psia
p_i = initial pressure, psia
p_l = liquid pressure, psia
p_m = mean or average pressure, psia
p_{matrix} = matrix pressure, psia
p_{nwt} = pressure of nonwetting phase, psia
p_o = oil pressure, psia
p_{ob} = overburden pressure, psia
p_w = water pressure, psia
p_{wf} = bottomhole flowing pressure, psia
p_{wt} = pressure of wetting phase, psia
p_2 = pressure at conditions below initial pressure, psia
q = flow rate, STB/D
q_n = flow rate of layer n or flow rate at time n, STB/D
q_o = flow rate of oil, STB/D
q_t = total flow rate, STB/D
r = average radius of capillary, ft
r_d = radius of damaged or altered zone around wellbore, ft
r_e = radius to edge of reservoir, ft
r_l = large radius, ft
r_m = medium radius, ft
r_s = small radius, ft
r_w = radius to edge of borehole, ft
s = skin, dimensionless
S_g = gas saturation, fraction or percent
S_{gc} = critical gas saturation, fraction or percent
S_i = initial fluid saturation, fraction or percent
S_{irr} = irreducible water saturation, fraction or percent
S_L = liquid saturation, fraction or percent
S_o = oil saturation, fraction or percent
S_{or} = residual oil saturation, fraction or percent
S_{org} = residual oil saturation from oil-gas relative permeability measurements, fraction or percent
S_{orw} = residual oil saturation from oil-water relative permeability measurements, fraction or percent
S_w = water saturation, fraction or percent
S_{wc} = connate saturation, fraction or percent
S_{wi} = initial water saturation, fraction or percent
S_{wp} = wetting-phase saturation, fraction or percent
v = flux velocity, L/t, ft/D
V_{grain} = grain volume, ft^3
V_p = pore volume, acre-ft, ft^3, or bbl
V_t = total unit volume, acre-ft, ft^3, or bbl
V_w = water volume, bbl
β = gas turbulence factor
Δp = pressure applied across L, psi
θ = angle, degrees
θ_l = contact angle at laboratory conditions, degrees
θ_r = reservoir angle at laboratory conditions, degrees
λ = mobility
μ = fluid viscosity, cp or Pa·s
μ_o = oil viscosity, cp or Pa·s
μ_w = water viscosity, cp or Pa·s
ρ = fluid density, lbm/ft^3
ρ_{nwt} = density of nonwetting phase, lbm/ft^3
ρ_o = oil density, lbm/ft^3
ρ_w = water density, lbm/ft^3
ρ_{wt} = density of wetting phase, lbm/ft^3
σ = interfacial tension, dynes/cm
σ_{aw} = interfacial tension between air and water, dynes/cm
σ_{os} = interfacial tension between oil and solid, dynes/cm
σ_{ow} = interfacial tension between oil and water, dynes/cm
σ_{so} = interfacial tension between solid and oil, dynes/cm
σ_{sw} = interfacial tension between solid and water, dynes/cm
σ_{ws} = interfacial tension between water and solid, dynes/cm
ϕ = porosity, percent
ϕ_i = initial porosity, percent
ϕ_1, ϕ_2, and ϕ_3 = porosity at different pressure conditions, percent
ϕ = potential energy

2.11 References

- Bass, D. M., Jr. 1987. Properties of Reservoir Rocks. In *Petroleum Engineering Handbook*, ed. Howard B. Bradley, Chap. 26. Richardson, Texas: Society of Petroleum Engineers.
- Bennion, B. 2010. The Impact of Wettability, Capillary Pressure and Relative Permeability on Recovery Efficiency in Oil and Bitumen Reservoirs. Presentation for SPE Canada, 27 April.
- Berg, R. R. and Avery, A. H. 1995. Sealing Properties of Tertiary Growth Faults, Texas Gulf Coast. *AAPG Bulletin* **79** (3): 375–392.
- Buckley, S. E. and Leverett, M. C. 1942. Mechanism of Fluid Displacement in Sands. In *Transactions of the Society of Petroleum Engineers*, Vol. 146, SPE-942107-G, 107–116. Richardson, Texas: Society of Petroleum Engineers. https://doi.org/10.2118/942107-G.
- Cossé, R. 1993. *Basics of Reservoir Engineering*. Houston, Texas: Gulf Publishing Company.
- Craft, B. C. and Hawkins, M. F. 1959. *Applied Petroleum Reservoir Engineering*. Englewood Cliffs, New Jersey: Prentice-Hall.
- Forchheimer, P. 1901. Wasserbewegung durch Boden. *Zeit Ver Deutsch Ing* **45**: 1782–1788.
- Hall, H. N. 1953. Compressibility of Reservoir Rocks. *J Pet Technol* **5** (1): 17–19. SPE-953309-G. https://doi.org/10.2118/953309-G.
- Klinkenberg, L. J. 1941. The Permeability of Porous Media to Liquids and Gases. In *Drilling and Production Practice*, API-41-200. New York, New York: API.
- Leverett, M. C. and Lewis, W. B. 1940. Steady Flow of Gas-Oil-Water Mixtures Through Unconsolidated Sands. Issued as T. P. 1206, SPE-941107-G. https://doi.org/10.2118/941107-G.
- Newman, G. H. 1973. Pore-Volume Compressibility of Consolidated, Friable, and Unconsolidated Reservoir Rocks Under Hydrostatic Loading. *J Pet Technol* **25** (2): 129–134. SPE-3835-PA. https://doi.org/10.2118/3835-PA.
- Tiab, D. and Donaldson, E. C. 1996. *Petrophysics: Theory and Practice of Measuring Reservoir Rock and Fluid Transport Properties*, first edition, 706. Waltham, Massachusetts: Gulf Professional Publishing.
- van der Knaap, W. 1959. Nonlinear Behavior of Elastic Porous Media. In *Transactions of the Society of Petroleum Engineers*, Vol. 216, SPE-1124-G, 179–187. Richardson, Texas: Society of Petroleum Engineers. https://doi.org/10.2118/1124-G.

Chapter 3

Reservoir Fluid Properties

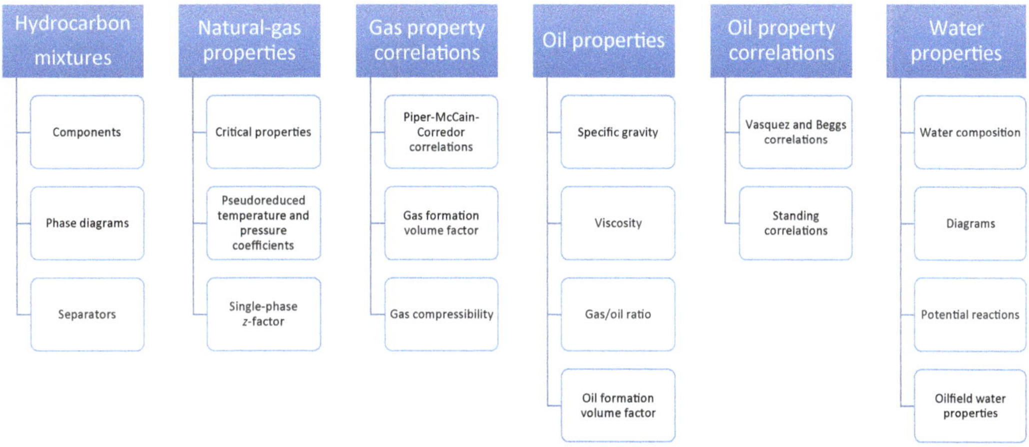

The physical properties of reservoir fluids have pronounced effects on the suitability of a given reservoir for further development by means of waterflooding. Petroleum crude oils and natural gases are mixtures of hydrocarbon and nonhydrocarbon components. Compositional and phase behavior studies involving petroleum reservoirs generally assume a constant temperature process with a declining reservoir pressure. However, the effects of separation temperature and pressure on the relative gas/liquid volumes of the produced fluids must be considered when developing reservoir-to-surface fluid property relationships.

Hydrocarbon gases often contain quantities of carbon dioxide (CO_2), hydrogen sulfide (H_2S), and nitrogen (N_2). CO_2 and H_2S are toxic and can cause significant problems in producing equipment, while nitrogen is inert and reduces the heat content of the produced gas.

3.1 Hydrocarbon Mixtures

3.1.1 Components. Hydrocarbon reservoirs are a mixture of hydrocarbon and nonhydrocarbon components. Compositions range from the lightest mixture of carbon/hydrogen, hydrocarbon methane (CH_4) gas, to complex, long-chained hydrocarbon components that are solids. **Table 3.1** provides an example of the compositional makeup of a moderate-gravity crude oil. Very heavy molecules are included in the heptanes (C_{7+}) fraction. Minor fractions of H_2S indicate the possibility of steel embrittlement, while the minor presence of CO_2 indicates scale formation should not be a problem. Nitrogen is inert and a nonhydrocarbon.

Natural gases and crude oils are classed as sweet or sour. A sour gas or oil contains nonhydrocarbon toxic and corrosive H_2S or mercaptans. CO_2 is often associated with natural gas and is usually the product of volcanic activity. The effect of CO_2 in the gas stream is a decrease of the heating value

Compound	mol %
Hydrogen sulfide	Nil
Carbon dioxide	0.9
Nitrogen	0.16
Methane	37.05
Ethane	9.66
Propane	6.94
Isobutane	1.4
n-Butane	3.5
Isopropane	1.44
n-Propane	1.4
Hexanes	4.3
Heptanes+	33.25
Total:	100.00

Table 3.1—Example fluid analysis sample.

of the gas while its corrosive capabilities are increased. The inert compound N_2 is sometimes found to be associated with gas and oil accumulations.

The relative volumes of oil and gas produced from a reservoir depend on the composition of the reservoir fluid. The relative weights of the components measured in a sample relegate the reservoir fluid to a particular category. Each component possesses distinctive physical characteristics. Most produced hydrocarbons form as a gas, oil, or water phase at the surface. The relative volume of each phase is a function of the mixture at reservoir conditions.

3.1.2 Phase Diagrams. Fig. 3.1 shows a subsurface/surface relationship for a saturated black-oil reservoir in the form of a pressure/temperature diagram.

The curve envelope possesses several points of interest:

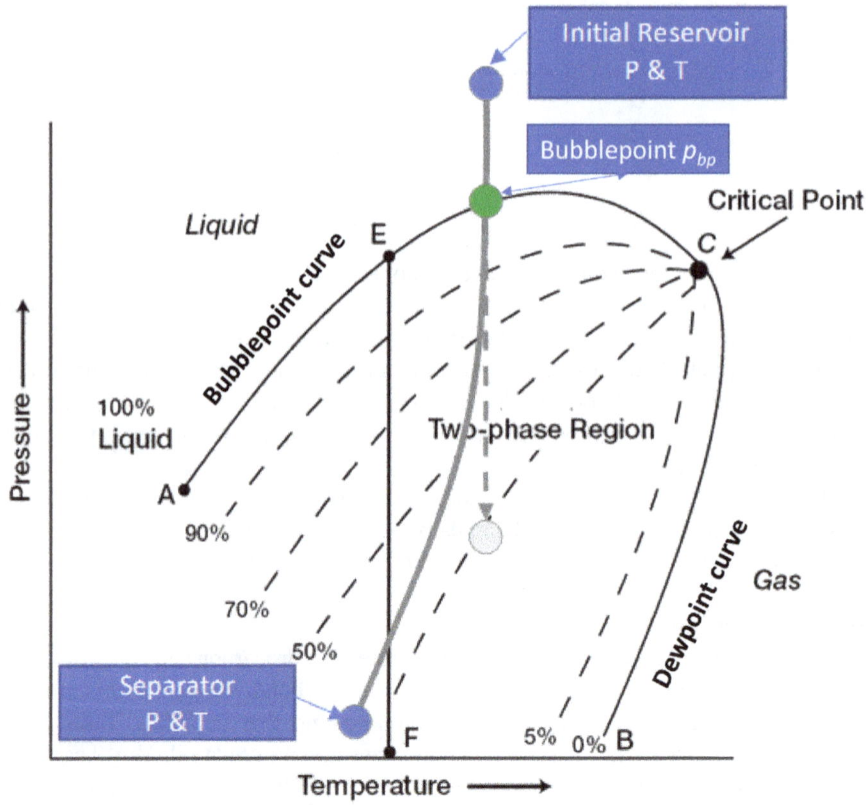

Fig. 3.1—Phase diagram of a multicomponent system (courtesy of Schlumberger NExT). The isovolume lines inside the envelope indicate the conversion of the liquid phase to gas as pressure declines in a constant temperature environment. Note that the temperature (T) and pressure (P) change as produced fluids travel from the reservoir to the separator at the surface. When does the temperature of the system begin to decline?

- Critical point: where the critical temperature (T_c) and pressure (p_c) meet. These points represent the highest values at which the two-phase mixture can exist.
- Bubblepoint pressure (p_{bp}): when the first gas evolves from the reservoir oil. The solid line extending from the left of the critical point defines the combination of temperature and pressure at which the bubblepoint is attained.
- Dewpoint pressure (p_{dp}): when the first oil separates from a reservoir gas. Conversely, the dewpoint line extends from the right of the critical point.

An undersaturated oil exists when reservoir pressures remain above the bubblepoint. No free gas is present. However, free gas and oil coexist at reservoir conditions, and the mixture is located in the two-phase region at separator conditions below the bubblepoint pressure (p_{bp}) and at lower reservoir temperatures.

Crude oils and natural gases are characterized by the proportions of light, medium, and heavy components present in the mixture. **Fig. 3.2** relates the five relative positions of dry gas, wet gas, condensate, volatile, and black-oil (light- and heavy-oil) systems on a pressure/temperature diagram.

The location of the dry-gas two-phase envelope is restricted to the low-temperature/low-pressure region because the mixture is composed primarily of methane or, at least, low-molecular-weight gases.

A gas can only be liquefied at very low temperatures. Conversely, a heavy-oil mixture contains a very high percentage of heavy C_{7+} hydrocarbons. The temperature must be high for the liquid to enter the two-phase envelope in this instance. Gas condensates and light oils inhabit intervening locations in Fig. 3.2, reflecting the relative quantities of light, medium, and heavy hydrocarbon components present in the mixture.

The location of the critical point defines the changing character of the oil or gas as reservoir pressure declines. The extreme locations of the critical points for the dry-gas and heavy-oil examples restrict the conditions in which either fluid can be transformed into the secondary phase.

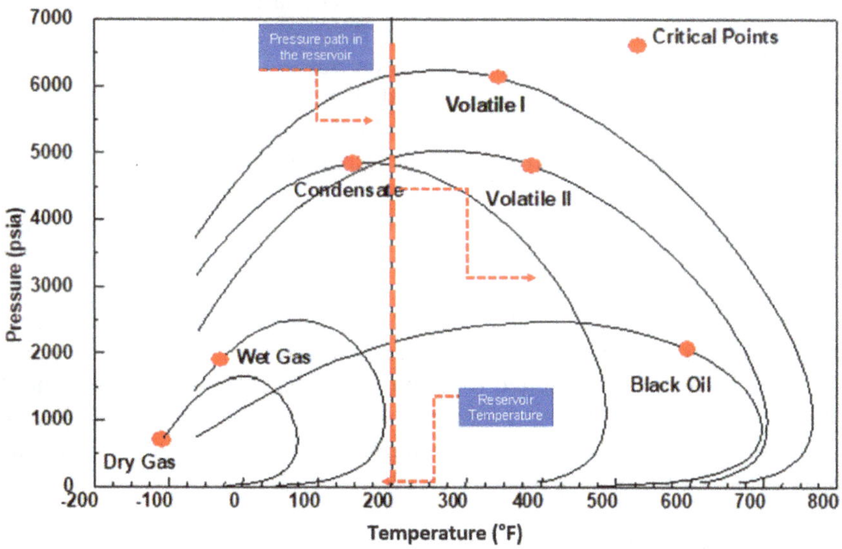

Fig. 3.2—Illustrating the varying locations of the two-phase envelopes on the pressure/temperature diagram for the five major fluid classifications. The locations of the critical points are defined by the volume of light or heavy hydrocarbon molecules present in the sample.

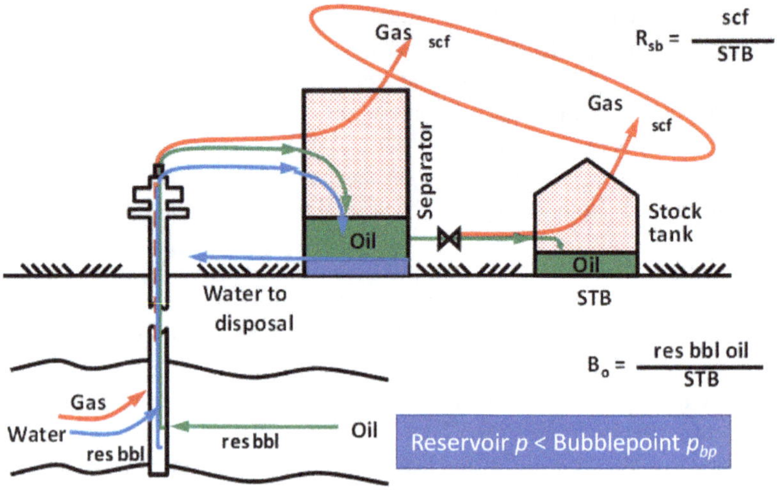

Fig. 3.3—An example separation system (courtesy of Schlumberger NExT). Reservoir fluids travel under controlled conditions to the wellhead, where the flow rate and pressure are reduced to a manageable level. Fluids are then routed to a multiwell header or a separator of sufficient strength to further reduce fluid pressure to near-standard conditions. Separation continues at close to ambient conditions in the stock tank.

3.1.3 Separators. Produced reservoir fluids rise to the surface under controlled conditions, but oil and water eventually must be reduced to standard pressure and temperature conditions and gas must be at a significantly reduced pressure. Separators are included in the flow stream to mitigate excessive pressure and temperature. **Fig. 3.3** shows a basic example of an oil, gas, and water separation system.

Gravity causes the separation of the gas, oil, and water phases. Gas is drawn off at the top of the receptacle, while water drains out at the bottom. Pressure and temperature reductions occur as reservoir fluids are produced and processed at the surface. Surface separators provide the process that maximizes liquids recovery. Produced gas is either flared or sold, oil is sold, and water is disposed of or cleaned for injection. The separation of oil and water in a waterflood is often a problem because the largest fraction of the produced fluids is often water, which must be separated from the oil.

In an oil field, different types of separators (horizontal and vertical) are used depending on the operations (onshore vs. offshore) and the gas and flow rates to be managed. Generally, a vertical configuration permits the separation of fluids when slugging occurs and is more efficient when accumulated sand, mud, and sediment must be removed. On the other hand, the small cross-sectional area inhibits the ability of fluids to separate. Vertical separators are often installed offshore because of the limited footprint and when fluids readily separate. Horizontal separators are most commonly used for the separation of large volumes of gas from small volumes of liquids. This type of unit is used on gas and gas-condensate wells and as an inlet separating device at gas processing plants.

3.2 Natural-Gas Properties

Natural gases are generally divided into dry, wet, and retrograde condensate categories. Methane (CH_4) remains in the gaseous state at reservoir and separator conditions and is the primary component of a dry-gas reservoir. Often, a small portion of the reservoir gas will be composed of intermediate- or higher-molecular-weight hydrocarbons whose dewpoint is close to the separator temperature and pressure setting or field pipeline pressure. Therefore, some liquid fallout will occur in the reservoir as the pressure declines.

The volume of liquid produced in conjunction with a gas is defined at yield with units of STB/MMscf. The yield for a dry gas typically exhibits a range of 0–2 STB/MMscf. However, this is an economic criterion because, strictly speaking, the yield for a dry gas should be zero. These insignificant produced-liquid volumes typically do not materially add to the income stream of the well.

A wet gas contains an increasingly larger fraction of intermediate-weight to heavier hydrocarbons. Reservoir gas does not enter the two-phase envelope until the temperature and pressure are substantially lowered in the production and separator system. Yields typically range from 2 to 20 STB/MMscf, and the condensate gravity is generally < 60 °API.

Natural gases are a mixture of pure hydrocarbon components and, to a lesser extent, nonhydrocarbon components. The composition of a natural gas can run the gamut, from essentially all methane with negligible condensate production to as much as 13% C_{7+} with a yield as high as 200 STB/MMscf (Moses and Donohoe 1987). Molecular weight, compressibility, and phase relations for each component often differ to a considerable degree. A weighted average of the properties of each component is assumed to reflect the average properties of a particular reservoir gas.

The reference volume for all gas- and oil-volume calculations is defined at a specified temperature and pressure. Reference or base conditions are promulgated by local governing bodies and vary according to geography and government edict. The z-factor accounts for the effects of molecular attraction, interference, and changes in temperature and pressure. It measures the divergence of a natural gas from ideal behavior. This nonlinear compensating expression relates the gas volume measured at surface conditions to the volume at reservoir conditions. A pressure-dependent relationship can be obtained from a laboratory fluid-analysis study or from published correlations. Fluid-analysis studies are seldom available, but published correlations usually produce acceptable results. **Fig. 3.4** illustrates a typical z-factor history as a function of pressure.

3.2.1 Critical Properties. Compressibility curves for pure-component paraffin hydrocarbon gases possess essentially the same shape. These curves can be coalesced into a single compressibility curve for a hydrocarbon gas figure. Application of the corresponding-states principle provides an approximation technique relating pressure/temperature/z-factor relationships for hydrocarbon gas mixtures of unknown composition but of a similar chemical nature. The point of reference is the critical temperature and pressure.

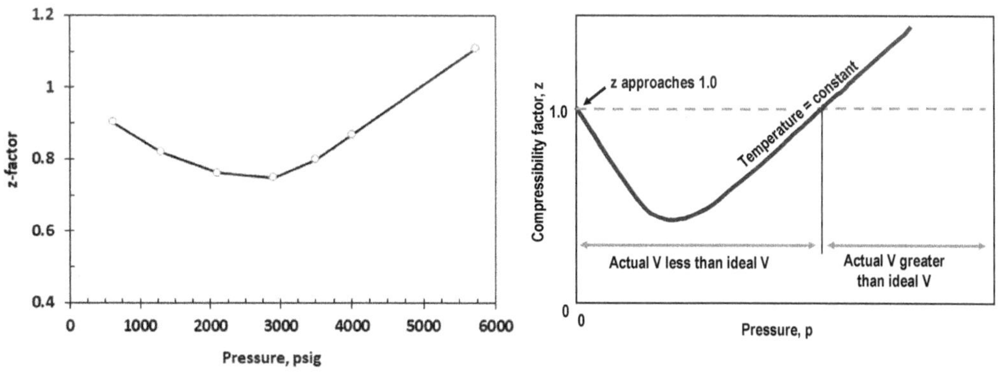

Fig. 3.4—Example *z*-factor plots. Note that there is minor variation of the *z*-factor at low pressures as *z* approaches a value of unity, but there is more variation at higher pressures. V = volume.

3.2.2 Pseudoreduced Temperature and Pressure Coefficients. Field parameters relate a specific field study to a z-factor correction. Reduced properties are ratios: the reservoir temperature measured in °R and pressure measured in psia divided by their respective critical properties. The "pseudo" prefix applies to instances when the composition of the gas mixture is unknown and specific gravity approximates the character of the gas. Pseudocritical properties are assumed to be the sum of the weighted averages of the critical properties of the individual components that compose the mixture. Eqs. 3.1 and 3.2, respectively, define the pseudocritical temperature and pressure:

$$T_{pr} = \frac{T}{T_{pc}}. \qquad (3.1)$$

$$p_{pr} = \frac{p}{p_{pc}}. \qquad (3.2)$$

Sutton (1985) published quadratic equations relating pseudocritical pressure and temperature to gas gravity. The correlations are based on 634 compositions developed from 275 laboratory studies.

$$p_{pc} = 756.8 - 131.0\gamma_g - 3.6\gamma_g^2. \qquad (3.3)$$

$$T_{pc} = 169.2 + 349.5\gamma_g - 74.0\gamma_g^2. \qquad (3.4)$$

Gas gravity is typically a known value because of its relation to the heating value of gas, which usually determines the product sales price. Gas specific gravity is referenced to the density of air in Eq. 3.5a. The molecular weight of air MW_a = 28.9647 lbm/lbm-mol.

$$\gamma_g = \frac{MW_g}{MW_a} = \frac{MW_g}{28.9647}, \qquad (3.5a)$$

$$MW_g = 28.9647\gamma_g. \qquad (3.5b)$$

Example 3.1—Calculating Gravity. What is the gravity of methane? MW_{CH4} = 16.04 lbm/lbm-mol.
Solution. Use Eq. 3.5a to calculate the gravity of methane:

$$\gamma_g = \frac{16.04}{28.9647} = 0.55.$$

As would be expected, the specific gravity of methane is at the low end of the scale for natural gases. The addition of heavier hydrocarbons can increase gas gravities to 0.8 or greater.

3.2.3 Single-Phase z-Factor. The real-gas law describes pressure/composition/volume relationships for a natural gas:

$$pV = znRT. \qquad (3.6)$$

The z-factor term adjusts for the difference between a gas volume measured at real and at ideal conditions. Gas composition and ambient temperature and pressure define a unique relationship for equating a gas volume to standard conditions. Eq. 3.7 expresses the z-factor relationship as a ratio of gas volume measured at nonstandard reservoir conditions to that measured at ideal conditions:

$$z = \frac{V_{\text{actual}}}{V_{\text{ideal}}}. \quad (3.7)$$

All components are assumed to remain as gas over the pressure and temperature range of interest. Estimating the gas deviation factor is complicated for retrograde gas condensate reservoirs because a portion of the reservoir gas changes to a liquid as reservoir pressure declines. Changes in the reference volume must be accounted for. Application of a two-phase z-factor compensates for the effects of liquid fallout.

3.3 Gas Property Correlations

The widely accepted Standing and Katz correlation (Standing and Katz 1942) reduces the z-factor for all sweet hydrocarbon gases to a single figure, as shown in **Fig. 3.5.** The correlating parameters are pseudoreduced temperature and pressure. Critical temperature and pressure values included in Eqs. 3.8 and 3.9 are found through laboratory analysis of reservoir fluids or from published correlations. Sutton (1985) developed equations (Eqs. 3.3 and 3.4) that can be used to estimate critical properties if the gas specific gravity is known.

The effects of impurities can change the z-factor estimate to a minor degree and should be considered. Wichert and Aziz (1971) introduced a manual procedure to account for the presence of H_2S, CO_2, and N_2 gases.

Abou-Kassem et al. (1990) developed an iterative technique for calculating the gas deviation factor. The solution reproduces the Standing and Katz calculation procedure to within a 0.6% accuracy of real laboratory data, but it requires a computer program.

3.3.1 Piper-McCain-Corredor Correlations. A more-intuitive method accounting for impurities was developed by Piper, McCain, and Corredor (Piper et al. 1993). Their starting point was also the Sutton method, which used the Stewart-Burkhardt-Voo mixing solutions shown in Eqs. 3.8 and 3.9 to develop critical-property equations for sweet gases. The J and K coefficients represent the adjustment factors for the presence of N_2, H_2S, and CO_2 mole fractions that are needed when calculating critical properties.

$$T_{pc} = \frac{K^2}{J}. \quad (3.8)$$

$$p_{pc} = \frac{T_{pc}}{J}. \quad (3.9)$$

Piper et al. (1993) extended this work by developing correlations to determine J and K coefficients (Eqs. 3.10 and 3.11) that could be directly applied to Eqs. 3.8 and 3.9.

$$J = 0.12 + 0.71\gamma_g - 0.10\gamma_g^2 - 0.24\gamma_{\text{H2S}} - 0.46\gamma_{\text{CO2}} - 0.30\gamma_{\text{N2}}. \quad (3.10)$$

$$K = 3.82 + 17.44\gamma_g - 3.22\gamma_g^2 - 1.22\gamma_{\text{H2S}} - 7.05\gamma_{\text{CO2}} - 9.34\gamma_{\text{N2}}. \quad (3.11)$$

In turn, the z-factor is determined from the Standing and Katz chart (Fig. 3.5).

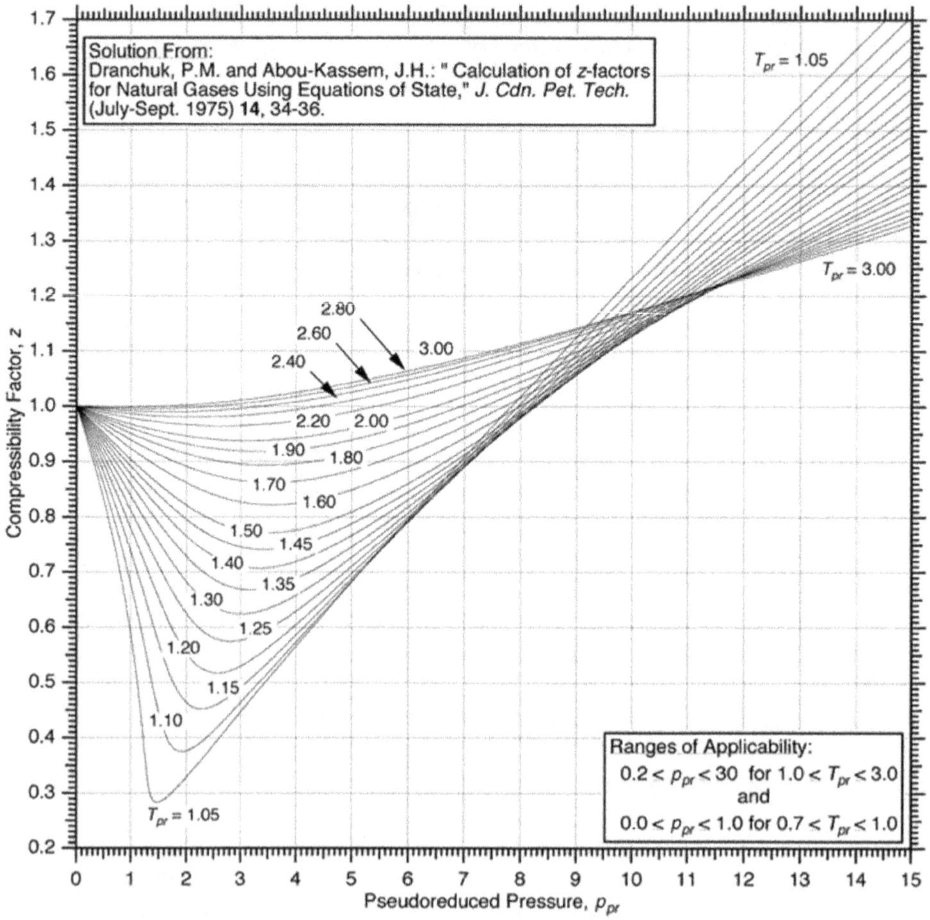

Fig. 3.5—The Standing and Katz *z*-factor chart for sweet gases. Note that the correlation values are pseudoreduced temperature and pressure.

Piper et al. (1993) included an analysis of 896 data points from 134 pressure/volume/temperature reports. The correlations work equally well for sweet and acid gases, and specific objectives were included to account for the presence of nonhydrocarbon impurities (H_2S, CO_2, and N_2).

Example 3.2—Calculating the *z*-Factor. Calculate the *z*-factor for a 0.68-gravity sour natural gas when the reservoir temperature is 200°F and the pressure is 2,000 psia. The natural-gas sample contains 2% N_2, 2% CO_2, and 14% H_2S.

Solution. Apply the Piper et al. (1993) equations (Eqs. 3.10 and 3.11) to calculate the coefficients *J* and *K*:

$$J = 0.12 + 0.71(0.68) - 0.10(0.68)^2 - 0.24(0.14) - 0.46(0.02) - 0.30(0.02) = 0.5077.$$

$$K = 3.82 + 17.44(0.68) - 3.22(0.68)^2 - 1.22(0.14) - 7.05(0.02) - 9.34(0.02) = 13.6917.$$

Apply the Stewart-Burkhardt-Voo mixing rules shown in Eqs. 3.8 and 3.9 to calculate the pseudocritical properties of the gas:

$$T_{pc} = \frac{K^2}{J} = \frac{13.69^2}{0.50} = 375°\text{R}.$$

$$p_{pc} = \frac{T_{pc}}{J} = \frac{375}{0.50} = 750 \text{ psia}.$$

Adjusted pseudocritical temperature and pressure correlating parameters (pseudoreduced properties) are calculated using Eqs. 3.1 and 3.2 and are then applied to the Standing and Katz chart shown in Fig. 3.5.

$$T_{pr} = \frac{T}{T_{pc}} = \frac{660}{375} = 1.76.$$

$$p_{pr} = \frac{p}{p_{pc}} = \frac{2{,}000}{750} = 2.67.$$

The z-factor for the contaminated natural gas is 0.889.

3.3.2 Gas Formation Volume Factor. The gas formation volume factor (FVF) (B_g) describes the difference between the gas volume measured at some temperature and pressure at reservoir conditions and the volume measured at base conditions. The following z-factor/gas and FVF relationships developed from the general gas equation assume a 14.65-psia base pressure and a base temperature of 60°F (base temperature = 460 + 60°F = 520°R). The base coefficients change if the base conditions are different.

$$B_g = \frac{V_{\text{res}}}{V_b} = \frac{p_b z_{\text{res}} T_{\text{res}}}{p_{\text{res}} z_b T_b} = \frac{14.65 z_{\text{res}} T_{\text{res}}}{p_{\text{res}}(1)(520)} = 0.0282 \frac{z_{\text{res}} T_{\text{res}}}{p_{\text{res}}}, \text{ rcf/scf.} \quad (3.12a)$$

B_g expressed in terms of reservoir barrels per standard cubic foot is

$$B_g = 0.00504 \frac{zT}{p}, \text{ RB/scf.} \quad (3.12b)$$

The reciprocal gas FVF ($1/B_g$) is sometimes called the gas expansion factor (E_g):

$$\frac{1}{B_g} = E_g = 35.46 \frac{p_{\text{res}}}{z_{\text{res}} T_{\text{res}}}, \text{ scf/RB.} \quad (3.12c)$$

Example 3.3—Calculating the FVF. Calculate the FVF for the sour natural gas described in Example 3.2. Reservoir temperature = 200°F (660°R), pressure = 2,000 psia, and z = 0.889.

Solution. Calculate the gas FVF:

$$B_g = 0.0282 \frac{z_{\text{res}} T_{\text{res}}}{p_{\text{res}}} = 0.0282 \frac{(0.889)(660)}{2{,}000} = 0.00827 \text{ rcf / scf}$$

$$= (0.00827 \text{ rcf / scf}) \frac{1}{5.615 \text{ ft}^3 / \text{bbl}} = 0.001473 \text{ RB / scf.}$$

The gas expansion factor is the reciprocal of the previous answer, and thus

E_g = 120.91 scf/rcf = 678 scf/RB.

3.3.3 Gas Compressibility. Knowledge of the gas compressibility at initial conditions and how that value changes with declining pressure is important when characterizing a reservoir gas. The general modified law for real gases was given in Eq. 3.6: $pV = znRT$.

The partial derivative of the pressure- and temperature-dependent general gas law is shown in Eq. 3.13. Assume that reservoir temperature remains constant.

$$\frac{dV}{dp} = \frac{nRT}{p}\frac{dz}{dp} - \frac{znRT}{p^2} \quad (3.13)$$

Substituting the real-gas law into Eq. 3.13 and solving for V results in

$$\frac{dV}{Vdp} = \frac{1}{z}\frac{dz}{dp} - \frac{1}{p} \quad (3.14)$$

The definition of compressibility relates a change in volume with a change in pressure to some reference volume:

$$c = -\frac{1}{V}\frac{dV}{dp} \quad (3.15)$$

Combining Eqs. 3.14 and 3.15 results in a relationship for determining gas compressibility:

$$c_g = \frac{1}{p} - \frac{dz}{zdp} \quad (3.16)$$

The slope of the z/pressure curve remains essentially constant for pressures > 4,000 psi. At low pressures, real gases behave as ideal gases, and the value of z is close to unity ($z \approx 1$) and the partial derivative of a constant will be zero $\left(\frac{dz}{dp} = 0\right)$. Therefore, the relationship between the z-factor and gas compressibility observed for low- to moderate-pressure gas reservoirs can be expressed as follows:

$$c_g = \frac{1}{p} \quad (3.17)$$

3.4 Oil Properties

Reservoir temperature and fluid composition position a crude oil on the left side of the critical point in the pressure/temperature diagram. Crude oils are classified as either black or volatile, but there is no sharp division between the two categories. However, volatile oils tend to lie close to the critical point, while moderate- to low-GOR black oils are located farther from the critical point.

The assumption is that the composition of black oil remains constant with a loss of reservoir pressure. This is not the case for near-critical or volatile crudes that contain a greater proportion of light- and intermediate-molecular-weight hydrocarbons. The evolution of these lighter-weight materials from the reservoir fluid with declining pressure causes the composition of the reservoir liquid to change. Therefore, the compositional material-balance approach should be applied to study the behavior of volatile-oil reservoirs.

Black-oil reservoir fluids are characterized by a stock-tank oil gravity < 35 °API, GOR < 1,500 scf/STB, and FVF < 1.5 RB/STB. The GOR for volatile oils ranges from 1,500 to 2,000 scf/STB, and the gravity ranges from 35 to 45 °API (Moses 1986). Gas evolution causes oil to exhibit high shrinkage characteristics, with the FVF between 1.5 and 2.0 RB/STB. Moses (1986) states that reservoir fluids exhibiting > 12.5% C_{7+} are classed as volatile oils, while those with < 12.5% C_{7+} are retrograde gas condensates.

A saturated crude oil contains a maximum-possible volume of dissolved gas. Reservoirs are defined as undersaturated when the initial pressure is greater than the bubblepoint pressure of the oil. This fact implies that more gas would go into solution in the reservoir oil if additional gas were introduced into the system. The presence of an initial gas cap implies the reservoir oil is fully saturated. That type of reservoir is typically defined as saturated.

Knowledge of how FVF, GOR, and for some cases viscosity vary with pressure for a reservoir oil aids in the development of insight into reservoir performance. Reservoir volumetric and performance studies include this information within the calculation procedure.

Laboratory fluid analyses of reservoir fluids provide the best means for determining these properties. However, fluid analysis reports are seldom available, and published correlations often form the basis for estimates of fluid property/pressure relationships.

The primary properties defining the character of a crude oil are viscosity, density or specific gravity, GOR, and FVF. The viscosity of crude oil is considered the most-important fluid property affecting the degree of success of a waterflooding project. Oil viscosity has the important effect of determining the mobility ratio, which, in turn, controls the sweep efficiency. In general, better-quality oils (those with lower viscosity) result in better waterflooding recovery. Specific gravity relates the crude-oil density to the density of fresh water when $\rho_w = 62.4$ lbm/ft^3.

3.4.1 Specific Gravity. The 10 °API base reference for pure water forms the basis of Eq. 3.18. The equation indicates there is an inverse relationship between liquid specific gravity and API gravity. The sales price of oil is based on gravity and the presence or absence of contaminants. The higher the API gravity, the greater the quantity of intermediate-sized molecules contained in the sample and, consequently, the higher the price of the oil.

$$\text{API gravity} = \frac{141.5}{\gamma_o} - 131.5. \quad (3.18)$$

Example 3.4—Calculating API Gravity. What is the API gravity of water when its specific gravity is 1.0?

Solution. Use Eq. 3.18:

$$\text{API gravity} = \frac{141.5}{\gamma_o} - 131.5 = \frac{141.5}{1.0} - 131.5 = 10\ ^\circ\text{API}.$$

Note the inverse relationship between specific gravity and API gravity.

API gravity is a measure of the relative density of a petroleum liquid with respect to the density of water. It has no units but is referred to using degrees.

API gravity is also used to compare the relative densities of petroleum liquids. For example, if one petroleum liquid floats on another, it is therefore less dense and has a greater API gravity. The property is measured on a hydrometer instrument designed so that most values would fall between 10 and 70 °API.

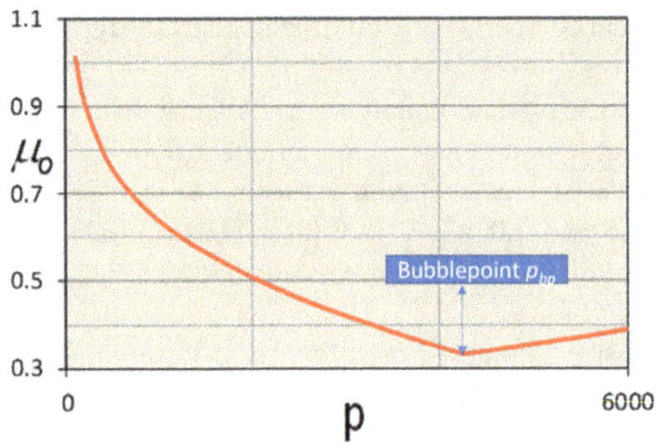

Fig. 3.6—A light-oil viscosity as a function of pressure measured at a constant temperature.

3.4.2 Viscosity. Oil viscosity μ_o is a measurement of the internal resistance of oil to flow (i.e., to shear).

Viscosity is defined as the force per unit area required to maintain a unit velocity gradient in a fluid. Newtonian fluids are typically assumed. Units are centipoise (cp) and Pa·s, where 1 mPa·s = 1 cp.

Viscosity governs the viscous forces. Gas viscosity is approximately 0.02 cp, water viscosity is approximately 1.3 cp (at 15°C), and most oils exhibit viscosities in the range of 0.2–30 cp or greater. The lower the oil viscosity, the greater the ability of the fluid to flow.

Figs. 3.6 through 3.8 show the viscosities for different oils. Viscosity is strongly dependent on temperature and solution GOR and moderately dependent on pressure (except at very high pressure), and it should be measured in the laboratory. Fig. 3.6 shows laboratory measurements of light-oil viscosity as a function of pressure measured at a constant temperature.

Fig. 3.7 shows oil viscosity as a function of API gravity measured at reservoir temperature for different oil-producing regions around the world.

Fig. 3.8 shows oil viscosity ranges as functions of values of density, API gravity, and specific gravity using a generic classification from conventional oils to heavy crude to bitumen.

3.4.3 Gas/Oil Ratio. The quantity of dissolved gas present in a reservoir crude oil is a function of

- Composition and number of C_{7+} molecules present
- Bubblepoint pressure
- Reservoir temperature and pressure

The volume of dissolved gas in a unit volume of crude oil is defined as the solution GOR, expressed as R_s in scf/STB units. The producing GOR is denoted as R. **Fig. 3.9** shows how the solution GOR behaves differently for conditions above and below the bubblepoint pressure. The GOR remains

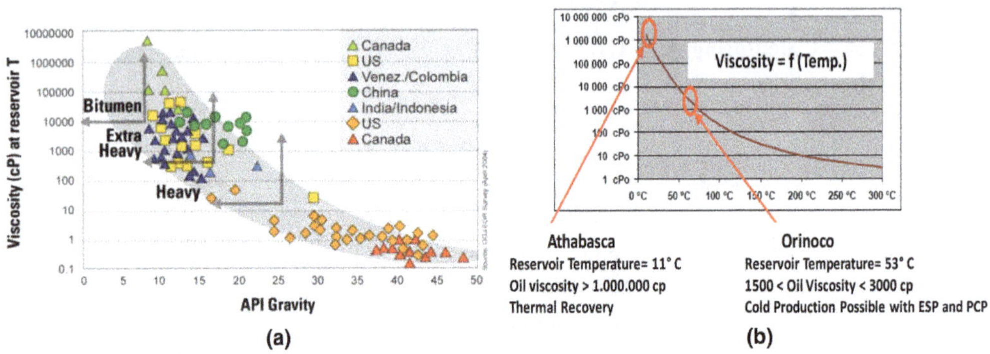

Fig. 3.7—Typical oil viscosity values (after *Oil & Gas Journal* 2004). (a) Viscosity (semilog scale) shown as a function of API gravity and (b) viscosity (semilog scale) shown as a function of temperature.

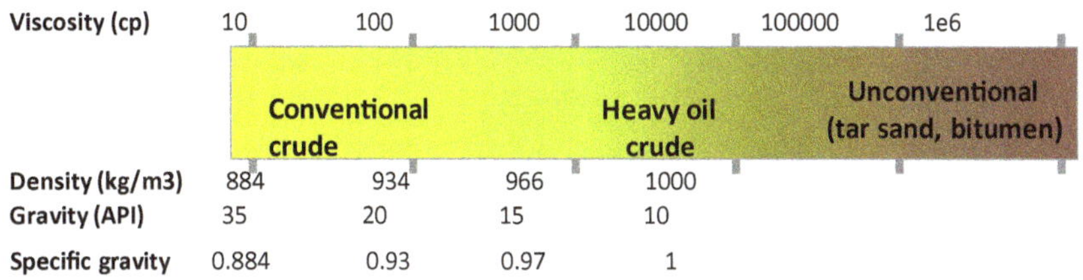

Fig. 3.8—Typical ranges of oil properties, relating viscosity to values of specific gravity, API gravity, and density. Courtesy of Schlumberger NExT.

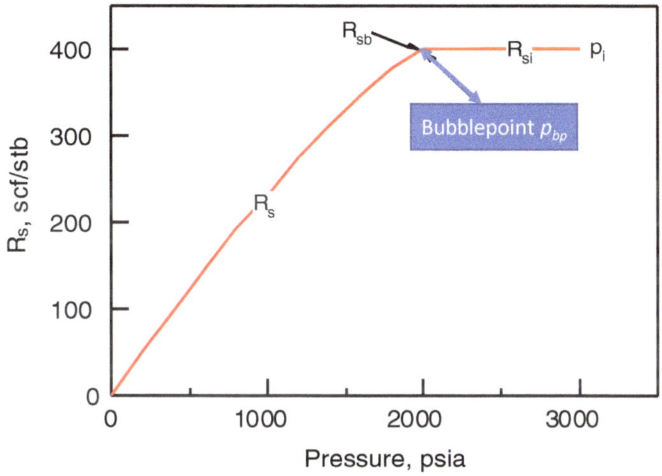

Fig. 3.9—Illustrating solution GOR as a function of pressure. Gas evolves from reservoir oil as conditions fall below the bubblepoint, but gas does not evolve as long as pressure remains above the bubblepoint over the p_i to p_{bp} range. Note the location of R_{si} at initial conditions, R_{sb} at the bubblepoint pressure, and variable R_s in the figure.

constant above the bubblepoint because gas is unable to evolve. However, gas evolves from reservoir oil when pressure declines below the bubblepoint. Evolved gas is either produced in conjunction with oil or migrates updip to form a secondary gas cap.

3.4.4 Oil Formation Volume Factor. Oil has a different volume at reservoir vs. surface conditions. Volume changes can be caused by

- The evolution of solution gas, causing reservoir oil to shrink
- A loss of temperature at the surface, which causes the reservoir oil to shrink
- A loss of pressure, which causes an expansion of reservoir oil

Solution gas evolution is an important consideration when reservoir conditions are below the bubblepoint. On the other hand, the expansion of reservoir liquids typically provides the primary energy force for conditions above the bubblepoint. The units for these volume changes between reservoir and surface conditions are RB/STB.

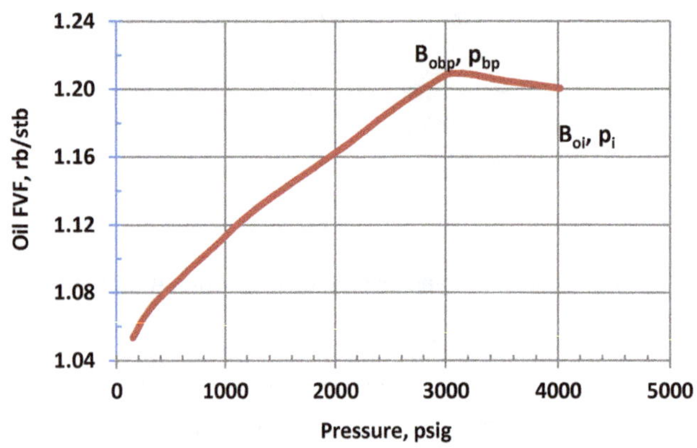

Fig. 3.10—Depletion history of a p_{bp} = 2,620-psi oil. The bubblepoint forms the nexus between liquid expansion and contracting volume caused by gas evolution from the reservoir oil when conditions fall below the bubblepoint.

Fig. 3.10 illustrates the effect of declining reservoir pressure on the oil FVF. Reservoir oil expands as pressure declines from some initial condition (B_{oi}) to bubblepoint conditions (B_{obp}). This effect causes the oil FVF to increase because the reservoir liquid volume expands. A loss of reservoir volume caused by gas evolution at conditions below the bubblepoint pressure causes the unit volume to shrink, which in turn reduces the oil FVF.

3.5 Oil Property Correlations

Correlations are a widely accepted method for estimating crude-oil properties, and they work reasonably well for reservoir crudes displaying a moderate gravity and GOR. However, care should be exercised when studying volatile oils or crude oils displaying some particular characteristic.

The sections that follow discuss the Vasquez and Beggs (1980) and Standing (1947) black-oil correlations. **Table 3.2** compares the range of data included in those correlations.

3.5.1 Vasquez and Beggs Correlations. Vasquez and Beggs (1980) developed a set of empirical correlations to estimate viscosity, API gravity, FVF, GOR, and bubblepoint pressure for crude oils. More than 600 laboratory analyses and 6,000 data points were analyzed. Correlations for estimating viscosity have been published, and in general, all of them consider that $\mu = f(T, p, \gamma_o, \gamma_g, R_s)$.

The correlation for undersaturated oils by Vasquez and Beggs (1980) is

$$\mu_o = \mu_{ob}\left(\frac{p}{p_{bp}}\right)^m, \qquad (3.19)$$

where $m = 2.6\,p^{1.187} \times 10^a$ and where $a = -3.9(10^{-5})\,p - 5$.

Condition	Vasquez and Beggs Minimum Value	Vasquez and Beggs Maximum Value	Standing Minimum Value	Standing Maximum Value
R_{sb} (scf/STB)	20	2,070	20	1,425
°API	16	58	16.5	63.8
γ_g	0.56	1.18	0.59	0.95
T_{res} (°F)	70	295	100	258
p_{bp} (psia)	50	5,250	130	7,000

Table 3.2—Range of conditions in Vasquez and Beggs (1980) and Standing (1947) correlations.

The data used to develop Eq. 3.19 ranged as follows:

- Pressure = 141–9,151 psia
- Gas solubility = 9.3–2,199 scf/STB
- Viscosity = 0.117–148 cp
- Gas gravity = 0.511–1.351
- API gravity = 15.3–59.5 °API

Table 3.3 lists coefficients for the Vasquez and Beggs correlations divided into < 30 and > 30 °API oil gravity.

Gas gravity was normalized to a 100-psig reference pressure to provide a starting point for the calculation procedure. Field gas gravity is assumed to be at a 100-psig reference pressure if the field separator pressure and the temperature are unknown.

Eq. 3.20 converts the gas gravity measured at some other separator condition to the 100-psig reference:

$$\gamma_{gc} = \gamma_{gs}\left[1.0+5.912\times10^{-5}\gamma_{\text{API}}T_s\log\left(\frac{p_s}{114.7}\right)\right], \quad (3.20)$$

where γ_{gc} is the gas gravity corrected to a 100-psig separator pressure, γ_{gs} is the gas gravity at a sample separator temperature and pressure, γ_{API} is the oil gravity (°API), T_s is the sample separator temperature (°F), and p_s is the sample separator pressure (psia).

Eq. 3.21 provides the correlation for solution GOR:

$$R_s = \text{C}1\gamma_{gc}p_{\text{res}}{}^{\text{C}2}\exp^{\left[\text{C}3\gamma_{\text{API}}/(T_{\text{res}}\,°\text{F}+460)\right]}, \text{ scf/STB}, \quad (3.21)$$

where γ_{gc} is determined from the separator gravity correction, T_{res} is the reservoir temperature (°F), and p_{res} is the reservoir pressure (psia).

Eq. 3.22 provides the correlation for the oil FVF (B_o) when $p_{\text{res}} < p_{bp}$ at saturated conditions:

$$B_o = 1+\text{C}1R_s+\text{C}2\left(T_{\text{res}}-60\right)\left(\frac{\gamma_{\text{API}}}{\gamma_{gc}}\right)+\text{C}3R_s\left(T_{\text{res}}-60\right)\left(\frac{\gamma_{\text{API}}}{\gamma_{gc}}\right), \text{ RB/STB}. \quad (3.22)$$

The equation for the $p_{\text{res}} > p_{bp}$ condition is less cumbersome because of the absence of free gas within the system:

$$B_o = B_{obp}\exp\left[c_o\left(p_{bp}-p\right)\right], \text{ RB/STB}. \quad (3.23)$$

Coefficient	Bubblepoint and R_s		FVF	
	°API ≤ 30	°API > 30	°API ≤ 30	°API > 30
C1	0.0362	0.0178	4.68×10^{-4}	4.67×10^{-4}
C2	1.0937	1.1870	1.75×10^{-5}	1.10×10^{-5}
C3	25.7240	23.9310	-1.81×10^{-8}	1.34×10^{-9}

Table 3.3—Vasquez and Beggs correlations coefficients.

A rewritten form of Eq. 3.23 can be used to calculate oil compressibility when system pressures are greater than the bubblepoint pressure:

$$c_o = \frac{-1{,}433.00 + 5.0R_s + 17.20T_{\text{res}} - 1{,}180.00\gamma_{gs} + 12.61\gamma_{\text{API}}}{10^5 p_{\text{res}}}, \text{ psi}^{-1}. \qquad (3.24)$$

To calculate the bubblepoint pressure, $R_{sb} = R_{si}$ = GOR at bubblepoint pressure. Eq. 3.24 can be rewritten as

$$p_{bp} = \left\{ \frac{R_{sb}}{\text{C1}\gamma_{gc}\exp^{\left[\text{C3}\gamma_{\text{API}}/(T_{\text{res}}+460)\right]}} \right\}^{1/1.187} \qquad (3.25)$$

Example 3.5—Using the Vasquez and Beggs Correlations. Production tests from an oil well indicate the perforated zone at 8,742–8,760 ft has exhibited the following properties:

- Presence of a gas/oil contact, which indicates saturated reservoir conditions.
- GOR was 380 scf/STB.
- Flowing tubinghead pressure was 890 psia.

The test rate was 137 BOPD.
Additional data are

- $\gamma_{\text{API}} = 45$ °API, $\gamma_{gs} = 0.82$.
- Static bottomhole pressure = p_{res} = 1,000 psia.
- T_{res} = 200°F (T_{res} = 200°F + 460 = 660°R).
- Separator conditions: T_s = 80°F and p_s = 60 psia.

Apply the Vasquez and Beggs correlations to calculate solution GOR (R_s), oil FVF (B_o), bubblepoint pressure (p_{bp}), and oil compressibility (c_o).

Table 3.3 lists coefficients for the Vasquez and Beggs correlations divided into ≤ 30 and > 30 °API oil gravity.

Solution. First, adjust the gas gravity to 100-psig conditions using Eq. 3.20:

$$\gamma_{gc} = \gamma_{gs}\left[1.0 + 5.912\times10^{-5}\gamma_{\text{API}}T_s\log\left(\frac{p_s}{114.7}\right)\right] = 0.82\left[1.0 + 5.912\times10^{-5}(45)(80)\log\left(\frac{60}{114.7}\right)\right] = 0.7709.$$

Then, apply the $\gamma_{\text{API}} > 30$ °API coefficients from Table 3.3 to calculate R_s using Eq. 3.21:

$$R_s = \text{C1}\gamma_{gc}p_{\text{res}}{}^{\text{C2}}\exp^{\left[\text{C3}\gamma_{\text{API}}/(T_{\text{res}}\,°\text{F}+460)\right]} = 0.0178(0.77)(1{,}000)^{1.187}\exp^{\left[(23.931)(45)/(660)\right]} = 255.289 \text{ scf/rcf}.$$

Apply the $\gamma_{\text{API}} > 30$ °API coefficients from Table 3.3 to estimate B_o using Eq. 3.22:

$$B_o = 1 + C1R_s + C2\left(T_{\text{res}} - 60\right)\left(\frac{\gamma_{\text{API}}}{\gamma_{gc}}\right) + C3R_s\left(T_{\text{res}} - 60\right)\left(\frac{\gamma_{\text{API}}}{\gamma_{gc}}\right) = 1 + \left(4.67\times10^{-4}\right)\left(255\right)$$

$$+\left(1.10\times10^{-5}\right)\left(200-60\right)\left(\frac{45}{0.77}\right) + \left(1.337\times10^{-9}\right)\left(255\right)\left(200-60\right)\left(\frac{45}{0.77}\right) = 1.2119 \text{ RB/STB.}$$

Next, calculate the bubblepoint pressure (p_{bp}) using Eq. 3.25:

$$p_{bp} = \left\{\frac{R_{sb}}{C1\gamma_{gc}\exp^{\left[C3\gamma_{\text{API}}/(T_{\text{res}}+460)\right]}}\right\}^{1/1.187} = \left\{\frac{255.289}{(0.0178)(0.77)\exp\left[\frac{(23.931)(45)}{660}\right]}\right\}^{1/1.187} = 1{,}000.0 \text{ psia.}$$

Then, calculate the oil compressibility using Eq. 3.24:

$$c_o = \frac{-1{,}433.00 + 5.0R_s + 17.20T_{\text{res}} - 1{,}180.00\gamma_{gs} + 12.61\gamma_{\text{API}}}{10^5 p_{\text{res}}}$$

$$= \frac{-1{,}433.00 + 5.0\left(255.288\right) + 17.20\left(200.0\right) - 1{,}180.00\left(0.77\right) + 12.61\left(45\right)}{10^5\left(1{,}000.0\right)}$$

$$= 1.67012\times10^{-5} \text{ psi}^{-1}.$$

3.5.2 Standing Correlations. Standing (1947) developed correlations for California crude oils. Ranges of correlation data from Standing (1947) and Vasquez and Beggs (1980) were compared in Table 3.2; note that both data sets span a wide range of conditions. The Standing correlations seem to work best for lower-API-gravity, California-type oils.

The bubblepoint pressure correlation can be used to estimate the bubblepoint pressure and gas solubility (solution GOR):

$$p_{bp} = 18\left(\frac{R_{sb}}{\gamma_g}\right)^{0.83} \times 10^{\gamma_g}, \qquad (3.26)$$

where γ_g is the gas gravity and R_{sb} is the solution GOR at the bubblepoint pressure (scf/STB).

$\gamma_{g(mf)}$ is the mole fraction of gas calculated as

$$\gamma_{g(mf)} = 0.00091T_{\text{res}} - 0.0125\gamma_{\text{API}}, \qquad (3.27)$$

where γ_{API} is the oil gravity (°API) and T_{res} is the reservoir temperature (°F).

Recall, the solution GOR (gas solubility) remains constant as long as pressure remains above the bubblepoint pressure. Thus, Eq. 3.26 can be rewritten to solve for the solution GOR (gas solubility):

$$R_s = \gamma_g\left(\frac{p}{18\times10^{\gamma_g}}\right)^{1.204} \qquad (3.28)$$

Example 3.6—Calculating Solution GOR. Preliminary data for an infield well indicate a reservoir pressure of 1,800 psia and a temperature of 200°F. Oil gravity = 34 °API, while the gas gravity = 0.75. Apply Eq. 3.28 to calculate the expected solution GOR for this well.

Solution. In the first step of the calculation procedure, apply Eq. 3.27 to determine the mole fraction of gas included in the sample:

$$\gamma_{g(mf)} = 0.00091(200) - 0.0125(34) = -0.243.$$

Then apply Eq. 3.28 to calculate the solution GOR:

$$R_s = \gamma_g \left(\frac{p}{18 \times 10^{\gamma_g}} \right)^{1.204} = 0.75 \left(\frac{1{,}800}{18 \times 10^{0.75}} \right)^{1.204} = 376 \text{ scf/STB}.$$

The Standing correlation to calculate the oil FVF is shown in Eq. 3.29, where F is a correlation factor given by a subsidiary relation between oil and gas gravity, which is shown in Eq. 3.30.

$$B_o = 0.972 + 0.000147 F^{1.175}, \quad (3.29)$$

$$F = R_s \left(\frac{\gamma_g}{\gamma_{API}} \right)^{0.5} + 1.25T, \quad (3.30)$$

where R_s is the solution GOR (gas solubility) (scf/STB) and T is the temperature of interest (°F).

Example 3.7—Calculating Oil FVF. Apply the information provided in Example 3.6 to calculate the oil FVF.

Solution. The first step is to calculate F:

$$F = 376 \left(\frac{0.75}{34} \right)^{0.5} + 1.25(200) = 305.8.$$

Then apply Eq. 3.29 to calculate the oil FVF:

$$B_o = 0.972 + 0.000147(305.8)^{1.175} = 1.06 \text{ RB/STB}.$$

3.6 Water Properties

Most oilfield waters contain a variety of inorganic and organic material. The most common inorganic constituents are the cations Na^+ (sodium), K^+ (potassium), Li^+ (lithium), Ca^{++} (calcium), Mg^{++} (magnesium), Ba^{++} (barium), Fe^{++} (ferrous iron), and Sr^{++} (strontium) and the anions $Cl=$ (chloride), $SO_4=$ (sulfate), $CO_3=$ (carbonate), HCO_3^- (bicarbonate), $NO_3=$ (nitrate), Br^- (bromide), and I^- (iodide). Injected waterflood water can contain organic material including plant material, plankton, and microscopic animals; produced fine-grained rock material; and water chemically different from the in-situ reservoir water. There are important environmental regulations that determine the need for antiscaling additives and limit water options.

3.6.1 Water Composition. Water-composition analysis is a must for estimating the makeup of produced and injected water for waterflooding, steam injection, or disposal. In addition, depending on the composition of the oilfield waters, management may need to be implemented to avoid problems such as corrosion, formation plugging, and scale formation. Oilfield waters are generally described using some of the following properties:

- pH: This is a measure of the hydrogen concentration, which affects the potential for corrosion. An increasing pH indicates a less-acidic atmosphere and a difficulty in holding iron and calcium in solution. Both minerals increase the presence of sulfides and scale formation.
- Alkalinity: This measures the acid-neutralizing capacity of a water sample. It is primarily the result of the carbonate, bicarbonate, and hydroxide content.
- Turbidity: It is typically composed of organic material and is an important test for determining water quality. Filters as well as biocides are required to inhibit the growth or activity of living organisms (bacteria downhole), which are detected by turbidity tests.
- Total dissolved solids (TDS): A measure of the residue after evaporation of the sample. Both sock and media filters are often required to strain out materials that can potentially plug pore spaces.
- Conductivity (electrical): This is routinely measured in siemens per meter, S/m. The most commonly used units are μS (microsiemens) and mS (millisiemens). The siemens has also been referred to as mho, and it is the reciprocal of electrical resistivity, whose unit is ohm.
- Redox potential (*Eh*): A negative *Eh* potential indicates ready oxidation. It is measured in volts (V) or millivolts (mV). It is also known as the reduction/oxidation potential.
- Hardness: This measures the concentration of cations, particularly calcium and magnesium because hard water is high in those dissolved minerals. It is typically expressed as mg/L calcium carbonate ($CaCO_3$).

Chang (1985) provided the properties of a typical seawater sample from the South China Sea. Note that **Table 3.4** lists both the organic and inorganic components present in the sample.

3.6.2 Diagrams. Diagrams illustrating the distribution of the anions (negatively charged ions) and cations (positively charged ions) for a particular sample are actively used when describing field character. The presence of various cations and anions can change solubility, acidity, and redox (oxidation) reactions. Core flow studies and laboratory water analyses of field water samples are a requirement for a thorough assessment of water quality.

The relative amounts of the materials present in a sample can be expressed in tabular as well as graphical form. Numerous techniques are available, such as pie and bar charts and Stiff (1951), Piper (1944), and fence diagrams. The Stiff diagram seems to be the most useful plot for the analysis of reservoir-water flow systems. **Fig. 3.11** provides an example of a Stiff diagram.

Comparing the plotted outlines of adjacent wells provides an indicator of the chemical content of the flow system occurring at that point in an active waterflood. Of course, there must be a sufficient chemical difference between the injected water and in-situ water to provide a reasonable estimate of the mixing at that point. The effectiveness of the water-injection process can be estimated by noting changes in the shape of well diagrams included in the injection/production system.

Calcium	400 mg/L
Magnesium	1100 mg/L
Iron	0.05 mg/L
Sodium	11 300 mg/L
Barium	0 mg/L
Chloride	19 600 mg/L
Sulfate	2400 mg/L
Bicarbonate	90 mg/L
Hydroxyl and sulfide	0 mg/L
TDS	34,890 ppm
pH	8.2
Temperature	29°C
Suspended solids (ppm)	0.2–0.9

Table 3.4—An example of the typical properties of a produced-water sample (after Chang 1985). 1 mg/L = 1 part per million (ppm).

3.6.3 Potential Reactions. The following potential chemical and physical reactions can restrict the flow capacity of surface equipment and injection rates:

- Montmorillonite clay particles swell when contacted by fresh and low-salinity water. Reactive

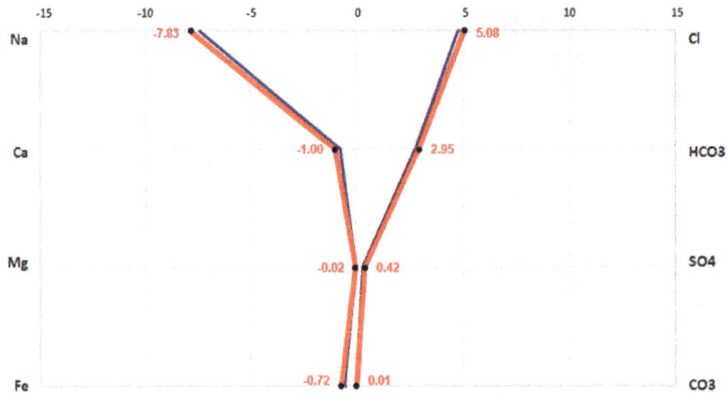

Fig. 3.11—An example Stiff diagram. The reaction values of ions are plotted as a function of concentration onto a polygon system. Cations are plotted to the left [Na^+ (sodium), Ca^{++} (calcium), Mg^{++} (magnesium), and Fe^{++} (ferrous iron)] and anions to the right [Cl= (chloride), HCO_3^- (bicarbonate), SO_4= (sulfate), and CO_3= (carbonate)] of a vertical zero line. Endpoints are connected by straight lines.

forces present within the clay particles expand them into the surrounding pore spaces, thus reducing the permeability of the system.
- Oxygen reacts with dissolved chemicals to precipitate oxides and scales. The ability of water to act as a corrosion agent is also increased.
- Entrained oxygen promotes bacteria growth and can carry sulfate-reducing bacteria.
- The deposition of precipitates provides a haven for anaerobic bacteria.
- Changes in dissolved-gas volumes can cause mineral precipitation.

The majority of water sources are derived from waters contaminated with oxygen and organic and inorganic material. The eradication of these impurities from a water source is a prerequisite before the water can be injected into a reservoir. Treating to rid a candidate water source of specific properties is of primary concern when designing a waterflood project, although sand quality is a good indicator of expected production problems.

Potential scaling and rock and fluid incompatibility problems can be determined from laboratory and field testing. Mitchell and Finch (1981) felt that most preliminary field testing should be treated qualitatively. However, the actual response should be closely monitored to compare to the original findings.

The following are concentrations of material found in water samples: total suspended solids, soluble hydrocarbons and organic material, carbonates, iron oxides/sulfides, total acid, and soluble and insoluble material. The measurement of the type and quantity of particulate matter in a water sample is typically carried out with membrane filters or Coulter counters:

- Membrane filtration: A membrane filter collects residue from a measured flow stream volume for a measured time interval. Trapped particulates remaining on the filter face represent the degree of contamination contained in that sample. The dried filter is weighed to measure the particulate quantity (in mL). Specific solvents are passed sequentially through the contaminated filter cake to determine the quantity of solute dissolved by that particular solvent.
- Coulter counter: This apparatus counts the number and size of particulates present in an electrolyte by giving off a charge as the solution moves through an aperture.

3.6.4 Oilfield Water Properties. Water properties are commonly used in petroleum engineering calculations to predict reservoir behavior by applying material balance, multiphase modeling, and simulation under conditions such as natural depletion, water injection, and water influx. The bubblepoint pressure p_{bp} of oilfield water is equal to the p_{bp} value of the coexisting oil as a result of thermodynamic equilibrium. The water FVF B_w depends on reservoir conditions of pressure and temperature while the amount of gas in solution has a negligible effect. Other properties such as density, viscosity, and specific gravity are commonly measured in the laboratory. **Fig. 3.12** shows an example of the typical shape of the water FVF.

Injecting water into the reservoir to replace the volumes of oil, gas, and water produced can be expressed by the voidage replacement ratio (VRR):

VRR = injected reservoir pore volumes/produced reservoir pore volumes;

$$\text{Instantaneous } VRR=\left(B_w \times i_w\right)/\left[\left(B_o \times q_o\right)+\left(B_w \times q_w\right)+(GOR-R_s)\times B_g \times q_o\right], \quad (3.31)$$

where B_w, B_o, and B_g are the FVFs of the water, oil, and gas phases, respectively (in RB/STB, RB/STB, and rcf/STB, respectively); q_w and q_o are the producing flow rates of the water and oil phases, respectively (in STB/D); i_w is the injection rate for the water phase (STB/D); GOR is in scf/STB; and R_s is the dissolved gas in the oil (scf/STB). The third term in the denominator, $(GOR-R_s)\times B_g \times q_o$, accounts for "free gas" produced that is in excess of gas in the reservoir that is dissolved in the oil.

The VRR can be calculated on an instantaneous basis using injected and produced fluids over any specific time period (flow rates measured or allocated on a daily basis) or calculated on a cumulative basis using the injected and produced fluids cumulative production numbers at the commencement of waterflooding.

Example 3.8—Calculating Pore Volumes of Oil and Gas Produced Fluids. Determine the pore volume on a per-day basis resulting from produced fluids in a reservoir from a well that is producing near the gas/oil contact.

Data: Oil rate = 50 STB/D, GOR = 2,000 scf/STB, R_s = 1,500 scf/STB, B_g = 0.00827 rcf/STB, and B_o = 1.5 RB/STB. No water production is evident, and water production is considered to be zero.

Solution. Calculate the oil produced reservoir pore volumes using a portion of Eq. 3.31:

$$\left(B_o \times q_o\right)=\left(1.5 \text{ RB / STB}\right)(50 \text{ STB / D})=75 \text{ RB/D}.$$

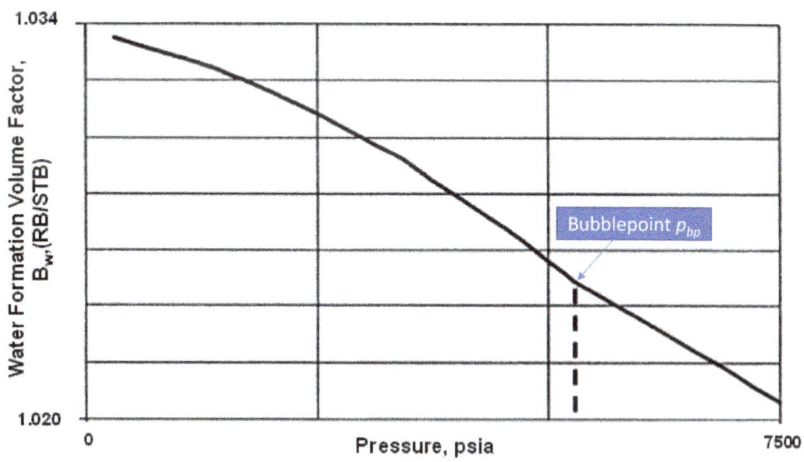

Fig. 3.12—An example of the typical shape of the water FVF.

Calculate the produced-free-gas pore volumes using a portion of Eq. 3.31:

$$\left(GOR - R_s\right)\times B_g \times q_o = \left(2{,}000 \text{ scf / STB} - 1{,}500 \text{ scf / STB}\right)\left(0.00827 \text{ rcf / STB}\right)\left(50 \text{ STB / D}\right)$$

$$\frac{1}{5.615 \text{ ft}^3 / \text{bbl}} = 37 \text{ RB/D}.$$

The total pore volume of the produced fluids = 112 RB/D.

Compare the void spaces calculated in Example 3.8 that were created by produced oil and gas. Liquid expansion drive is inefficient when compared to gas expansion. The displacement of 1 bbl of produced oil by means of injected water requires the injection of two times the volume of water to compensate for the depleted pore space per each bbl of water required to replace the gas void space.

3.7 Summary

Petroleum crude oils and natural gases are mixtures of hydrocarbons that are contained within the system. Hydrocarbons are blends of carbon and hydrogen atoms ranging from methane gas (CH_4) to long-chained liquid and solid rock.

Pressure, temperature, and composition define the gas/liquid relationship at reservoir conditions. Phase relationships change as oil and gas are produced. Reservoir pressure declines as a result of produced fluids passing through a separation system and a pressure drop overcoming hydraulic head as the fluids move to the surface. This process changes the liquid/gas proportion of the hydrocarbons. These relationships can be obtained from laboratory analyses of a reservoir sample or from fluid-property correlations.

Produced-liquid volumes are measured at the surface while all fluid interactions typically occur in the subsurface at some elevated temperature and pressure. Reservoir analysis and material-balance studies relating the production history to the reservoir depletion history require knowledge of these gas/liquid interactions.

The Standing and Katz z-factor chart (Fig. 3.5) shows changing gas compressibility as a function of temperature and pressure. The gas specific gravity value provides a starting point for applying equations to determine pressure/volume relationships. Correlations are available to compensate for the effects of impurities.

GOR and FVF account for the volume of gas in solution with the oil at reservoir conditions. These effects must be accounted for when calculating oil-in-place and reserves. The Standing, Standing and Katz, Vasquez and Beggs, and Piper-McCain-Corredor correlations (to name a few) serve for the majority of reservoir situations.

3.8 Questions

- Why must continuous field monitoring be performed when producing an oil or gas field containing significant volumes of nonhydrocarbon components such as H_2S or CO_2? (Answer: Because CO_2 increases corrosion problems while H_2S causes steel embrittlement, and both can increase the deterioration of producing equipment. In addition, H_2S reduces human sensitivity to odors, and it is lethal at moderate concentrations.)
- List at least two reasons why recovery is less from volatile-oil reservoirs than in black-oil reservoirs. [Answer: (1) There is less stock-tank oil in volatile-oil reservoirs as a result of high concentrations of solution gas in the reservoir oil. (2) Excessive gas production often controls the oil rate in high-productivity wells. Waterflooding such a reservoir requires more capital because additional volumes of water are required to fill up the void space created by the evolution of the excessive volumes of solution gas.]
- Why do the condensate API gravity and GOR producing from a retrograde gas condensate reservoir generally increase with time, as seen in **Fig. 3.13?** (Answer: Heavier hydrocarbon components tend to be the first to condense in the reservoir. The GOR increases with time

because the produced gas contains increasing volumes of lighter gas molecules. High pressure keeps more liquids in the gas stream, which is why pressure maintenance projects are often recommended.)

- How does the loss of solution gas affect the residual oil saturation in an oil reservoir that is being waterflooded? (Answer: The API gravity of the reservoir oil decreases and the oil viscosity increases as a result of blowing off the lighter hydrocarbons first. These effects decrease the efficiency of the waterflood process, and the residual oil saturation increases as a result of having to displace a higher-viscosity oil.)

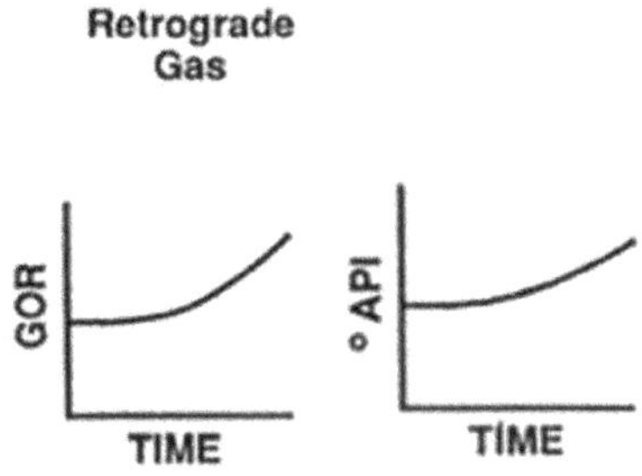

Fig. 3.13—Fluid property behavior of a retrograde gas condensate. GOR and API gravity are plotted as a function of time. Courtesy of Schlumberger NExT.

3.9 Nomenclature

a	=	constant for undersaturated oils (Vasquez and Beggs 1980)
B_g	=	gas formation volume factor, rcf/scf, RB/scf, or rcf/STB
B_o	=	oil formation volume factor, RB/STB
B_{obp}	=	oil formation volume factor at bubblepoint conditions, RB/STB
B_{oi}	=	initial oil formation volume factor, RB/STB
B_w	=	water formation volume factor, RB/STB
c	=	compressibility, psi^{-1}
c_g	=	gas compressibility, psi^{-1}
c_o	=	oil compressibility, psi^{-1}
E_g	=	gas expansion factor, scf/RB
Eh	=	redox potential, V or mV
f	=	function of a certain number of variables
F	=	Standing (1947) correlation factor to calculate oil formation volume factor
i_w	=	injection rate for water phase, STB/D
J	=	parameter in Stewart-Burkhardt-Voo mixing rules, °R/psia
K	=	parameter in Stewart-Burkhardt-Voo mixing rules, °R/psia
m	=	exponent factor used in correlations for undersaturated oils by Vasquez and Beggs (1980)
MW_a	=	molecular weight of air, lbm/lbm-mol
MW_{CH4}	=	molecular weight of methane, lbm/lbm-mol
MW_g	=	molecular weight of gas, lbm/lbm-mol
n	=	amount of substance, number of moles of gas measured in lbm-mol, used in the gas law equation $pV = znRT$
p	=	pressure, psia
p_b	=	base pressure, psia
p_{bp}	=	bubblepoint pressure, psia
p_c	=	critical pressure, psia
p_{dp}	=	dewpoint pressure, psia
p_i	=	initial pressure, psia
p_{pc}	=	pseudocritical pressure, psia
p_{pr}	=	pseudoreduced pressure, psia
p_{res}	=	reservoir pressure, psia
p_s	=	sample separator pressure, psia
q_o	=	producing flow rate of oil phase, STB/D
q_w	=	producing flow rate of water phase, STB/D
R	=	universal gas constant, which has a value of 8.314 J/mol-K or 10.730 psia ft^3/lbm-mol °R, used in the gas law equation $pV = znRT$
R	=	producing gas/oil ratio (gas solubility in oil), scf/STB
R_s	=	solution gas/oil ratio, scf/STB
R_{sb}	=	solution gas/oil ratio at bubblepoint pressure, scf/STB

R_{si} = solution gas/oil ratio at initial conditions, scf/STB
T = temperature, °F
T_b = base temperature, °F
T_c = critical temperature, °F
T_{pc} = pseudocritical temperature, °R
T_{pr} = pseudoreduced temperature, dimensionless
T_{res} = reservoir temperature, °F
T_s = sample separator temperature, °F
V = gas volume, ft³ or L
V_{actual} = gas volume measured at actual conditions, ft³ or L
V_b = base volume, ft³ or L
V_{ideal} = gas volume measured at ideal conditions, ft³ or L
V_{res} = reservoir volume, acre-ft, ft³, or bbl
z = z-factor, dimensionless. Corrections (z-factor) must be made to account for the deviation of real-gas behavior from that of the ideal gas.
z_b = compressibility factor, dimensionless at 14.65-psia base pressure and base temperature of 60°F
z_{res} = compressibility factor, dimensionless at reservoir pressure (psia) and reservoir temperature (°R)
γ_{API} = oil gravity, °API
γ_{CO2} = carbon dioxide impurity, mol fraction
γ_g = gas specific gravity, dimensionless
γ_{gc} = gas gravity corrected to 100-psig separator pressure, dimensionless
$\gamma_{g(mf)}$ = mole fraction of gas
γ_{gs} = gas gravity at sample separator temperature and pressure, dimensionless
γ_{H2S} = hydrogen sulfide impurity, mol fraction
γ_{N2} = nitrogen impurity, mol fraction
γ_o = liquid specific gravity, dimensionless
μ = viscosity, cp or Pa·s
μ_o = oil viscosity, cp or Pa·s
μ_{ob} = oil viscosity at bubblepoint, cp or Pa·s
ρ_w = density of water, lbm/ft³

3.10 Problems

Exercise 3.9—Calculating Pore Volumes of Oil and Gas Produced Fluids. Apply the gas FVF to calculate unit production as reservoir pressure declines from 5,000 to 1,000 psi. The following fluid properties are available for the pressure-depleting reservoir: At 5,000 psi, *zres* = 1.06, while at 1,000 psi, z_{res} = 0.978, and T_{res} = 230°F (690°R).

Solution. Apply the reciprocal gas FVF equation (Eq. 3.12c) to determine the volume change.

At 5,000 psi, $\frac{1}{B_g}=35.46\frac{p_{res}}{z_{res}T_{res}}=35.46\frac{5{,}000}{(1.06)(690)}=242.4$ scf / RB; at 1,000 psi, $\frac{1}{B_g}=$ 52.5 scf/RB.

Therefore, 1 ft³ of reservoir gas should yield 189.9 scf of expanded gas that was either produced or migrated to the gas cap. Fill-up of the evolved gas stored in a reservoir must occur before a response to water injection is achieved.

Exercise 3.10—Determining the True Oil Volume at Surface Conditions. Geologists have mapped a prospect and calculated that the reservoir could contain 86,900,000 RB of oil. Determine the true oil volume at surface conditions so that an economical field development plan can be developed.

Reservoir data: γ_{gc} = 0.81, γ_{API} = 22.5 °API, p_{res} = 1,834 psia, and T_{res} = 118°F.

Solution. Use the Vasquez and Beggs correlations to find the oil FVF (B_o). However, because the solution GOR (R_s) is included in Eq. 3.22, this value must be determined first.

Calculate the solution GOR using Eq. 3.21:

$$R_s = \mathrm{C1}\gamma_{gc} p_{\mathrm{res}}{}^{\mathrm{C2}} \exp^{\left[\mathrm{C3}\gamma_{\mathrm{API}}/(T_{\mathrm{res}}\,{}^\circ\mathrm{F}+460)\right]} = (0.0362)(0.81)1{,}834^{1.0937}\exp^{\left[(25.724\times 22.5)/(118+460)\right]} = 295$$

scf/STB.

Then, calculate B_o:

$$B_o = 1 + \mathrm{C1}R_s + \mathrm{C2}(T_{\mathrm{res}} - 60)\left(\frac{\gamma_{\mathrm{API}}}{\gamma_{gc}}\right) + \mathrm{C3}R_s(T_{\mathrm{res}} - 60)\left(\frac{\gamma_{\mathrm{API}}}{\gamma_{gc}}\right) = 1 + (4.68\times 10^{-4})(295) +$$

$$(1.75\times 10^{-5})(118-60)\left(\frac{22.5}{0.81}\right) + (-1.81\times 10^{-8})(295)(118-60)\left(\frac{22.5}{0.81}\right) = 1.158 \text{ RB/STB.}$$

OOIP = (86,900,000 RB)/(1.158 RB/STB) = 75,043,178 BO.

The true oil volume at surface conditions is 75,043,178 BO.

Exercise 3.11—Analyzing Plotted Values of Oil FVF and Solution GOR. Using **Fig. 3.14** (developed using the Vasquez and Beggs correlations), identify the location of p_{bp} and indicate the different temperatures at which the calculations were performed.

Solution. The answer is shown in **Fig. 3.15.**

Exercise 3.12—Calculating the Adjusted *z*-Factor. A natural-gas well tested a 0.7-gravity gas. The shut-in pressure was 4,680 psia, while the reservoir temperature was 240°F. Preliminary indications are that the gas contains 20% CO_2, and neither H_2S nor N_2 were observed in the tests. Apply the Piper-McCain-Corredor correlations to calculate an adjusted *z*-factor.

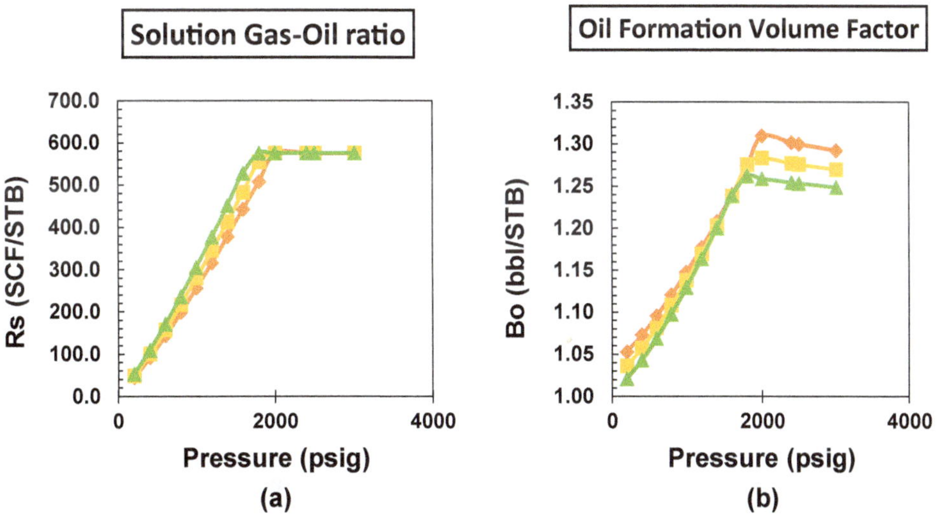

Fig. 3.14—Statement of problem. (a) Solution GOR as a function of pressure at different temperatures. (b) Oil FVF as a function of pressure at different temperatures.

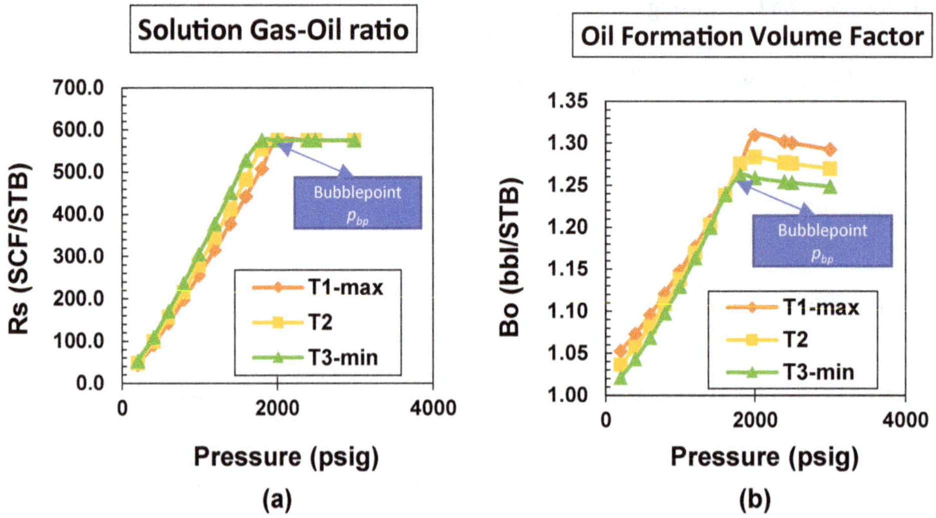

Fig. 3.15—Solution to problem. (a) Solution GOR as a function of pressure at different temperatures. (b) Oil FVF as a function of pressure at different temperatures.

Solution. Use the Sutton correlations to determine the *z*-factor from the gas gravity and calculate the pseudocritical properties using Eqs. 3.3 and 3.4:

$$p_{pc} = 756.8 - 131.0\gamma_g - 3.6\gamma_g^2 = 756.8 - 131.0(0.7) - 3.6(0.7)^2 = 663.33 \text{ psia}.$$

$$T_{pc} = 169.2 + 349.5\gamma_g - 74.0\gamma_g^2 = 169.2 + 349.5(0.7) - 74.0(0.7)^2 = 377.59°\text{R}.$$

Calculate the coefficients using Eqs. 3.10 and 3.11:

$$J = 0.12 + 0.71\gamma_g - 0.10\gamma_g^2 - 0.24\gamma_{\text{H2S}} - 0.46\gamma_{\text{CO2}} - 0.30\gamma_{\text{N2}} = 0.12 + 0.71(0.7) - 0.10(0.7)^2 - 0.46(0.2) = 0.476.$$

$$K = 3.82 + 17.44\gamma_g - 3.22\gamma_g^2 - 1.22\gamma_{\text{H2S}} - 7.05\gamma_{\text{CO2}} - 9.34\gamma_{\text{N2}} = 3.82 + 17.44(0.7) - 3.22(0.7)^2 - 7.05(0.2) = 13.042.$$

Adjust the pseudocritical properties for the presence of impurities using Eqs. 3.8 and 3.1 and Eqs. 3.2 and 3.9:

$$T_{pc} = \frac{K^2}{J} = \frac{13.042^2}{0.476} = 290.88°\text{R},$$

$$T_{pr} = \frac{T}{T_{pc}} = \frac{700}{290.88} = 2.41,$$

and

$$p_{pc} = \frac{T_{pc}}{J} = \frac{290.88}{0.476} = 611.09 \text{ psia},$$

$$p_{pr} = \frac{p}{p_{pc}} = \frac{4{,}680}{611.09} = 7.65.$$

The calculated pseudoreduced temperature (2.41) and pressure (7.65) correlating parameters are applied to the Standing and Katz chart shown in Fig. 3.5 to determine the z-factor for the contaminated natural gas.

Thus, $z = 1.08$.

3.11 References

- Abou-Kassem, J. H., Mattar, L., and Dranchuk, P. M. 1990. Computer Calculations of Compressibility of Natural Gas. *J Can Pet Technol* **29** (5): 105–108. PETSOC-90-05-10. https://doi.org/10.2118/90-05-10.
- Chang, C. K. 1985. Water Quality Considerations in Malaysia's First Waterflood. *J Pet Technol* **37** (9): 1689–1698. SPE-12387-PA. https://doi.org/10.2118/12387-PA.
- Dranchuk, P. M. and Abou-Kassem, J. H. 1975. Calculation of Z Factors for Natural Gases Using Equations of State. *J Can Pet Technol* **14** (3): 34–36. PETSOC-75-03-03. https://doi.org/10.2118/75-03-03.
- Mitchell, R. W. and Finch, E. M. 1981. Water Quality Aspects of North Sea Injection Water. *J Pet Technol* **33** (6): 1141–1152. SPE-8069-PA. https://doi.org/10.2118/8069-PA.
- Moses, P. L. 1986. Engineering Applications of Phase Behavior of Crude Oil and Condensate Systems. *J Pet Technol* **38** (7): 715–723. SPE-15835-PA. https://doi.org/10.2118/15835-PA.
- Moses, P. L. and Donohoe, C. W. 1987. Gas-Condensate Reservoirs. In *Petroleum Engineering Handbook*, ed. H. B. Bradley, Chap. 39. Richardson, Texas: Society of Petroleum Engineers.
- *Oil & Gas Journal*. 2004. Worldwide EOR Survey, https://www.ogj.com/ogj-survey-downloads/worldwide-eor/document/17299692/2004-worldwide-eor-survey.
- Piper, A. M. 1944. A Graphic Procedure in the Geochemical Interpretation of Water Analyses. *Eos, Transactions American Geophysical Union* **25**: 914–923. https://doi.org/10.1029/TR025i006p00914.
- Piper, L. D., McCain, W. D., Jr., and Corredor, J. H. 1993. Compressibility Factors for Naturally Occurring Petroleum Gases. Paper presented at the SPE Annual Technical Conference and Exhibition, Houston, Texas, USA, 3–6 October. SPE-26668-MS. https://doi.org/10.2118/26668-MS.
- Standing, M. B. 1947. A Pressure-Volume-Temperature Correlation for Mixtures of California Oils and Gases. In *Drilling and Production Practice*. New York, New York: API. API-47-275.
- Standing, M. B. and Katz, D. L. 1942. Density of Natural Gases. In *Transactions of the Society of Petroleum Engineers*, Vol. 146, SPE-942140-G, 140–144. Richardson, Texas: Society of Petroleum Engineers. https://doi.org/10.2118/942140-G.
- Stiff, H. A., Jr. 1951. The Interpretation of Chemical Water Analysis by Means of Patterns. *J Pet Technol* **3** (10): 15–17. SPE-951376-G. https://doi.org/10.2118/951376-G.
- Sutton, R. P. 1985. Compressibility Factors for High-Molecular-Weight Reservoir Gases. Paper presented at the SPE Annual Technical Conference and Exhibition, Las Vegas, Nevada, USA, 22–25 September. SPE-14265-MS. https://doi.org/10.2118/14265-MS.
- Vasquez, M. and Beggs, H. D. 1980. Correlations for Fluid Physical Property Prediction. *J Pet Technol* **32** (6): 968–970. SPE-6719-PA. https://doi.org/10.2118/6719-PA.
- Wichert, E. and Aziz, K. 1971. Compressibility Factor of Sour Natural Gases. *The Canadian Journal of Chemical Engineering* **49** (2): 267–273. https://doi.org/10.1002/cjce.5450490216.

Chapter 4

Well and Reservoir Models

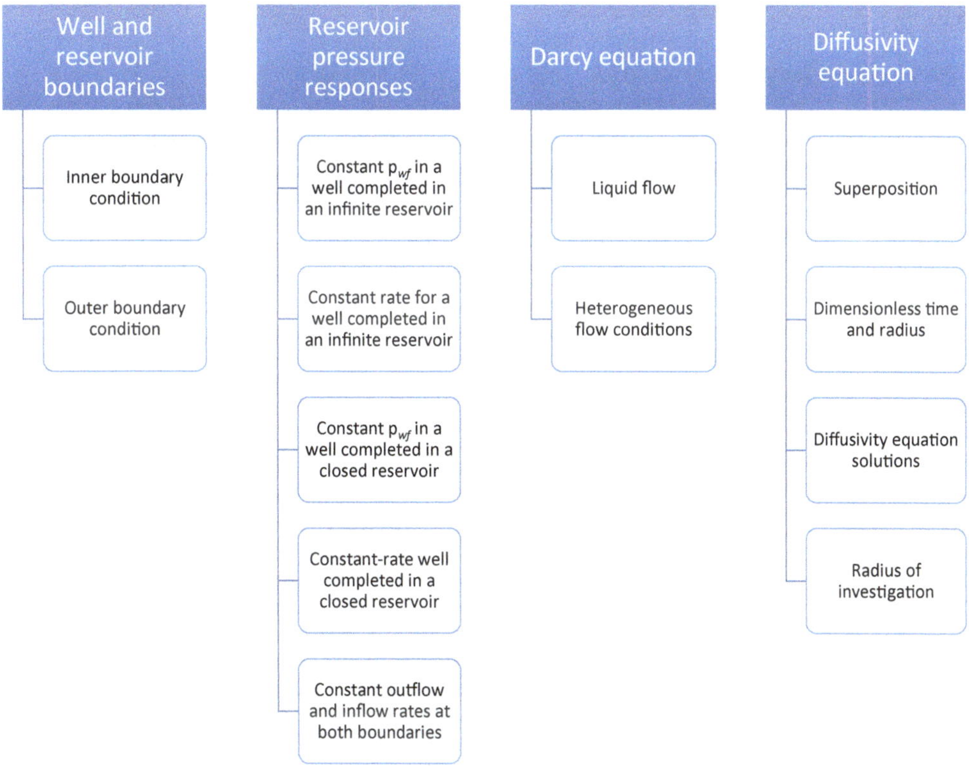

The displacement of fluids in a porous medium in a waterflood project involves the simultaneous flow of two or more immiscible phases between an inner and an outer boundary. In this chapter, we introduce the mathematical models that have been developed for forecasting fluids migration and pressure performance within the drainage area. The diffusivity equation is derived by combining the law of conservation of mass and Darcy's law for the isothermal flow of slightly compressible fluids, and the solutions are used to model the physics of the diffusion of pressure and/or fluid dynamics in accordance with a specific set of limiting or boundary conditions. One must be aware of the assumptions and boundary conditions included in the development of a particular flow equation to ensure its proper application when solving field problems. Rate equations form the basis of the oil, gas, and water inflow-performance relationships. Water-influx calculation methods, on the other hand, are founded on changing rates and pressures as a function of reservoir architecture.

Solutions to field problems assume an instantaneous change in conditions. However, changes occurring in the reservoir are usually gradual, and they are typically represented by a smooth curve, which indicates stable oil and water flow in channels. Compensating assumptions accounting for these discrepancies must be included when applying the diffusivity equation to solve field problems.

As shown in **Fig. 4.1,** fluids migrating toward the wellbore can assume a parallel or series flow geometry instead of unidirectional flow. Therefore, combinations of flow equations are often applied to study fluid flow in layered reservoirs and the effects of formation damage on well productivity. Reservoir geometry, drainage limits, and the time interval over which a disturbance is operating affect fluid flow rates.

The following discussion develops the different types of reservoir boundaries included in the solutions to the Darcy and diffusivity equations.

4.1 Well and Reservoir Boundaries

Wells are drilled to target hydrocarbon saturated intervals, and perforations are located at the inner limit of a system, defined as the wellbore radius (r_w), which is typically defined at the center of a circle (i.e., the drainage area of the well is circular, which implies radial flow). Wellbore pressure or rate changes are measured at this wellbore, which is considered mathematically to be a point of infinitesimally small size when compared to the outer radius. Pressure and/or rate changes induced at the perforations initiate a disturbance that migrates outward into the reservoir as a function of time and distance to the outer boundary.

Certain definitions and assumptions are needed to develop the appropriate solutions:

- Distance to the outer reservoir boundary (r_e).
- Initial conditions at $t = 0$ and $q = 0$ define the starting point of any reservoir depletion study.
- Initial reservoir pressure (p_i) and average static reservoir pressure ($\overline{p}$).

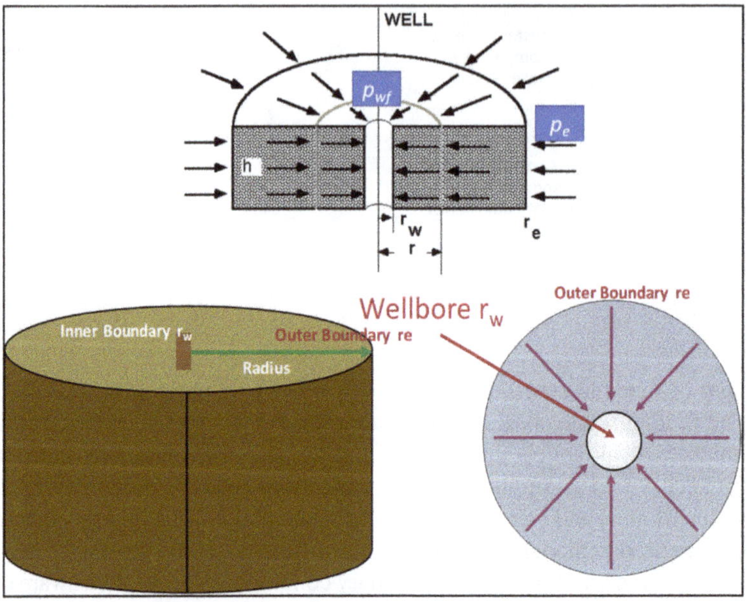

Fig. 4.1—Most cases are modeled assuming radial flow conditions.

Inner and outer reservoir boundary conditions describe the physical arrangement of the flow system as well as the type of disturbance induced at the perforations. Solutions to either the Darcy or the diffusivity equation are developed with this arrangement in mind. These solutions can be expressed in terms of (1) pressure/time changes or (2) rate/time changes. The types of boundary conditions commonly applied to the study of petroleum reservoirs are summarized in the sections that follow.

4.1.1 Inner Boundary Condition. Two different disturbances can be applied at the perforations. They are:

- Constant rate case: The production rate remains constant while the temporal variation between the static reservoir pressure and the measured pressure at the wellbore is measured. van Everdingen and Hurst (1949) set the producing rate to zero to develop their transient pressure testing solution to the diffusivity equation.
- Constant pressure case: This is often called the pseudopressure solution to the diffusivity equation. However, that is a misnomer because pressure at the perforations (p_{wf}) declines simultaneously with reservoir pressure loss. Pressure at the perforations is held constant to study the producing rate variation as a function of time. van Everdingen and Hurst (1949) applied a constant pressure solution to the diffusivity equation to calculate water influx. The constant pressure case was extended to study water influx (Fetkovich 1971) and production decline curves (Fetkovich 1980; Fetkovich et al. 1987). For most cases, a constant flowing pressure implies a declining reservoir pressure and production rate.

4.1.2 Outer Boundary Condition. Reservoir geology defines three possible outer boundary conditions:

- An infinite-acting reservoir: Rate or pressure disturbances migrate outward from the wellbore in an ever-increasing radial form and infinite manner into an unbounded reservoir. This infinite-acting-behavior assumption is present for only a limited time, and the pressure disturbance will diffuse in the porous medium until some outer boundary is reached. In reality, this reservoir condition is seldom present in nature. However, very large reservoirs and aquifers act for all practical purposes in an infinite manner.
- Finite reservoir with a partially sealing outer boundary: Partially permeable faults or juxtapositioned sands of different characters can form "leaky," or partially sealing, boundaries. Steady-state conditions result when the volume of fluid inflow from across the boundary replaces the volume of produced reservoir fluid. The pressure at the inner and outer boundaries remains constant for this particular case, in addition to a well responding to a waterflood or to any inflow of fluids from the outer boundary of the drainage area. Production or pressure loss occurs if the volume of produced fluids is only partially replaced by fluid inflow from across the outer boundary. A material balance between what is going out and what is coming in is not maintained for this particular circumstance.
- Closed reservoir boundary: The reservoir being drained is of a finite size, with a sealed outer limit. Examples of closed-reservoir behavior include (1) a reservoir defined by sealing faults, (2) a producing sand encased in an impermeable shale section, and (3) a well producing from a developed reservoir where each well is draining its own drainage area. The presence of the permeability seal prevents fluids from being transmitted into or out of the reservoir. This geological model provides the foundation for the formation of overpressured reservoirs. Fluids are unable to escape from the sand as continued burial causes the reservoir pressure to increase. The pressure buildup could eventually exceed the normal pressure gradient, creating overpressured conditions.

The Darcy equation and solutions to the diffusivity equation form the basis for studying production and pressure responses in petroleum reservoirs. Specific boundary conditions are applied to either general equation to develop a specific solution that represents the behavior of a particular reservoir configuration and flow situation.

4.2 Reservoir Pressure Responses

The discussion that follows presents the different well-testing and production decline conditions that could be applied to a particular well and to reservoir studies.

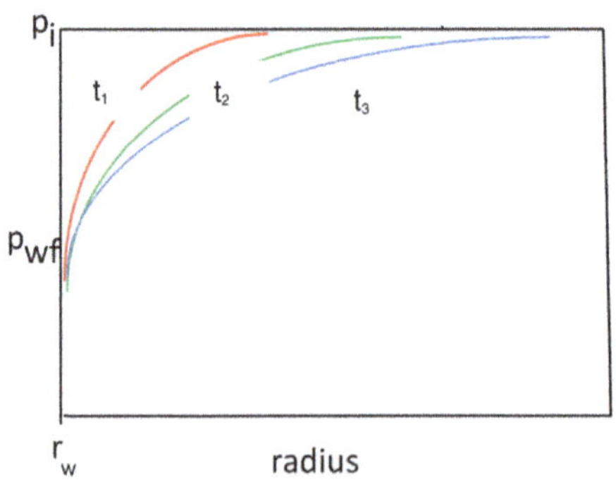

Fig. 4.2—Infinite-acting constant pressure case. The combination of the constant bottomhole flowing pressure and the increasing drainage radius forces the production rate to decline.

4.2.1 Constant p_{wf} in a Well Completed in an Infinite Reservoir. The constant pressure condition implies a declining production rate because the pressure at the continuously expanding outer boundary remains equal to initial pressure. The constant pressure level at the inner boundary with an increasing outer reservoir boundary causes the production rate to decline as a function of time. Theoretically, a reservoir boundary will never be encountered, and the reservoir is considered an infinite medium. **Fig. 4.2** indicates that the expanding pressure profiles are anchored at the constant pressure level located at the perforations. The production rate must decline because the drainage radius is constantly expanding.

4.2.2 Constant Rate for a Well Completed in an Infinite Reservoir. The bottomhole flowing pressure must decline to maintain the constant rate condition. A declining bottomhole flowing pressure creates a greater pressure differential, which helps to compensate for the constantly expanding reservoir radius. **Fig. 4.3** presents changing pressure profiles over time. These curves constantly expand until they reach the outer boundary.

The constant rate and increasing reservoir limit force the bottomhole flowing pressure to decline to compensate for the effects of the increasing drainage radius.

4.2.3 Constant p_{wf} in a Well Completed in a Closed Reservoir. The producing rate continuously decreases to compensate for a loss of reservoir pressure. The lack of fluid flow across the no-flow outer boundary causes a pressure differential between the inner and outer reservoir limits, as shown

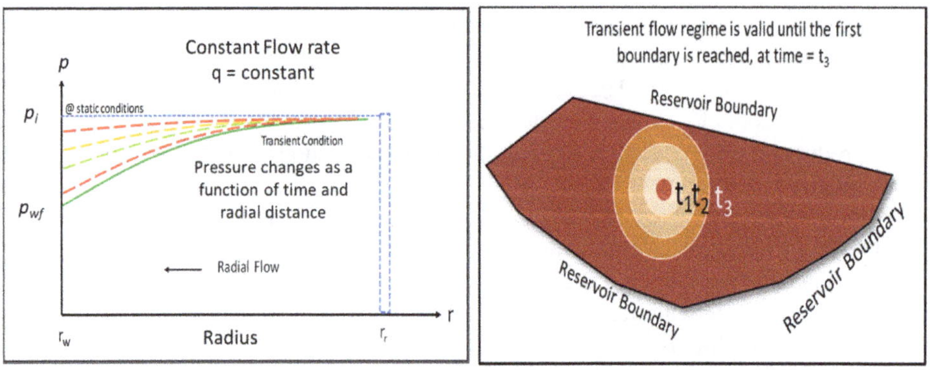

Fig. 4.3—Constant rate/pressure transient for the infinite-acting case.

in **Fig. 4.4.** This particular type of pressure response has been designated the post-transient condition. However, many authors define this case as a pseudosteady state. The name sometimes causes confusion because a closed reservoir with a constant rate case is also sometimes designated a pseudo-steady state.

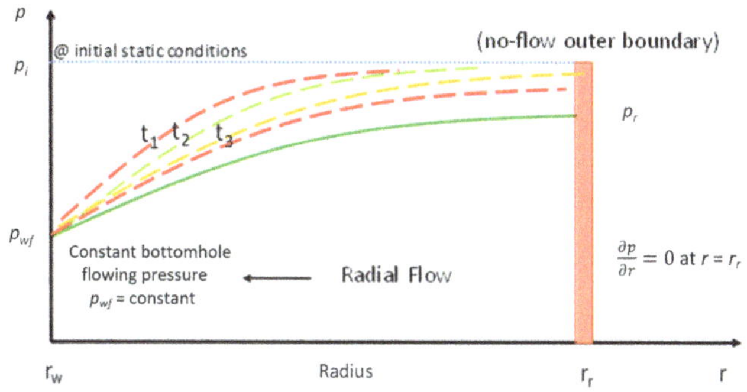

Fig. 4.4—Post-transient case. Reservoir pressure must decline because the sealed reservoir condition prohibits inflow to the reservoir. A declining production rate results because the complete container has been investigated by pressure response.

4.2.4 Constant-Rate Well Completed in a Closed Reservoir. **Fig. 4.5** shows that the difference between the wellbore pressure and the constant outer-boundary pressure must remain the same to maintain a constant production rate. Therefore, the constantly decreasing pressure profiles are parallel.

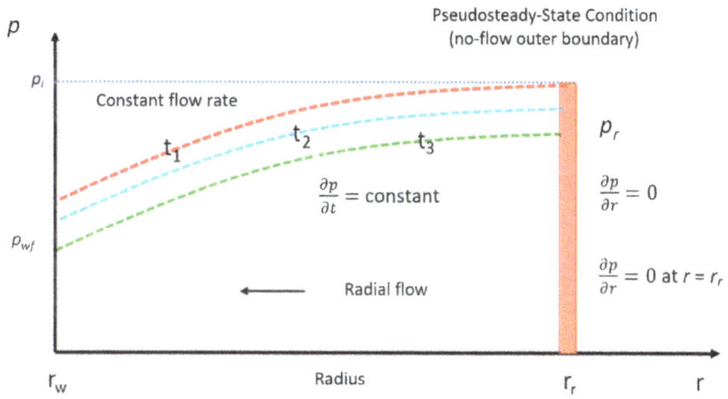

Fig. 4.5—Pseudosteady-state case. The constant rate and pressure drop across the reservoir cause the pressure profiles to remain constant.

4.2.5 Constant Outflow and Inflow Rates at Both Boundaries. Steady-state conditions are the third and most restrictive type of reservoir-depletion condition. Fluid inflow across the outer boundary is equal to the voidage rate caused by the production of reservoir fluids. The producing rate and pressure profiles remain constant and are independent of time. Steady-state conditions are rarely encountered in field operations. **Fig. 4.6** shows the single pressure profile across the reservoir for the steady-state case.

4.3 Darcy Equation

The steady-state condition is applied when developing the Darcy equation, while the diffusivity equation can be solved for a wider range of boundary conditions.

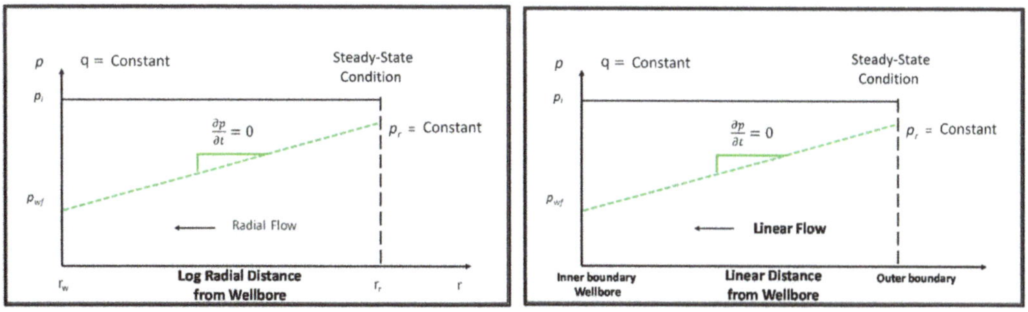

Fig. 4.6—Steady-state case. The pressures and producing rate remain constant because the fluid volume migrating across the partially sealing fault replaces the produced-fluid volume. The result is a single pressure profile during the life of the condition for flow of an incompressible fluid through homogeneous and isotropic porous media. The left side shows the pressure profile for radial flow where the *x*-axis is in log scale. The right side shows the pressure profile for linear flow where the *x*-axis is in Cartesian/linear scale.

4.3.1 Liquid Flow. Liquid and gas radial flow equations can be presented in terms of steady-state and pseudosteady-state depletion conditions. Steady-state solutions assume boundary pressures remain constant, while pseudosteady-state solutions assume the well is producing from a closed reservoir with a constantly declining reservoir pressure but a constant flowing bottomhole pressure.

4.3.1.1 Liquid Flow—Steady-State Flow Equation—Radial Flow Conditions. The generalized Darcy equation introduced in Chapter 2 (see Eq. 2.4) can be rearranged into an integral form to represent radial and horizontal liquid flow occurring between an inner and outer reservoir boundary. Permeability and viscosity are assumed to remain constant and are located outside of the integral for this case.

$$\frac{q}{2\pi h}\int_{r_w}^{r_e}\frac{\mathrm{d}r}{r}=\frac{k}{\mu}\int_{p_w}^{p_e}\mathrm{d}p, \qquad (4.1)$$

where r is the radial distance (ft), p_w is the pressure (psia) at the wellbore, q is the rate at which the well is flowing and is in surface units (RB/D) rather than reservoir units, p_e is the pressure (psia) maintained constant at the external radius or drainage radius r_e (this condition can be observed if the reservoir has a large surrounding aquifer), h is thickness (ft), r_e is the external radius (ft) and is calculated from well spacing, r_w is the wellbore radius (ft), and μ is the viscosity of the fluid (cp).

Integrating Eq. 4.1 results in Eq. 4.2. Note that Eq. 4.2 is expressed in the oilfield units of STB/D. The formation volume factor is included to relate the reservoir flow rate to stock-tank conditions where the produced volumes are measured. The formation volume factor and viscosity values are evaluated at the static reservoir pressure.

$$q=\frac{0.00708\ kh(p_e-p_w)}{\mu B[\ln(r_e/r_w)]}, \text{ STB/D}, \qquad (4.2)$$

where μ is the viscosity of the oil (cp) and B is the formation volume factor (RB/STB).

4.3.1.2 Liquid Flow—Steady-State Flow Equation—Radial Flow. This is expressed in terms of the average reservoir pressure ($\bar{p}=p_e$) in the drainage area:

$$q=\frac{0.00708\ kh(p_e-p_w)}{\mu B[\ln(r_e/r_w)-0.50]}, \text{ STB/D}. \qquad (4.3)$$

4.3.1.3 Liquid Flow—Pseudosteady-State Conditions. The Darcy equation can be expressed in reference to the outer boundary at average reservoir pressure. The change in the formulation from the steady state to the pseudosteady state causes the constant in the denominator to change from 0.50 to 0.75 for the average static pressure case. The general solution for a circular drainage area in oilfield units is

$$q=\frac{0.00708\ kh(\bar{p}-p_{wf})}{\mu B[\ln(r_e/r_w)-0.75]}, \text{ STB/D} \qquad (4.4)$$

When we have pseudosteady-state conditions in the reservoir, the primary assumptions are

- Porous medium: isotropic and homogeneous.
- Slightly compressible fluid.
- Laminar/radial flow.

- Pressure at the outer boundary is not constant. This is often because of natural limits (e.g., faults, pinchouts, or the result of artificial boundaries induced by adjoining production wells).
- Pressure declines at a constant rate with time (assuming the well rate is constant).

4.3.2 Heterogeneous Flow Conditions. Homogeneous and isotropic reservoirs are seldom encountered in actuality. Therefore, heterogeneous flow conditions must be assumed to occur in most instances. The discussion that follows applies the Darcy equation to develop ideas about specific flow conditions that are often observed.

The method used to average permeability depends on the permeability distribution in the reservoir. One must know the vertical and lateral permeability distributions to describe a reservoir thoroughly. Lateral permeability distributions can be estimated from individual-well and multiwell pressure transient tests. Vertical permeability distributions can be estimated from core analysis in individual wells.

As mentioned in Chapter 2, the geometric average is used when permeability varies randomly. The arithmetic average is applicable to flow parallel to depositional bedding boundaries and when layer permeabilities are similar over a relatively large areal extent. The harmonic average is applicable to flow perpendicular to depositional bedding boundaries and when layer permeabilities are similar over a relatively large areal extent.

4.3.2.1 Layered or Parallel Flow—Horizontal Flow Parallel to Depositional Bedding Planes. Layered, multizone reservoirs are often encountered. Two geological models feature (1) sediment layers of unequal thickness and permeability, caused by a variation in the energy level during the time of deposition, and (2) widespread shale layers that often divide sand bodies into layered reservoirs.

Production is commingled at the wellbore when the well is perforated across two or more layers. The discussion that follows assumes a layered system, with each zone possessing different thickness and permeability rock properties and no vertical communication between layers. The total flow capacity is the sum of the individual capacity of each layer in the system. Many reservoirs have depositional layers that are parallel to each other, and in some cases, they are near horizontal. The overall permeability of this system is given by Eq. 4.5, where the flow through the three rock layers shown in **Fig. 4.7** is similar to an electrical current flowing through three parallel resistors.

As seen in Fig. 4.7, $q_t = q_A + q_B + q_C$ and $h_t = h_A + h_B + h_C$, and flow area $A = wh_t$.

The arithmetic-average permeability of a layered system is given by

$$\bar{k} = \frac{\sum_{i=1}^{n} k_i h_i}{h_t}. \qquad (4.5)$$

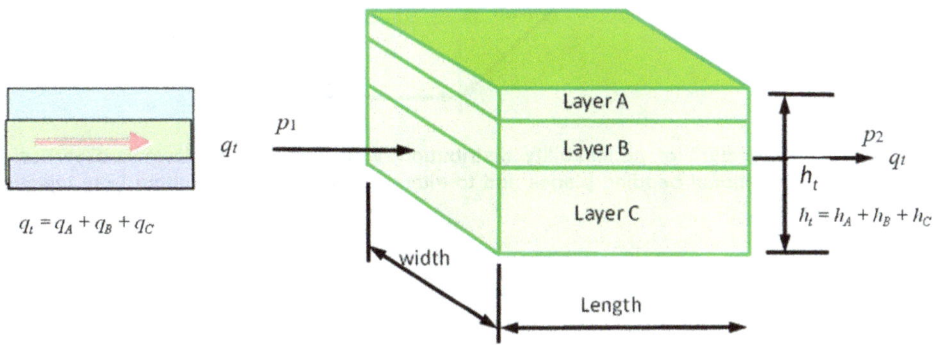

Fig. 4.7—Horizontal flow parallel to depositional bedding planes.

4.3.2.2 Horizontal Flow Perpendicular to Depositional Bedding Planes—Series Flow. Flow in the vertical direction with near-horizontal depositional layers is relatively common. Gravity typically must be included when considering vertical flow. **Fig. 4.8** shows the hypothetical arrangement of permeabilities in series where the following parameters are specified: total flow rate is $q_t = q_A = q_B = q_C$; pressure drop is $p_1 - p_2 = \Delta p_A + \Delta p_B + \Delta p_C$; and total length is $L = L_A + L_B + L_C$.

The harmonic-average system permeability is given by

$$\bar{k} = \frac{L}{\sum_{i=1}^{n} \frac{L_i}{k_i}} \quad (4.6)$$

4.3.2.3 Layered or Parallel Flow—Composite Radial Flow System. Skin effect is a measure of near-wellbore permeability damage or stimulation. Fluids introduced during drilling and completion operations often cause shale particles to expand or force very fine-grained material in the reservoir rock to the near-wellbore region. The reduction of near-wellbore permeability is the result of these operations. On the other hand, acidizing or hydraulic fracturing increases the permeability adjacent to the wellbore.

Fig. 4.9 shows the near-wellbore permeability alteration as a function of the change in the reservoir permeability, which can act in a positive or negative manner, where $q_t = q_A = q_B = q_C$, $h_t = h_A = h_B = h_C$, and $p_e - p_w = \Delta p_A + \Delta p_B + \Delta p_C$.

Eq. 4.7 shows the results of rearranging, simplifying, and developing an expression for calculating the average permeability displayed by an affected well:

$$\bar{k} = \frac{\ln(r_e / r_w)}{\sum_{i=1}^{n} \frac{\ln(r_i / r_{i-1})}{k_i}}, \quad (4.7)$$

which is the average permeability of a composite system.

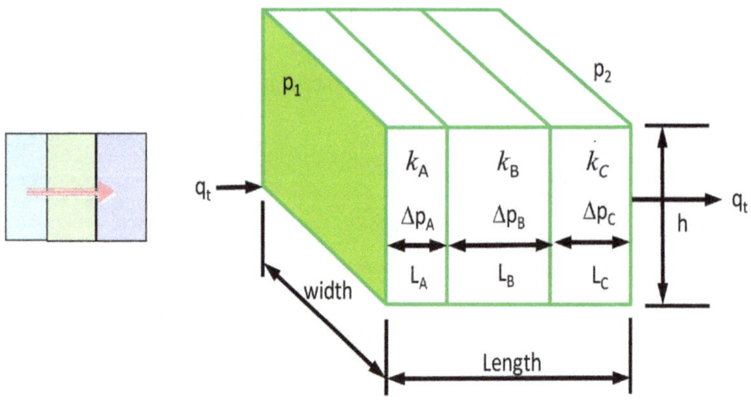

Fig. 4.8—Series arrangement of permeability distribution, which applies to downward vertical flow perpendicular to depositional bedding planes and to altered permeabilities resulting from injection or production operations.

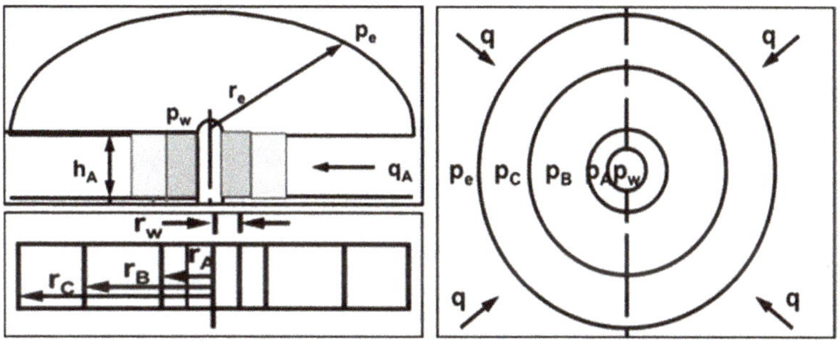

Fig. 4.9—Theoretical visualization of horizontal radial flow, showing zones with different permeabilities and the pressures at the different boundaries. Courtesy of Schlumberger NExT.

Fig. 4.10 illustrates how stimulation reduces the pressure drop close to the wellbore $[p_{wf(s)}]$ while formation damage increases the pressure drop, shown as $p_{wf(d)}$. The formation damage effect extends out to the distance r_d. Hawkins (1956) developed a relation between the reservoir permeability and skin permeability and the radius of damage, which is expressed in Eq. 4.8.

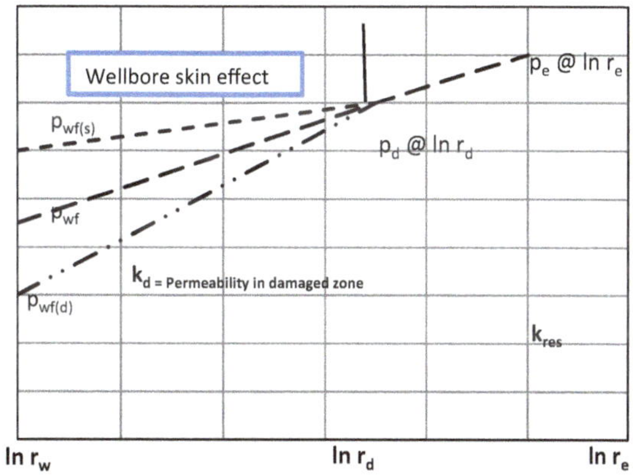

Fig. 4.10—Pressure distribution as a function of the natural log of the radius from the wellbore, illustrating the effect of wellbore damage or stimulation on pressure drawdown. Both alterations extend out some distance r_d into the reservoir. Assume a constant rate process.

Skin s can be defined as a change in permeability close to the wellbore:

$$s = \frac{\left(k_f - k_d\right)}{k_d} \ln \frac{r_d}{r_w}. \quad (4.8)$$

Eq. 4.8 indicates that a negative skin will be exhibited by stimulated wells and a positive skin will be exhibited by wells suffering formation damage near the wellbore. In hydraulically fractured wells, s is often > -3.

The combination of reservoir permeability and skin permeability alters the pressure response and inflow performance relationships. The ratio of the normal pressure drop to the pressure drop created by near-wellbore alteration was defined by Matthews and Russell (1967) as the effective wellbore radius (r_{wa}):

$$\ln \frac{r_e}{r_{wa}} = \ln \frac{r_e}{r_w} + s. \quad (4.9)$$

$$r_{wa} = r_w \exp(-s). \quad (4.10)$$

The total system pressure drop is the sum of the individual pressure drops across the damaged zone and the unaffected reservoir:

$$p_e - p_{wf} = \left(p_d - p_{wf}\right) + \left(p_e - p_d\right). \quad (4.11)$$

The Darcy equation can be applied to develop an expression to calculate the average permeability displayed by a damaged well:

$$q_d = \frac{0.00708\bar{k}h(p_e - p_{wf})}{\mu B[\ln(r_e / r_w)]} = \frac{0.00708k_d h(p_d - p_{wf})}{\mu B[\ln(r_d / r_w)]} + \frac{0.00708k_{\text{res}} h(p_e - p_d)}{\mu B[\ln(r_e / r_d)]}. \quad (4.12)$$

Eq. 4.12 can be rearranged and simplified to develop an expression for calculating the average permeability displayed by an affected well, which is the particular case shown in Eq. 4.7.

$$\bar{k} = \frac{\ln(r_e / r_w)}{\frac{\ln(r_d / r_w)}{k_d} + \frac{\ln(r_e / r_d)}{k_{\text{res}}}}. \quad (4.13)$$

Series flow coupled with the logarithm of the radius shows why near-wellbore formation damage or hydraulic fracturing can greatly affect the well producing rate.

Exercise 4.1—Creating Pressure Profiles. Rewrite Eq. 4.4 for liquid flow to calculate and plot the pressure profile as a function of radial distance. Use **Fig. 4.11** in Cartesian coordinates and in semilog (radius on the log scale) for a well draining a 40-acre circular reservoir when r_e = 745 ft, r_w = 0.25 ft, and p_{wf} = 1,200 psia. The constant reservoir parameters have been collected and were found to be equal to 306 psia. Calculate the reservoir pressure and plot these values as a function of the radius of investigation.

Solution. Rewrite Eq. 4.4 and observe that $\bar{p} = p$ at each assumed value of r in the established range of r = 10 ft to r_e = 745 ft:

$$q = \frac{0.00708\ kh(\bar{p} - p_{wf})}{\mu B[\ln(r_e / r_w) - 0.75]}, \text{ STB/D.}$$

$$p = p_{wf} + \frac{q\mu B\left[\ln\left(\frac{r_e}{r_w} - 0.75\right)\right]}{0.00708kh} = p_{wf} + 306\left[\ln\left(\frac{r_e}{0.25} - 0.75\right)\right].$$

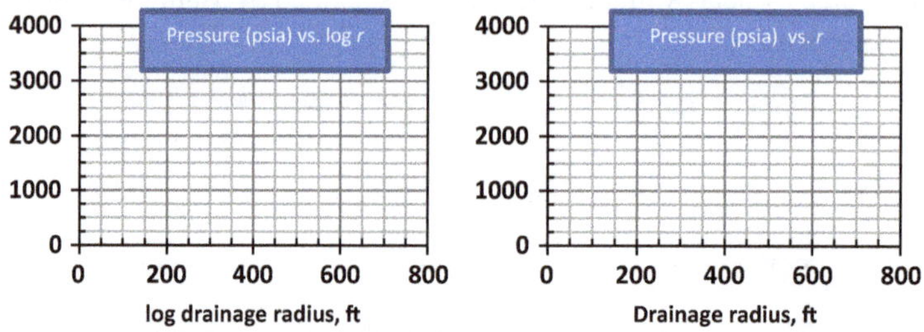

Fig. 4.11—Plot the pressure profile for a well draining a 40-acre circular reservoir when r_e = 745 ft, r_w = 0.25 ft, and p_{wf} = 1,200 psia. Using Cartesian graph paper is suggested for two purposes: plotting pressure as a function of the logarithm of radial distance and plotting pressure as a function of radial distance.

Fig. 4.12 shows the graphical results from the calculations reported in **Table 4.1.**

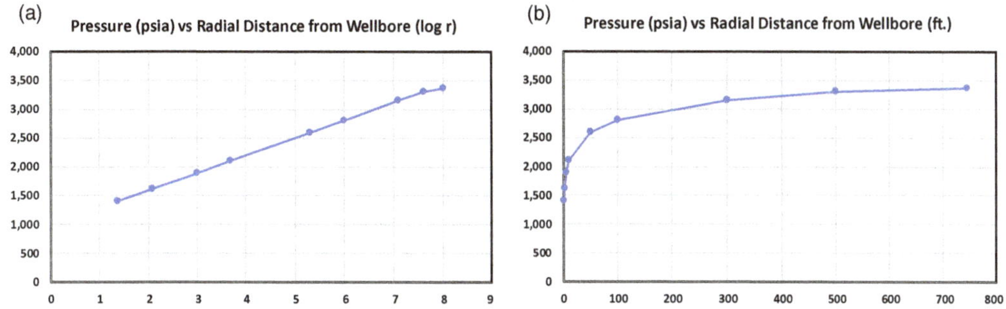

Fig. 4.12—(a) Pressure distribution as a function of radial distance (natural log) from the wellbore; (b) pressure vs. radial distance in Cartesian coordinates. Both graphs show the very large pressure drop, of approximately 20%, within 10 ft of the wellbore.

r (ft)	$\ln r_e/r_w$	p_e (psia)	r (ft)	$\ln r_e/r_w$	p_e (psia)	r (ft)	$\ln r_e/r_w$	p_e (psia)
1	1.386	1,395	**10**	3.689	2,099	**300**	7.090	3,140
2	2.080	1,607	**50**	5.300	2,592	**500**	7.600	3,296
5	3.000	1,887	**100**	5.991	2,804	**745**	8.00	3,357

Table 4.1—Reservoir pressures as a function of drainage radius.

4.4 The Diffusivity Equation

Darcy's law, the law of conservation of mass, and an equation of state are included in the formulation of the diffusivity equation. The rock is assumed to be 100% saturated with a single fluid, and the well is assumed to be producing from the entire vertical section of an isotropic homogeneous reservoir.

The diffusivity equation expresses a mathematical relation of multivariable functions between pressure, distance, and the time of the flow of a slightly compressible fluid in a radial and porous system in terms of viscosity, system compressibility, and reservoir permeability. Any of the five boundary conditions discussed in Section 4.1 can be imposed on the diffusivity equation to derive an expression applicable to that particular set of operating parameters. Refer to Dake (1978) and Lee et al. (2003) for a more comprehensive development of the equation.

A variety of boundary conditions can be applied to solve the diffusivity equation to approximate different reservoir flow/pressure situations. The best-known solutions were provided by van Everdingen and Hurst (1949), and they form the basis to study the effects of pressure-transient and water-influx changes in petroleum reservoirs. The results were presented in dimensionless tabular and graphical forms. Dimensioned and dimensionless relationships were also presented to permit the dimensionless solutions to be applied to solve "real world" problems.

4.4.1 Superposition. Reservoir pressures and flow rates typically change in a smooth manner in a reservoir. However, applying either the Darcy or the diffusivity equation to study field problems assumes any pressure or rate change occurs instantaneously. The principle of superposition gets around this problem by dividing a smooth curve into a series of instantaneous pressure declines arranged as step functions, and the effect of a number of pressure waves can be found by adding up the disturbances that would have been produced separately by the individual waves. A step function is defined as the difference in average pressures. **Fig. 4.13** illustrates the procedure for calculating the step function from a pressure decline curve.

The relationship given by the field curve representing measured pressures at the average rates is calculated as the midpoints between the measured rates. The process converts the smooth curve to a step function.

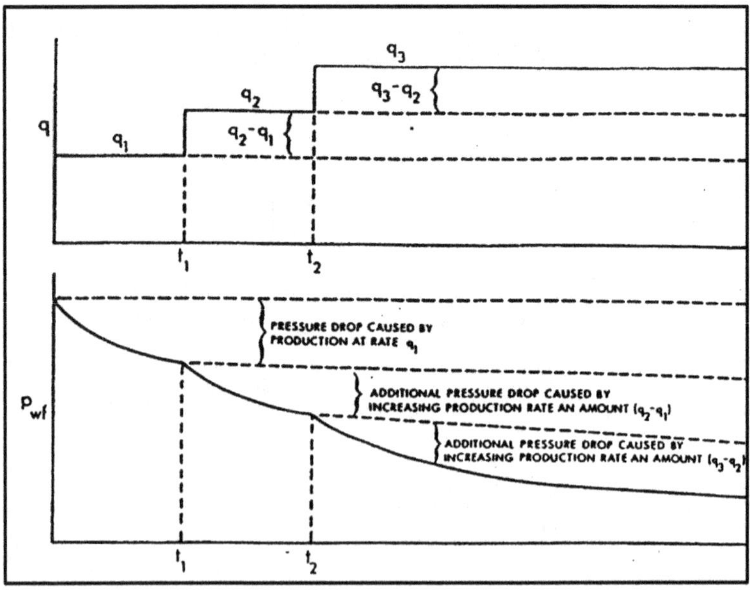

Fig. 4.13—Relating the effect of a change in flow rate to the gradual flowing pressure loss (after Matthews and Russell 1967).

Applying the step-function concept to the smooth field curve divides the performance history into a series of approximately rectangular-shaped layers, with each incremental step reflecting the performance behavior over a definite time period. The time of influence or effective time for Layer 1 in Fig. 4.13 covers the entire period. However, Layer 2 begins at the time when the rate changes and influences production until the end of the prediction.

Applying the step-function and effective time concepts rearranges the smooth but irregular field curve into a series of geometrically regular shapes. The rectangular-shaped rate/time curves are easily integrated to calculate the cumulative produced volume. Fig. 4.13 relates the effects of rate/pressure-response relationships to the step function and effective time over which each rate change/pressure response occurs.

Eqs. 4.14a through 4.14e provide example calculations of the average pressure drop at different effective times: at t_1 between p_i and p_1; at t_2 between p_1 and p_2; at t_3 between p_2 and p_3; at t_4 between p_3 and p_4; and at a time period t_j between sequential recorded pressure drops p_j and p_{j-1}:

$$\Delta p_1 = p_i - \frac{(p_i + p_1)}{2} = \frac{(p_i - p_1)}{2}, \qquad (4.14a)$$

$$\Delta p_2 = \left(\frac{p_i + p_1}{2}\right) - \left(\frac{p_1 + p_2}{2}\right) = \frac{(p_i - p_2)}{2}, \qquad (4.14b)$$

$$\Delta p_3 = \left(\frac{p_1 + p_2}{2}\right) - \left(\frac{p_2 + p_3}{2}\right) = \frac{(p_1 - p_3)}{2}, \qquad (4.14c)$$

$$\Delta p_4 = \left(\frac{p_2 + p_3}{2}\right) - \left(\frac{p_3 + p_4}{2}\right) = \frac{(p_2 - p_4)}{2}, \qquad (4.14d)$$

$$\Delta p_j = \left(\frac{p_{j-2} - p_{j-1}}{2}\right) - \left(\frac{p_{j-1} - p_j}{2}\right) = \frac{(p_{j-2} - p_j)}{2} \quad (4.14e)$$

4.4.2 Dimensionless Time and Radius. Terms were developed by van Everdingen and Hurst to relate pressure, producing time, and drainage-volume size to dimensionless-value correlation variables (Lee et al. 2003). Eq. 4.15 expresses the dimensionless time vs. real time relationship when the time units are days. These scaling factors allow field characteristics to be directly related to complex solutions. Therefore, the analyst does not have to derive the diffusivity equation to calculate a specific solution when solving a field problem.

$$t_D = \frac{0.00633kt}{\phi \mu c_t r_w^2} \quad (4.15)$$

System compressibility (c_t) was defined by Craft and Hawkins (1959) when oil, gas, and water saturation values are expressed as fractions:

$$c_t = S_o c_o + S_g c_g + S_w c_w + c_f \quad (4.16)$$

For many cases, water and formation compressibility values in hard-rock areas can be assumed to be c_w and $c_f = 3 \times 10^{-6}$ psi^{-1} and oil compressibility is $c_o = 3 \times 10^{-5}$ psi^{-1}; gas compressibility can vary widely, but at pressures below 2,000 psi, it can be approximated by using $c_g = \frac{1}{p}$. If at all possible, fluid and rock property studies should be used to calculate compressibility values instead of using rules of thumb or correlations. The previous set of values should only be applied as a last resort.

Eq. 4.17 defines the dimensionless drainage volume relationship between the outer reservoir boundary and the wellbore radius. This dimensionless expression provides an estimate of well drainage area or reservoir or aquifer size.

$$r_{eD} = \frac{r_e}{r_w} \quad (4.17)$$

Analytical solutions to the diffusivity equation are developed with Bessel functions. However, the complexity of the resulting equation and calculation procedure usually renders these solutions impractical. Some type of mathematical function is commonly assumed to approximate a portion of a rate or pressure history curve. The cases are roughly divided into (1) an early time when the flow or pressure response is infinite acting and (2) a late time when the flow or pressure response is boundary dominated. The multiplicity of influences in the transitional region precludes the application of approximation techniques to study reservoir behavior.

As noted previously, the two broad solution categories that form the basis for characterizing petroleum reservoirs are (1) well-test analysis when $q = 0$ for constant rate solutions and (2) decline curve analysis when $p_{wf} = C$ (constant value). Each of these subjects as they are related to specific solutions to the diffusivity equation are discussed in the section that follows.

4.4.3 Diffusivity Equation Solutions. The diffusivity equation forms the basis for describing petroleum engineering fluid flow in a reservoir and in transient pressure analysis studies.

The diffusivity equation includes all physical properties required to describe fluid flow through a porous medium. Recall, the Darcy equation measures a constant flow rate through a porous medium, and fluid and rock properties are considered to be constant and homogeneous.

The diffusivity differential equation includes Darcy's law, conservation of mass, and the equation of state to provide a basis for altering possible combinations of reservoir conditions to study fluid flow. Physical boundary conditions are imposed at the inner and outer boundary of the system. To build the reservoir model, the outer boundary is defined as a reservoir limit that can be classified for a number of conditions:

- Finite or closed: a sealing fault or reservoir encased by a shale envelope.
- Finite with a partially sealing boundary: a leaky fault or a juxtapositioned sand from across a fault. This specific case is classified as a pseudosteady state, and p_{wf} = a constant value.
- Infinite acting: For practical purposes, an aquifer volume that is five times greater than the reservoir volume can be treated as infinite.

Either the constant bottomhole flowing pressure or the constant rate conditions must be set at the inner boundary to solve the diffusivity equation. The constant rate solution for the study of pressure responses in a reservoir and the constant pressure solution for the study of production decline curves are included in tables provided in van Everdingen and Hurst (1949). Pressure responses are defined as the pseudosteady-state condition for this case. Modeling transient flow conditions for well-test analysis implies that the well is shut in and the pressure response is monitored.

Eq. 4.18 defines the dimensionless flow rate as a quantity to which no physical dimension is assigned:

$$q_D = \frac{141.2qB\mu}{kh\left(p_i - p_{wf}\right)} \quad (4.18)$$

Multiplying the dimensionless flow rate by dimensionless time results in a dimensionless estimate of cumulative water inflow, as shown by Eq. 4.19. The same is true for the dimensioned values, shown by Eq. 4.20.

$$Q_{pD} = q_D t_D \quad (4.19)$$

$$Q_w = q_w t \quad (4.20)$$

4.4.3.1 Constant Pressure Solution. Constant pressure solutions are presented as a series of dimensionless rate (q_{pD}) and dimensionless cumulative production (Q_{pD}) values. These solutions are arranged according to dimensionless reservoir size and the dimensionless time that has elapsed since the initiation of the disturbance.

Eq. 4.21 compares the ratio of the radius of the outermost reservoir boundary (r_e) to that of the innermost boundary, which for this case is the wellbore radius (r_w). This expression provides a size estimate for a particular study case—aquifer influx or a well case. In this case, the wellbore radius is assumed to be the distance to the oil/water contact (r_o).

$$r_{eD} = \frac{r_e}{r_w} \quad (4.21)$$

Eq. 4.22 presents the dimensionless time/real time relationship. The time units are days.

$$t_D = \frac{0.00633kt}{\phi\mu c_t r_w^2} \quad (4.22)$$

One must first apply reservoir size, rock, and fluid data to determine the dimensionless size (r_{eD}) and time frame (t_D) to determine the solution to the problem. Use solution tables, such as **Table 4.2,** to select the proper solution after the first task has been completed. Table 4.2 is an amended listing of the tables provided in van Everdingen and Hurst (1949).

The applied-research interest of van Everdingen and Hurst was oriented in modeling natural water influx. The dimensionless rate expression was integrated over time to develop a dimensionless water-influx term, Q_{pD} or W_D. The results of these infinite and boundary-dominated solutions mirror those shown in **Fig. 4.14.**

Ramey and Cobb (1971) showed that the constant pressure solution for a semicompressible fluid producing from a homogeneous reservoir can be expressed as Eq. 4.23. This equation was also presented in the form of a straight-line equation when the logarithm of the rate is plotted as a function of time; the plot that results is equivalent to an exponential decline curve.

$$\ln q = \ln\left\{\frac{kh\left(p_i - p_{wf}\right)}{141.2B\mu\left[\ln\left(\frac{r_e}{r_w}\right)-0.75\right]}\right\} - \left\{\left(\frac{0.00633kt}{\phi\mu c_t r_w^2}\right)\frac{2}{\left(r_e / r_w\right)^2\left[\ln\left(r_e / r_w\right)-0.75\right]}\right\} \quad (4.23)$$

r_{eD} = 20			r_{eD} = 50			r_{eD} = 100		
t_D	q_D	Q_{PD}	t_D	q_D	Q_{pD}	t_D	q_D	Q_{pD}
100	0.3394	42.91	600	0.2652	189.0	2,000	0.2304	532
130	0.3174	52.76	800	0.2915	241	3,000	0.2179	757
160	0.2975	61.98	1,000	0.2393	290	4,000	0.2070	969
200	0.2728	73.38	1,300	0.2220	359	5,000	0.1967	1,171
240	0.2502	83.83	1,600	0.2060	473	6,000	0.1869	1,363
300	0.2197	97.91	2,000	0.1865	502	8,000	0.1686	1,718
400	0.1770	117.7	2,400	0.1682	573	1×10^4	0.1536	2,000
500	0.1426	133.6	3,000	0.1543	667	1.3×10^4	0.1304	2.46×10^3
600	0.1146	146.4	4,000	0.1133	795	1.6×10^4	0.1118	2.82×10^3
700	0.1148	156.7	5,000	0.0833	895	2×10^4	0.0910	3.23×10^3
800	0.0925	165.1	6,000	0.0682	974	2.4×10^4	0.0741	3.56×10^3
1,000	0.0745	177.1	8,000	0.0418	1,082	3×10^4	0.0645	3.94×10^3
1,300	0.0483	187.8	1×10^4	0.0254	1,148	4×10^4	0.0326	4.37×10^3
1,600	0.0132	193.4	1.3×10^4	0.0210	1,201	5×10^4	0.0195	4.62×10^3
2,000	0.0056	196.9	1.6×10^4	0.0056	1,227	6×10^4	0.0117	4.77×10^3
3,000	0.0006	199.2	2×10^4	0.0021	1,241	8×10^4	0.0042	4.92×10^3
			2.4×10^4	0.0006	1,246	1×10^5	0.0015	4.97×10^3
			3×10^4	0.0002	1,249	1.1×10^5	0.0009	4.98×10^3

Table 4.2—Dimensionless values for a finite radial system with a closed outer boundary and constant-pressure inner boundary (after Lee 1982).

This expression was also recognized by Fetkovich (1980) and Fetkovich et al. (1987) when developing Fetkovich's water-influx model and type curve approaches to study reservoir behavior.

4.4.3.2 Constant Rate Solution. Eq. 4.24 expresses the van Everdingen and Hurst (1949) constant rate solution to the diffusivity equation:

$$p_D = \frac{kh\left(p_i - p_{wf}\right)}{141.2qB\mu}. \quad (4.24)$$

Solutions are presented in tabular and graphical form and expressed in terms of dimensionless time and reservoir size. The constant rate solution shown in Fig. 4.14 is divided into the infinite-acting, constantly increasing value solid curve and flat, finite-acting curves at different values of r_{eD}. The uncertain transitional region divides these two boundaries. The producing rate drops to zero when the well is shut in at some initial time.

Fig. 4.14 also shows that there is a continual pressure buildup as time elapses as long as the reservoir acts in an unbounded or infinite manner. However, the pressure-buildup history for a finite dimensioned reservoir takes place in two stages. Initially, the pressure history is infinite-acting until the onset of outer boundary effects. The curve bends over and becomes essentially horizontal when the now-constant reservoir volume is the primary influence, and reservoir pressure must level off at this condition.

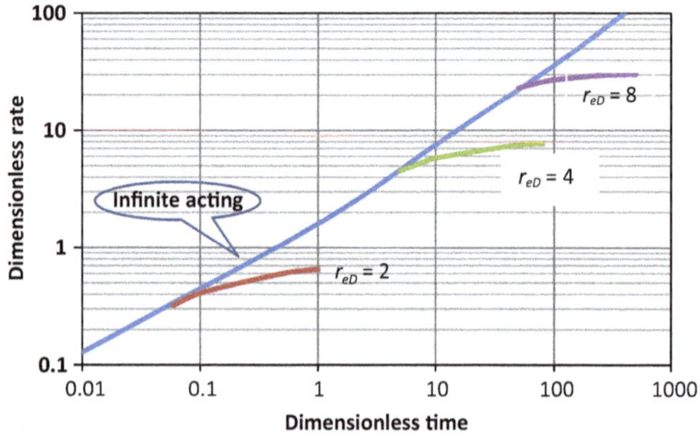

Fig. 4.14—Results of the van Everdingen and Hurst (1949) constant flowing pressure infinite- and finite-acting solutions. Note the transitioning from the smooth upward-trending infinite- to finite-acting flow rate curve to completely boundary-dominated conditions.

The time required to attain boundary-dominated flow can be calculated with a rewritten form of the diffusivity equation:

$$t = 948\left(\frac{\phi\mu c_t r_e^2}{k}\right), \text{ hours}. \quad (4.25)$$

Example 4.2—Calculating the Time Required to Reach Pseudosteady-State Conditions. Calculate the time required to attain pseudosteady-state flow in an aquifer when reD = 5 and the radius to the oil/water contact is 2,000 ft. The area of the oil reservoir is 288 acres. Other pertinent properties: $\mu_w = 0.6$ cp, $\phi = 12\%$, $c_f = 3 \times 10^{-6}$ psi^{-1}, $c_w = 3 \times 10^{-6}$ psi^{-1}, and $k = 100$ md.

Solution. Note that the radius to the oil/water contact (r_o) has been substituted for the wellbore radius (r_w):

$$r_e = r_o\left(r_{eD}\right) = 2{,}000(5) = 10{,}000 \text{ ft}.$$

Calculate system compressibility using Eq. 4.16:

$$c_t = S_o c_o + S_g c_g + S_w c_w + c_f.$$

When $S_o = 0$, $S_g = 0$, and $S_w = 1$,

$$c_t = c_w + c_f = 3\times10^{-6} + 3\times10^{-6} = 6\times10^{-6}\ \text{psi}^{-1}.$$

The approximate time for the onset of boundary-dominated conditions is

$$t = 948\left(\frac{\phi\mu c_t r_e^2}{k}\right) = \frac{948(0.12)(0.6)(6\times10^{-6})(10{,}000)^2}{100} = 410 \text{ hours or 17 days.}$$

Pseudosteady-state conditions begin only 17 days after a change is induced at the reservoir.

Table 4.3 shows an analysis of the effect of aquifer volume on the time of onset of pseudosteady-state flow. The extraordinarily large aquifer (Case 1), of 1,125 sq miles, stabilizes within 4.7 years. Stabilization times are less than 2.5 months for the smaller aquifers.

van Everdingen and Hurst (1949) provide an exact solution to the problem of a constant-rate radial flow condition in a homogeneous system. The rather-cumbersome analysis method provides a standard for comparison to approximate solutions that are much easier to apply and provide more information about reservoir character.

The exponential integral models the line source solution to study the characteristics of an infinite-acting cylindrical reservoir. A well with an infinitely small (assumed zero) radius is chosen as the line source. Eq. 4.26 approximates the pressure buildup in the infinite-acting region. Pressure is measured at some radius out in the reservoir at any time. Lee (1982) provides mathematical tables for the *Ei* function values.

$$p = p_i + 70.6\frac{qB\mu}{kh} Ei\left(-948\frac{\phi\mu c_t r^2}{kt}\right), \quad (4.26)$$

under the following assumptions:

- Initial condition: The reservoir is at a uniform pressure p_i before production begins.
- Outer boundary: The well drains an infinite area ($p \rightarrow p_i$ as $r \rightarrow \infty$).
- Inner boundary: The well produces at a constant rate qB.
- The well radius is considered to be approximately zero when compared to a drainage radius of infinite size.
- Homogeneous and isotropic rock properties.
- Slightly compressible fluids, single-phase flow.

Slider (1976) states that the line source assumption of a wellbore size of zero limits the minimum practical application of the solution to $t = 3.79\times10^5\ \phi\mu c_t r^2 / kh$, while reservoir boundary effects begin to alter the infinite-acting curve for limits when $t = 948\frac{\phi\mu c_t r_w^2}{k}$.

CASE	r_e (ft)	Area (acres)	Time (days)
1	10,000	7,209	17
2	20,000	28,546	68
3	100,000	720,556	1,706

Table 4.3—Comparing aquifer size to time of stabilization.

The basic pressure-buildup-analysis equation was developed from the exponential integral solution by Horner (1951). Pressure drawdown effects caused by a prior production and pressure buildup that occurred when the well was shut in are considered with the principle of superposition, and are expressed by the Horner equation in terms of oilfield units:

$$p_{ws} = p_i - 162.6\frac{q\mu B}{kh}\log\left(\frac{t_p + \Delta t}{\Delta t}\right), \quad (4.27)$$

where t_p is the producing time (hours), Δt is the shut-in time (hours), q is the average rate before the well is shut in (STB/D), and p_{ws} is the shut-in pressure (psia) at Δt.

Fig. 4.15 shows an idealized model with zero skin and no presence of wellbore storage during shut-in. When a semilog straight line equation is applied to the Horner equation (Eq. 4.27), *kh* is the slope.

The line source equation can be changed to the following form when boundary-dominated flow conditions occur:

$$p_{wf} = p_i - 141.2\frac{qB\mu}{kh}\left(\frac{0.000527kt}{\phi\mu c_t r_e^2} + \ln\frac{r_e}{r_w} - 0.75\right). \quad (4.28)$$

Eq. 4.28 indicates that plotting p_{wf} vs. t should result in a straight line **(Fig. 4.16).**

Eq. 4.29 shows the slope of the straight line during the late-time period under boundary-dominated flow conditions:

$$\frac{\mathrm{d}p(r,t)}{\mathrm{d}t} = -\frac{1.79qB}{h\phi c_t r_e^2}, \quad (4.29)$$

where h is reservoir thickness (ft), c_t is total compressibility (psi^{-1}), r_e is the drainage radius (ft), and q is the flow rate (STB/D).

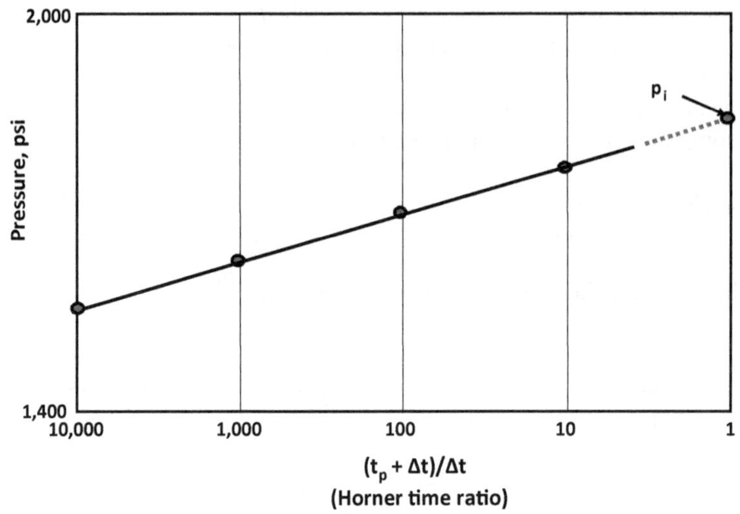

Fig. 4.15—Ideal pressure response after a well is shut in. There is a constant flowing rate before shut-in, and the reservoir is infinite acting with no boundary-dominated conditions.

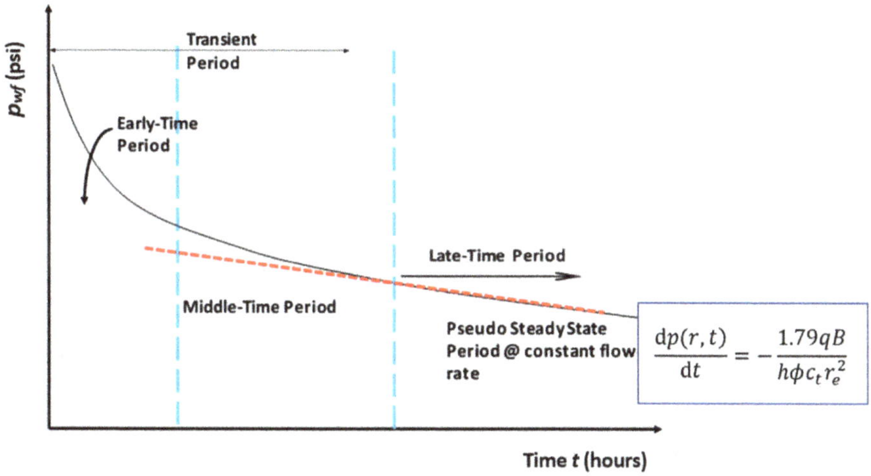

Fig. 4.16—Pressure at the wellbore. There is a constant flowing rate, and the reservoir is finite acting with boundary-dominated conditions. The slope of the line results in an equation defining the reservoir pore volume.

This drawdown test is often applied to estimate reservoir limits, and it is important for calculating porosity/thickness (ϕh) and permeability/thickness (kh) from the pseudosteady state (Eq. 4.29).

4.4.4 Radius of Investigation. Pressure/rate/time relationships are often used to study reservoir behavior under boundary-dominated conditions when analyzing production decline curves and developing water influx models. Matthews and Russell (1967) state that the time/radius relationship for the onset of pseudosteady-state conditions can be calculated with a rearranged version of Eq. 4.25. Notice that the radius of investigation is a function of time (hours) and permeability (md), and neither flow rate nor the size of the pressure differential appear in Eq. 4.30.

$$r_i = \sqrt{\frac{kt}{948\phi\mu c_t}}, \quad (4.30)$$

under the following assumptions:

- Radial flow
- Homogeneous reservoir rock properties
- Slightly compressible fluid

Example 4.3—Radius of Investigation. A 36-hour pressure buildup test was conducted in an oil reservoir. An analysis of the well-test information indicated a 1,588-psia initial pressure. Declining reservoir pressure might have caused the establishment of a 20% gas saturation after 360 days of production time. Assume radial flow conditions exist. **Table 4.4** lists the reservoir rock and fluid properties.

ϕ = 25%	c_w = 3 × 10^{-6} psi^{-1}	k = 48.4 md
S_{wi} = 21%	c_f = 3 × 10^{-6} psi^{-1}	μ = 0.8 cp
	c_o = 15 × 10^{-6} psi^{-1}	S_o = 79%

Table 4.4—Example rock and fluid properties.

Answer the following questions:

- How far out into the reservoir did the test investigate?
- What is the gas compressibility?
- Would the change in the system compressibility value affect the radius of investigation when free gas is present in the drainage area of the well?

Solution. Use Eq. 4.30 to calculate the radius of investigation with the 20% free-gas saturation and compare with the radius of investigation under the assumption of undersaturated conditions.

The first step involves calculating the radius of investigation at the initial undersaturated conditions ($S_g = 0$). Initial saturation = $S_{wi} + S_o$.

$$c_t = S_{wi}c_w + S_o c_o + c_f = 10^{-6}\left[(0.21)(3)+(0.79)(15)+3\right] = 15.5 \times 10^{-6} \text{ psi}^{-1}.$$

$$r_i = \left(\frac{kt}{948\phi\mu c_t}\right)^{0.5} \left[\frac{(48.4 \text{ md})(36 \text{ hours})}{948(0.25)(0.8 \text{ cp})\left(15.5\times10^{-6} \text{ psi}^{-1}\right)}\right]^{0.5} = 770 \text{ ft.}$$

Adjust the fluid saturation to calculate the new oil saturation value after the 360-day production period:

$$S_{or} = 1.0 - S_g - S_{wi} = 1.0 - 0.20 - 0.21 = 0.59.$$

Thus, $S_o = 0.59$.
Then, apply the equation approximating the gas compressibility at low to moderate pressures:

$$c_g = \frac{1}{p} = \frac{1}{1{,}588 \text{ psi}} = 63\times10^{-6} \text{ psi}^{-1}.$$

Calculate the new system compressibility:

$$c_t = S_o c_o + S_g c_g + S_w c_w + c_f = 10^{-6}\left[(0.59)(15)+(0.20)(63)+(0.21)(3)+3.0\right] = 25.1\times10^{-6} \text{ psi}^{-1}.$$

Applying the radius of investigation equation (Eq. 4.30),

$$r_i = \left(\frac{kt}{948\phi\mu c_t}\right)^{0.5} = \left[\frac{(48.4 \text{ md})(36 \text{ hours})}{948(0.25)(0.8 \text{ cp})\left(25.1\times10^{-6} \text{ psi}^{-1}\right)}\right]^{0.5} = 609 \text{ ft.}$$

The calculated radius of investigation in the presence of a 20% gas saturation is 609 ft. This value is smaller than the original estimate by approximately 13%. These calculations indicate that gas saturation at moderate to low pressures will increase the time to reach the boundaries of the investigated reservoir volume. Shutting in a well to attain static pressure would take considerably longer when a 20% gas saturation is present in a reservoir system.

4.5 Summary

Rate equations form the basis of the study of oil, gas, and water flow performance relationships. In addition, water-influx calculation methods can be used with aquifer architecture data to study water encroachment effects.

This chapter introduces the mathematics and calculation procedures for the study of reservoir fluid flow. Solutions to the diffusivity equation model the physics of pressure and fluid dynamics. The Darcy equation models a specific flow system, while the more-complex diffusivity equation describes a variety of flow situations. One must be aware of the reservoir architecture and the rock and fluid parameters related to the particular flow system to solve these equations. The pressure drop between the inner boundary (wellbore) and the outer limit must be included to calculate the solution. Darcy avoided this problem by assuming steady-state flow (i.e., that the quantity of fluid migrating into the drainage volume is equal to that being produced). The assumptions and boundary conditions pertaining to a particular field problem must be included for every flow situation. The development of a flow equation modeling reservoir and fluid flow is used for solving the field problem.

van Everdingen and Hurst (1949) solved the diffusivity equation and presented the results in dimensioned and dimensionless forms. The translation allows the input field values to be converted to obtain the results of their dimensionless solution. These results are then translated back to a specific field application.

Fluids migrating toward the wellbore can assume a reservoir with parallel or series flow geometry instead of unidirectional flow. Therefore, combinations of flow equations are often applied to study fluid flow in layered reservoirs and the effects of formation damage on well productivity. Reservoir geometry, drainage limits, and the time interval over which a disturbance is operating affect fluid flow rates.

4.6 Questions

- Is bottomhole flowing pressure or the production rate held constant when solving the diffusivity equation to study production decline curves? (Answer: Bottomhole flowing pressure is held constant. Generally, there is a loss of bottomhole flowing pressure as a result of declining reservoir pressure.)
- How does one translate the complicated solutions to the diffusivity equation to answer relatively straightforward field problems? [Answer: The van Everdingen and Hurst solutions to the diffusivity equation are presented in terms of ratios of the radius of the outermost reservoir boundary (r_e) to that of the innermost boundary (r_w) and of real time to dimensionless time. Values are provided in terms of dimensionless rate, pressure, and cumulative water influx.]
- What two reservoir conditions are applied in the solution to the diffusivity equation? (Answer: Infinite and boundary-dominated conditions.)
- Which particular flow condition of the diffusivity equation does the Darcy equation present? (Answer: Steady-state flow.)
- Why is it difficult to apply the Darcy equation to solve real-world field problems? (Answer: It is difficult to determine the correct reservoir parameters to apply.)

4.7 Nomenclature

A = area, acres
B = formation volume factor, RB/STB
B_o = oil formation volume factor, RB/STB
C = constant value
c_f = formation compressibility, psi^{-1}
c_g = gas compressibility, psi^{-1}
c_o = oil compressibility, psi^{-1}
c_t = total system compressibility, psi^{-1}
c_w = water compressibility, psi^{-1}
Ei = exponential integral function
h = thickness, ft
h_A, h_B, h_C = thickness of various layers, ft
h_i = layer i thickness, ft
h_t = total thickness, ft
k = permeability, md
$\bar{k}$ = average permeability, md
k_A, k_B, k_C = permeability of various layers, md
k_d = permeability of damaged region, md
k_f = formation permeability, md
k_i = layer i permeability, md
k_{res} = reservoir permeability, md
L = length, ft
L_A, L_B, L_C = length of various layers, ft
L_i = layer i length, ft
n = number
p = pressure, psia

$\overline{p}$ = average pressure, psia
p_A, p_B, p_C = pressure of various layers, psia
p_d = pressure at a distance d to damaged zone (indicates extent of damaged zone), psia
p_D = pressure, dimensionless
p_e = pressure at external radius or drainage radius r_e, psia
p_i = initial reservoir pressure, psia
p_j = pressure at timestep j, psia
p_r = reservoir pressure, psia
p_w = pressure at wellbore, psia
p_{wf} = pressure at perforations or bottomhole flowing pressure, psia
$p_{wf(d)}$ = bottomhole flowing pressure at perforations resulting from damaged zone, psia
$p_{wf(s)}$ = bottomhole flowing pressure at perforations resulting from stimulated zone, psia
p_{ws} = shut-in pressure at Δt, psia
p_1, p_2, p_3, p_4 = pressure at a certain point or pressure at a certain time, psia
q = flow rate, RB/D or STB/D
q_A, q_B, q_C = flow rate at various layers, RB/D or STB/D
q_d = flow rate at a distance d to damaged zone (indicates extent of damaged zone), RB/D or STB/D
q_D = flow rate, dimensionless
Q_{pD} or W_D = cumulative water influx, dimensionless
q_t = total flow rate, RB/D or STB/D
q_w = water flow rate, RB/D or STB/D
Q_w = cumulative water inflow, STB
q_1, q_2, q_3 = flow rate at a certain time, RB/D or STB/D
r = radius, ft
r_A, r_B, r_C = radius at a certain point, ft
r_d = radius of damaged zone, ft
r_e = external radius or distance to outer reservoir boundary, ft
r_{eD} = radius, dimensionless
r_i = radius of investigation, ft
r_o = distance to the oil/water contact, ft
r_r = reservoir radius, ft
r_w = wellbore radius, ft
r_{wa} = effective wellbore radius, ft
$r_{w(\text{damaged zone})}$ = true inner radius that is damaged, ft
s = skin, dimensionless
S_g = gas saturation, fraction or percent
S_o = oil saturation, fraction or percent
S_{or} = residual oil saturation, fraction or percent
S_w = water saturation, fraction or percent
S_{wi} = initial water saturation, fraction or percent
t = time, hours, days, months, or years
t_D = time, dimensionless
t_j = time at period j, hours, days, months, or years
t_p = producing time, hours
t_1, t_2, t_3, t_4 = times of interest, days
w = width, ft
W_D or Q_{pD} = cumulative water influx, dimensionless
Δt = shut-in time, hours
μ = viscosity, cp
μ_w = water viscosity, cp
ϕ = porosity, percent

4.8 Problems

Exercise 4.4—Calculating Flow Rate. You are the completions engineer in charge of a recently drilled oil well. The well was drilled with a brackish contaminated water, and formation damage is expected. Formation damage effects will reduce formation permeability by a factor of 5 close to the wellbore and are expected to extend out 3 ft. The maximum-allowable bottomhole flowing pressure (p_{wf}) because of potential sanding problems is 2,250 psi. **Table 4.5** lists pertinent reservoir and fluid data.

B_o = 1.23 RB/STB	h = 52 ft	r_w = 0.33 ft	k_{res} = 500 md
p_e = 2,644 psia	μ = 1.17 cp	r_e = 1,000 ft	$r_{w(\text{damaged zone})}$ = 3 ft

Table 4.5—Reservoir and fluid data.

Apply the Darcy equation to calculate the expected flow rate for the damaged condition and a new flow rate if perforations were washed with acid to eliminate the skin problem.

Hint

- Calculate the permeability of the damaged portion.
- Apply the Darcy equation to calculate the change in flow rate.

Solution. Because the permeability is five times less than the original permeability, therefore $k_d = 500 \div 5 = 100$ md.

First, calculate the radius of damage:

$$r_d = r_{w(\text{damaged zone})} + r_w = 3.0 + 0.33 = 3.33 \text{ ft}.$$

Then, apply the series flow equation (Eq. 4.13) to calculate the average permeability displayed by the system after that:

$$\bar{k} = \frac{\ln(r_e / r_w)}{\dfrac{\ln(r_d / r_w)}{k_d} + \dfrac{\ln(r_e / r_d)}{k_{\text{res}}}} = \frac{\ln(1{,}000 / 0.33)}{\dfrac{\ln(3.33 / 0.33)}{100} + \dfrac{\ln(1{,}000 / 3.33)}{500}} = 236 \text{ md}.$$

Calculate the flow rate under damaged conditions around the wellbore using Eq. 4.12:

$$q_d = \frac{0.00708\bar{k}h(p_e - p_{wf})}{\mu B[\ln(r_e / r_w)]} = (0.00708) \times \frac{236 \times 52(2{,}644 - 2{,}250)}{1.23 \times 1.17[\ln(1{,}000 / 0.33)]} = 2{,}968 \text{ BOPD}.$$

After treatment, $k = 500$ md.
Calculate the flow rate after treatment conditions around the wellbore using Eq. 4.12:

$$q = (0.00708) \times \frac{k_{\text{res}} h(p_e - p_{wf})}{B_o \mu[\ln(r_e / r_w)]} = 0.00708 \frac{500 \times 52(2{,}644 - 2{,}250)}{1.23 \times 1.17[\ln(1{,}000 / 0.33)]} = 6{,}288 \text{ BOPD}.$$

4.9 References

- Craft, B. C. and Hawkins, M. F. 1959. *Applied Petroleum Reservoir Engineering*. Englewood Cliffs, New Jersey: Prentice-Hall.
- Dake, L. P. 1978. *Fundamentals of Reservoir Engineering*, first edition, Vol. 8. New York: Developments in Petroleum Science, Elsevier.
- Fetkovich, M. J. 1971. A Simplified Approach to Water Influx Calculations—Finite Aquifer Systems. *J Pet Technol* **23** (7): 814–828. SPE-2603-PA. https://doi.org/10.2118/2603-PA.
- Fetkovich, M. J. 1980. Decline Curve Analysis Using Type Curves. *J Pet Technol* **32** (6): 1065–1077. SPE-4629-PA. https://doi.org/10.2118/4629-PA.
- Fetkovich, M. J., Vienot, M. E., Bradley, M. D. et al. 1987. Decline Curve Analysis Using Type Curves: Case Histories. *SPE Form Eval* **2** (4): 637–656. SPE-13169-PA. https://doi.org/10.2118/13169-PA.
- Hawkins, M. F., Jr. 1956. A Note on the Skin Effect. *J Pet Technol* **8** (12): 65–66. SPE-732-G. https://doi.org/10.2118/732-G.
- Horner, D. R. 1951. Pressure Build-up in Wells. *Proc.*, Third World Petroleum Congress, The Hague, The Netherlands, 28 May–6 June.
- Lee, J. 1982. *Well Testing*, Vol. 1. Richardson, Texas: Textbook Series, Society of Petroleum Engineers.
- Lee, J., Rollins, J. B., and Spivey, J. P. 2003. *Pressure Transient Testing*, Vol. 9. Richardson, Texas: Textbook Series, Society of Petroleum Engineers.
- Matthews, C. S. and Russell, D. G. 1967. *Pressure Buildup and Flow Tests in Wells*, Vol. 1. Richardson, Texas: Monograph Series, Society of Petroleum Engineers.
- Ramey, H. J., Jr. and Cobb, W. M. 1971. A General Pressure Buildup Theory for a Well in a Closed Drainage Area. *J Pet Technol* **23** (12): 1493–1505. SPE-3012-PA. https://doi.org/10.2118/3012-PA.
- Slider, H. C. 1976. *Practical Petroleum Reservoir Engineering Methods*, 70. Tulsa, Oklahoma: Petroleum Pub. Co.
- van Everdingen, A. F. and Hurst, W. 1949. The Application of the Laplace Transformation to Flow Problems in Reservoirs. *J Pet Technol* **1** (12): 305–324. SPE-949305-G. https://doi.org/10.2118/949305-G.

Chapter 5

Oil Reservoirs

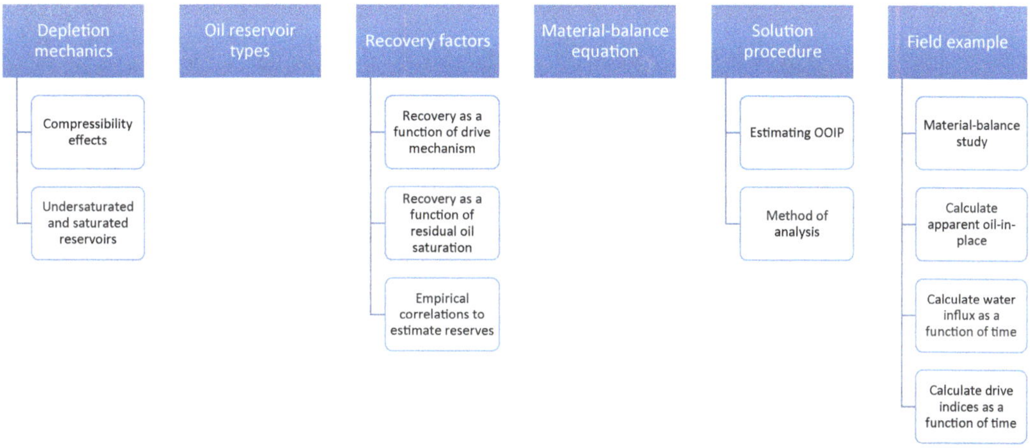

This chapter introduces the concept of material-balance modeling, emphasizing the fundamental energy mechanisms and the use of pressure and production data to understand the performance of oil reservoirs when considered a zero-dimensional tank. The discussion shows how to apply this physics-based model to estimate original oil-in-place (OOIP) and identify drive mechanisms.

Ultimate recovery for black-oil reservoirs is predicated on two primary concerns:

- The ability of reservoir oil to migrate toward the wellbore. Relatively homogeneous dipping sandstone reservoirs with low oil viscosity display high recoveries for this reason.
- Maintaining sufficient bottomhole flowing pressure to overcome hydrostatic head in the producing string. Artificial-lift pumps are often installed to reduce hydrostatic head. Reducing the flowing bottomhole pressure often enhances producing life and ultimate recovery.

The effectiveness of the properties and depletion mechanisms present in the reservoir forms the physical basis for determining reserves. Recovery is highest in low-viscosity, relatively homogeneous moderate- to high-permeability reservoirs. The maximum recovery that can be expected for a good-quality black-oil reservoir with an efficient waterdrive is approximately 50–60% of the OOIP.

Expanding gas caps or aquifer drive mechanisms tend to force oil to migrate toward the wellbore and to reduce the oil saturation to residual values in the invaded zone. Invading water or gas displaces oil ahead of the flood front. Maintaining reservoir pressure allows natural repressuring forces to overcome the effects of the hydrostatic head within the producing string. Water influx is generally the more efficient displacement mechanism because water reduces oil saturation to a minimum value and forces oil to migrate updip. Conversely, gas-cap expansion forces oil outward to a constant narrowing of the annular-ring oil zone.

5.1 Reservoir Depletion Mechanisms

The production of reservoir fluids with a resultant loss of reservoir pressure activates the following depletion mechanisms, which might or might not affect the depletion history:

- The expansion of a primary or secondary gas cap is often a measurable influence augmenting reservoir pressure, and the effect can vary from negligible to significant. Efforts should be made to ensure that oil wells adjacent to the expanding gas/oil are not allowed to produce at the high gas/oil ratio (GOR) condition, which unnecessarily expends reservoir energy.
- The reduction of a liquid-phase hydrocarbon pore volume usually has a minor effect on ultimate recovery, except in the case of a high GOR or volatile-oil reservoirs where considerable gas evolves from the reservoir oil.
- Residual oil saturation is established as pressure depletion continues and as the oil leg expands into a primary gas cap. Water encroachment or gas-cap expansion will not mobilize this residual oil at a later date.
- Encroaching aquifer water can be a very effective drive mechanism. Unfortunately, the repressuring effect can vary from strong to negligible.
- Gravity segregation is the most efficient of all displacement mechanisms provided that the mobility term (kh/μ) is sufficiently high. Naturally occurring segregation can improve ultimate recovery when the proper conditions occur. Gravity segregation is present in all reservoirs but acts as an effective production accelerant only when both low-oil-viscosity and very-high-permeability conditions exist. Gravity segregation can recover up to 80–90% of the OOIP but is typically too long lived for consideration as an economic alternative to waterflooding or other possible enhanced oil recovery solutions.
- Viscous forces: Gas from an overlying gas cap or water from an underlying water leg can be commingled with the producing stream by means of the magnified pressure differential adjacent to the wellbore. Gas cusping or water coning is the result. The well typically displays a rapidly increasing water/oil ratio (WOR) or GOR, with a resultant decrease in the oil production rate. Production of the unwanted fluid is a function of the well mechanical arrangement plus the oil column thickness.
- Capillary forces: Formation rock typically has a greater affinity for water than oil, while gas has the least wetting action of all three fluids. Therefore, water resides in the smallest pore spaces while gas resides in the largest pore spaces. Oil is trapped in the intervening region between the maximum and minimum pore-size distributions. An increasing gas saturation can dramatically affect the relative permeability to oil because migrating gas automatically resides in the largest pore spaces. On the other hand, encroaching water will imbibe into the smaller pore spaces and displace oil. Capillary displacement is a complicated, time-dependent, and difficult process to model. Capillary forces can help or hinder oil production.

5.1.1 Compressibility Effects. Formation, oil, gas, and water compressibility values are considered when evaluating performance and by using the material-balance equation. The discussion that follows compares the gas, oil, water, and rock compressibility relative effectiveness as a function of changing pressure.

As shown in Chapter 2, Fig. 2.1, the Hall rock compressibility correlation is often presented and generally indicates a 3 to 4 $\times$ 10^{-6} psi^{-1} average formation compressibility for a relatively well-consolidated rock. However, rock compressibility combines the effects of expanding rock material and formation compaction as a function of pressure. A loss of pore pressure causes an expansion of the rock grains and cementing material comprising the reservoir framework. The framework can fail structurally and compact when the increased stress becomes greater than the matrix strength. These collapsing effects are prevalent in friable to unconsolidated sediments. In this case, a true average formation compressibility value for a producing interval is difficult to determine because of variations in the degree of cementing and the internal rock character.

Water compressibility effects are related to the expansion of interstitial or connate water residing in the oil reservoir. Connate-water expansion, except in the case of overpressured conditions, typically exerts a minimal influence on performance history. Reservoir-water compressibility remains essentially constant over the pressure range shown in **Fig. 5.1** and is approximately 3×10^{-6} psi^{-1}. Water compressibility forms the basis of aquifer-water influx models.

Oil compressibility, as shown in Fig. 5.1, is at least three times greater than that of water and generally varies from 3 to 11×10^{-6} psi^{-1} for undersaturated conditions and up to 180×10^{-6} psi^{-1} for pressures below the bubblepoint at 2,000 psi (18 °API oil at T = 220°F and GOR = 200 scf/STB). It is evident that this value increases when pressures are below the bubblepoint, even for this heavy oil.

Fig. 5.1 illustrates that the gas compressibility value can vary 25-fold over a 0- to 10,000-psi pressure range. (Gas properties: specific gravity = 0.8, T = 220°F, air specific gravity = 1, nitrogen = 2%, carbon dioxide = 3%, and hydrogen sulfide = 1%.) Fig. 5.1 shows the predominance of the gas compressibility term in relation to formation, water, and oil compressibilities at normal pressures. An oil-reservoir depletion mechanism is a combination of low-value liquid- and formation-expansion terms in the absence of free gas. Water encroachment provides an effective energy for the oil reservoir because low-compressibility water is multiplied by an extremely large aquifer volume, particularly when viewed in the context of oil to aquifer volume ratios.

5.1.2 Undersaturated and Saturated Reservoirs. The effects on the performance history of an oil reservoir of changing from undersaturated to saturated conditions are compared in **Fig. 5.2.** The rather-steep pressure loss at early time becomes less significant after reaching the bubblepoint

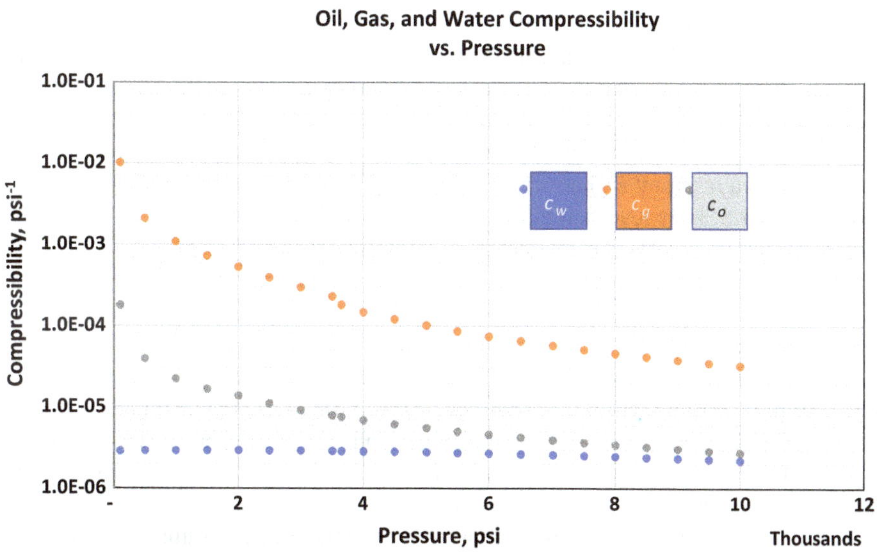

Fig. 5.1—Comparing relative trends for the compressibility values of oil, gas, and formation water. Note the significant response of the gas compressibility value as a function of changing pressure. In fact, gas compressibility is close to that of a liquid at very high pressures. Oil compressibility changes dramatically at the bubblepoint pressure of 2,000 psi when gas begins to evolve.

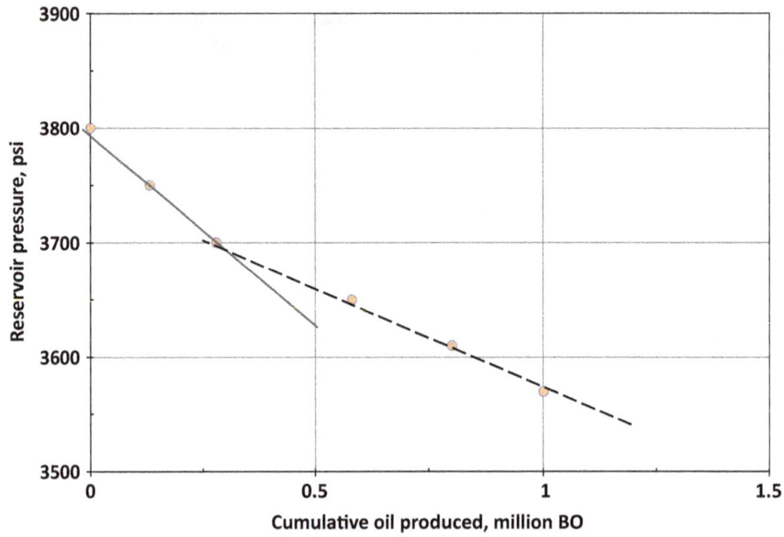

Fig. 5.2—The evolution of gas below the bubblepoint causes the reservoir energy sources to change from liquid expansion to gas formation and solution gas drive. The bubblepoint pressure = 3,705 psi where the two lines intersect.

because gas compressibility effects are now able to influence the pressure history. Increasing liberated gas volumes increases the solution-gas-drive effects and unit recoveries. The presence of free gas more efficiently augments reservoir pressure and production efficiency.

5.2 Oil Reservoir Types

Oil reservoirs can be classified by oil composition as black or volatile. Volatile oil typically contains a greater proportion of lighter and intermediate-weight carbon/hydrogen molecules. Generally, a black-oil reservoir is characterized by a stock-tank oil gravity of < 35 °API, GOR < 1,500 scf/STB, and B_{oi} < 1.5 RB/STB. The GOR of volatile oils ranges from 1,500–2,000 scf/STB, and gravity ranges from 35–45 °API. Excessive gas evolution causes the reservoir oil to exhibit high shrinkage characteristics, with a resultant B_{oi} between 1.5 and 2.0. The result is that the reserves for a volatile-oil reservoir are significantly less than those expected for a black-oil reservoir.

An undersaturated oil reservoir is characterized by an absence of free gas ($p_{res} > p_{bp}$). Oil, water, and formation compressibility values can become important because the more-significant reservoir energies are typically not present. The recovery efficiency typically ranges from 1–5% if no free gas has evolved.

In a saturated oil reservoir, the reservoir oil is at or above the bubblepoint pressure. A primary gas cap may or may not be present at initial conditions p_i ($p_{res} < p_{bp}$). A continuous gas saturation thread (i.e., equilibrium gas saturation S_{eq}) must be established within the reservoir confines before an increasing GOR is achieved. Evolving gas can migrate upward to form a secondary gas cap or can migrate to a producing well to form a solution gas drive.

Fig. 5.2 compares the effects of transitioning from undersaturated to saturated conditions on the performance history of an oil reservoir.

5.3 Recovery Factors

Estimated ultimate recovery is one of the primary considerations when evaluating a reservoir. More-detailed computer simulator models typically provide higher-quality ultimate recovery estimates.

However, one seldom has the luxury of applying such an exhaustive investigation, and in certain cases, correlations will be sufficient.

Oil-in-place obtained from an isopach map multiplied by a recovery factor results in an ultimate recovery estimate when only volumetric data are available for analysis. Recovery factors can be developed from a variety of sources, but one should keep in mind that the results are qualitative estimates and should be viewed only in a general sense. This type of estimate can be developed using prior knowledge of the recovery characteristics of the geographic and geologic province in which the property is located.

5.3.1 Recovery as a Function of Drive Mechanism. It is important to determine drive mechanisms to estimate recovery efficiencies from oil and gas reservoirs. One often determines the remaining potential for a workover or project using knowledge gained from previous experience working in the same region by using analogies, but each reservoir displays characteristics common to that region. For instance, near-shore Louisiana sandstone, Texas Austin Chalk, and west Texas Permian sand and shale producing regions differ in rock and well-completion character as well as in expected recovery.

Fundamental guidelines have been established within each region that are based on the types of fluids present in the reservoir, initial pressure and temperature, reservoir rock and fluids characterization and description (including information on rock permeability and compressibility), relative permeabilities, aquifer strength, formation-dip permeability, gas-cap size and presence, and producing rates. All of the analysis will depend on whether the oil reservoir is above the bubblepoint pressure, whether the potential for a primary or secondary gas cap exists, or whether an oil reservoir with solution gas experiences water influx from an aquifer. In gas reservoirs, the characterization and analysis are simplified and depend on whether there is an expansion of gas as pressure declines or the reservoir experiences water influx from an aquifer.

Other factors to be considered are economics, surface facilities, market constraints, and the number of wells needed. **Table 5.1** presents the expected recovery ranges for the primary drive mechanisms. Note the degree of latitude for each category. Relatively clean sands present along the US Gulf Coast should trend on the high side of each range, while midcontinent, low-permeability, moderate- to high-viscosity reservoirs should trend on the low side of each category. However, it should be recognized that most reservoirs are depleted by some combination of the primary drive mechanisms.

Drive Mechanism	% Recovery of OOIP
Rock and fluid compressibility	3–5
Solution gas drive	5–30
Gas-cap drive	15–40
Waterdrive	30–60
Gravity drainage	40–85

Table 5.1—Recovery factors by drive mechanism for oil reservoirs.

Ultimate recovery factors are affected by certain physical constraints:

- Degree of sand/shale stratification: This profoundly affects the flow system established in the reservoir.
- Offset production and well locations: Field development wells are drilled in anticipation of becoming productive but also to further knowledge of the geological makeup of the prospect. However, one often finds that these wells were not placed in an optimal arrangement, and provisions for these irregularities must be included in any ultimate recovery estimate. Cross-lease drainage is often a problem.
- Thickness of the oil column in relation to a gas cap or water leg: Ultimate recovery in these cases is a tradeoff between gravity effects and viscous effects caused by inflow rates.

Specific recovery factors should be assigned to an operating area according to reservoir and drainage characteristics.

5.3.2 Recovery as a Function of Residual Oil Saturation (Eq. 5.1). The unit recovery factor assumes a 100% volumetric sweep efficiency. The maximum possible recovery is a function of the oil saturation change occurring between two periods within the reservoir history. Either water or gas is assumed to sweep all movable oil from the volume.

$$N_p = (7{,}758 \text{ bbl/acre-ft})(\phi)\left(\frac{S_{oi}}{B_{oi}} - \frac{S_{or}}{B_o}\right), \text{ STB/acre-ft.} \quad \text{(5.1)}$$

The residual oil saturation can be developed by means of material-balance calculations, correlations, and coreflood studies. Eq. 5.1 should be applied with caution because displacement efficiencies are assumed to be 100%.

5.3.3 Empirical Correlations To Estimate Reserves. The American Petroleum Institute (API) Subcommittee on Recovery Efficiency (1967) performed a statistical study on 70 sandstone reservoirs, and they developed empirical relationships estimating recovery factors for solution-gas-drive and waterdrive sandstone reservoirs. (Work was also conducted on multiple carbonate reservoirs, but it was statistically imprecise.)

These estimates are reasonably good for moderate- to low-oil-gravity reservoirs composed of a fairly uniform sand. A carbonate correlation was also presented, but it should be viewed as a qualitative estimate at best because of the often highly irregular nature of carbonate reservoirs. However, Eqs. 5.2 and 5.3, originally from the API Subcommittee on Recovery Efficiency (1967), provide reasonably good recovery estimates for sandstone reservoirs.

Equations for solution-gas-drive and waterdrive reservoirs are presented in Eqs. 5.2 and 5.3, respectively. The solution-gas-drive equation (Eq. 5.2) is a function of hydrocarbon pore volume, absolute permeability, fluid properties at the bubblepoint pressure, and the relative difference between the bubblepoint and the abandonment pressure. The ability to maintain a reservoir close to initial conditions is measured by the negative exponent term in the waterdrive equation (Eq. 5.3).

$$RE_{SGD} = 41.8\left\{\left[\frac{\phi(1-S_{wi})}{B_{obp}}\right]^{0.1611}\left(\frac{k}{\mu_{obp}}\right)^{0.0979}\left(S_{wi}\right)^{0.3722}\left(\frac{p_{bp}}{p_a}\right)^{0.1741}\right\}, \% \quad \text{(5.2)}$$

$$RE_{WD} = 54.9\left\{\left[\frac{\phi(1-S_{wi})}{B_{oi}}\right]^{0.0422}\left(\frac{k\mu_w}{\mu_{oi}}\right)^{0.077}\left(S_{wi}\right)^{-0.19}\left(\frac{p_i}{p_a}\right)^{-0.2159}\right\}, \% \quad \text{(5.3)}$$

where RE_{SGD} is the solution-gas-drive recovery efficiency (percent), RE_{WD} is the waterdrive recovery efficiency (percent), k is the permeability (darcies), S_{wi} is the initial water saturation (fraction), μ_{oi} is the initial oil viscosity (cp), μ_{obp} is the oil bubblepoint viscosity (cp), μ_w is the water viscosity (cp), and ϕ is the porosity (fraction). In addition, p_a is the abandonment pressure (psig), p_i is the initial reservoir pressure (psig) or the current reservoir pressure (psig) when a waterflood project is being planned, p_{bp} is the bubblepoint pressure (psig), B_{obp} is the oil formation volume factor at bubblepoint pressure (RB/STB), and B_{oi} is the oil formation volume factor at initial pressure (RB/STB).

Note that abandonment pressures for solution gas drive and waterdrive are usually assumed to be different.

5.4 Material-Balance Equation

The material-balance approach can model reservoir rock and fluid volume changes as a function of performance and reservoir pressure. The effects of gravity segregation and capillary forces are a function of the reservoir rock, fluid character, and transfer rate and cannot be modeled with the material-balance approach. Rate equations are excluded from the material-balance equation because the tank-like model monitors only pressure/volume relationships. Including transfer rate exchanges in a reservoir model requires the application of the more intricate and accurate computer reservoir simulation method. A much more accurate reservoir static model description and technical competencies are required to develop and apply this dynamic model.

Performance analysis of constant-composition black-oil reservoirs is a relatively straightforward analysis process because only the changes in reservoir volumes must be monitored to determine reservoir character. However, a compositional material balance must be applied to study volatile-oil reservoirs to account for changing oil composition. Evolving gas draws off the lighter-weight compounds to change the molecular arrangement of the reservoir oil in this case.

Comparing volumetric oil and gas-in-place estimates calculated with a material-balance study provides a useful starting point for characterizing and modeling water encroachment effects. This type of analysis is particularly useful for developing a fundamental understanding of reservoir mechanics before conducting a more expensive and time-consuming computer reservoir simulator study.

Oil-reservoir performance histories are typically influenced by some combination of formation and fluid compressibility, solution gas drive, gas-cap expansion, and waterdrive depletion mechanisms. A material balance includes all these possible reservoir drives in a general equation and provides a relationship between reservoir pore volume, reservoir pressure, and cumulative production/injection. The technique has important applications for evaluating volumes of hydrocarbons-in-place, average reservoir pressure, and average fluid saturations in the reservoir.

Fig. 5.3 compares changes in the volume of materials originally present in the reservoir after modification by the production of reservoir fluids. The effects of an expanding gas cap and encroaching water leg are included in the final configuration.

The reservoir is considered to be a tank at average reservoir pressure filled with fluid (oil, gas, water) at all times; therefore, as fluids are produced, the change in reservoir pore volume = the change in reservoir oil volume + the change in reservoir free-gas volume + the change in reservoir water volume.

The reservoir remains filled with fluids during production because of the energy loss caused by pressure drops, and as the remaining fluids expand, pore volume compresses and water from adjacent aquifers flows into the reservoir. Equating production values with changes in rock and fluid properties permits one to calculate changes in the fluid volumes remaining in the reservoir.

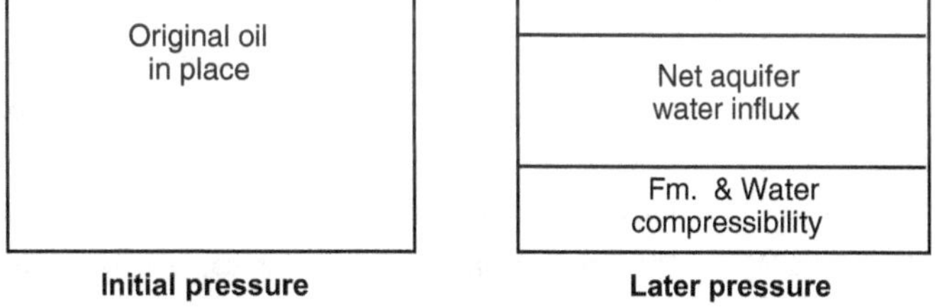

Fig. 5.3—Volumetric changes of materials within the reservoir are caused by the production of fluids. Remember that the reservoir is represented by a unidirectional tank. Rate equations are not included in any of the calculations.

The data requirements for performing calculations are as follows and are also summarized in **Table 5.2:**

- Representative and reliable volume of fluids produced and injected, average reservoir pressure, reservoir rock properties, and fluid pressure/volume/temperature (PVT) property data
- OOIP (*N*, STB)
- Original gas-cap volume (*m*, gas reservoir volume/oil reservoir volume)
- Water influx (W_e, STB)

Havlena and Odeh (1963) provided a material-balance equation accounting for the effects of water influx, gas cap, and solution-gas-drive mechanisms and the expansion of rock and formation water on reservoir performance. McEwen (1962) developed a statistical method to simultaneously determine the best answer for OOIP (*N*) and aquifer description.

The development of Eq. 5.4 follows the Havlena and Odeh (1963) model. (Neither water nor gas injection are considered.) The interpretation of the results is predicated on a thorough review of reservoir geology, volumetric results, and performance.

$$N_p B_o + \left(G_p - N_p R_s\right) B_g + W_p B_w = NB_{oi}\left[\frac{\left(B_o - B_{oi}\right) + \left(R_{si} - R_s\right) B_g}{B_{oi}} + m\left(\frac{B_g}{B_{gi}} - 1\right)\right] +$$

$$NB_{oi}\left[\left(1+m\right)\left(\frac{c_w S_{wi} + c_f}{1 - S_{wi}}\right)\left(p_i - p\right)\right] + W_e B_w, \quad \ldots\ldots\ (5.4)$$

where the B_g units are RB/scf, not rcf/scf.

Eq. 5.5 simplifies the rather lengthy and confusing Eq. 5.4. Eq. 5.5 represents oil, gas, and water production and oil, gas, and water expansion and is expressed at surface conditions:

$$F = N\left(E_o + E_g + E_{fw}\right) + E_w, \quad \ldots\ldots\ (5.5)$$

B_o, B_g, and B_w	Oil, gas, and water formation volume factor, respectively (volume at reservoir conditions/unit volume at STB)
B_{oi}, B_{gi}, and B_{wi}	Initial oil, gas, and water formation volume factor, respectively (volume at initial reservoir conditions/unit volume at STB)
c_w and c_f	Water and formation compressibility, respectively (psi^{-1})
G_i	Cumulative gas injection (measured at stock-tank conditions)
m	Original gas-cap volume (gas reservoir volume/oil reservoir volume)
N and G	In-place oil and gas volumes, respectively (measured at stock-tank conditions)
N_p, G_p, and W_p	Produced oil, gas, or water volumes, respectively (measured at stock-tank conditions)
R_p	Produced-fluid GOR (scf/STB)
R_s	Solution GOR (scf/STB)
R_{si}	Initial solution GOR (scf/STB)
S_{wi}	Initial water saturation (fraction or percent)
W_e	Water encroachment volume (RB)
W_i	Cumulative water injection (RB)

Table 5.2—Data used in the different forms of material-balance application.

where NE_o is the expansion of oil and originally dissolved gas, NE_g is the expansion of the original gas cap, and NE_{fw} is the reduction in hydrocarbon pore volume resulting from connate-water expansion and pore volume compression.

F is the underground withdrawal and is calculated as shown in Eq. 5.6:

$$F = N_p B_o + \left(G_p - N_p R_{si}\right) B_g + W_p B_w, \text{ RB.} \qquad (5.6)$$

The general form of Eq. 5.4 can be rewritten as Eq. 5.6 and compacted using the defined terms in Eqs. 5.7 through 5.11a.

Solution gas effects are expressed as the volume change of the oil and that of its originally dissolved gas:

$$E_o = \left(B_o - B_{oi}\right) + \left(R_{si} - R_s\right) B_g, \text{ RB/STB..} \qquad (5.7)$$

Gas-cap-drive expansion is expressed as the volume change of the original gas cap:

$$E_g = mB_{oi}\left(\frac{B_g}{B_{gi}} - 1\right), \text{ RB/STB.} \qquad (5.8)$$

Formation- and connate-water volume changes are expressed as

$$E_{fw} = B_{oi}\left[\left(1+m\right)\left(\frac{c_w S_{wi} + c_f}{1 - S_{wi}}\right)\left(p_i - p\right)\right], \text{ RB/STB..} \qquad (5.9)$$

Water-influx effects are expressed as

$$E_w = W_e B_w, \text{ RB..} \qquad (5.10)$$

Adding up the expansion terms represented by Eqs. 5.7 through 5.9, the total expansion is represented by Eq. 5.11a:

$$E_{\text{total}} = E_o + E_g + E_{fw}, \qquad (5.11a)$$

and the material-balance equation can be written in a straight-line form as Eq. 5.11b:

$$F = NE_{\text{total}}. \qquad (5.11b)$$

A plot of F/E_{total} vs. cumulative oil production should yield a horizontal straight line representing the value of the oil in place. Remember, in Eq. 5.11b, F is measured, E_{total} is calculated from known fluid data, and N must increase with any pressure/production history study because the presence of water influx and an incorrect gas-cap size are assumed to be absent. An increasing N in a pressure/production history study indicates the presence of an additional energy source because the OOIP must remain constant.

If the general data trend is not a straight, horizontal line, then either the incorrect reservoir model (water influx) has been chosen or the fluid PVT relationships are not correct.

In reality, some combination of depletion mechanisms can affect production continuously or intermittently. Eq. 5.4 can be altered to account for the performance of a specific reservoir type by deleting or adding the appropriate components. In fact, other authors have included expanding water and rock formation conditions for the overpressured gas reservoir case (Poston and Berg 1997).

5.5 Solution Procedure

This section describes a process to apply the material balance to calculate OOIP, water influx effects, and the drive indices supplying energy to the reservoir. The procedure involves analyzing the results of the calculation concurrently with the producing history, geologic maps, and isopach volumes.

5.5.1 Estimating OOIP. Assume the absence of water influx in the material-balance equation. Eq. 5.5 changes to Eq. 5.12a when $W_e = 0$ and E_{fw} is assumed to be negligible:

$$N_a = \frac{F}{\left(E_o + E_g + E_{fw}\right)} \quad \text{(5.12a)}$$

The OOIP term N has been replaced with an apparent oil-in-place term N_a that signifies that this is an approximation because of the assumption that the water influx effect is absent.

A backward extrapolation to zero time of the calculated OOIP (N_a) as a function of oil production (N_p) affords an OOIP estimate because any repressuring effect from water influx is absent. **Fig. 5.4** shows how the extrapolation procedure works. This procedure is not as easy as the example solution shows, but including production and geologic information with the interpretation increases the likelihood of determining a correct value.

Eq. 5.12b includes all reservoir, fluid, and production parameters:

$$N_a = \frac{N_p B_o + \left(G_p - N_p R_s\right) B_g + W_p B_w}{\left[\left(B_o - B_{oi}\right) + \left(R_{si} - R_s\right) B_g + m B_{oi}\left(\frac{B_g}{B_{gi}} - 1\right)\right] + B_{oi}\left[\left(1+m\right)\left(\frac{c_w S_{wi} + c_f}{1 - S_{wi}}\right)\left(p_i - p\right)\right]}, \text{ STB.} \quad \text{(5.12b)}$$

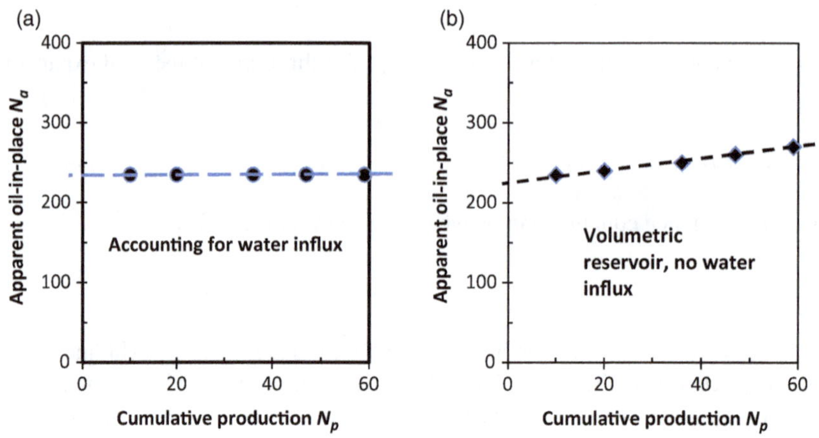

Fig. 5.4—Comparing the results of apparent oil calculations when the effects of water influx are (a) included and (b) not included. (a) The solution accounts for water influx influence on the OOIP estimate, and a straight, horizontal line is observed. (b) Extrapolation back to zero time when one would expect negligible water influx should approximate the true OOIP.

5.5.2 Method of Analysis. The following workflow procedure is applied to perform the analysis:

- For each timestep, apply production, pressure, and fluid property data to solve the material-balance equation.
- Construct an N_a vs. N_p plot from these calculations.
- Extrapolate backward to $N_p = 0$ to determine the apparent oil-in-place N_a.

5.5.2.1 Water Influx Calculations. Rewrite Eq. 5.5 to solve for water influx because the OOIP (N) has been calculated:

$$E_w = W_e B_w = F - N\left(E_o + E_g\right). \qquad (5.13a)$$

The expanded form of this equation is

$$E_w = W_e B_w = \left[N_p B_o + \left(G_p - N_p R_{si}\right) B_g + W_p B_w\right]$$

$$N_a\left[\left(B_o - B_{oi}\right) + \left(R_{si} - R_s\right) B_g + mB_{oi}\left(\frac{B_g}{B_{gi}} - 1\right)\right] \qquad (5.13b)$$

Include the previously determined N value to calculate a water influx value for each timestep.

5.5.2.2 Drive Indices. The material-balance equation can be rewritten to comprise the effects of all possible drive mechanisms in an oil reservoir. The general material-balance equation (Eq. 5.5) can be subdivided into component parts, with each component expressing the contribution of formation and connate water (*RFWDI*), solution gas (*SGDI*), gas-cap expansion (*GCDI*), and waterdrive (*WDI*) depletion mechanisms:

$$SGDI + GCDI + RFWDI + WDI = 1, \qquad (5.14)$$

where

$$SGDI = \frac{NE_o}{F} = \frac{N\left[\left(B_o - B_{oi}\right) + \left(R_{si} - R_s\right) B_g\right]}{N_p B_o + \left(G_p - N_p R_s\right) B_g + W_p B_w}, \qquad (5.15)$$

$$GCDI = \frac{NE_g}{F} = \frac{mNB_{oi}\left(\dfrac{B_g}{B_{gi}} - 1\right)}{N_p B_o + \left(G_p - N_p R_s\right) B_g + W_p B_w}, \qquad (5.16)$$

$$RFWDI = \frac{NE_{fw}}{F} = \frac{NB_{oi}(1+m)\left(\dfrac{c_w B_w + c_f}{1 - S_{wi}}\right)\mathrm{d}p}{N_p B_o + \left(G_p - N_p R_s\right) B_g + W_p B_w}, \qquad (5.17)$$

$$WDI = \frac{NE_w}{F} = \frac{W_e B_w}{N_p B_o + \left(G_p - N_p R_s\right) B_g + W_p B_w}. \qquad (5.18)$$

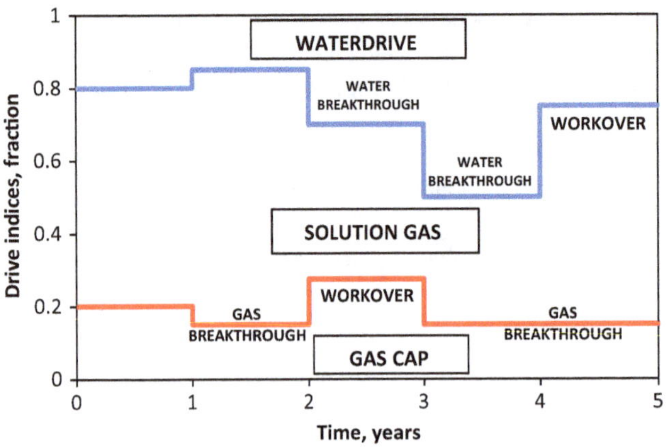

Fig. 5.5—Illustrating a typical performance plot. Excess water and gas production reduces the WDI and GCDI values.

5.5.2.3 Analysis of Past Performance. Reservoir performance history reflects the potential influences of the formation and connate-water expansion, solution gas, gas-cap, and waterdrive mechanisms. Each of these drive mechanisms can be expressed in component form by rewriting the general material-balance equation, and the sum of the fractional strengths of each mechanism should be set equal to unity, as seen in Eq. 5.14. This formulation provides a method to quantitatively compare the influence of each drive mechanism over the life of the depletion history. An example drive-indices illustration is shown in **Fig. 5.5.** The waterdrive index (WDI) is the top layer, while the gas-cap drive index (GCDI) represents the bottom layer. The solution gas drive index (SGDI) is sandwiched between the two more-efficient drive mechanisms. Reservoir formation-water drive index (RFWDI) was assumed to be ineffectual and was excluded from Fig. 5.5.

In Fig. 5.5, the WDI increased until $t = 2$, when water breakthrough occurred in some of the downdip wells. Increasing water production decreased the water influx effect at this point.

Meanwhile, an expanding secondary gas cap caused the GOR in some of the more-updip wells to go to high GOR. Workovers during Year 3 shut in some high-GOR and -water-cut wells. The effect was an increase in the relative values of the WDI and GCDI. An increasing GOR caused the GCDI to decrease during Year 4. Water breakthrough again occurred during Year 3.

In Fig. 5.5, the SGDI is sandwiched between the two more-efficient depletion mechanisms and generally decreases over time. This fact demonstrates that the field was apparently operating under good oilfield practices. A final analysis indicates that gas-cap depletion has remained at a moderate level and water influx improved performance. Should a workover program be instituted to shut off the water from the high-water-cut wells?

Reservoir pressure must be maintained at as high a level as is practical to maximize oil recovery. A moderate loss in reservoir pressure increases recovery efficiency in two ways: Flowing wells last longer because a loss of flowing bottomhole pressure is minimized and encroaching aquifer water or gas-cap gas displaces otherwise-unaffected reservoir oil toward the wellbore. In either case, ultimate recovery is increased.

Ranked according to their relative displacement efficiencies for an oil reservoir, from most to least effective, the drive mechanisms are water influx, gas-cap expansion, solution gas expansion, and formation and interstitial-water compressibility. Gravity segregation and capillary effects are not included in the material-balance equation because of their time-dependent, rather than pressure-dependent, nature. However, their influence should be at least considered in any reservoir analysis because under proper conditions, these effects can profoundly influence performance history.

Another example of the use of drive indices for the rapid visualization of drive mechanisms and reservoir management evaluation is shown in **Fig. 5.6.** In this case example, waterdrive is the dominant energy mechanism, with a WDI value of approximately 90%.

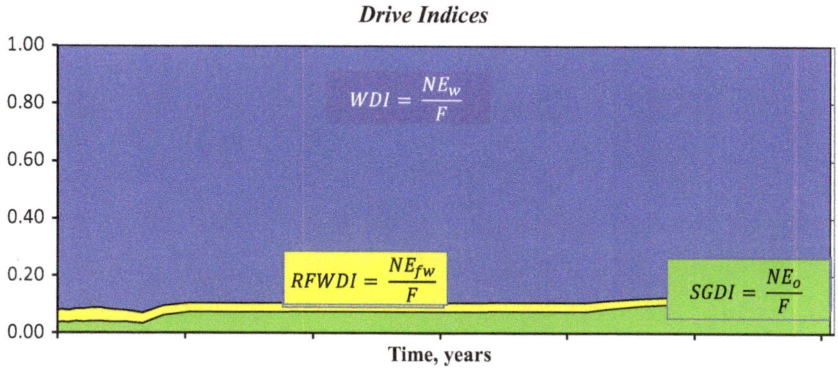

Fig. 5.6—A drive-indices performance plot. WDI values indicate that 90% of the energy mechanism has been waterdrive during the 20 years of production history.

5.6 Field Example

Apply the material-balance solution process to perform the following calculations:

- Verify the OOIP (N) and gas-cap size (G).
- Calculate water influx as a function of time.
- Analyze production performance by calculating the solution gas drive, gas-cap expansion, and waterdrive effects.

Volumetric calculations have indicated OOIP = 220 million BO and the ratio of the gas-cap volume to the reservoir volume (m) = 0.92.

Fig. 5.7 shows the dichotomy of the performance history. All values are credited to the end of the year. Production for the first 5 years was very erratic, while the 6- or 7-year history that followed remained relatively stable.

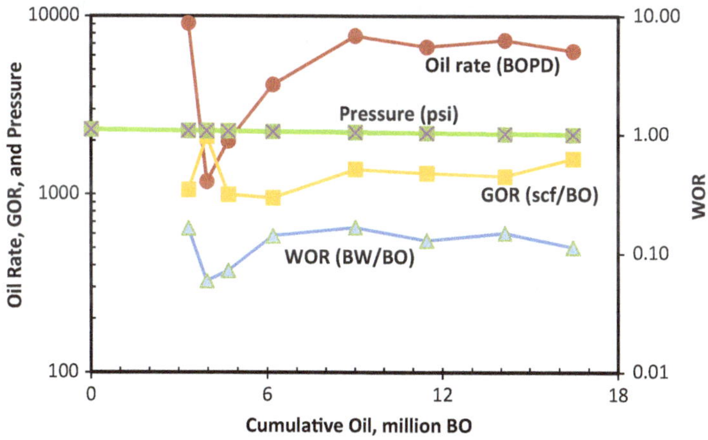

Fig. 5.7—Field performance. Reservoir pressure remained close to the original pressure. A significant portion of this pressure maintenance resulted from the presence of a very large primary gas cap. The oil rate, GOR, and WOR remained remarkably constant after the initial variability.

A	B	C	D	E	F	G	H	I
t (years)	p (psig)	Average Flow Rate (BOPD)	N_p (million BO)	W_p (million BW)	G_p (Bscf)	B_o (RB/STB)	R_s (scf/STB)	B_g (RB/scf)
0	2,310	0	0	0	0	1.400	656	0.00119
1	2,275	9,110	3.325	0.543	3.511	1.396	647	0.00121
2	2,268	1,175	3.954	0.775	4.413	1.396	646	0.00122
3	2,259	1,981	4.677	1.110	5.129	1.394	643	0.00122
4	2,245	4,123	6.182	1.981	6.561	1.392	639	0.00123
5	2,220	7,745	9.009	3.470	10.443	1.390	633	0.00124
6	2,198	6,707	11.457	4.943	13.639	1.387	627	0.00126
7	2,174	7,323	14.130	7.047	16.998	1.386	621	0.00127
8	2,155	6,367	16.454	8.903	20.665	1.373	612	0.00129

Table 5.3—Yearly production, pressure, and fluid properties.

Pressure/fluid property relationships and the calculation modules for each yearly timestep are listed in **Table 5.3.** Note that the gas formation volume factor (B_g) is expressed in RB/scf units to facilitate the calculation process. B_w = 1.0 RB/STB.

5.6.1 The Material-Balance Study. Assume formation and connate-water compressibility effects are negligible for this problem. Therefore, apply Eq. 5.12a in the following form:

$$N_a = \frac{F}{\left(E_o + E_g\right)}.$$

The equation has been divided into three calculation modules to simplify the calculation procedure.

Modules and calculated apparent-oil-in-place values are listed for each timestep in **Table 5.4.** The calculation procedure is illustrated for the t = 3 years period:

$$N_a = \frac{(4.677)(1.394)+\left[5.129-(4.677)(643)\right]0.00122+1.11}{\left[(1.394-1.400)+(656-643)(0.00122)\right]+\left[(0.92)(1.4)\left(\frac{0.00122}{0.00119}-1\right)\right]} = 241.2 \text{ million STB.}$$

5.6.2 Calculate Apparent Oil-in-Place. Assume negligible formation and connate-water (E_{fw}) and water-influx (W_e) effects when calculating apparent oil-in-place N_a. Therefore, continue to use Eq. 5.12a in the following form:

$$N_a = \frac{F}{\left(E_o + E_g\right)}.$$

The equation has been divided into three calculation modules to simplify the calculation procedure.

$$N_a = \frac{N_p B_o + \left(G_p - N_p R_s\right) B_g + W_p B_w}{\left[\left(B_o - B_{oi}\right) + \left(R_{si} - R_s\right) B_g + m B_{oi}\left(\frac{B_g}{B_{gi}} - 1\right)\right]}.$$

A	J	K	L	M	N	O	P	Q	R
t (years)	F (million RB)	E_o (RB/STB)	E_g (RB/STB)	N_a (million STB)	W_e (million STB)	SGDI	GCDI	WDI	Σ
0	0	0	0	0	0	0	0	0	
1	6.830	0.00689	0.0217	239	0.55	0.22	0.70	0.08	1.00
2	8.562	0.00820	0.0325	210	−0.35	0.21	0.84	−0.04	1.01
3	10.218	0.00986	0.0325	241	0.90	0.21	0.70	0.09	1.00
4	13.798	0.01291	0.0433	245	1.46	0.21	0.69	0.11	1.01
5	21.870	0.01852	0.0541	299	5.89	0.19	0.54	0.27	1.00
6	28.968	0.02354	0.0758	292	7.14	0.18	0.58	0.25	1.01
7	37.075	0.03045	0.0866	317	11.31	0.18	0.51	0.31	1.00
8	45.162	0.02976	0.1082	327	14.91	0.15	0.53	0.33	1.01

Table 5.4—Modules and results of calculations. (Values have been rounded off.)

Calculation modules and N_a values are listed in columns J, K, L, and M in Table 5.4 and plotted in **Fig. 5.8.** Which extrapolation is correct—this one, or the one provided in Section 5.6.1?

A straight-line extrapolation of the data backward to initial time results in an estimate of $N_a = 217$ million STB. However, the early-time data points tend to flatten. Extrapolation of the early-time flat-lying data results in an estimate of $N_a = 234$ million STB. The question has been raised: Which one is correct?

Interpretation of the performance history indicated that the early-time data were very erratic, with a very low producing rate during Years 3 and 4. Therefore, a reasonable interpretation would be OOIP = 220 million STB with a ±10 million STB range in possible answers.

5.6.3 Calculate Water Influx as a Function of Time. The OOIP value ($N_a = 220$ million STB) represents one of the two significant unknowns in the general material balance. Use Eq. 5.13a to solve for water influx because we now know OOIP:

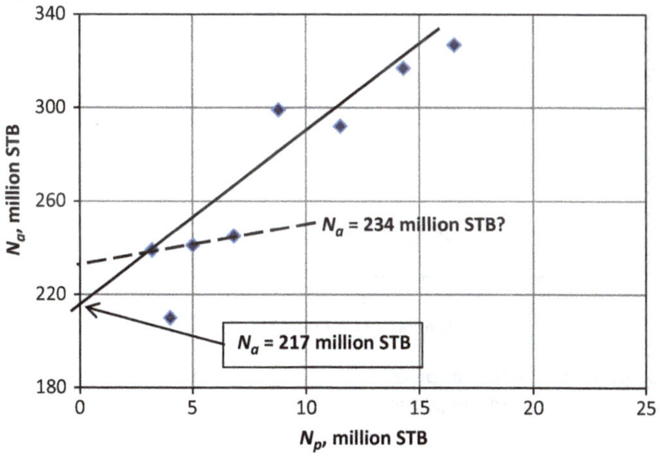

Fig. 5.8—Interpreting the apparent oil-in-place. Note that the sharply increasing line after approximately 6 million STB of oil was produced. Could the OOIP value calculated by the geologists help in your decision? Should N_a = 234 million STB be chosen?

$$E_w = W_e B_w = F - N\left(E_o + E_g\right).$$

Recall that the expanded form of Eq. 5.13a was given in Eq. 5.13b:

$$E_w = W_e B_w = \left[N_p B_o + \left(G_p - N_p R_{si}\right) B_g + W_p B_w\right] - N_a\left[\left(B_o - B_{oi}\right) + \left(R_{si} - R_s\right) B_g + m B_{oi}\left(\frac{B_g}{B_{gi}} - 1\right)\right].$$

Cumulative water influx for each timestep was calculated. The sharply increasing curve after 6 million STB production seems to indicate that water influx became an important energy source only after 6 million STB of oil had been produced.

The results of these calculations are shown in Table 5.4 and plotted in **Fig. 5.9.** The essentially linear cumulative water influx curve after the first three data points indicates that steady-state conditions existed during the majority of reservoir history. The sharply increasing curve also indicates water influx became an important energy source after 6 million STB of oil had been produced.

5.6.4 Calculate Drive Indices as a Function of Time. Section 5.6.1 showed how the general material-balance equation can be subdivided into its component parts. Each component expresses the contribution of solution gas, gas-cap expansion, and waterdrive depletion mechanisms.

Recall Eq. 5.14, with formation and interstitial-water effects excluded:

$$SGDI + GCDI + WDI = 1,$$

where

$$SGDI = \frac{NE_o}{F} = \frac{N\left[\left(B_o - B_{oi}\right) + \left(R_{si} - R_s\right) B_g\right]}{N_p B_o + \left(G_p - N_p R_s\right) B_g + W_p B_w},$$

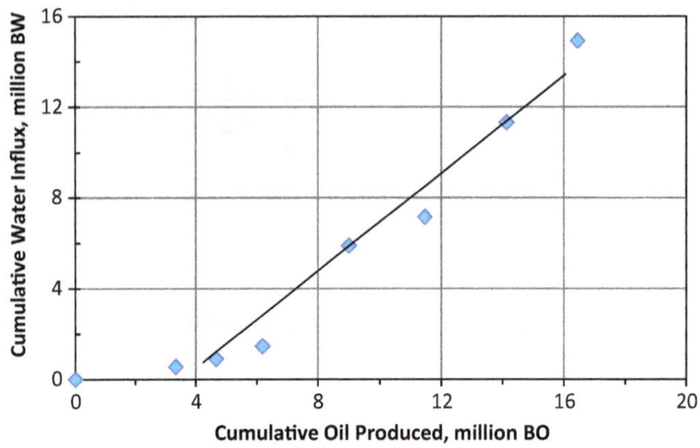

Fig. 5.9—Water influx plot. The history appears to be divided into two episodes, just as in Fig. 5.8. Does this information help when selecting the proper N_a estimate?

$$GCDI = \frac{NE_g}{F} = \frac{mNB_{oi}\left(\frac{B_g}{B_{gi}} - 1\right)}{N_pB_o + \left(G_p - N_pR_s\right)B_g + W_pB_w},$$

$$WDI = \frac{NE_w}{F} = \frac{W_eB_w}{N_pB_o + \left(G_p - N_pR_s\right)B_g + W_pB_w}.$$

The drive indices are plotted in **Fig. 5.10.** The most-significant indices are the WDI, which increases with the production history, and the GCDI.

Example calculations using the Year 3 values in Table 5.4:

$$SGDI = \frac{NE_o}{F} = \frac{220 \times 0.00986}{10.218} = 0.21.$$

$$GCDI = \frac{NE_g}{F} = \frac{220 \times 0.0325}{10.218} = 0.70.$$

$$WDI = \frac{NE_w}{F} = \frac{220 \times 0.9}{10.218} = 0.21.$$

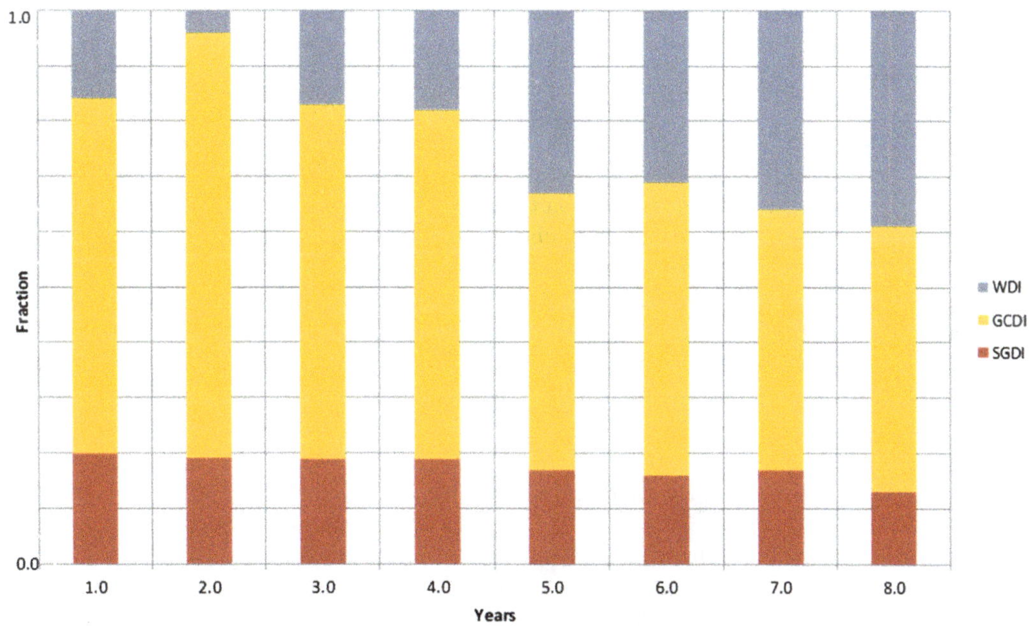

Fig. 5.10—The drive-indices plot. Note the increasing water-influx influence on production history. As one would expect, the gas cap was always a significant influence on performance.

5.7 Summary

The significant principles underlying the depletion of an oil reservoir are (1) rock and fluid expansion, which physically displaces oil; (2) gas or water invasion, which displaces reservoir oil; and (3) the overcoming of gravity forces to drive reservoir oil to the surface. Recovery estimates that account for these principles provide an incentive to expend the time and resources needed to investigate whether a waterflood should be added to a candidate reservoir.

A study of the reservoir production history offers clues to the reactions occurring during depletion. Applying performance history and/or material-balance studies to the analysis procedure provides information about the effects of the dynamic forces present during production.

Comparing oil-in-place values obtained from API correlations, analysis of production decline curves, and material-balance estimations provides a foundation for estimating primary and secondary recovery. Decline curve analysis has been extensively discussed in Poston and Poe (2008) and is not reviewed here. The API equations are useful for clastic reservoirs displaying rather-uniform properties.

Poston, Laprea-Bigott and Poe (2019) and material-balance modeling couples fundamental energy mechanisms with pressure and production data to diagnose the performance of an oil reservoir. Unfortunately, the model represents a zero-dimensional tank and does not track actual intra-reservoir fluid flow. However, it does provide the foundation for applying the more-complicated reservoir simulator model, which divides the reservoir into a grid. Each cell in the grid operates as a unit, so the effects of well location can be determined by coupling intra-grid fluid flow with the material balance.

The material-balance model can be easily set up on a spreadsheet. Appropriate reservoir fluid properties and production information are entered as a function of production and reservoir pressure. Rearranging the material-balance equation into various forms allows one to evaluate the most likely oil-in-place and water-influx history and determine the relative effects of solution gas, gas-cap expansion, and waterdrive mechanisms. Capillary effects and gravity segregation are not included in these calculations.

5.8 Questions

- Why does a material-balance study not include production rate calculations? (Answer: Because a material-balance study represents the monitoring of rock and fluid volume changes into or out of a reservoir volume as a function of changing pressure.)
- Are the effects of gravity included in the material-balance equation? (Answer: No.)
- What is the difference between a saturated and an undersaturated reservoir? (Answer: The bubblepoint pressure is below the reservoir pressure in an undersaturated reservoir. It can also be said that the reservoir pressure is below the bubblepoint pressure in a saturated reservoir.)

5.9 Nomenclature

B_g = gas formation volume factor, RB/scf or rcf/scf
B_{gi} = initial gas formation volume factor, RB/scf or rcf/scf
B_o = oil formation volume factor, RB/STB
B_{obp} = oil formation volume factor at bubble-point pressure, RB/STB
B_{oi} = initial oil formation volume factor, RB/STB
B_w = water formation volume factor, RB/STB
B_{wi} = initial water formation volume factor, RB/STB
c_f = formation compressibility, psi^{-1}
c_g = gas compressibility, psi^{-1}
c_o = oil compressibility, psi^{-1}
c_w = water compressibility, psi^{-1}
E_{fw} = formation-water effects (in-situ water expansion), RB/STB
E_g = gas expansion factor, RB/STB
E_o = oil expansion term, RB/STB
E_{total} = total fluid expansion, RB/STB
E_w = water influx, RB
F = underground withdrawal, RB

G = in-place gas volume or gas-cap size, scf, MMscf, or Bscf or Bscf/acre-ft
G_i = cumulative gas injection, scf, MMscf, or Bscf or Bscf/acre-ft
G_p = produced-gas volume, scf, MMscf, or Bscf or Bscf/acre-ft
h = thickness, ft
k = permeability, md (or darcies when using API correlations)
m = original gas-cap volume, fraction, gas reservoir volume (acre-ft)/oil reservoir volume (acre-ft)
N = original oil-in-place, million BO, million STB, or STB/acre-ft
N_a = apparent oil-in-place, million BO, million STB, or STB/acre-ft
N_p = produced-oil volume, million BO, million STB, or STB/acre-ft
p = pressure, psig
p_a = abandonment pressure (API correlations), psig
$p_{a(SGD)}$ = abandonment pressure in solution-gas-drive case (API correlations), psig
$p_{a(WD)}$ = abandonment pressure in waterdrive case (API correlations), psig
p_{bp} = bubblepoint pressure (API correlations), psig
$p_{current}$ = reservoir pressure at actual conditions, psig
p_i = initial pressure, psig
p_{res} = reservoir pressure, psig
q_o = oil flow rate, STB/D or RB/D
RE_{SGD} = solution-gas-drive reservoir recovery, %
RE_{WD} = waterdrive reservoir recovery, %
R_p = produced-fluid gas/oil ratio, scf/STB
R_s = solution gas/oil ratio, scf/STB
R_{si} = initial solution gas/oil ratio, scf/STB
S_{eq} = equilibrium gas saturation, fraction or %
S_{oi} = initial oil saturation, fraction or %
S_{or} = residual oil saturation, fraction or %
S_{wi} = initial water saturation, fraction or %
t = time, years
T = temperature, °F
V_p = reservoir volume, RB
W_e = water influx or water encroachment volume, STB or RB
W_i = cumulative water injection, RB
W_p = produced-water volume, million BW
μ = viscosity, cp
μ_{obp} = oil bubblepoint viscosity, cp
μ_{oi} = initial oil viscosity, cp
μ_w = water viscosity, cp
ϕ = porosity, fraction or %

5.10 Problems

Exercise 5.1—Estimating Secondary Reserves. You are the reservoir engineer in charge of the single-well Boggy Hollow Unit. The production response seems to indicate the existence of pressure-depletion conditions. The installation of a waterflood is being contemplated at the current reservoir pressure conditions, specified in **Table 5.5.**

A shut-in well located in a good downdip location can be converted to water injection. Assume a 70% areal sweep efficiency and a 50% vertical sweep factor because the sand is heterogeneous. Reservoir volume = 5,400 acre-ft. Reservoir and rock properties can be found in Table 5.5.

Pressure (psig)	p_i = 5,000	$p_{current}$ = 2,700	p_{bp} = 2,000	$p_{a(WD)}$ = 2,000	$p_{a(SGD)}$ = 1,000
Reservoir	Volume = 5,400 acre-ft	h = 28 ft	ϕ = 28%	S_{wi} = 25%	k = 180 md
Fluid Properties	B_{oi} = 1.205 RB/STB	B_{obp} = 1.234 RB/STB	μ_w = 0.6 cp	μ_{oi} = 0.81 cp	μ_{obp} = 0.56 cp

Table 5.5—Reservoir and rock properties for Exercise 5.1.

The objective of this exercise is to recognize the production response for above-bubblepoint and below-bubblepoint depletion conditions. Apply the API equations to qualitatively estimate secondary reserves for a contemplated waterflood.

Analysis procedure:

1. Determine OOIP at p_i and at the bubblepoint.
2. Apply the API equations (Eqs. 5.2 and 5.3) to estimate the solution-gas-drive pressure depletion and waterdrive recovery factors if a waterflood project were installed at current pressure conditions.
3. Determine the following recovery estimates:
   - Oil to be produced from p_i to p_{bp}
   - Possible primary recovery by means of solution gas drive

Solution.

1. Calculate OOIP at the initial and bubblepoint pressures.

Calculate pore volume (V_p) and OOIP using Eqs. 1.5 through 1.7b (see Chapter 1, Reservoir Rocks and Framework):

$V_p = PV$ = reservoir volume × ϕ = 5,400 acre-ft × 7,758 RB/acre-ft × 0.28 = 11,730,096.00 RB.

$$OOIP \text{ at } p_i = \frac{(V_p)(1-S_{wi})}{B_{oi}} = \frac{(11{,}730{,}096.00 \text{ RB})(1-0.25)}{1.205 \text{ RB / STB}} = 7{,}300{,}890 \text{ STBO.}$$

$$OOIP \text{ at } p_{bp} = \frac{(V_p)(1-S_{wi})}{B_{obp}} = \frac{(11{,}730{,}096.00 \text{ RB})(1-0.25)}{1.234 \text{ RB / STB}} = 7{,}129{,}313 \text{ STBO.}$$

The values of OOIP at p_i and OOIP at p_{bp} are different based on the values of the formation volume factor at initial and bubblepoint conditions and the fact that oil has been produced by the effect of rock and fluid compressibility.

2. Compare the oil recoveries by means of solution gas drive (Eq. 5.2) and waterdrive (Eq. 5.3) at current pressure conditions to estimate the potential recovery range if a waterflood were actually installed:

$$RE_{SGD} = 41.8\left\{\left[\frac{\phi(1-S_{wi})}{B_{obp}}\right]^{0.1611}\left(\frac{k}{\mu_{obp}}\right)^{0.0979}(S_{wi})^{0.3722}\left(\frac{p_{bp}}{p_a}\right)^{0.1741}\right\} =$$

$$41.8\left\{\left[\frac{0.28(1-0.25)}{1.234}\right]^{0.1611}\left(\frac{0.18}{0.56}\right)^{0.0979}0.25^{0.3722}\left(\frac{2{,}000}{1{,}000}\right)^{0.1741}\right\} = 18.9\%.$$

$$RE_{WD} = 54.9\left\{\left[\frac{\phi(1-S_{wi})}{B_{oi}}\right]^{0.0422}\left(\frac{k\mu_w}{\mu_{oi}}\right)^{0.077}(S_{wi})^{-0.19}\left(\frac{p_{\text{current}}}{p_a}\right)^{-0.2159}\right\} =$$

$$54.9\left\{\left[\frac{0.28(1-0.25)}{1.205}\right]^{0.0422}\left(\frac{0.18(0.6)}{0.81}\right)^{0.077}0.25^{-0.1903}\left(\frac{2{,}715}{2{,}015}\right)^{0.2159}\right\} = 60.63\%.$$

The resulting values are then rounded off because they are really qualitative and should be viewed as such. Using the estimated values calculated above, the solution-gas-drive pressure depletion and water-influx recovery factors can be approximated:

- RE_{SGD} = 18.9 can be rounded up to 20%.
- RE_{WD} = 60.63% can be rounded down to 60% as a conservative assumption.

These calculations indicate that approximately 2.5 to 3 times of additional oil would be obtained by developing a waterflood project.

4. Determine the following recovery estimates:
   - OOIP at p_{bp} = oil to be produced from p_i to p_{bp} = 7,300,890 – 7,129,313 = 171,577 STBO.
   - OOIP at $p_{bp} \times RE_{SGD}$ = possible primary recovery by means of solution gas drive = 7,129,313 × 0.20 = 1,425,862.6 STBO.

Exercise 5.2—Calculate Secondary Gas-Cap Volume for Hardin Field. Apply the information in **Table 5.6** and the knowledge gained during the discussion on the material-balance equation in Section 5.4 to make the specified calculations.

Complete the following tasks:

1. Calculate pore volume and OOIP at p_i.
2. Calculate the original gas-in-place (OGIP). Then, calculate the remaining gas after 2.8 Bscf of gas is produced.
3. Calculate the volume of liberated solution gas remaining in the reservoir after the pressure declines to 1,500 psig. Then, calculate the reservoir volume of the secondary gas cap (calculate hydrocarbon pore volume and the reservoir volume).

Solution.

1. Calculate pore volume and OOIP at p_i:

$V_p = PV$ = reservoir volume × ϕ = 7,652 acre-ft × 7,758 RB/acre-ft × 0.31 = 18,402,907 RB.

$$OOIP \text{ at } p_i = \frac{(V_p)(1-S_{wi})}{B_{oi}} = \frac{(18{,}402{,}907)(1-0.25)}{1.33 \text{ RB / STB}} = 10{,}377{,}579 \text{ STB.}$$

2. Calculate OGIP:

$$OGIP = (10{,}377{,}579 \text{ STB})(1{,}008 \text{ scf / STB}) = 10.4606 \text{ Bscf.}$$

Calculate the remaining gas after 2.8 Bscf of gas is produced:

$$NR_{si} - G_p = NR_{si} - N_pR_p = (10.4606 \text{ Bscf}) - (1.0 \text{ million STB})(2{,}800 \text{ scf / STB}) = 7.6605 \text{ Bscf.}$$

Reservoir volume = 7,652 acre-ft	R_{si} = 1,008 scf/STB	S_{wi} = 25%	S_{or} = 20%
N_p = 1.0 million STB	G_p = 2.8 Bscf	m = 0	R_s = 630 scf/STB
B_{oi} = 1.33 RB/STB	B_g = 0.0107 rcf/scf	ϕ = 31%	R_p = 2,800 scf/STB

Table 5.6—Reservoir rock and fluid properties and production data for Exercise 5.2.

3. Calculate the volume of liberated solution gas remaining in the reservoir after the pressure declines to 1,500 psig:

$$\text{Free gas} = NR_{si} - \left(N - N_p\right)R_s - G_p = 10.4606\ \text{Bscf} - \left(10{,}377{,}579 - 1{,}000{,}000\ \text{STB}\right)\times$$
$$\left(630\ \text{scf / STB}\right) - 2.8\ \text{Bscf} = 1.75\ \text{Bscf.}$$

Calculate the reservoir volume of the secondary gas cap.

Secondary gas cap occupies a volume = free gas × B_g;

SGIP = secondary gas-in-place volume = 1.75 Bscf (0.0107 rcf/scf) = 18.88 MMscf.

The reservoir volume occupied by the secondary free-gas cap:

$$\text{Reservoir volume} = \frac{(SGIP)}{(\phi)\left(1 - S_{or} - S_{wi}\right)(43{,}560\frac{\text{ft}^2}{\text{acre}-\text{ft}})} =$$

$$\frac{18.88\ \text{MMscf}}{(0.31)(0.55)\left(43{,}560\frac{\text{ft}^2}{\text{acre}-\text{ft}}\right)} = 2{,}542\ \text{acre-ft.}$$

Is this value absolutely correct? Not completely because an equilibrium gas saturation was established in the oil reservoir and remains immovable.

Exercise 5.3—Underground Withdrawal. Oil production from the Burton Field has resulted in a rapid pressure decline with a resultant increasing GOR. These characteristics suggest a dominant solution-gas-drive mechanism. A waterflood project should possibly be installed.

Pressure and production data and reservoir-fluid properties for a 1-year period are listed in **Table 5.7.** Assume the bubblepoint pressure p_{bp} is the same as the initial pressure p_i. Apply material-balance concepts to calculate the minimum water-injection requirements. q_o = 31,000 STB/D = 41,702 RB/D.

Hint:

Compare the oil producing rate to the reservoir voidage rate (underground withdrawal) F in RB/D units. Do they differ? Why?

Assume average reservoir properties for the calculations by interpolating between the two pressures. Rearrange the underground withdrawal equation (Eq. 5.6) into the following form:

$$F = N_pB_o + \left(G_p - N_pR_s\right)B_g,\ \text{RB.}$$

Solution. Compare the oil producing rate to the reservoir voidage rate to determine the minimum water-injection requirements:

Time (years)	Pressure (psig)	N_p (million STB)	G_p (Bscf)	B_o (RB/STB)	B_g (RB/scf)	R_s (scf/STB)
1	2,453	9.873	10.940	1.364	0.001204	564
2	2,318	11.250	13.163	1.353	0.001276	540

Table 5.7—Reservoir producing history for Exercise 5.3.

$$F = N_pB_o + (G_p - N_pR_s)B_g = (21.123 \text{ million STB})(1.353 \text{ RB/STB}) +$$
$$[24{,}103 \text{ MMscf} - (21.123 \text{ million STB} \times 540 \text{ scf/STB})] \times (0.001276 \text{ RB/scf}) =$$
$$44.780 \text{ million RB in 2 years} = 61{,}343 \text{ RB/D}.$$

The water-injection rate must be at least two times greater than the current producing rate because of the production of what is probably excess gas. The high-GOR wells should be shut in.

Exercise 5.4—Oil- and Gas-In-Place Estimates. Well control for the determination of the gas-cap size in the D3 Oil Reservoir Unit is not good. Originally, geologists estimated a gas-cap size of $G = 21$ Bscf.

Alfred Newman, the field production geologist, has determined that there could be a considerable extension of the gas cap in the northwest corner of the field. Newman's new estimate of volume occupied by the gas cap is 19,215 acre-ft. The isopach oil reservoir volume estimate is 57,300 acre-ft.

Use the pressure, production, and fluid property information provided in **Table 5.8** with a rewritten form of the material-balance equation (Eq. 5.5) to prove or disprove Newman's hypothesis.

Analysis procedure:

Rearrange the material-balance equation (Eq. 5.5) into the following form to determine oil- and gas-in-place from production records: $F = N(E_o + E_g)$. Assume water influx and rock and water expansion are absent.

Rearrange to the equation of a straight line, $y = mx + b$:

$$N_pB_o + (G_p - N_pR_s)B_g = NB_{oi}\left[\frac{(B_o - B_{oi}) + (R_{si} - R_s)B_g}{B_{oi}} + m\left(\frac{B_g}{B_{gi}} - 1\right)\right].$$

Dividing through N, substituting m for G/N, and rearranging to replicate the expression of a straight line result in

$$\frac{N_pB_o + (G_p - N_pR_s)B_g}{(B_o - B_{oi}) + (R_{si} - R_s)B_g} = G\frac{B_o\left(\dfrac{B_g}{B_{gi}} - 1\right)}{(B_o - B_{oi}) + (R_{si} - R_s)B_g} + N.$$

Pressure (psig)	N_p (million STB)	G_p (Bscf)	B_o (RB/STB)	B_g (rcf/scf)	B_g (RB/scf)	R_s (scf/STB)
2,500	0	0	1.498	0.00588	0.00105	721
2,300	3.654	2.616	1.463	0.00648	0.00116	669
2,100	6.929	6.693	1.429	0.00718	0.00128	617
1,900	9.425	12.224	1.395	0.00808	0.00144	565
1,700	10.719	17.397	1.361	0.00917	0.00164	513
1,500	12.631	24.668	1.327	0.01056	0.00188	461

Table 5.8—Reservoir pressure and production data and fluid properties for Exercise 5.4.

The OOIP (N) is the b-intercept, and the slope of the line represents the gas-cap size in the same units. The plotting functions are

$$y = \frac{F}{E_o} = \frac{N_p B_o + \left(G_p - N_p R_s\right) B_g}{\left(B_o - B_{oi}\right) + \left(R_{si} - R_s\right) B_g},$$

$$x = \frac{E_g}{E_o} = \frac{B_o\left(\frac{B_g}{B_{gi}} - 1\right)}{\left(B_o - B_{oi}\right) + \left(R_{si} - R_s\right) B_g}.$$

Table 5.9 breaks down the calculation procedure into individual modules. The last two columns show the plotting variables, which should be plotted and interpreted using **Fig. 5.11.**

Compare the calculated estimates with Newman's estimates using **Table 5.10.** However, because Newman redrew the structure and isopach maps, his values are in acre-ft, but reserves and material-balance calculations are calculated in STB. Common units must be used when comparing the two estimates. Convert Newman's numbers, calculated in acre-ft, to STB and Bscf units and use the estimated values of $\phi = 30\%$ and $S_{wi} = 18\%$.

Then, calculate the estimated values of OOIP and OGIP:

$$V_p = PV = \text{reservoir volume} \times \phi = 57{,}300 \text{ acre-ft} \times 7{,}758 \text{ RB/acre-ft} \times 0.30 = 133{,}360{,}020 \text{ RB}.$$

p (psig)	$(B_o - B_{oi}) + (R_{si} - R_s)B_g$ E_o (RB/STB)	$N_pB_o + (G_p - N_pR_s)B_g$ F (million RB)	$B_o\left(\frac{B_g}{B_{gi}} - 1\right)$ (RB/STB)	x	y (million STB)
2,300	0.0253	5.5446	0.1569	6.2	219.0
2,100	0.0642	12.9962	0.3282	5.1	202.7
1,900	0.1213	23.0821	0.5564	4.6	189.8
1,700	0.2040	34.1020	0.8416	4.1	167.1
1,500	0.3178	52.1906	1.1842	3.7	164.2

Table 5.9—Calculation modules for Exercise 5.4.

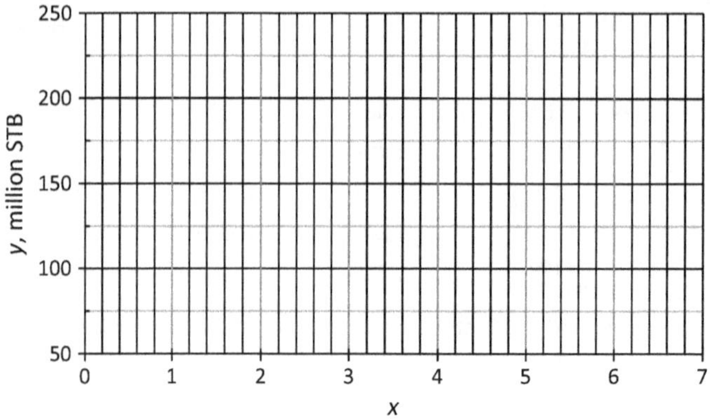

Fig. 5.11—Use the graph to plot the OOIP and gas-cap size estimates.

	OGIP (Bscf)	OOIP (million STB)
Calculated volumes		
Newman's estimated volumes	35	73

Table 5.10—Use the table to compare gas-cap and oil volume estimates.

$$OOIP \text{ at } p_i = \frac{(V_p)(1-S_{wi})}{B_{oi}} = \frac{(133{,}360{,}020)(1-0.18)}{1.498 \text{ RB / STB}} = 73{,}000{,}812 \text{ STB} \approx 73 \text{ million STB.}$$

$$\text{OGIP} = [\text{gas-cap volume} \times \phi \times (1 - S_{wi})]/B_{gi} = \frac{(19{,}215)(43{,}560 \text{ rcf / acre} - \text{ft})(0.30)(1-0.18)}{0.00588 \text{ rcf/scf}} = 35$$
Bscf.

Solution. As seen in **Fig. 5.12,** the intercept indicates $N = 77.4$ million STB, while the slope of the line indicates the gas-cap size is 23.4 million STB, which must be converted to scf units. Compare with original values.

Converting OGIP from STB to scf gives

$$OGIP = \frac{23.4 \text{ million STB} \times 1.498 \text{ RB / STB}}{0.00105 \text{ RB / scf}} = 33.4 \text{ Bscf.}$$

Table 5.11 compares the results to the original hydrocarbon volumes estimates.

There could be a field extension. Newman's estimates were probably correct.

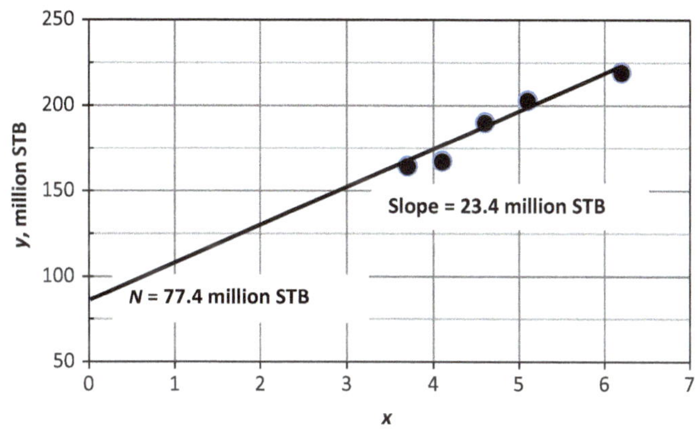

Fig. 5.12—OOIP and gas-cap size estimates for Example 5.4.

	OGIP (Bscf)	OOIP (million STB)
Calculated volumes	33.4	77.4
Newman's estimated volumes	35	73
Absolute difference between Newman's estimates and calculated values (%)	5%	6%

Table 5.11—Comparing gas-cap and oil volume estimates.

5.11 References

- API Subcommittee on Recovery Efficiency. 1967. A Statistical Study of Recovery Efficiency. API Bulletin D14, American Petroleum Institute, New York (October 1967).
- Havlena, D. and Odeh, A. S. 1963. The Material Balance as an Equation of a Straight Line. *J Pet Technol* **15** (8): 896–900. SPE-559-PA. https://doi.org/10.2118/559-PA.
- McEwen, C. R. 1962. Material Balance Calculations with Water Influx in the Presence of Uncertainty in Pressures. *J Pet Technol* **2** (2): 120–128. SPE-225-PA. https://doi.org/10.2118/225-PA.
- Poston, S. W. and Berg, R. R. 1997. *Overpressured Gas Reservoirs*. Richardson, Texas: Society of Petroleum Engineers.
- Poston, S. W. and Poe, B. D., Jr. 2008. *Analysis of Production Decline Curves*. Richardson, Texas: Society of Petroleum Engineers.

Chapter 6

Inflow Performance

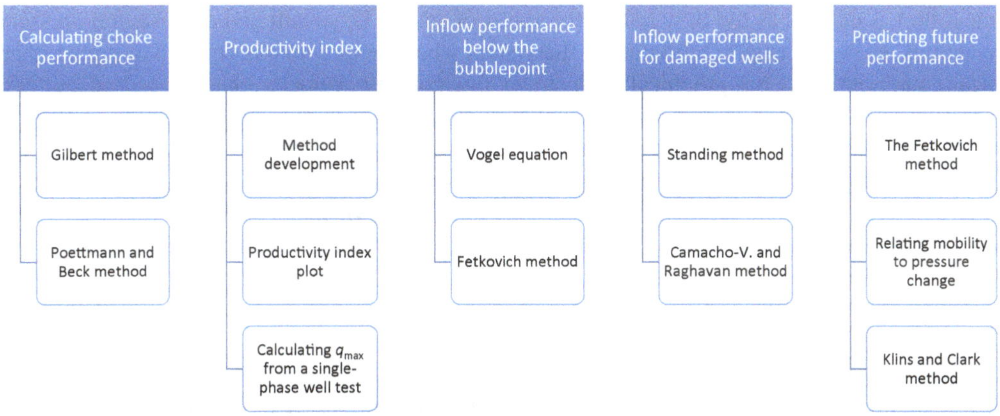

Estimating future well producing rates is one of the most-important tasks when evaluating an oil reservoir. Hydrostatic head controls the relationship between the backpressure at the wellhead and the bottomhole flowing pressure in the production tubing. Gas or water phases introduced into the flow stream can change the hydrostatic head while chokes control backpressure, both of which affect the surface producing rate.

Water entering the flow stream increases the weight of the hydrostatic head, which must exceed the differential between the bottom and surface flowing pressure. Chokes are used to limit the flow rates at the surface and subsurface. Excessive quantities of commingled gas with the produced oil can dramatically reduce the oil production rate because of the ability of the gas phase to flow in preference to the oil phase. Equations describing the relation between the bottomhole flowing pressure and the oil producing rate are defined as inflow performance relationships. Differences between the bottom and surface flowing pressure caused by changes in the hydrostatic head and the effects of turbulence in the tubing are classified as well performance characteristics.

The productivity-index (PI) concept applies the Darcy or the pseudosteady-state flow equations to model single-phase fluid flow. However, relative permeability and viscosity effects must be included in most two- or three-phase flow systems. The inability to exactly replicate reservoir and near-wellbore behavior usually means this model is impractical for use. Simulation studies combined with well-test information seem to provide the best approach for studying the inflow problem.

6.1 Calculating Choke Performance

Wellhead and downhole chokes exert a backpressure on the flow system to control flow rates. Maintaining the correct allowable backpressure to restrict the production of excessive sand and preventing, or at least inhibiting, gas and water coning are some of the primary reasons for controlling well performance. Excessive sand production can present a serious operational problem when producing from unconsolidated to poorly cemented reservoirs. Entrained sand erodes and degrades surface and production equipment and can pose a significant environmental hazard. Casing failure can occur when rock material collapses into the voids surrounding the perforated interval created by excessive sand production. Restricting production by introducing a smaller-sized choke into the flow system increases the bottomhole flowing pressure, which in turn decreases the pressure drop between the inner and outer drainage boundaries.

Chokes are constructed from ceramic to inhibit the abrasive influence of the sand and formation material often entrained in the producing stream. Choke boxes or manifolds are designed to provide ready entry so that chokes can be easily replaced. The upstream side of the orifice should have a knife edge. Rounding of the knife edge by means of erosion alters the pressure drop profile, which in turn reduces the accuracy of the pressure-drop/rate calculations.

All rate/choke-size calculations assume the presence of critical flow conditions within the system. Critical flow is defined as the flow rate at the speed of sound. Upstream and downstream pressure fluctuations are not transmitted throughout the system when the flow rate is equal to or greater than critical conditions. **Fig. 6.1** graphically illustrates this concept. Velocity changes caused by fluid flowing through a restriction are solely a function of the pressure drop caused by the restriction when the flow rates are greater than or equal to the critical flow conditions. As a rule of thumb, the velocity of the flow stream through a choke is greater than the critical flow rate when the upstream pressure is at least approximately two times greater than the downstream pressure.

Fig. 6.2 illustrates the nonlinearity of the pressure-drop/rate relationship measured across a choke when the flow rate is less than the critical flow rate. Downstream flow restrictions influence the pressure drop across the choke and alter the pressure-drop/rate relationship when the well is flowing at less than this critical value.

An oil well seldom flows in a smooth, continuous stream. Fluids passing through a choke often surge in an erratic and discontinuous manner because of fluid segregation within the producing string. In addition, produced formation fines can alter viscosity and density relationships. In a practical sense, analytical solutions to derive equations relating the oil flow rate and the pressure drop

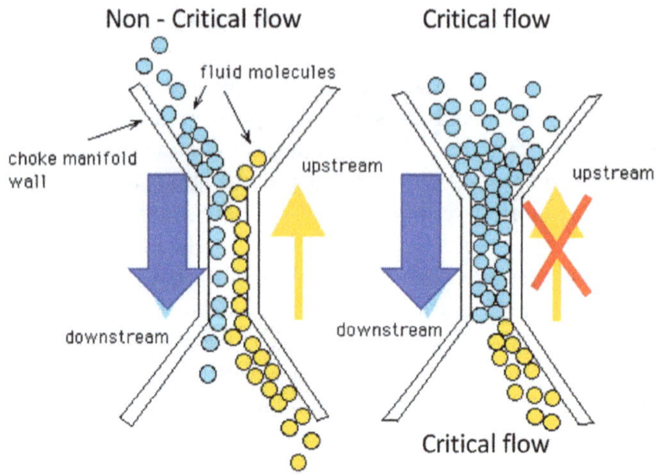

Fig. 6.1—Critical flow rate. This shows flow conditions occurring across a choke manifold valve. Pressure fluctuations downstream of the valve cannot affect conditions above the valve if the downstream pressure is less than approximately 0.5–0.6 times the upstream pressure. Courtesy of Schlumberger NExT.

across a choke are essentially impossible to develop because of the wide variety of poorly defined variables that must be considered.

Gilbert (1954) statistically analyzed well-test information obtained from a field to develop a pressure-drop/choke-size/rate relationship particular to those field characteristics. Poettmann and Beck (1963) developed equations describing oil and gas flow through a choke when oil is assumed to be dispersed as droplets in the carrier gas. The Gilbert (1954) equation is probably the most practical of the two methods.

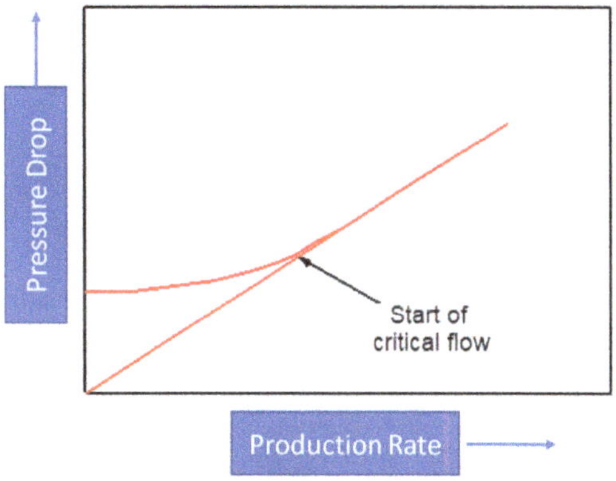

Fig. 6.2—The rate/pressure-drop relationship is linear when the system operates at or above the critical flow region, and the downstream pressure does not affect the upstream pressure.

6.1.1 Gilbert Method. Gilbert (1954) presented an empirical equation relating the flowing tubinghead pressure, (oil plus water) liquid flow rate, gas/liquid ratio, and choke-size properties to calculate the oil flow rate across a choke:

$$p_t = \frac{Cq_t R^{0.5}}{S^2}, \quad (6.1)$$

where p_t is the flowing tubinghead pressure (psia); C is the coefficient particular to a field (typically equal to 600); $q = q_t$, which is the gross liquid flow rate, oil plus water (BFPD); S is the choke size expressed in 1/64 in.; and R is the gas/liquid ratio (Mscf/bbl).

A large number of well tests are statistically analyzed to calculate the combination of variables and coefficients, providing the least standard deviation from the field data. The values of the gas/liquid ratio exponent and coefficient C are varied to provide a best-fit answer. Gilbert (1954) refined the basic equation to develop a rate/tubinghead pressure relationship for the Tensleep Field. Eq. 6.2 contains three constants that were deemed to provide the best fit for the Tensleep Field data:

$$p_t = \frac{435q_t R^{0.546}}{S^{1.89}}. \quad (6.2)$$

Example 6.1—Calculating Flowing Tubinghead Pressure. A well produces 60 BOPD and 40 BWPD with a 700 scf/bbl gas/liquid ratio through a ¼-in. choke. Determine the flowing tubinghead pressure.

Solution. Apply the generalized Gilbert (1954) equation (Eq. 6.1) to calculate p_t:

$$p_t = \frac{Cq_t R^{0.5}}{S^2} = \frac{600q_t R^{0.5}}{S^2} = \frac{(600)(100)(0.7)^{0.5}}{16^2} = 196 \text{ psia.}$$

6.1.2 The Poettmann and Beck Method. Poettmann and Beck (1963) used an equation derived by Ros (1961) describing single-phase flow through pipe to develop a correlation for (oil plus gas) flow/pressure-drop relationships measured across a choke. Solutions were presented in chart and equation forms. Water production invalidates the method, and the system must flow at critical conditions. Gas is the continuous phase, with liquid dispersed in the gas/liquid phase, and the gas expands polytropically.

$$q = \frac{88{,}992A}{5.61\rho_o + 0.0765\gamma_g R_{st}} \left[\frac{9{,}273.6p_t}{V_1(1+0.5m_1)}\right]^{0.5} \left[\frac{0.4513(R_1+0.766)^{0.5}}{R_1+0.5663}\right]. \quad (6.3)$$

Grouping constant terms, the R_1, m_1, and V_1 constant values are determined as follows:

$$R_1 = \frac{0.00504T_t z_t (R - R_{st})}{p_t B_{ot}}, \quad (6.4)$$

$$m_1 = \frac{1}{1 + R_1 \dfrac{\rho_{gt}}{\rho_{ot}}}, \quad (6.5)$$

$$V_1 = \frac{m_1}{\rho_{ot}}, \quad (6.6)$$

where q is the flow rate (STB/D), A is the choke cross-sectional area (ft^2), ρ_o is the oil density measured at 60°F and 14.7 psia (lbm/ft^3), p_t is the flowing tubinghead pressure (psia), ρ_{ot} is the oil density at upstream T and p (lbm/ft^3), ρ_{gt} is the gas density at upstream T and p (lbm/ft^3), ρ_g is the gas gravity measured at 60°F and 14.7 psia (lbm/ft^3), T_t is the upstream temperature (°R), z_t is the dimensionless gas compressibility factor at upstream T and p, R is the producing gas/oil ratio (scf/STB), R_{st} is the gas solubility at upstream T and p (scf/STB), and B_{ot} is the oil formation volume factor at upstream T and p (bbl/STB).

The R_1, m_1, and V_1 values for a particular set of conditions are calculated with Eqs. 6.4, 6.5, and 6.6, respectively. These answers are used in Eq. 6.3 to calculate the flow rate measured across the choke. Gas and oil specific gravity and density values referenced to surface conditions must be adjusted to the temperature and pressure at the choke. Assume temperature and pressure at the choke are the same as those measured at the tubing head.

Example 6.2—Calculating Flow Rate. Determine the flow rate for the Zenon No. 2 well producing under the following conditions: flowing tubinghead pressure = 1,264 psia, flowing tubinghead temperature = 85°F, choke size = 6/64 in., gas/oil ratio = 2,250 scf/STB, $\gamma_g = 0.6$, and oil gravity = 44.4 °API.

Solution. The first step is to calculate the physical and fluid properties referenced to conditions at the choke.

Convert the choke area to ft^2 units:

$$A = \left(\frac{6}{64}\right)^2 \left(\frac{\pi}{4}\right)\left(\frac{1}{12}\right)^2 = 4.794 \times 10^{-5} \text{ ft}^2.$$

Convert the oil API gravity measured at 60°F and 14.7 psia to specific gravity:

$$\gamma_o = \frac{141.5}{131.5 + °\text{API}} = \frac{141.5}{131.5 + 44.4} = 0.804(62.4 \text{ lbm/ft}^3) = 50.2 \text{ lbm/ft}^3.$$

Apply the tables in *API 852-27045* (1980) to find the oil density at the 1,264-psia and 85°F tubinghead conditions. The answer is $\rho_{ot} = 49.4$ lbm/ft^3.

Apply the Sutton (1985) correlations to calculate the gas critical properties and Fig. 3.5 (Standing and Katz 1942) in the Fluid Properties chapter to determine the *z*-factor at the 1,264-psia and 85°F tubinghead conditions, z_t:

$$p_{pc} = 756.8 - 131.0(0.6) - 3.6(0.6)^2 = 677 \text{ psia.}$$

$$T_{pc} = 169.2 + 349.5(0.6) - 74.0(0.6)^2 = 352°\text{R.}$$

The dimensionless pseudoreduced properties are ratios of the reservoir temperature and pressure to their pseudocritical properties:

$$T_{pr} = \frac{545}{352} = 1.55.$$

$$p_{pr} = \frac{1,264}{677} = 1.87.$$

Thus, using Fig. 3.5, $z_t = 0.86$.

Calculate the gas density at the flowing tubinghead conditions. Apply the gas molecular weight equation (Eq. 3.5b) and a rearranged version of the general gas law (Eq. 3.6):

$$M = 28.9647\gamma_g = (28.9647)(0.6) = 17.38 \text{ lbm/lbm-mol.}$$

$$\rho_g = \frac{pM}{zRT} = \frac{(1,264)(17.38)}{(0.86)(10.73)(545)} = 4.37 \text{ lbm/ft}^3.$$

Apply the Vasquez and Beggs (1980) correlation (Eq. 3.21) and the requisite coefficients found in Table 3.3 (see the Fluid Properties chapter) to calculate the solution gas/oil ratio at flowing tubinghead conditions R_{st} for °API > 30:

$$R_{st} = \text{C1}\gamma_{gc} p_{\text{res}}{}^{\text{C2}} \exp^{[\text{C3}\gamma_{\text{API}}/(T_{\text{res}}\,°\text{F}+460)]} = 0.0178(0.6)(1,264)^{1.187} \exp^{[(23.931\times 44.4)/(85+460)]} = 339 \text{ scf/STB.}$$

Apply the Vasquez and Beggs (1980) correlation (Eq. 3.22) and coefficients found in Table 3.3 to calculate B_o at flowing tubinghead conditions, B_{ot}:

$$B_o = 1 + \text{C1}R_s + \text{C2}(T_{\text{res}} - 60)\left(\frac{\gamma_{\text{API}}}{\gamma_{gc}}\right) + \text{C3}R_s(T_{\text{res}} - 60)\left(\frac{\gamma_{\text{API}}}{\gamma_{gc}}\right),$$

$$B_{ot} = 1 + 4.67\times 10^{-4}(449) + 1.10\times 10^{-5}(85-60)\left(\frac{44.4}{0.6}\right) + 1.34\times 10^{-9}(449)(85-60)\left(\frac{44.4}{0.6}\right) =$$

$$1.1809 \text{ RB/STB.}$$

Calculate the R_1 term:

$$R_1 = \frac{0.00504 T_t z_t (R - R_{st})}{p_t B_{ot}} = \frac{0.00504(545)(0.86)(2,250-339)}{(1,264)(1.1809)} = 2.899.$$

Calculate m_1:

$$m_1 = \frac{1}{1+R_1\dfrac{\rho_{gt}}{\rho_{ot}}} = \frac{1}{1+2.899\left(\dfrac{4.37}{49.4}\right)} = 0.7958.$$

Calculate V_1:

$$V_1 = \frac{m_1}{\rho_{ot}} = \frac{0.7958}{49.4} = 0.016.$$

Now calculate the flow rate. The flow rate expected at the specified conditions can now be determined because all of the terms included in Eq. 6.3 have been calculated:

$$\begin{aligned} q &= \frac{88{,}992A}{5.6\rho_O + 0.0765\gamma_g R_{st}}\left[\frac{9{,}273.6p_t}{V_1(1+0.5m_1)}\right]^{0.5}\left[\frac{0.4513(R_1+0.766)^{0.5}}{R_1+0.5663}\right] \\ &= \frac{(88{,}992)(4.794\times10^{-5})}{(5.61)(49.4)+0.0765(0.6)(339)}\left\{\frac{(9{,}273.6)(1{,}264)}{0.016[1+0.5(0.7958)]}\right\}^{0.5}\left[\frac{0.4513(2.899+0.766)^{0.5}}{2.899+0.5663}\right] \\ &= (0.01)(22{,}892.84)(0.25) = 65\text{ BOPD}. \end{aligned}$$

6.2 Productivity Index

The PI measures the ability of a well to produce oil. Evinger and Muskat (1942) showed that a well producing water flowing through a cross-sectional area of 1 ft^2 and a 1-darcy-permeability sand with a 1-psi pressure drop displays a PI of 0.9. The study indicated that only wells with an extraordinarily high kh/μ value display PI $\geq$ 1 values. Flow tests relate a series of q vs. p_{wf} data to measure the ability of a well to flow oil. Most oil wells display a PI substantially less than unity.

6.2.1 Method Development. A homogeneous producing interval with uniform fluid saturations, negligible formation compressibility effects, and single-phase and laminar flow is assumed. The assumption of pseudosteady-state flow conditions forms the basis for estimating the PI, and Eq. 4.4 (see the Well and Reservoir Models chapter) is used. The skin term can be deleted from the denominator under the assumption that neither damage nor stimulation is present and s = 0. Eq. 6.7 shows the skin factor in the denominator.

$$q = \frac{0.00708\ kh(\bar{p} - p_{wf})}{\mu B[\ln(r_e/r_w) - 0.75 - s]},\ \text{BOPD}. \qquad (6.7)$$

PI has been also defined in the literature as J and is expressed as

$$J = \frac{q}{p - p_{wf}},\ \text{BOPD/psi}. \qquad (6.8)$$

Eq. 6.8 expresses the PI as the ratio of the producing rate to the drawdown pressure.

The oil relative permeability concept, $k_{ro} = \left(\dfrac{k_o}{k}\right)$, can be conveniently introduced into Eq. 6.7:

$$q = \frac{0.00708\ kh(\bar{p} - p_{wf})}{[\ln(r_e/r_w) - 0.75 - s]}\left(\frac{k_{ro}}{\mu_o B_o}\right) \qquad (6.9)$$

Combining Eqs. 6.7 and 6.8 and assuming $s = 0$ result in

$$J = \frac{0.00708 k_o h}{\mu_o B_o \left[\ln\left(\frac{r_e}{r_w} \right) - 0.75 \right]} \quad (6.10)$$

The following observations and conclusions can be drawn from Eq. 6.8:

- The PI rate solution gathers all of the Darcy components into a single value.
- When $\Delta p = 0$, the reservoir and flowing pressures are equal, and therefore $q = 0$. When $p_{wf} = 0$ and the pressure drop is at its maximum, $q = q_{max}$. The q_{max} term is defined as the theoretical maximum rate at which the well could produce. At initial conditions, the original reservoir pressure is p_i and $q = 0$. These two endpoints anchor the straight-line plot shown in **Fig. 6.3.**

6.2.2 Productivity Index Plot. This plot can be expressed in the form of a straight-line equation: $p_{wf} = p - q/J$. Plotting and outcome variables are defined in **Table 6.1.**

6.2.2.1 Test and Analysis Procedure. The well is shut in at the beginning of the test to attain static conditions. Tests are then conducted by means of opening the well and measuring several different rates at stabilized flowing bottomhole pressure combinations. Stabilized flow rate and bottomhole flowing pressure data are plotted as shown in Fig. 6.3. The slope of the interpreted best-fit straight line drawn through the data points determines the PI particular to that well.

6.2.3 Calculating q_{max} from a Single-Phase Well Test. Applying the similar-triangles concept (in Euclidean geometry, the corresponding sides of similar triangles always have the same ratio) provides a relationship between q_{max}, p_{res}, and a $q - p_{wf}$ plot, where q_{max} is the maximum flow rate at an assumed value of bottomhole flowing pressure, $p_{wf} = 0$, and $p_{res} = p$ = reservoir pressure.

$$\frac{q}{q_{max}} = \frac{p - p_{wf}}{p - 0} \text{ or } q_{max} = \frac{q}{\left(1 - \frac{p_{wf}}{p} \right)} \quad (6.11)$$

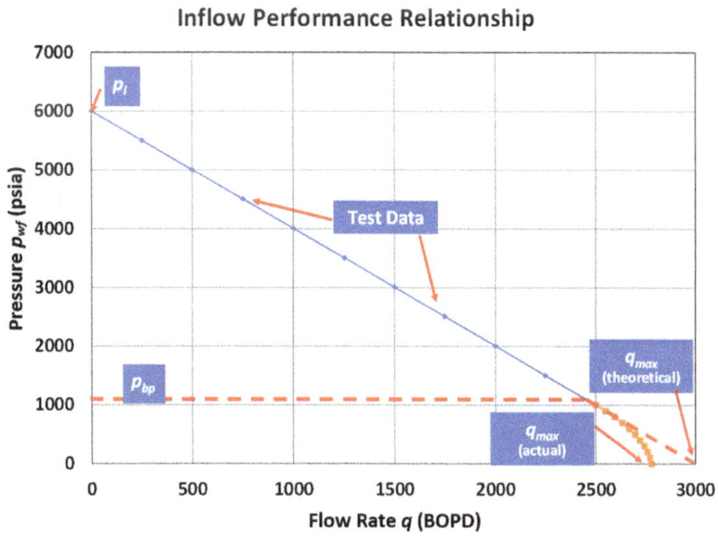

Fig. 6.3—The PI plot. The endpoints of the straight line are at ($p_i \rightarrow q = 0$) and ($p_{wf} = 0 \rightarrow q_{max}$).

Plotting Variables	Outcome Variables
x-axis: *q* (BOPD)	*y*-intercept: *p* (psia)
y-axis: p_{wf} (psia)	Slope of the line: −1/*J* (scf/psi)

Table 6.1—PI-plot plotting and outcome variables.

The relationship permits the calculation of any rate/bottomhole flowing pressure combination as long as q_{max} and a p_{res} value are known.

Substituting Eq. 6.11 into Eq. 6.9 results in Eq. 6.12:

$$q_{max} = Jp. \qquad (6.12)$$

Example 6.3—Calculating Flow Rate. Calculate the expected flow rate for the SL 1772 No. 21 well when conditions are p_{wf} = 1,000 psia and p_{res} = 2,000 psia. The well previously tested q = 65 BOPD and p_{wf} = 1,500 psia when p_{res} = = 2,000 psia.

Solution. Calculate the q_{max} term:

$$q_{max} = \frac{q}{\left(1 - \frac{p_{wf}}{\bar{p}}\right)} = \frac{65}{(1 - \frac{1{,}500}{2{,}000})} = 260 \text{ BOPD.}$$

Then, calculate the flow rate when p_{wf} =1,000 psia:

$$q = q_{max}\left(1 - \frac{p_{wf}}{p}\right) = 260\left(1 - \frac{1{,}000}{2{,}000}\right) = 130 \text{ BOPD.}$$

Thus, the expected flow rate for the SL 1772 No. 21 well is 130 BOPD.

Example 6.4—Evaluating Remediation Efforts. Two cases of tested wells are shown in **Fig. 6.4.** Case A averaged 100 BOPD before an acid treatment was performed, and the well tested 110 BOPD after the acid job. Compare this with Case B. Was the Case A remedial effort successful? How does it compare with Case B?

Solution. From Fig. 6.4, it can be observed that for Case B, an estimated flow rate of 130 BOPD is obtained after treatment, indicating a 30% increase. In comparison, Case A experienced only a 10% increase after treatment.

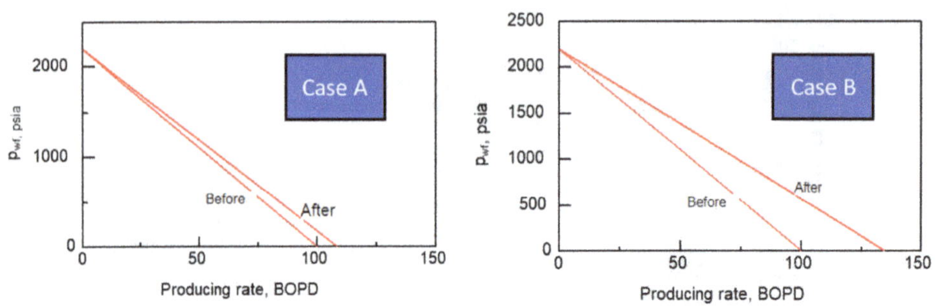

Fig. 6.4—Before and after PI tests. In Case A, on the left, the two tests essentially overlap. Therefore, the remedial efforts did not solve the problem. In Case B, on the right, there is a significant difference between the before and after q_{max} value; thus, the remedial work was deemed to be effective.

Example 6.5—Estimating the PI. A well with 80-acre spacing produces 137 BOPD at p_{wf} = 1,827 psia. Estimate the PI for the well and the formation permeability when the following properties are known: h = 10 ft, r_w = 0.25 ft, p_{res} = 2,000 psia, r_e = 1,053 ft, m_o = 0.5 cp, B_o = 1.5 RB/STB, and s = 0.

Solution. Include test data to calculate the PI:

$$J = \frac{q}{p_{res} - p_{wf}} = \frac{137}{2{,}000 - 1{,}827} = 0.79 \text{ BOPD/psi.}$$

Then, rearrange the pseudosteady-state flow equation (Eq. 6.7) to solve for permeability:

$$k_o = (J)\left\{\frac{\mu_o B_o\left[\ln(r_e - r_w) - 0.75 - s\right]}{0.00708h}\right\} = (0.79)\left\{\frac{(0.5)(1.5)\left[\ln\left(\frac{1{,}053}{0.25}\right) - 0.75\right]}{0.00708(10)}\right\} = 63 \text{ md.}$$

6.3 Inflow Performance below the Bubblepoint

The straight-line pressure-drawdown/rate relationship typically does not apply to two-phase flow conditions, which can occur when any portions of the reservoir are below the bubblepoint pressure. This problem is particularly prevalent in low-permeability, high gas/oil ratio wells because the presence of significant pressure drawdowns near the wellbore often reduces the reservoir pressure below the bubblepoint. Factors that can cause this inhibiting effect include

- An increasing gas saturation decreases the relative permeability to oil.
- A loss of solution gas increases the oil viscosity.
- A loss of solution gas causes shrinkage of the reservoir oil.
- An increasing gas saturation increases turbulence effects.
- Formation damage is present.

The total effect can reduce the effective permeability to oil in the portion of the reservoir located closest to the wellbore and introduce an apparent skin effect to the flow system. Recall that the pressure drop between reservoir boundaries is exaggerated close to the wellbore because the radius is presented as a logarithm in the Darcy equation.

Fig. 6.5a illustrates the effect of gas liberation on a pressure profile sketched between the inner and outer flow boundaries of an oil reservoir. The PI inflow performance relationship will appear as a straight line for single-phase conditions prevalent above the bubblepoint pressure. However, a decreased effective permeability to oil occurs as evolved gas enters the flow system. This new addition to the flow stream causes the pressure profile to diverge in a nonlinear manner. The location of the bubblepoint pressure in the profile defines the intersection of the parallel and divergent lines.

An equivalent inflow performance relationship curve is shown in Fig. 6.5b. An increasing pressure drop per unit of production causes the flow process to become inefficient. Therefore, the straight line becomes curved. The location of the bubblepoint pressure defines the point of inflection between the linear and curved conditions.

6.3.1 The Vogel Equation. Vogel (1968) performed simulation studies on oil reservoirs operating above and below the bubblepoint to develop a dimensionless relationship for two-phase oil and gas inflow. The assumptions are as follows:

- Reservoir pressure is less than or equal to the initial saturation pressure.
- There is two-phase and pseudosteady-state flow.

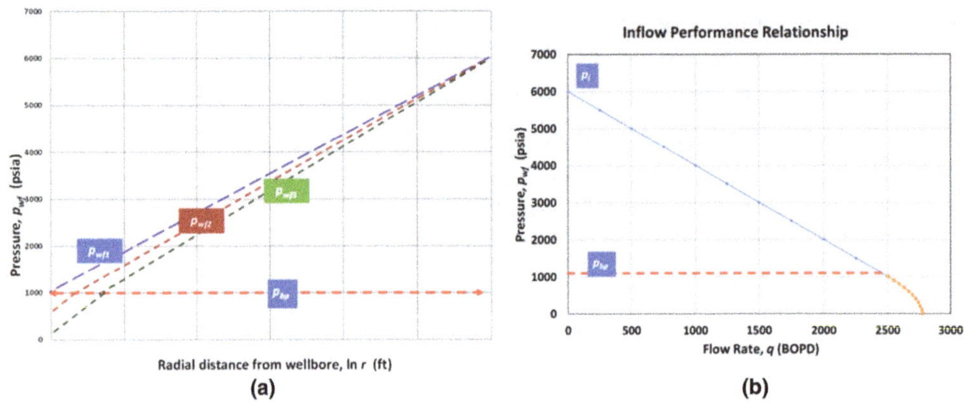

Fig. 6.5—The effect of bubblepoint pressure on the inflow performance profile. (a) Pressure profile as a function of radial distance from the wellbore, showing the effect of an additional pressure drop below the bubblepoint pressure and (b) pressure profile as a function of flow rate, showing the deviation effect from a straight line when pressure reaches the bubblepoint.

- Wells are neither damaged nor stimulated.
- Oil and gas saturations and pressure vary with distance from the wellbore.

The results of these efforts produced a quadratic relationship relating two-phase oil and gas flow to oil well inflow performance:

$$\frac{q}{q_{\max}}=1-0.2\left(\frac{p_{wf}}{p}\right)-0.8\left(\frac{p_{wf}}{p}\right)^2. \quad (6.13)$$

The Vogel equation contains four unknowns:

- q and p_{wf} data, which are obtained from a well test
- The static reservoir pressure p
- The single constant $q_{\max}$ term, which anchors one end of the curve at a maximum theoretical pressure drop

The Vogel relationship can be applied to damaged wells as long as the skin value does not change. Simulation studies by both Vogel (1968) and Klins and Majcher (1989) indicated that skin either adds or subtracts from the bottomhole flowing pressure in a constant manner. **Fig. 6.6** shows the effect of a skin factor that is not zero. Damage can cause a pressure drop that results from changes in the reservoir permeability around the bottom hole of the well (s = 8). However, stimulation or fracturing can improve the permeability in the wellbore (s = −5).

Some practical observations about the Vogel method:

- It is useful for two-phase but not three-phase flow.
- The $q_{\max}$ term is rate dependent to a certain degree.
- It has a 10–20% error rate, while the PI equation might have a 70–80% error rate.
- It can be used with wells producing up to a 40% water cut.
- It can be used for damaged wells when the mechanical skin does not change.

Example 6.6—Using the Vogel Equation. A well tested 65 BOPD at p_{wf} = 1,500 psia. Reservoir pressure p_{res} = 2,000 psia. Bubblepoint pressure p_{bp} = 1,000 psia. Positive apparent skin (s = 1.5)

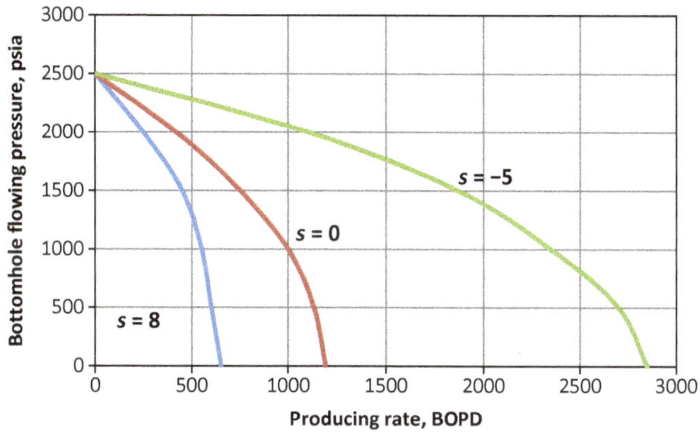

Fig. 6.6—Illustrating the effect of formation damage on the inflow performance profile.

indicates formation damage. Apply the Vogel equation (Eq. 6.13) to construct an inflow performance relationship curve.

Solution. First, rearrange the equation and use the test data to calculate q_{max}:

$$q_{max} = \frac{q}{1-0.2\left(\frac{p_{wf}}{p}\right)-0.8\left(\frac{p_{wf}}{p}\right)^2} = \frac{65}{1-0.2\left(\frac{1,500}{2,000}\right)-0.8\left(\frac{1,500}{2,000}\right)^2} = 163 \text{ BOPD}.$$

Then, construct an inflow performance relationship curve by assuming different p_{wf} values and calculating q.

Assume $p_{wf} = 500$ psia and calculate q:

$$\frac{q}{163} = 1-0.2\left(\frac{500}{2,000}\right)-0.8\left(\frac{500}{2,000}\right)^2 = 147 \text{ BOPD}.$$

Assume $p_{wf} = 1,000$ psia and calculate the equivalent flow rate q:

$$\frac{q}{163} = 1-0.2\left(\frac{1,000}{2,000}\right)-0.8\left(\frac{1,000}{2,000}\right)^2 = 114 \text{ BOPD}.$$

The calculated values are plotted in **Fig. 6.7,** which shows the performance of a well that has a positive skin (i.e., formation damage around the wellbore). It can also be observed that the behavior of the curve below the bubblepoint pressure indicates that the PI cannot be considered constant below the bubblepoint pressure.

In conclusion, it can be said that

- The Vogel equation accounts for the effects of two-phase flow and turbulence at bubblepoint conditions.
- Vogel calculations require well test and reservoir data, similar to the PI method.

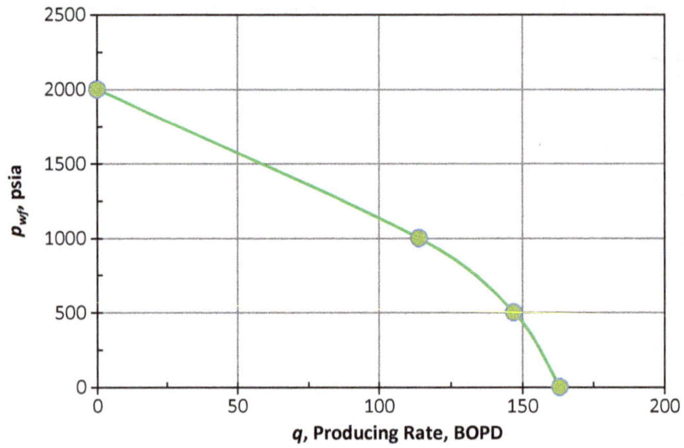

Fig. 6.7—The resulting inflow performance relationship curve for the example well. Note how the straight line bends when $p_{wf} < p_{bp}$ (in this case, at p = 1,000 psia) at high production rates.

6.3.2 Fetkovich Method. Fetkovich (1973) used well tests to develop an empirical method to properly study wells producing from saturated oil reservoirs. Production tests were conducted on saturated to highly undersaturated wells at a variety of rates. Reservoir permeability ranged from 6 to 1,000 md. The well tests indicated that wells flowing commingled gas and oil were better characterized using the gas-well backpressure test equation than the single-phase flow PI equation (Eq. 6.13).

Fetkovich (1973) rearranged the gas-well test equation to develop an empirical oilwell inflow performance equation producing at two-phase flow conditions below the bubblepoint:

$$q = J'\left(p_{\text{res}}^2 - p_{wf}^2\right)^n, \quad (6.14)$$

where J' is the PI for a well producing below the bubblepoint and n is the turbulence term, which should vary from 0.5 to 1.0.

Taking logs of Eq. 6.14, $\ln q = \ln J' + n\ln\left(\bar{p}^2 - p_{wf}^2\right)$, and reversing the axis results in **Fig. 6.8.** The slope of the line is $1/n$ and the intercept is.

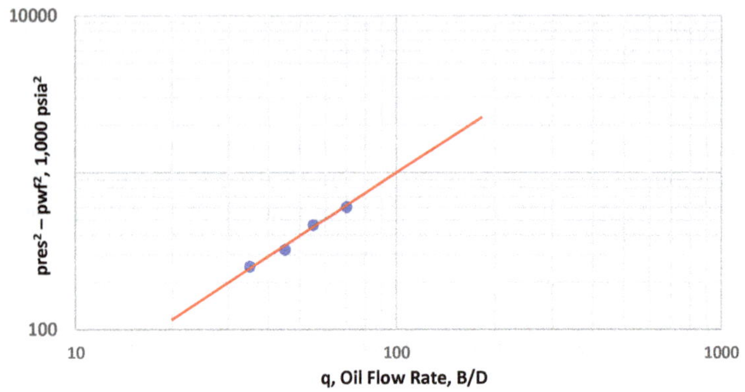

Fig. 6.8—The Fetkovich backpressure plot for oil wells.

A relation between q_{max} and a lesser flow rate can be developed:

$$\frac{q}{q_{max}} = \frac{J'(\bar{p}^2 - p_{wf}^2)^n}{J'(\bar{p}^2 - 0)^n} \text{ or } \frac{q}{q_{max}} = \left(1 - \frac{p_{wf}^2}{\bar{p}^2}\right)^n \quad (6.15)$$

Eq. 6.14 can be expressed in terms of q_{max}:

$$q_{max} = J' \times (\bar{p}^2)^n \quad (6.16)$$

The analysis procedure is as follows:

- Apply the results of a four-point backpressure test to calculate the inverse slope of n and J'.
- Include these two values in the backpressure equation (Eq. 6.14) to calculate the flow characteristics for the well.

Example 6.7—Using the Fetkovich Method. A four-point backpressure test for the Kent well is available. The test information is listed in **Table 6.2.** Apply the Fetkovich method to determine the inflow performance equation for the current reservoir pressure p_{res} = = 1,345 psia.

Solution. First, use the test data to calculate $^2 - p_{wf}^2$, which is the pressure difference. The answers are given in **Table 6.3** and are plotted and interpreted in **Fig. 6.9.**

$$\text{Slope} = \frac{\log 3{,}300 - \log 220}{\log 1{,}000 - \log 100} = \frac{3.48 - 2.34}{3 - 2} = 1.18,\ n = \frac{1}{m} = 0.85.$$

Next, rearrange Eq. 6.14 to determine the J' and q_{max} terms using test data specified in Table 6.3:

$$J' = \frac{q}{(\bar{p}^2 - p_{wf}^2)^n} = \frac{342}{(961{,}000)^{0.85}} = 0.0028.$$

$$q_{max} = J' \times (\bar{p}^2)^n = 0.0028 \times (1{,}345^2)^{0.85} = 586 \text{ BOPD}.$$

Then, calculate the inflow performance equation for the Kent well:

$$q = J'(p_{res}^2 - p_{wf}^2)^n = 0.0028(1{,}345^2 - p_{wf}{}^2)^{0.85}.$$

Rate (BOPD)	BHFP (psia)
215	1,123
342	921
450	719
477	638

Table 6.2—Test data for the Kent well. BHFP = bottomhole flowing pressure.

Rate (BOPD)	BHFP (psia)	$p^{-2} - p_{wf}^2$ (1,000 psia²)
215	1,123	548
342	921	961
450	719	1,292
477	638	1,402

Table 6.3—Data reduction for the Kent well test. BHFP = bottomhole flowing pressure.

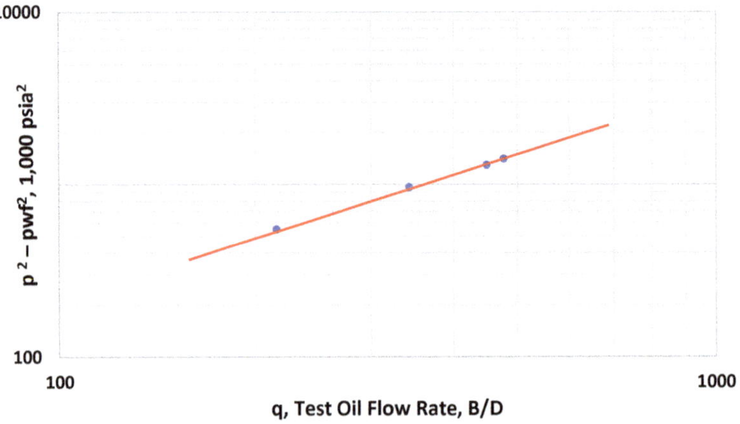

Fig. 6.9—Backpressure plot for the Kent well. The slope of the log-log plot = 1.18.

6.4 Inflow Performance for Damaged Wells

The effect of wellbore damage or stimulation close to the wellbore is included in flow equations as the skin effect term s. The skin effect is assumed to be a function of only the mechanical change in permeability around the wellbore.

The basic definition for skin (Craft and Hawkins 1959) is

$$s=\left(\frac{k-k_s}{k_s}\right)\ln\frac{r_s}{r_w} \quad (6.17)$$

Eq. 6.17 indicates that stimulated wells exhibit a negative skin ($-s$) while wells suffering formation damage near the wellbore exhibit a positive skin ($+s$). Therefore, for the constant rate solution, the following relationships hold:

- Unaffected wellbore: $s = 0$, $k_s = k_r = k$, $s = 0$, and $p_{wf} = p_{wfr}$.
- Stimulation around the wellbore annulus: $s < 0$, $k_s > k_r$, and p'_{wf} (also termed p_{wfs}) $< p_{wf}$; radius of skin is r_s.
- Damage around the wellbore annulus: $s > 0$, $k_s < k_r$, and $p'_{wf} > p_{wf}$; radius of skin is r_s.

Formation-damaged wells are not able to produce at a rate equal to that of an unaffected well. A flow efficiency term FE, which is also called E_f in the literature, should be defined to account for the flow rate differences caused by altering the formation permeability near the wellbore. An ideal drawdown occurs when the pressure drop incurred near the wellbore is not affected by changes in the reservoir permeability and in the wellbore, and reservoir permeability is considered uniform while the actual drawdown occurs at the test rate.

$$FE=E_f=\frac{\text{Ideal drawdown}}{\text{Actual drawdown}}=\frac{p-p'_{wf}}{p-p_{wf}} \quad (6.18)$$

6.4.1 Standing Method. Standing (1970) performed essentially the same simulation studies as those that were used in the development of the Vogel equation. However, these studies were conducted on damaged and stimulated wells. **Fig. 6.10** presents the results of the Standing (1970) studies.

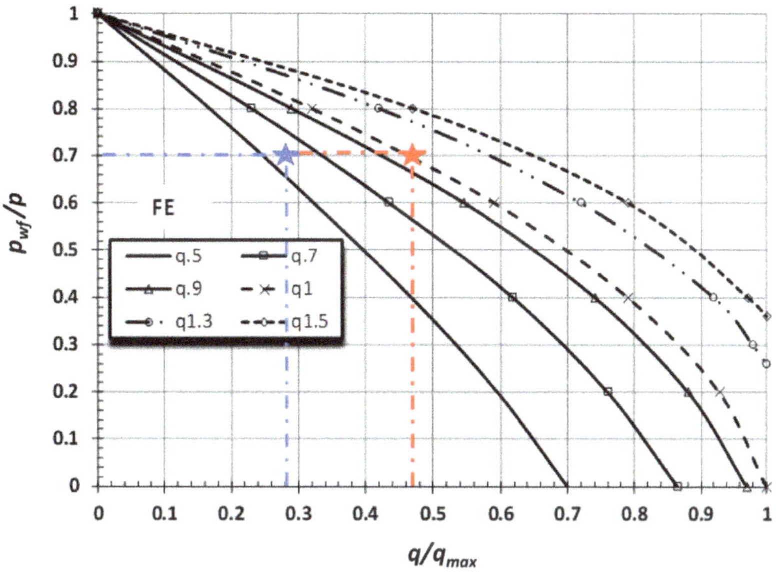

Fig. 6.10—The Standing chart (after Standing 1970).

The dimensionless pressure and rate terms are exactly the same as those presented by Vogel (1968). In Standing (1970), the q_{max} term was evaluated at FE = 1. The manipulation of the data allowed Standing (1970) to include all variables in a single figure. These conclusions might be incorrect at high flow efficiencies and at high rates. The objective is to adjust the damaged q_{max} value to q_{max} when FE = 1.

The analysis procedure is as follows:

- Estimate the change in productivity caused by formation damage using a bottomhole flowing pressure/producing rate flow test. Data needed to perform the calculations:
  - Tests from a damaged and an undamaged well
  - Reservoir pressure
- Use the difference in drawdown pressures between the damaged and undamaged wells to calculate the flow efficiency (Eq. 6.18).
- Use the Standing (1970) chart (Fig. 6.10) to determine the equivalent (q/q_{max}) value, which in turn allows one to calculate q_{max}. Recall that q_{max} is defined when FE = 1.
- Use the p_{wf} value obtained with the initial test to obtain a new p_{wf}/p value, and use the FE = 1 value to obtain q/q_{max}. Calculate the new rate.

Example 6.8—Evaluating Damaged Wells. A well suffered formation damage during completion. p_{res} = 1,000 psia. What would the new flow rate be if the damage were removed? The following test results are available:

- The damaged well tested 375 BOPD with a p_{wf} of 700 psia.
- Normal wells test 375 BOPD with a p_{wf} of 820 psia.

Solution. First, calculate the flow efficiency:

$$FE = E_f = \frac{\text{Ideal drawdown}}{\text{Actual drawdown}} = \frac{p - p'_{wf}}{p - p_{wf}} = \frac{1{,}000 - 820}{1{,}000 - 700} = 0.6.$$

When $\frac{p_{wf}}{p} = \frac{700}{1{,}000} = 0.7, \frac{q}{q_{max}} = 0.29$ can be read from the Standing (1970) chart (blue star in Fig. 6.10).

Next, calculate q_{max} using the $\frac{q}{q_{max}}$ value read from the blue star in Fig. 6.10 at $FE = 1$:

$$q_{max} = \frac{375}{0.29} = 1{,}293 \text{ BOPD at } FE = 1.$$

Then, calculate the new flow rate q at the $FE = 1$ value. When $\frac{p_{wf}}{p} = \frac{700}{1{,}000} = 0.7$ from Fig. 6.10 at $FE = 1$, $\frac{q}{q_{max}} = 0.47$ can be read from the Standing (1970) chart (red star).

$$q = (1{,}293 \text{ BOPD})(0.47) = 608 \text{ BOPD}.$$

The calculations indicate that investing in a workover to remedy the well formation damage will increase production from 375 BOPD to 608 BOPD.

6.4.2 Camacho-V. and Raghavan Method. The Standing method assumed that flow efficiency was a linear difference between damaged and undamaged drawdown. Camacho-V. and Raghavan (1989) applied basic principles and simulation modeling to show that the application of a linear equation to the Vogel quadratic equation (Eq. 6.13) can result in an inaccurate relationship.

A new quadratic expression, shown as Eq. 6.19, defined an alternative flow efficiency term E_f (also called FE in the literature). Simulation studies showed that the Vogel (Eq. 6.13) and Standing graphical solutions produced slightly different inflow curves when compared to the new definition. An adjustment term V was introduced to permit the use of the flow efficiency term E_f in both cases.

$$E_f = FE = \frac{\left(1 + \frac{Vp'_{wf}}{p}\right)\left(1 - \frac{p'_{wf}}{p}\right)}{\left(1 + \frac{Vp_{wf}}{p}\right)\left(1 - \frac{p_{wf}}{p}\right)} = \frac{\text{Ideal conditions}}{\text{Actual conditions}}, \quad \text{(6.19)}$$

where $V = 0.8$ for Vogel and $V = 1.0$ for Fetkovich and $p'_{wf} = p_{wf}$ when the well is undamaged. Recall the Vogel equation (Eq. 6.13),

$$\frac{q}{q_{max}} = 1 - 0.2\left(\frac{p_{wf}}{p}\right) - 0.8\left(\frac{p_{wf}}{p}\right)^2,$$

and the Fetkovich backpressure relationship (Eq. 6.15),

$$\frac{q}{q_{max}} = \left(1 - \frac{p_{wf}^2}{\overline{p}^2}\right)^n.$$

Well-test data can be entered into Eq. 6.19 to calculate q_{max} or to calculate a new p_{wf} vs. rate relationship:

$$\frac{q}{q_{max}} = E_f\left[1 + V\left(\frac{p_{wf}}{p}\right)\right]\left[1 - V\left(\frac{p_{wf}}{p}\right)\right] \quad \text{(6.20)}$$

Eq. 6.20 can then be included in either the Vogel or Fetkovich equation (Eq. 6.13 or 6.15, respectively) to calculate the inflow performance relationship if the adjustment term V is included in the calculation process.
The analysis procedure is as follows:

- Obtain tests for the damaged well and a comparable undamaged well. An estimated value of reservoir pressure is essential to perform the calculations.
- Determine the correct V term to apply in the calculation procedure.
- Calculate E_f using Eq. 6.19.
- Calculate q_{max} at the test conditions.
- Use either the Vogel or the Fetkovich equation to calculate the remedial inflow performance.

Example 6.9—Evaluating Damaged Wells with the Vogel Equation. A well suffered formation damage during completion when p_{res} = 1,000 psia. What would the new flow rate be if the damage were removed? Apply the Vogel equation (Eq. 6.13) in the calculation procedure. The following test results are available:

- The damaged well tested 375 BOPD with a p_{wf} of 700 psia.
- Normal wells test 375 BOPD with a p_{wf} of 820 psia.

Solution. First, select $V = 0.8$ because the Vogel equation will be applied to the solution:

$$E_f = FE = \frac{\left(1+\frac{Vp'_{wf}}{p}\right)\left(1-\frac{p'_{wf}}{p}\right)}{\left(1+\frac{Vp_{wf}}{p}\right)\left(1-\frac{p_{wf}}{p}\right)} = \frac{\left[1+\frac{(0.8)(820)}{1,000}\right]\left(1-\frac{820}{1,000}\right)}{\left[1+\frac{(0.8)(700)}{1,000}\right]\left(1-\frac{700}{1,000}\right)} = 0.64.$$

Next, apply the modified Vogel equation (Eq. 6.20) to calculate the q_{max} term:

$$\frac{q}{q_{max}} = E_f\left[1+V\left(\frac{p_{wf}}{p}\right)\right]\left[1-V\left(\frac{p_{wf}}{p}\right)\right];$$

$$\frac{375}{q_{max}} = 0.64\left[1+0.8\left(\frac{700}{1,000}\right)\right]\left[1-0.8\left(\frac{700}{1,000}\right)\right];\ q_{max} = 854 \text{ BOPD}.$$

Then, apply the Vogel equation to calculate the rate when p_{wf} = 700 psia:

$$\frac{q}{q_{max}} = 1-0.2\left(\frac{p_{wf}}{p}\right)-0.8\left(\frac{p_{wf}}{p}\right)^2;\ \frac{q}{854} = 1-0.2\left(\frac{820}{1,000}\right)-0.8\left(\frac{820}{1,000}\right)^2. = 400 \text{ BOPD}.$$

6.5 Predicting Future Performance

The inflow performance relationships discussed up to this point have been concerned with determining the effect of changing bottomhole flowing pressure on the producing rate at some constant reservoir pressure. However, declining reservoir pressure changes the oil mobility term, which in turn can affect the well performance coefficient.

Changing oilwell inflow performance relations as pressure declines are primarily caused by the increasing presence of free gas in the reservoir. Changing gas/liquid relationships affect the relative permeability to oil and the oil viscosity and oil formation volume factor values.

Recall that the PI, Vogel, and Fetkovich inflow performance relationships relate changing bottom-hole flowing pressure to the producing rate at some reservoir pressure. Future performance models must account for the effect of declining reservoir pressure on fluid property relationships and how that effect alters performance.

The oil mobility term λ_o can be defined as the ratio of oil effective permeability to the product of oil viscosity and the oil-phase formation volume factor:

$$\lambda_o = \frac{k_o}{\mu_o B_o}, \quad (6.21)$$

where k_o is the effective permeability to oil, which decreases as a gas saturation is established; μ_o is the oil viscosity, which increases as gas evolves from the reservoir oil; and B_o is the oil formation volume factor, which decreases as gas evolves from the reservoir oil, causing the volume to shrink.

These effects should be taken into account when assessing the future, with the assumption that

$\lambda_o = \dfrac{k_o}{\mu_o B_o}$ is a linear function of p and $p_e \approx p_i$.

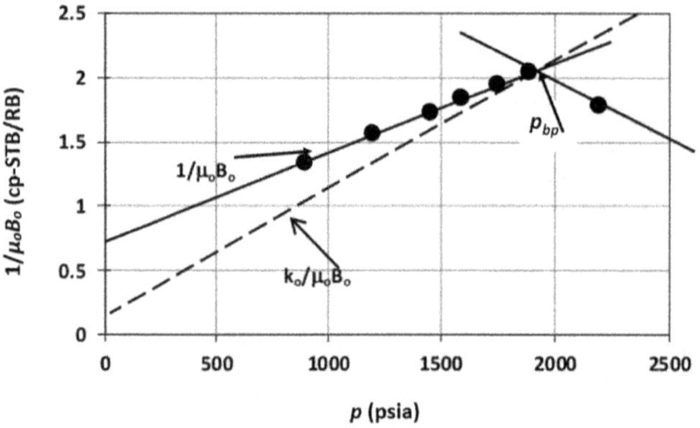

Fig. 6.11—Adapting the straight-line approach both above and below bubblepoint conditions to changes in the mobility term.

Fig. 6.11 illustrates how the evolution of solution gas changes the mobility term. Straight lines intersect at the bubblepoint pressure, with each representing a different mobility response. The solid lines represent the mobility with no free gas present. Because gas saturation is based on fluid properties, the dashed line represents an approximation of the mobility term, which intersects the y-axis near zero. Note that the divergence between the two lines increases at very low pressure, near exhaustion of the property.

A useful prediction method must account for changes in the mobility term as a function of declining reservoir pressure. The methods that follow circumvent the need to apply Darcy or diffusivity-based flow equations by substituting well-test data to provide a reasonable model for predicting future production for undersaturated oil reservoirs.

6.5.1 Fetkovich Method. Fetkovich (1973) applied the results of an earlier study by Evinger and Muskat (1942) to develop a method for tracking the well mobility term with a change in pressure. Recall that the term is the non-Darcy PI when operating below the bubblepoint.

Eq. 6.22 represents a steady-state, radial, multiphase model. Notice that the oil mobility term remains inside the integral, expressed using the relative permeability definition $k_{ro} = k_o / k$ and the effective permeability term $k_o = kk_{ro}$.

$$q = \frac{0.00708kh}{\ln r_e / r_w} \int_{p_{wf}}^{p_e} \frac{k_{ro}}{\mu_o B_o} dp \quad (6.22)$$

The integral can be divided into above and below the bubblepoint areas:

$$q=\frac{0.00708kh}{\ln r_e / r_w}\left(\int_{p_{wf}}^{p_{bp}}\frac{k_{ro}}{\mu_o B_o}\mathrm{d}p+\int_{p_{bp}}^{p_e}\frac{k_{ro}}{\mu_o B_o}\mathrm{d}p\right) \quad (6.23)$$

The integrals can be divided into below the bubblepoint in the two-phase region,

$$\int_{p_{wf}}^{p_{bp}}\frac{k_{ro}}{\mu_o B_o}\mathrm{d}p\text{''} \quad (6.24a)$$

and above the bubblepoint in the single-phase region,

$$\int_{p_{bp}}^{p_e}\frac{k_{ro}}{\mu_o B_o}\mathrm{d}p. \quad (6.24b)$$

Integration of Eqs. 6.24a and 6.24b results in

$$q=J'\left(p_{bp}^2-p_{wf}^2\right)^n+J\left(\bar{p}-p_{bp}\right), \quad (6.25)$$

with the assumption that $\bar{p}=p_e$ and where J is the single-phase PI operating above the bubblepoint pressure and J' is the two-phase non-Darcy PI operating below the bubblepoint pressure.

6.5.2 Relating Mobility to Pressure Change. Fetkovich (1973) demonstrated that the mobility term can be assumed to change in a linear manner with a change in pressure:

$$\frac{\left(\dfrac{k_{ro}}{\mu_o B_o}\right)_{p_{test}}}{\left(\dfrac{k_{ro}}{\mu_o B_o}\right)_{p_e}}=p_{test}/p_e. \quad (6.26)$$

The two-phase portion of Eq. 6.14, $q=J'\left(p_{res}^2-p_{wf}^2\right)^n$, developed by Fetkovich (1973), is included to produce the following equation:

$$q=J'\frac{\left(\dfrac{k_{ro}}{\mu_o B_o}\right)_{p_{test}}}{\left(\dfrac{k_{ro}}{\mu_o B_o}\right)_{p_e}}\left(p_{res}^2-p_{wf}^2\right)^n=(p_{test}/p_e)J'\left(p_{res}^2-p_{wf}^2\right)^n. \quad (6.27)$$

The analysis procedure is as follows:

- Obtain a backpressure test and use the four-point backpressure analysis procedure to obtain the well-description coefficients J' and n.
- Use the equation $q=J'\left(p_{bp}^2-p_{wf}^2\right)^n$.
- Construct inflow performance relationships as a function of changing reservoir pressure.

Example 6.10—Constructing Inflow Performance Relationship Curves. Recall Example 6.7, regarding the Kent well. $\bar{p} = p_{res} = 1{,}345$ psia.

Complete the following tasks:

1. Prepare inflow performance relationship curves for different shut-in future static pressures: 1,200, 1,000, and 800 psia.
2. Plot the calculated flow rate as a function of declining reservoir pressure at a constant bottomhole flowing pressure. $p_{wf} = 791$ psia.

Solution.

1. The general inflow equation for the Kent well modified to account for changing reservoir pressure below the bubblepoint is

$$J' = \frac{q}{\left(\bar{p}^2 - p_{wf}^2\right)^n} = \frac{342}{(961{,}000)^{0.85}} = 0.0028.$$

$$q = \frac{p_{test}}{p} J'\left(p^2 - p_{wf}^2\right)^n = \frac{p_{test}}{p} 0.0028\left(1{,}345^2 - p_{wf}{}^2\right)^{0.85}.$$

Rewrite the equation for $p = 1{,}200$ psia:

$$q = \frac{1{,}200}{1{,}345} 0.0028\left(1{,}200^2 - p_{wf}{}^2\right)^{0.85} = 0.0025\left(1{,}200^2 - p_{wf}{}^2\right)^{0.85}.$$

The inflow performance relationship data provided in **Table 6.4** were generated for the three reservoir pressure profiles using information provided in Table 6.2. **Fig. 6.12** illustrates the shapes of the three inflow performance relationship plots. Note the nonlinearity of the curve accentuated at high producing rates.

2. Using the calculated values from Table 6.4 to plot **Fig. 6.13,** it can be observed how declining reservoir pressure affects well production when the bottomhole flowing pressure is maintained constant at 719 psi.

6.5.3 Klins and Clark Method. Klins and Majcher (1989), in addition to others, showed how the bubblepoint pressure and the extent of reservoir depletion can significantly affect the shape of the inflow curve.

$p = 1{,}200$ psia		$p = 1{,}000$ psia		$p = 800$ psia	
p_{wf} (psia)	Rate (BOPD)	p_{wf} (psia)	Rate (BOPD)	p_{wf} (psia)	Rate (BOPD)
1,123	73				
921	202	921	53		
719	295	719	141	719	35
638	325	638	168	638	61
0	431	0	262	0	144

Table 6.4—Data reduction for the Kent well test for the three reservoir pressure profiles.

These studies reinforced the Camacho-V. and Raghavan (1989) concept of skin. The generation of a gas saturation effect close to the wellbore as a function of declining pressure was shown to have a particularly negative effect on the production rate.

The inflow performance relationship curve becomes more accurate as the two-phase flow effect becomes more pronounced with increasing bubblepoint pressure. However, the inflow performance relationship curve becomes more similar to the PI curves when single-phase flow conditions predominate at lower bubblepoint pressures **(Fig. 6.14).**

Production tests typically indicate $n < 1$. Camacho-V. and Raghavan (1989) showed that the difference between the test and simulated values was more than likely caused by the buildup of a gas saturation around the wellbore. An increased gas saturation decreases the relative permeability to oil and consequently reduces the oil producing rate.

As a matter of interest, applying the Vogel equation to the results of the simulation studies resulted in a value of $n = 1.24$. Camacho-V. and Raghavan (1989) also showed that increasing system compressibility in a pressure-depleting reservoir causes the exponent term to change. They also demonstrated that the exponent does not decline monotonically and $n > 1$ when the effect of near-wellbore gas saturation is accounted for. A value of $n < 1$ generally reflects the presence of a low gas saturation value.

Klins and Clark (1993) studied different solution-gas-drive

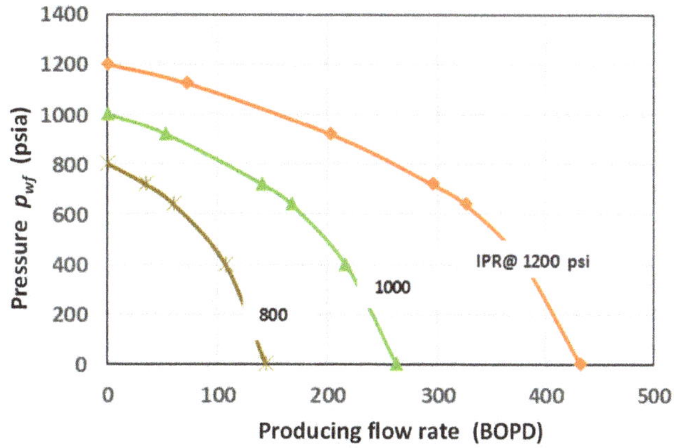

Fig. 6.12—Inflow performance relationship (IPR) plots for the Kent well. Note how the method accounts for the nonparallelism of the performance curve by adjusting for changing mobility.

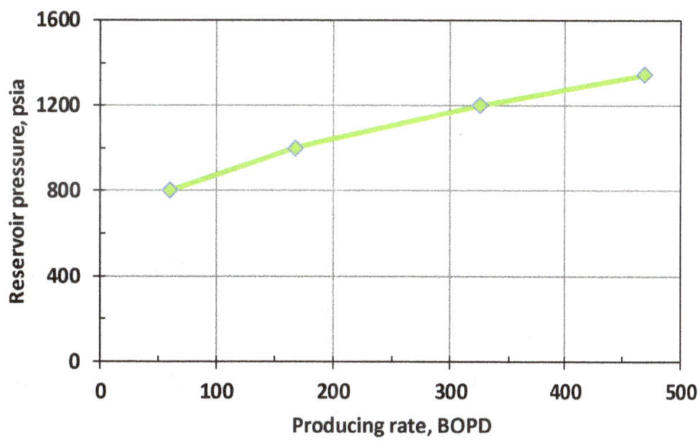

Fig. 6.13—Illustrating the nonlinear decline of the production rate as a result of the changing mobility factor in an undersaturated reservoir.

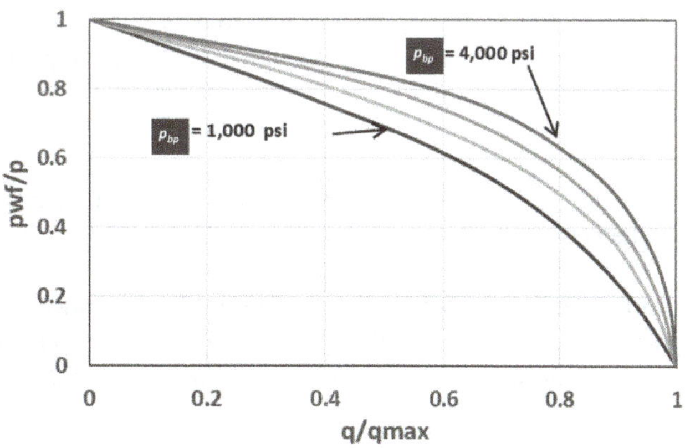

Fig. 6.14—Increasing dissolved gas increases nonlinear flow as the pressure is reduced (after Klins and Majcher 1989).

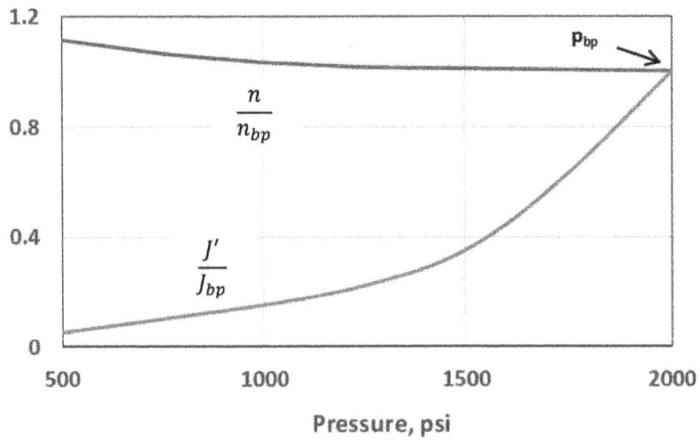

Fig. 6.15—The Klins and Clark (1993) variation of the exponent *n* and well coefficient as functions of declining reservoir pressure.

reservoirs to extend the Fetkovich concept. The results of these studies are presented in **Fig. 6.15** and show trends of and n as functions of reservoir pressure depletion.

Note that in Fig. 6.15, the dimensionless n term increased by approximately 10% and the dimensionless PI term could decrease by 100%.

The equations of the lines of these curves are shown in Eqs. 6.28 and 6.29. Note that the bubblepoint pressure p_{bp} now forms the anchor to the curve.

$$\frac{n}{n_{bp}}=1+0.06\left(1-\frac{p}{p_{bp}}\right)-0.25\left(1-\frac{p}{p_{bp}}\right)^2+0.50\left(1-\frac{p}{p_{bp}}\right)^3, \quad \text{(6.28)}$$

$$\frac{J'}{J_{bp}}=1-3.57\left(1-\frac{p}{p_{bp}}\right)+4.80\left(1-\frac{p}{p_{bp}}\right)^2-2.31\left(1-\frac{p}{p_{bp}}\right)^3. \quad \text{(6.29)}$$

The solution procedure is similar to the Vogel inflow performance evaluation method except that instead of determining the value of $q_{\max}$, it is the n and J_{bp} values at the bubblepoint pressure p_{bp} that must be determined.

These values form the basis for estimating the needed values n and J' to calculate the new rate at the reservoir pressure using Eq. 6.20.

The analysis procedure is as follows:

- Obtain a backpressure test and use the four-point backpressure analysis procedure to obtain the well-description coefficients J' and n.
- Use the equation $q=J'\left(p_{bp}^2-p_{wf}^2\right)^n$.
- Use a fluid analysis or correlation to determine the bubblepoint pressure. Include the n and J' values determined from the initial well test and p_{bp} to calculate n_{bp} and J_{bp} at bubblepoint conditions.
- Use Eq. 6.14, $q=J'\left(p_{\text{res}}^2-p_{wf}^2\right)^n$, to calculate the n and J' terms that are appropriate for any reservoir pressure.
- Apply the Fetkovich backpressure equation (Eq. 6.15) to calculate future performance using the new n and J' terms.

Example 6.11—Determining the Inflow Performance Equation. A four-point backpressure test was performed on the Willis No. 2 well. p_{bp} = 2,000 psia. Obtained values of n = 0.85 and J' = 0.0048 are specified in the following inflow performance equation:

$$q=0.0048\left(1{,}600^2-p_{wf}^2\right)^{0.85}.$$

Determine the inflow performance equation when the reservoir pressure has declined to $\bar{p} = p_{res} = 1{,}450$ psia.

Solution. First, calculate n_{bp} and J_{bp}:

$$\frac{n}{n_{bp}} = 1 + 0.06\left(1 - \frac{p}{p_{bp}}\right) - 0.25\left(1 - \frac{p}{p_{bp}}\right)^2 + 0.50\left(1 - \frac{p}{p_{bp}}\right)^3;$$

$$\frac{0.85}{n_{bp}} = 1 + 0.06\left(1 - \frac{1{,}600}{2{,}000}\right) - 0.25\left(1 - \frac{1{,}600}{2{,}000}\right)^2 + 0.50\left(1 - \frac{1{,}600}{2{,}000}\right)^3;\ n_{bp} = 0.8449.$$

$$\frac{J'}{J_{bp}} = 1 - 3.57\left(1 - \frac{p}{p_{bp}}\right) + 4.80\left(1 - \frac{p}{p_{bp}}\right)^2 - 2.31\left(1 - \frac{p}{p_{bp}}\right)^3;$$

$$\frac{0.0048}{J_{bp}} = 1 - 3.57\left(1 - \frac{1{,}600}{2{,}000}\right) + 4.80\left(1 - \frac{1{,}600}{2{,}000}\right)^2 - 2.31\left(1 - \frac{1{,}600}{2{,}000}\right)^3;\ J_{bp} = 0.0104.$$

Then, calculate n and J' for $\bar{p} = p_{res} = 1{,}450$ psia:

$$\frac{n}{0.8449} = 1 + 0.06\left(1 - \frac{1{,}450}{2{,}000}\right) - 0.25\left(1 - \frac{1{,}450}{2{,}000}\right)^2 + 0.50\left(1 - \frac{1{,}450}{2{,}000}\right)^3;\ n = 0.8517.$$

$$\frac{J'}{0.0104} = 1 - 3.57\left(1 - \frac{1{,}450}{2{,}000}\right) + 4.80\left(1 - \frac{1{,}450}{2{,}000}\right)^2 - 2.31\left(1 - \frac{1{,}450}{2{,}000}\right)^3;\ J' = 0.003465.$$

Thus, the rate equation for the Willis No. 2 well at 1,450 psia is

$$q = 0.003465\left(1{,}450^2 - p_{wf}^2\right)^{0.8517}.$$

6.6 Summary

Well producing rates are a function of both subsurface conditions and mechanical restrictions located in the tubing or wellhead. Controlling the surface pressure with chokes is the only option for monitoring well flow because the majority of flow variables are uncontrollable. Hydrostatic head defines the relationship between the backpressure at the wellhead and the bottomhole flowing pressure in the production tubing. Restricting the well production rate provides a measure of integrity for controlling the production of high-pressure reservoir fluid and is universally applied throughout the industry.

The PI method for monitoring single-phase Darcy flow in tubing provides the foundation for studying fluid flow through production tubing. Theoretically, plotting bottomhole flowing pressure vs. flow rate should result in a straight line for this condition. The effects of multiphase flow, mechanical restrictions in the tubing, changing pressure profiles, and establishing a wellbore skin can affect the production rate. Pressure/rate graphs follow a curved profile modeled by a quadratic equation for these cases.

This chapter reviews methods to predict how changes in a current well setup can affect performance predictions. For instance, what would be the new flow rate if the skin damage were removed and the well were produced unimpeded? Predicting performance for the following cases has been discussed:

- PI—single-phase flow—Darcy equation
- Vogel equation—accounts for the effect of simultaneous oil and free-gas flow
- Camacho-V. and Raghavan method—adjusts for the effects of formation damage
- Fetkovich concepts—predict future performance as a function of reservoir pressure decline

6.7 Questions

- Why is the PI method theoretically incorrect for the study of oil- and gas-well rate changes? (Answer: The effects of turbulence and two-phase flow are not considered. The PI method is theoretically correct for laminar, single-phase flow.)
- Why do flow studies of damaged wells compare their performance characteristics to those of an undamaged well? (Answer: A mathematical comparison between the two allows one to estimate the production increase if the damage were removed.)
- Into what two segments did Fetkovich divide well flow? (Answer: He divided well flow into conditions above the bubblepoint and below the bubblepoint.)
- Which three variables did Fetkovich assume have straight-line declines to reservoir exhaustion? (Answer: Oil viscosity, the formation volume factor, and relative permeability to oil.)

6.8 Nomenclature

A = choke cross-sectional area, ft^2
B = formation volume factor, RB/STB
B_o = oil formation volume factor, RB/STB
B_{ot} = oil formation volume factor at upstream T and p, bbl/STB or RB/STB
C = coefficient particular to a field (typically equal to 600)
E_f or FE = flow efficiency (ideal conditions/actual conditions), fraction
h = eight or thickness, ft
J = productivity index, BOPD/psi
= productivity index intercept, BOPD/psi
J_{bp} = productivity index at bubblepoint pressure, BOPD/psi
k = permeability, md
k_o = effective oil permeability, md
k_r = relative permeability, md
k_{ro} = relative permeability to oil, md
k_s = altered permeability as a result of skin effect, md
m = gas reservoir pore volume/oil reservoir pore volume (can also be defined as slope of straight line), fraction
m_1 = Poettmann and Beck method grouping constant term
M = gas molecular weight, lbm/lbm-mol
n = turbulence, also defined in the Fetkovich method as an exponent term, dimensionless
n_{bp} = Klins and Clark method well-description coefficientexponent term, dimensionless
p = pressure, psia
$\bar{p}$ = average pressure, psia
p_{bp} = bubblepoint pressure, psia
p_e = pressure at drainage radius of well, psia
p_i = initial pressure, psia
p_{pc} = pseudocritical pressure, psia
p_{pr} = pseudoreduced pressure, psia
p_{res} = reservoir pressure, psia
p_t = flowing tubinghead pressure, psia
p_{test} = pressure measured at well testing conditions, psia
p_{wf} = bottomhole flowing pressure, psia
p'_{wf} or p_{wfs} = bottomhole flowing pressure when there is stimulation or damage (altered permeability) around wellbore, psia
p_{wf1}, p_{wf2}, and p_{wf3} = bottomhole flowing pressure at different times, psia
p_{wfr} = p_{wf} = bottomhole flowing pressure at original conditions, psia
p_{wfs} or p'_{wf} = bottomhole flowing pressure when there is stimulation or damage (altered permeability) around wellbore, psia
q = flow rate, STB/D, BOPD, or B/D
q_{max} = theoretical maximum production rate, BOPD
q_o = oil flow rate, STB/D or BOPD
q_{omax} = theoretical maximum oil production rate, BOPD
q_t = total fluids flow rate (oil plus water), BFPD, STB/D, or BOPD
q_w = water flow rate, BWPD
r = radius, ft
r_e = radius to edge of reservoir, ft
r_s = radius of skin, ft
r_w = radius to edge of borehole, ft
R = gas/liquid ratio, Mscf/bbl or scf/STB
R = universal gas constant in , with a value of 8.314 J/mol-K or 10.730 psia ft^3/lbm-mol °R
R_s = solution gas/oil ratio, scf/STB
R_{sb} = solution gas/oil ratio at bubblepoint pressure, scf/STB
R_{st} = gas solubility at upstream T and p, scf/STB

R_1 = Poettmann and Beck method grouping constant term
s = skin, dimensionless
S = choke size expressed in 1/64 in.
T = temperature, °F
T_{pc} = pseudocritical temperature, °R
T_{pr} = pseudoreduced temperature, dimensionless
T_{res} = reservoir temperature, °F
T_t = upstream temperature, °R
V = adjustment term
V_1 = Poettmann and Beck method grouping constant term
z = z-factor, dimensionless. Corrections (z-factor) must be made to account for the deviation of real-gas behavior from that of the ideal gas
z_t = gas compressibility factor at upstream T and p, dimensionless
γ_{API} = oil gravity, °API
γ_g = gas gravity, at 60°F and 14.7 psia, lbm/ft³
γ_{gc} = gas gravity, ratio, dimensionless
γ_o = oil gravity, °API
γ_o = oil mobility term, ratio of oil effective permeability to product of oil viscosity and oil-phase formation volume factor, md/(cp × RB/STB)
λ_w = water mobility term, ratio of water effective permeability to product of water viscosity and water-phase formation volume factor, md/(cp × RB/STB)
μ = viscosity, cp
μ_o = oil viscosity, cp
μ_w = water viscosity, cp
ρ_g = gas density, lbm/ft³
ρ_{gt} = gas density at upstream T and p, lbm/ft³
ρ_o = oil density, at 60°F and 14.7 psia, lbm/ft³
ρ_{ot} = oil density at upstream T and p, lbm/ft³
ρ_w = water density, at 60°F and 14.7 psia, lbm/ft³

6.9 Problems

Exercise 6.12—Constructing Inflow Performance Relationship Curves. Construct an inflow performance relationship curve for the Burton No. 4 well.

The following production test data are available: p = 3,300 psia, p_{wf} = 1,500 psia, and q = 6,500 BOPD.

Use the data provided in **Table 6.5** to plot p_{wf} vs. q using the blank graph in **Fig. 6.16.** Does the curve follow the definition of the PI? Why or why not?

Solution. The resulting inflow performance relationship curve is shown in **Fig. 6.17.**

p_{wf} (psia)	q (BOPD)
3,300	0
2,500	3,403
2,000	5,112
1,500	6,500
1,000	7,667
500	8,314

Table 6.5—Data for the well test for the inflow performance relationship curve in Exercise 6.12.

Exercise 6.13—Using the Standing Method. Apply the Standing method to calculate the effect of repairing formation damage on the Moran No. 7 well. What would the new flow rate be if the formation

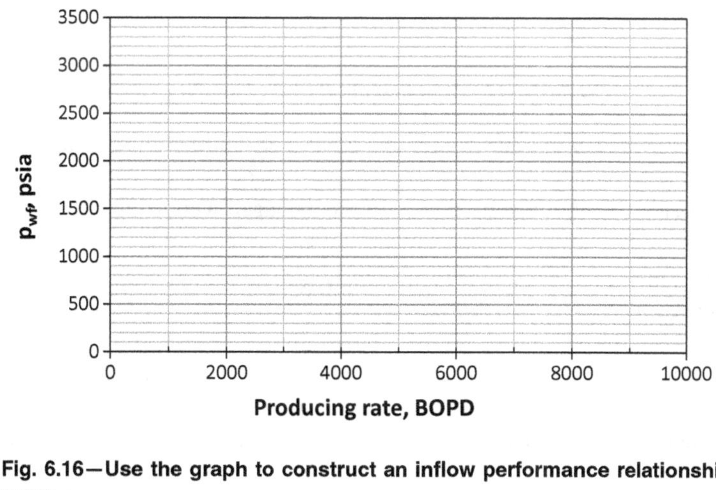

Fig. 6.16—Use the graph to construct an inflow performance relationship curve.

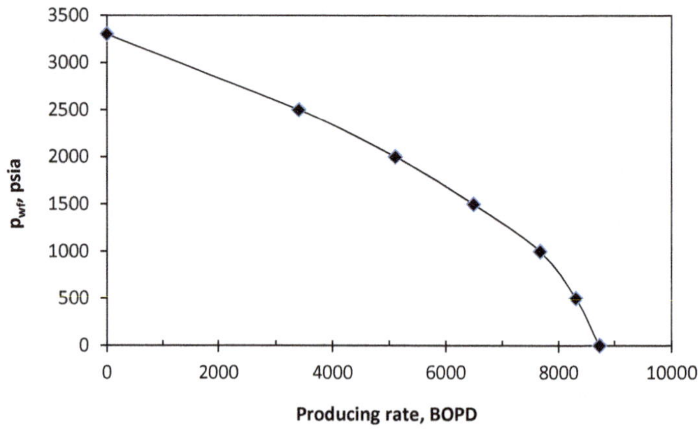

Fig. 6.17—The resulting inflow performance relationship curve using the well-test data.

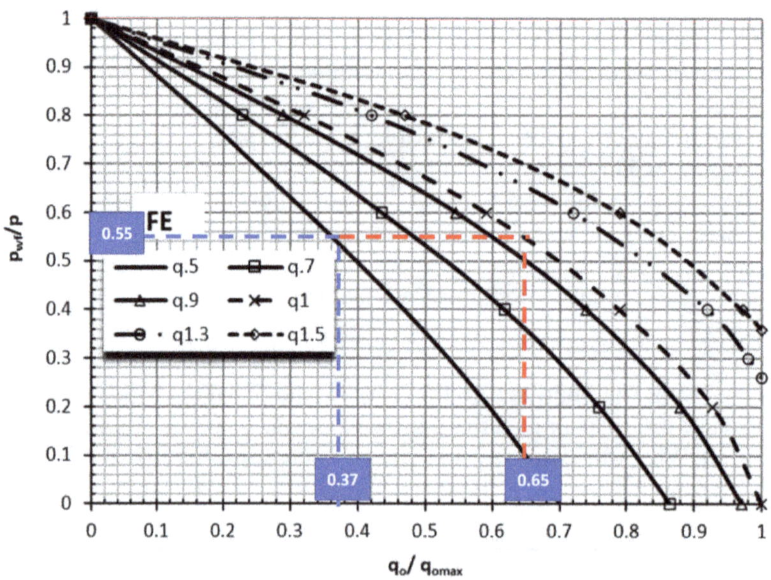

Fig. 6.18—The Standing chart (after Standing 1970).

damage were removed and p_{wf} = 300 psia? Test data: q = 2,300 BOPD at p_{wf} = 1,600 psia and p_{res} = 2,900 psia.

Solution. First, determine the flow efficiency of the damaged well using Eq. 6.18:

$$FE = \frac{\text{Ideal drawdown}}{\text{Test drawdown}} = \frac{p - p'_{wf}}{p - p_{wf}} = \frac{2{,}900 - 1{,}600}{2{,}900 - 300} = 0.5.$$

Using **Fig. 6.18,** which is a modified Cersion of Fig. 6.10, the value of $\frac{q}{q_{max}}$ can be read from the blue line.

Next, calculate and find the equivalent value. Remember, q_{max} was determined at $FE = 1$.

$$\frac{p_{wf}}{\overline{p}} = \frac{1{,}600}{2{,}900} = 0.55;\ \text{thus,}\ \frac{q}{q_{max}} = 0.37\cdot$$

Apply the well-test data to determine q_{max}:

$$q_{max} \frac{2{,}300}{0.37} \text{ BOPD when } FE = 1.$$

Remember, the Standing chart was constructed using the q_{max} term in the denominator when $FE = 1$. Then, calculate the new flow rate if the formation damage were removed (the red line in Fig. 6.18):

$$\text{When } FE = 1 \text{ and } \frac{p_{wf}}{\bar{p}} = \frac{1{,}600}{2{,}900} = 0.55, \ \frac{q}{q_{max}} = 0.65.$$

If the formation damage were removed from the Moran No. 7 well, the new flow rate would be

$$q = 0.65(6{,}216) = 4{,}041 \text{ BOPD}.$$

Exercise 6.14—Effect of Formation Damage. The Muleshoe No. 2 well was drilled and tested, with resulting values of q = 230 BOPD, p_{wf} = 500 psia, and = 1,450 psia. The low bottomhole flowing pressure was probably caused by formation damage, and under normal, undamaged conditions around the wellbore, it would likely be 1,300 psia.

The well-performance equation for the well is $q = 0.0034\left(1{,}450^2 - p_{wf}^2\right)^{0.85}$. What would the new flow rate be if the skin problem were removed and p_{wf} remains at 1,300 psia?

Solution. Apply the Camacho-V. and Raghavan concept to calculate the effect of formation damage on the Muleshoe No. 2 well. Remember, the V term is determined by the type of flow equation used in the calculations; here, V = 1 because the Fetkovich technique will be applied.

$$E_f = \frac{\left(1 + \frac{vp'_{wf}}{p}\right)\left(1 - \frac{p'_{wf}}{p}\right)}{\left(1 + \frac{vp_{wf}}{p}\right)\left(1 - \frac{p_{wf}}{p}\right)} = \frac{\left[1 + \frac{(1.0)(1{,}300)}{1{,}450}\left(1 - \frac{1{,}300}{1{,}450}\right)\right]}{\left[1 + \frac{(1.0)(500)}{1{,}450}\right]\left(1 - \frac{500}{1{,}450}\right)} = 0.222.$$

Next, estimate q_{max}:

$$\frac{q}{q_{max}} = E_f\left[1 + V\left(\frac{p_{wf}}{p}\right)\right]\left[1 - V\left(\frac{p_{wf}}{p}\right)\right]; \frac{230}{q_{max}} = 0.222\left[1 + 1\left(\frac{500}{1{,}450}\right)\right]\left[1 - 1\left(\frac{500}{1{,}450}\right)\right].$$

Thus, q_{max} = 1,186 BOPD.

Calculate the new rate with a rewritten form of the Fetkovich equation (Eq. 6.15):

$$\frac{q}{q_{max}} = \left(1 - \frac{p_{wf}^2}{\bar{p}^2}\right)^n; \frac{q}{1{,}186} = \left[1 - \left(\frac{1{,}300}{1{,}450}\right)^2\right]^{0.85}. \text{ Thus, } q = 297 \text{ BOPD.}$$

The workover could increase well production from 230 to 297 BOPD.

Exercise 6.15—Calculating Future Performance. Use the Fetkovich method to calculate future performance for the Buffalo No. 2 oil well. Isochronal test data are available in **Table 6.6.**

Apply the open flow concept to plot and interpret the data on **Fig. 6.19** to determine n and J values using the Fetkovich equation (Eq. 6.15).

Then, determine the general inflow performance equation when the average reservoir pressure = 1,230 psia.

p_{wf} (psia)	q (BOPD)	$\bar{p}^2 - p_{wf}^2$ (1,000 psia^2)
2,150	0	
2,065	860	358
1,968	1,390	750
1,724	3,100	1,650
1,274	5,000	2,999

Table 6.6—Isochronal test data for the Fetkovich method, Exercise 6.15.

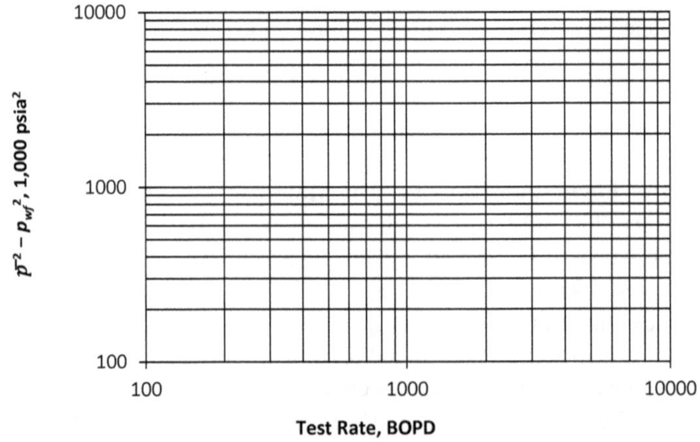

Fig. 6.19—Use the graph to plot the data using the Fetkovich method.

Solution. Apply the Fetkovich method to calculate future performance for the Buffalo No. 2 well.

$$m = \text{slope} = \frac{\log 3{,}000{,}000 - \log 358{,}000}{\log 5{,}000 - \log 860} = 1.21,\ n = \frac{1}{m} = 0.83.$$

Using the corresponding values from Table 6.6,

$$J' = \frac{q_o}{\left(\bar{p}^2 - p_{wf}^2\right)^n} = \frac{3{,}100}{1{,}650^{0.83}} = 0.0219.$$

Thus, the inflow performance equation is $q = 0.0219\left(\bar{p}^2 - p_{wf}{}^2\right)^{0.83}$.

Next, determine the general future inflow performance equation using Eq. 6.27:

$$q = \frac{p_{\text{test}}}{p} J'\left(p^2 - p_{wf}^2\right)^n.$$

$$q = \frac{\bar{p}}{2{,}150} 0.0219\left(\bar{p}^2 - p_{wf}{}^2\right)^{0.83}.$$

Determine the general inflow performance equation when $\bar{p}$ = 1,230 psia:

$$q = \frac{\bar{p}}{2{,}150} 0.0219\left(\bar{p}^2 - p_{wf}{}^2\right)^{0.83};\ q = 0.0125\left(1{,}230^2 - p_{wf}{}^2\right)^{0.83}.$$

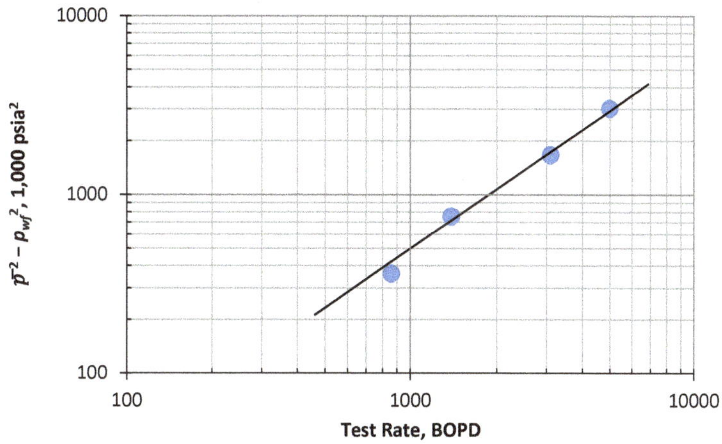

Fig. 6.20—Plotted data for the Fetkovich method, Exercise 6.15.

Fig. 6.20 shows a straight line on log-log coordinates using the values calculated using the Fetkovich method.

Exercise 6.16—Calculating the Inflow Performance Equation. Apply the Klins and Clark method to calculate the inflow performance equation for predicted conditions when average reservoir pressure declines to $\bar{p} = 1{,}915$ psia.

Initial reservoir pressure is $p_i = 3{,}310$ psia and the most-recent well test was = 2,815 psia, and the corresponding flow equation was calculated as

$$q = 0.0069\left(2{,}815^2 - p_{wf}^2\right)^{0.73}, \text{ BOPD}.$$

Gas gravity () = 0.73, oil gravity = 32.1 °API, and T_{res} = 185°F.

Hint:

The first order of business is to apply Eq. 3.25 from Vasquez and Beggs (1980) to calculate the bubblepoint pressure. **Table 6.7** lists the Vasquez and Beggs correlation parameters as a function of API gravity.

Assume that R_{sb} equals the current R_s value. Recall, T_{res} is in °F and is in psia.

$$\text{Solution gas/oil ratio} = R_s = \text{C1}\gamma_{gc}\bar{p}^{\text{C2}}\exp^{[\text{C3}\gamma_{\text{API}}/(T_{\text{res}}\,°\text{F}+460)]}, \text{ scf/STB}.$$

$$\text{Bubblepoint pressure} = p_{bp} = \left\{\frac{R_{sb}}{\text{C1}\gamma_{gc}\exp^{[\text{C3}\gamma_{\text{API}}/(T_{\text{res}}+460)]}}\right\}^{1/1.187}, \text{ psia}.$$

Coefficient	Bubblepoint and Gas/Oil Ratio		Formation Volume Factor	
	°API ≤ 30	°API > 30	°API ≤ 30	°API > 30
C1	0.0362	0.0178	4.68×10^{-4}	4.67×10^{-4}
C2	1.0937	1.1870	1.75×10^{-5}	1.10×10^{-5}
C3	25.7240	23.9310	-1.81×10^{-8}	1.34×10^{-9}

Table 6.7—Vazquez and Beggs correlation coefficients.

After determining the bubblepoint pressure, n_{bp} and J_{bp} can be calculated by rearranging Eqs. 6.28 and 6.29:

$$n_{bp} = \frac{n}{1+0.06\left(1-\frac{\overline{p}}{p_{bp}}\right)-0.25\left(1-\frac{\overline{p}}{p_{bp}}\right)^2+0.50\left(1-\frac{\overline{p}}{p_{bp}}\right)^3},$$

$$J_{bp} = \frac{J'}{1-3.57\left(1-\frac{\overline{p}}{p_{bp}}\right)+4.80\left(1-\frac{\overline{p}}{p_{bp}}\right)^2-2.31\left(1-\frac{\overline{p}}{p_{bp}}\right)^3}.$$

The two coefficients are calculated for the bubblepoint pressure. Then, rearrange Eqs. 6.28 and 6.29 again to solve for n and J' when $\overline{p}$ = 1,915 psia and calculate the inflow performance equation for p_{res} = 1,915 psia.

Solution. Apply the Klins and Clark method to calculate the inflow performance equation for the $\overline{p}$ = 1,915 psia case.

The previous flow equation at $\overline{p}$ = 2,815 psia was calculated as

$$q = 0.0069\left(2{,}815^2 - p_{wf}^2\right)^{0.73}, \text{ BOPD.}$$

Apply the Vasquez and Beggs equation (Eq. 3.25) to calculate the bubblepoint pressure using the correlation coefficients provided in Table 6.7.

Solution gas/oil ratio =

$$R_s = C1\gamma_{gc}\overline{p}^{C2}\exp^{[C3\gamma_{API}/(T_{res}\,°F+460)]} = 0.018(0.73)(3{,}310)^{1.187}\exp^{[(23.93)(32.1)/645]} = 651 \text{ scf/STB.}$$

Bubblepoint pressure =

$$p_{bp} = \left\{\frac{R_{sb}}{C1\gamma_{gc}\exp^{[C3\gamma_{API}/(T_{res}+460)]}}\right\}^{1/1.187} = \left\{\frac{651}{(0.018)(0.73)\exp^{[(23.93)(32.1)/(645)]}}\right\}^{1/1.187} = 3{,}308 \text{ psia.}$$

After determining the 3,308-psia bubblepoint pressure, which is approximately equal to $p_{i,}$ the requisite n_{bp} and J_{bp} values can be calculated by rearranging Eqs. 6.28 and 6.29:

$$n_{bp} = \frac{0.73}{1+0.06\left(1-\frac{2{,}815}{3{,}308}\right)-0.25\left(1-\frac{2{,}815}{3{,}308}\right)^2+0.50\left(1-\frac{2{,}815}{3{,}308}\right)^3} = 0.73.$$

$$J_{bp} = \frac{0.0069}{1-3.57\left(1-\frac{2{,}815}{3{,}308}\right)+4.80\left(1-\frac{2{,}815}{3{,}308}\right)^2-2.31\left(1-\frac{2{,}815}{3{,}308}\right)^3} = 0.012.$$

Solve for n and when reservoir pressure = 1,915 psia:

$$n_{1,915} = 0.73\left[1 + 0.06\left(1 - \frac{1,915}{3,308}\right) - 0.25\left(1 - \frac{1,915}{3,308}\right)^2 + 0.50\left(1 - \frac{1,915}{3,308}\right)^3\right] = 0.74.$$

$$J'_{1,915} = 0.012\left[1 - 357\left(1 - \frac{1,915}{3,308}\right) + 4.80\left(1 - \frac{1,915}{3,308}\right)^2 + 2.31\left(1 - \frac{1,915}{3,308}\right)^3\right] = 0.0021.$$

Thus, the inflow performance equation for = 1,915 psia is

$$q_{1,915} = 0.0021\left(1,915^2 - p_{wf}^2\right)^{0.74}, \text{ BOPD.}$$

Exercise 6.17—Case History Study. Fig. 6.21 shows the producing rate and water-cut history for a well producing from a rather flat-lying reservoir. The oil producing rate fluctuated but increased during the 3.5 years of history. Initially, production was water-free. Scale formation presents a constant problem. Washing the perforations with acid to reduce scaling effects is the preferred remedial treatment.

Water production occurred a few months before the start of Year 2 and became markedly more obvious with time. The pay zone consists of a highly variable interbedded carbonate/sand section. Water breakthrough in the highest-permeability streak was soon evident.

Downhole conditions were constantly monitored by running inflow performance tests. Both acid treatments and enlarging the pump capacity were considered as a means to increase performance.

The segments shown in Fig. 6.21, indicating the timing of four tests taken during the producing history, are described as follows:

- A: The initial loss in the oil production rate coupled with the PI from test T1 indicated declining productivity as the result of an ineffective drainage system. The productivity loss likely resulted from formation damage that occurred during completion. An acid wash was conducted, with a resultant production increase.

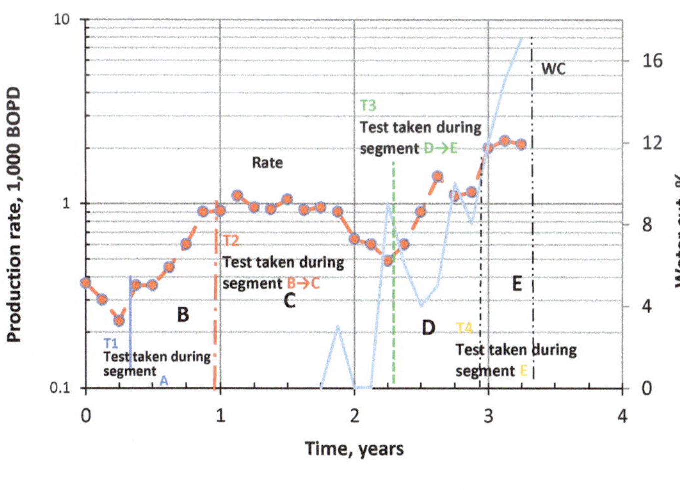

Fig. 6.21—Performance history of a well. The letters on the *x*-axis denote historical events described in the discussion that follows. WC = water cut.

- B→C: A submersible pump was installed, which significantly increased the production rate. The second PI test (T2) illustrates the reaction.
- C→D: The production rate declined as a result of gradual scale buildup.
- D→E: The T3 inflow performance test taken at Year 2.2 showed a reduced well potential when compared with T2, which resulted from the onset of water production. The acid wash conducted during this period resulted in a production increase.
- E: A larger-capacity pump was installed to handle oil + water flow. Note, the increased capacity did not materially change the oil production rate (T4) even though the total flow rate increased. The reservoir drainage volume did not increase even though the rate of depletion did.

Four production tests were run to monitor performance over the course of the well history. The results of these tests are shown in **Fig. 6.22.** Use these tests and the information provided to answer the questions that follow.

Questions and tasks

1. Calculate the water/oil ratio (WOR) at Year 3.25.
2. Did water breakthrough occur at Year 3.25? What stage of the waterflood cycle are we in?
3. Did the two acid treatments effectively clean the near-wellbore scale from the plugged pore spaces? The acid wash stimulation job had a positive effect by increasing the effective permeability around the wellbore, which caused increased water production. Installing a larger submersible pump also increased operating costs. Is this necessarily a bad thing?
4. Did the installation of the larger pump affect reservoir behavior? What is your evidence?
5. The installation of the larger submersible pump made it possible to produce more fluid. Did the oil and water production rates increase proportionately?

Solution.

1. From Fig. 6.21, it is observed that the water cut (WC) = 17% and q_o = 2,000 BOPD.

$$WC = \frac{q_w}{q_w + q_o} = 0.17;\ q_w = \frac{WC \times q_o}{(1-WC)} = \frac{0.17 \times 2{,}000}{(1-0.17)} = 410 \text{ BWPD.}$$

$$WOR = \frac{q_w}{q_o} = \frac{410}{2{,}000} = 0.2 \text{ BW / BO.}$$

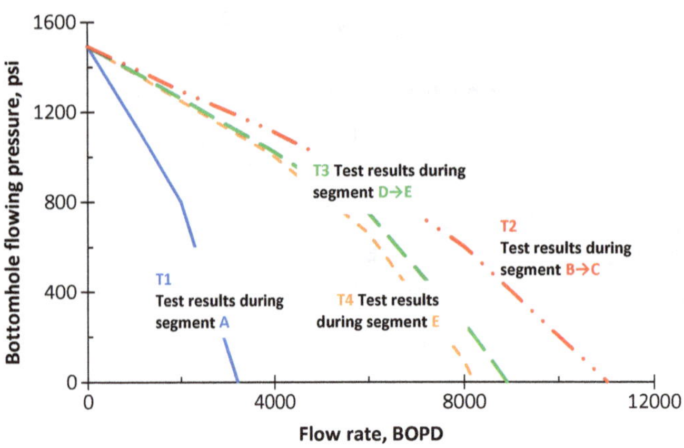

Fig. 6.22—The four productivity tests run on the problem well.

2. A WOR of 0.2 BW/BO is considered low, and it can be concluded that water breakthrough has started to occur at Year 3.25. Continuous monitoring is recommended to determine the evolution of water breaking through. It could be considered that the waterflood project is in a transition from the fill-up stage to water breakthrough.
3. Yes, the DE period indicates that the acid treatments had a positive effect. Increased operating costs

are offset by the increased oil production during Period D→E:, from approximately 500 BOPD to 2,000 BOPD, so the changes had a positive impact on the cash flow.

4. The installation of a larger pump did not affect reservoir behavior. One can tell because the q_{max} values of the T3 and T4 tests are similar, indicating the well drainage area was essentially unchanged.
5. No, the oil and water production rates did not increase proportionally. The higher mobility of water coursing through the high-permeability streaks caused bypassing of reservoir oil. Water production increased at the expense of oil production.

6.10 References

- *API 852-27045, Volume Correction Factors*. 1980. Washington, DC: API.
- Camacho-V., R. G. and Raghavan, R. 1989. Inflow Performance Relationships for Solution-Gas-Drive Reservoirs. *J Pet Technol* **41** (5): 541–550. SPE-16204-PA. https://doi.org/10.2118/16204-PA.
- Craft, B. C. and Hawkins, M. F. 1959. *Applied Petroleum Reservoir Engineering*. Englewood Cliffs, New Jersey: Prentice-Hall.
- Evinger, H. H. and Muskat, M. 1942. Calculation of Theoretical Productivity Factor. In *Transactions of the Society of Petroleum Engineers*, Vol. 146, SPE-942126-G, 126–139. Richardson, Texas: Society of Petroleum Engineers. https://doi.org/10.2118/942126-G.
- Fetkovich, M. J. 1973. The Isochronal Testing of Oil Wells. Paper presented at the Fall Meeting of the Society of Petroleum Engineers of AIME, Las Vegas, Nevada, USA, 30 September–3 October. SPE-4529-MS. https://doi.org/10.2118/4529-MS.
- Gilbert, W. E. 1954. Flowing and Gas-Lift Well Performance. In *Drilling and Production Practice*, API-54-126. New York, New York: API.
- Klins, M. A. and Clark, J. W., III. 1993. An Improved Method to Predict Future IPR Curves. *SPE Res Eng* **8** (4): 243–248. SPE-20724-PA. https://doi.org/10.2118/20724-PA.
- Klins, M. A. and Majcher, M. W. 1989. Inflow Performance Relationships for Damaged or Improved Wells Producing Under Solution-Gas Drive. Paper presented at the SPE Annual Technical Conference and Exhibition, San Antonio, Texas, USA, 8–11 October. SPE-19852.
- Poettmann, F. H. and Beck, R. L. 1963. New Charts Developed to Predict Gas-Liquid Flow Through Chokes. *World Oil*: 95-101.
- Ros, N. C. J. 1961. Simultaneous Flow of Gas and Liquid as Encountered in Well Tubing. *J Pet Technol* **13** (10): 1037–1049. SPE-18-PA. https://doi.org/10.2118/18-PA.
- Standing, M. B. 1970. Inflow Performance Relationships for Damaged Wells Producing by Solution-Gas Drive. *J Pet Technol* **22** (11): 1399–1400. SPE-3237-PA. https://doi.org/10.2118/3237-PA.
- Standing, M. B. and Katz, D. L. 1942. Density of Natural Gases. In *Transactions of the Society of Petroleum Engineers*, Vol. 146, SPE-942140-G, 140–144. Richardson, Texas: Society of Petroleum Engineers. https://doi.org/10.2118/942140-G.
- Sutton, R. P. 1985. Compressibility Factors for High-Molecular-Weight Reservoir Gases. Paper presented at the SPE Annual Technical Conference and Exhibition, Las Vegas, Nevada, USA, 22–25 September. SPE-14265-MS. https://doi.org/10.2118/14265-MS.
- Vasquez, M. and Beggs, H. D. 1980. Correlations for Fluid Physical Property Prediction. *J Pet Technol* **32** (6): 968–970. SPE-6719-PA. https://doi.org/10.2118/6719-PA.
- Vogel, J. V. 1968. Inflow Performance Relationships for Solution-Gas Drive Wells. *J Pet Technol* 20 (1): 83–92. SPE-1476-PA. https://doi.org/10.2118/1476-PA.

Acronyms used:

- BFPD = barrels of fluid per day
- BOPD = barrels of oil per day
- BWPD = barrels of water per day
- FE = flow efficiency
- PI = productivity index

Chapter 7

Water Influx

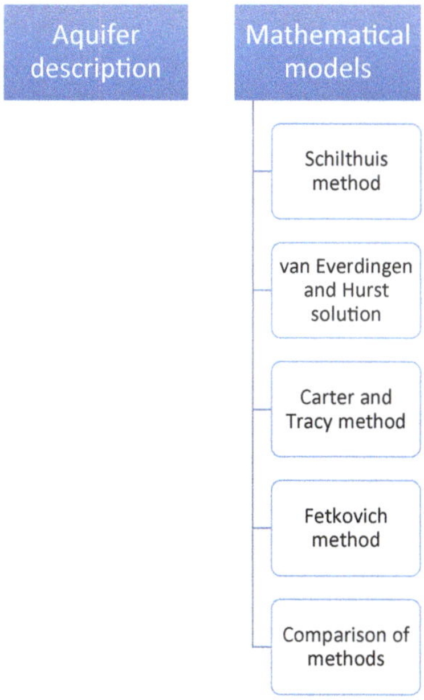

During the productive life of a reservoir, several possible primary mechanisms can be present. One of the most efficient of these is water influx, which influences the economic viability of a reservoir development and optimization project. The primary objective of managing water influx is to efficiently displace oil and maximize the economic recovery of oil from a reservoir. Water influx increases the amount of oil recovered from a reservoir in two ways: through pressure support and oil displacement by means of water.

Groundwater can migrate through a permeable aquifer in regions when there is a pressure differential established between the outcrop that is accepting water at a higher elevation and the outflow location at a much lower elevation. There have been cases of oil fields located near mountainous areas where the water flow through the associated aquifer was sufficiently strong to tilt the water contact away from the horizontal. Hydrocarbons migrated from the source rock, and because oil and gas are less dense than water, buoyant forces caused them to flow up to shallower depths and become trapped in porous reservoir rock, displacing the original water that saturated the permeable layer. Generally, the lack of quantifying information makes it very difficult, if not impossible, to model this type of aquifer.

Water movement can also be induced by rock and water expansion when oil and gas production establishes a pressure drop between the water contact and the outer limit of the aquifer. In general, the expanding water leg moves updip to displace movable hydrocarbons located near the oil/water contact (OWC). Modeling the expansion of aquifer water is the preferable method for determining the effects of water influx on the performance history of a petroleum reservoir, which can be very efficient in displacing naturally available energy.

Water influx can exert a strong to marginal influence on the performance history of an oil or gas reservoir. There are no reasonable methods available to measure water-influx effects because the physical characteristics of an aquifer are seldom known. Therefore, the effect of water influx on reservoir performance must be inferred by developing a mathematical model to approximate and characterize the aquifer, taking into consideration the hydrocarbon-zone response to changes in compressibility in the porous system.

The rate of water encroaching into a petroleum reservoir is dependent on the pressure drop measured as a function of time, aquifer volume, and rock and fluid properties. Physical characteristics must be determined from geological and performance data when developing a mathematical model to approximate the field conditions. Reservoir pressure responses observed in the field are sorted into regular geometric shapes. In this chapter, four different methods for predicting water influx are developed, and to smooth out the pressure irregularities, the principles of superposition and effective time are included in the Schilthuis (1936) and van Everdingen and Hurst (1949) analysis methods. The solution procedures for the Fetkovich (1971) and Carter and Tracy (1960) methods do not need these regularizing techniques because they are solved with the constant-pressure solution to the diffusivity equation (Eq. 4.23 in the Well and Reservoir Models chapter).

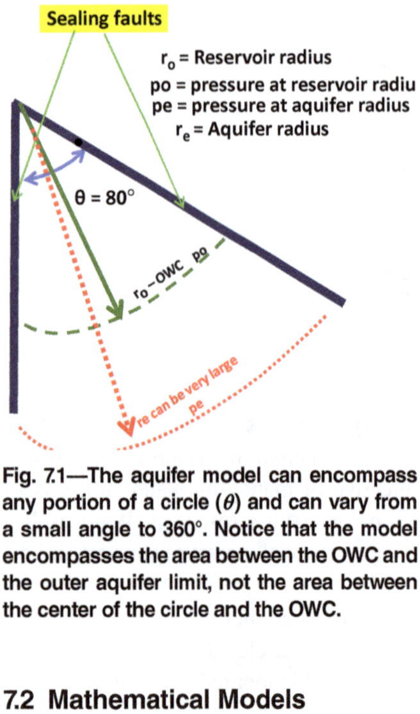

Fig. 7.1—The aquifer model can encompass any portion of a circle (θ) and can vary from a small angle to 360°. Notice that the model encompasses the area between the OWC and the outer aquifer limit, not the area between the center of the circle and the OWC.

7.1 Aquifer Description

Petroleum reservoirs are commonly irregular in shape and must be fitted to a regular geometric shape so that pressure/water-influx characteristics can be studied with a mathematical model. Geometric models can assume a radial or rectangular shape, but generally, the radial shape is deemed appropriate. **Fig. 7.1** presents an example of a radial segment approximating the shapes and relationships between an oil reservoir and an underlying aquifer bounded by two sealing faults.

In Fig. 7.1, r_o is the distance from the center of the circle to the edge of the reservoir at the OWC, r_e is the distance from the center of the circle to the aquifer limit, p_o is the pressure at the OWC, p_e is the pressure at the aquifer limit, and θ is the angle subtending the portion of the circle and can vary from 360° to < 90°.

Aquifer-water expansive effects are the sum of water and formation compressibility values:

$$c_t = c_w + c_f . \qquad (7.1)$$

7.2 Mathematical Models

7.2.1 Schilthuis Method. Schilthuis (1936) applied the Darcy flow equation (Eq. 4.2), assuming steady-state conditions and that the pressure at the aquifer limit and at the water contact remains constant and independent of time, to calculate water-influx volumes. These assumptions imply that the aquifer is very large and can be considered infinite in size. Step functions and effective time concepts are incorporated into the solution procedure to divide the analysis of the field pressure response curve into a series of instantaneous segments.

The steady-state solution to the Darcy equation (Eq. 4.3) for radial flow can be rewritten in the following form when θ is the portion of the circle encompassed by the reservoir and the aquifer. Notice that Eq. 7.2 relocates the conditions measured at the wellbore to the reservoir OWC.

$$q = \frac{\mathrm{d}W_e}{\mathrm{d}t} = \frac{2\pi kh(p_e - p_o)}{\mu B_w \ln(r_e / r_o)}\left(\frac{\theta}{360}\right). \quad (7.2)$$

Permeability, thickness, viscosity, formation volume factor, and reservoir limits are assumed to remain constant, and there is no pressure gradient across the aquifer. Therefore, these terms can be coalesced into a single coefficient, defined as the steady-state aquifer description term C_s. This term is similar to the definition of the productivity-index (PI) rate equation (Eq. 6.8 in the Inflow Performance chapter), which also assumes the existence of steady-state conditions.

An alternative method for the steady-state solution (Eq. 7.2) was proposed by Craft and Hawkins (1959):

$$C_s = \frac{0.00708kh}{\mu B_w \ln(r_e / r_o)}\left(\frac{\theta}{360}\right), \text{ STB/D-psi}. \quad (7.3)$$

The inclusion of the aquifer description coefficient C_s in the steady-state fluid flow Darcy equation results in Eq. 7.4, which is used to calculate the cumulative water influx as a function of expected field conditions. The pressure drop is the difference between pressure at the water contact and at the outer boundary, while the time interval over which the pressure difference acts is the effective time.

$$W_e = C_s \int_0^t (p_e - p_o)\mathrm{d}t = C_s \sum_0^t \Delta p \Delta t, \text{ STB}. \quad (7.4)$$

Using the principle of superposition transforms Eq. 7.4 into an expression consisting of a series of instantaneous rate changes applied over individual effective times. Recall that the principle of superposition, which was developed in Chapter 4, Well and Reservoir Models, deals with these rate changes. However, in this particular case, pressure changes are of interest. Cumulative water influx is the sum of the water influx of the individual layers defined by their particular influence on the instantaneous pressure drop and the effective time. Eq. 7.4 can be expressed in a more-generalized form as

$$W_{e(t-t_j)} = C_s \int_0^t (p_e - p_o)\mathrm{d}t = C_s \int_0^t \Delta p \, \mathrm{d}t_e = C_{sj-1}^{n} \, \Delta p_j (t - t_j). \quad (7.5)$$

7.2.1.1 Solution Procedure. A step function is created by dividing the pressure/time curve into a series of rectangular layers. The height of each step is represented by the average pressure drop between p_i and p_1 and p_1 and p_2, and the method consists of approximating the continuous pressure decline behavior by means of a series of sequential finite timesteps, as illustrated in **Fig. 7.2.**

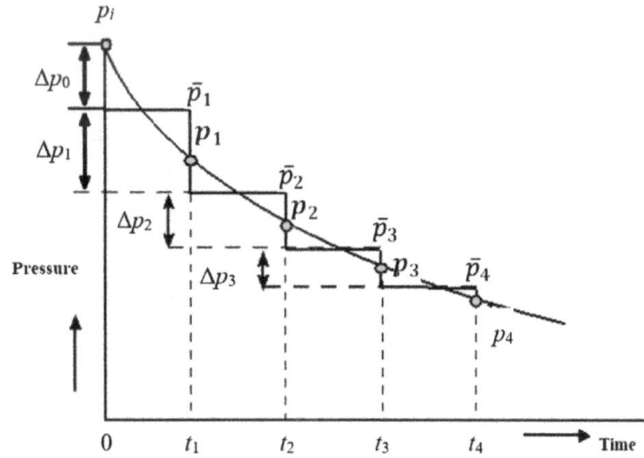

Fig. 7.2—The recommended method to calculate the cumulative water influx is to approximate the continuous pressure decline by means of a series of pressure steps.

Eqs. 7.6 through 7.8c are used to calculate the pressure drop of each effective timestep:

$$\Delta p_1 = p_i - \frac{(p_i + p_1)}{2} = \frac{(p_i - p_1)}{2}, \quad \text{(7.6)}$$

$$\Delta p_2 = \left(\frac{p_i + p_1}{2}\right) - \left[\frac{(p_1 + p_2)}{2}\right] = \frac{(p_i - p_2)}{2}, \quad \text{(7.7)}$$

$$\Delta p_3 = \left(\frac{p_1 + p_2}{2}\right) - \left[\frac{(p_2 + p_3)}{2}\right] = \frac{(p_1 - p_3)}{2}, \quad \text{(7.8a)}$$

$$\Delta p_4 = \left(\frac{p_2 + p_3}{2}\right) - \left[\frac{(p_3 + p_4)}{2}\right] = \frac{(p_2 - p_4)}{2}, \quad \text{(7.8b)}$$

$$\Delta p_j = \left(\frac{p_{j-2} - p_{j-1}}{2}\right) - \left[\frac{(p_{j-1} - p_j)}{2}\right] = \frac{(p_{j-2} - p_j)}{2}. \quad \text{(7.8c)}$$

The effective time is the length of time each pressure step influences the aquifer.

The analysis procedure is as follows:

- The most-likely values for all the aquifer-character terms in Eq. 7.3 are determined to calculate the aquifer description term C_s. Most aquifers can be modeled as radial in shape, with a form ranging from a complete to a partial circle. [A more-rigorous solution for determining the C_s value by identifying steady-state conditions from an analysis of production records was developed by Craft and Hawkins (1959) and will be discussed later in the chapter.]
- Use the step-function model with the aquifer description term to calculate cumulative water influx using Eq. 7.5.

Exercise 7.1—Calculating Steady-State Water-Influx Volume Using the Schilthuis Method. Use the information that follows to calculate the steady-state water-influx volume in a reservoir at the end of 1,100 days.

Determine the pressure step function and effective time values. The effective times represent the difference between the 1,100-day endpoint and the time of initiation of the pressure drop defined by the specific pressure step function.

The time/pressure history for the reservoir is shown in the first two columns in **Table 7.1** and as the smooth orange curve in **Fig. 7.3.** Aquifer characteristics are also listed in Table 7.1; note that the measured pressure at the OWC increases at 1,000 days and the complete pressure history extends to 1,200 days. The last column in Table 7.1 lists the effective time over which that particular disturbance influenced the 1,200-day history. The r_{eD} = 500 value indicates that the aquifer is extraordinarily large and in fact probably acts in an infinite manner.

k_w = 100 md, B_w = 1.03 RB/STB, θ = 95°, h = 30 ft, μ_w = 0.6 cp, μ = 12%, r_{eD} = 500, r_o = 2,000 ft, $c_w = 3 \times 10^{-6}$ psi^{-1}, $c_f = 3 \times 10^{-6}$ psi^{-1}

1	2	3	4	5	6 (Column 4 × Column 5)
Time (days)	Pressure (psia)	Average Pressure (psia)	Pressure Change Δ*p* (psia)	Effective Time t_e (days)	Δpt_e (1,000 psi-D)
0	5,000	5,000	–	–	–
300	4,700	4,850	150	1,100	165.0
600	4,500	4,600	250	800	200.0
1,000	4,750	4,625	−25	500	−12.5
1,100	4,300	4,525	100	100	10.0
1,200	4,300				Total: 362.5

Table 7.1—Aquifer character data and results of calculations showing the total Δpt_e component term = 362.5 (1,000 psi-D) for estimating the total water influx at t =1,100 days.

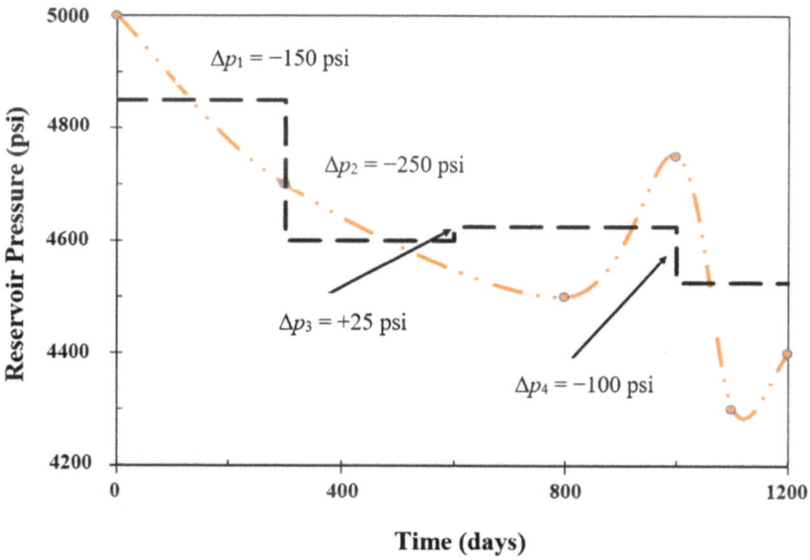

Fig. 7.3—Comparing the measured pressure history to the step pressure function calculation.

Solution. After calculating the value of r_e, use Eq. 7.3 to calculate the aquifer description coefficient defining the characteristics of this field:

$$r_{eD} = r_e / r_o;\ r_e = r_{eD} \times r_o = 500 \times 2{,}000 = 1{,}000{,}000 \text{ ft}.$$

$$C_s = \frac{0.00708kh}{\mu B_w \ln(r_e / r_o)}\left(\frac{\theta}{360}\right) = \frac{(0.00708)(100 \text{ md})(30 \text{ ft})}{(0.6 \text{ cp})(1.03)\left[\ln\left(\dfrac{1{,}000{,}000}{2{,}000}\right)\right]}\left(\frac{95°}{360°}\right) = 1.459 \text{ STB/D-psi}.$$

Then, calculate the step pressure function and effective times for the first two layers:

$$\text{Layer 1:}\Delta p_1 = \frac{(p_i - p_1)}{2} = \frac{(5{,}000 - 4{,}700)}{2} = 150 \text{ psia; } t_e = 1{,}100 - 0 = 1{,}100 \text{ days.}$$

$$\text{Layer 2: }\Delta p_2 = \frac{(p_i - p_2)}{2} = \frac{(5{,}000 - 4{,}500)}{2} = 250 \text{ psia, } t_1 = 1{,}100 - 300 = 800 \text{ days.}$$

Using Eq. 7.5 and including the principle of superposition and the aquifer description term (C_s = 1.459 STB/D-psi) results in the calculations given in **Table 7.2a.** Recall, we want to calculate cumulative water influx out to 1,100 days.

Or, we could apply Eq. 7.5 and the summation value determined in Table 7.1:

$$W_{e(t-t_j)} = C_{sj-1}^{n}\,\Delta p_j\left(t - t_j\right);\ W_e = 1.459 \text{ STB/D-psi} \times 362{,}500 \text{ psi-D} = 528{,}887 \text{ STB.}$$

No difference between the two methods of estimating water influx is observed.

However, some differences could appear as a result of rounding up numbers. In this case, if an approximation for C_s is considered valid, C_s = 1.459 ≈ 1.50 STB/D-psi. Water-influx values are shown in **Table 7.2b.**

Applying Eq. 7.5 and the summation value determined in Table 7.1,

$$W_{e(t-t_j)} = C_{sj-1}^{n}\,\Delta p_j\left(t - t_j\right);\ W_e = 1.5 \text{ STB/D-psi} \times 362{,}500 \text{ psi-D} = 543{,}750 \text{ STB.}$$

A difference of 14,862 STB (approximately 3%) between the calculated values can be observed by rounding up the values.

Exercise 7.2—Estimating Cumulative Water Influx Using the Schilthuis Method. Estimate cumulative water influx for the San Miguel Field for a 0.7- and 1.2-year time period assuming steady-state flow. Aquifer properties and the pressure history are given in **Table 7.3.** Assume a radial aquifer, θ = 360°.

This problem illustrates how applying the pressure step function requires rearranging the effective time and pressure drop values for each case. Table 7.3 provides the necessary aquifer properties and performance history assuming the pressure in the aquifer remains constant. Notice that the average pressures have already been determined.

$W_e = C_s(\Delta p_1)(t - t_1) = 1.459 \times 150 \times (1{,}100 - 0)$	240,735 STB
$W_e = C_s(\Delta p_2)(t - t_2) = 1.459 \times 250 \times (1{,}100 - 300)$	291,800 STB
$W_e = C_s(\Delta p_3)(t - t_3) = 1.459 \times -25 \times (1{,}100 - 600)$	−18,237 STB
$W_e = C_s(\Delta p_4)(t - t_4) = 1.459 \times 100 \times (1{,}100 - 1{,}000)$	14,590 STB
	Total: W_e = 528,887 STB

Table 7.2a—Water influx as a function of effective time.

$W_e = C_s(\Delta p_1)(t - t_1) = 1.5 \times 150 \times (1{,}100 - 0)$	247,500 STB
$W_e = C_s(\Delta p_2)(t - t_2) = 1.5 \times 250 \times (1{,}100 - 300)$	300,000 STB
$W_e = C_s(\Delta p_3)(t - t_3) = 1.5 \times -25 \times (1{,}100 - 600)$	−18,750 STB
$W_e = C_s(\Delta p_4)(t - t_4) = 1.5 \times 100 \times (1{,}100 - 1{,}000)$	15,000 STB
	Total: W_e = 543,750 STB

Table 7.2b—Water influx as a function of effective time.

$\phi = 27\%$	$k_w = 212$ md	$r_e = 10{,}800$ ft	$B_w = 1.0$ RB/STB
$h = 120$ ft	$\mu_w = 0.8$ cp	$r_o = 1{,}800$ ft	$c_f = 6.7 \times 10^{-6}$ psi^{-1}
Time (years)	p_{res} (psia)	$\bar{p}$ (psia)	Δp (psi)
0	2,500	2,500	–
0.3	2,400	2,450	50
0.7	2,300	2,350	100
1.2	2,200	2,250	100
1.7	2,100	2,150	100
2.8	1,800	1,950	200

Table 7.3—Aquifer properties and pressure history and calculations.

Solution. First, calculate the aquifer description coefficient. Include the 365 day/year multiplier because time is reported in years.

$$C_s = \frac{0.00708kh}{\mu B_w \ln(r_e / r_o)} \left(\frac{\theta}{360} \right) = \frac{0.00708(365)(212)(120)}{(0.8)(1.0)\left[\ln(10,800/1,800)\right]} \left(\frac{360}{360} \right) = 45,844 \text{ STB/yr-psia.}$$

Next, determine the effective time values and step pressure function (t_e and Δpt_e, respectively). **Table 7.4** illustrates the calculation procedure for the time frames of $t = 0.7$ years and $t = 1.2$ years.

Then, calculate cumulative water influx. For the $t = 0.7$-years timestep,

$W_e = 45{,}844$ STB/yr-psia $\times$ 75 psi-yr $= 3{,}438{,}300$ STB $\approx$ 3.4 million STB.

Calculate cumulative water influx for the $t = 1.2$-years timestep:

$W_e = 45{,}844$ STB/yr-psia $\times$ 200 psi-yr $= 9{,}168{,}800$ STB $\approx$ 9.2 million STB.

Time	W_e	Pressure	Δp	t_e	Δpt_e	W_e
(years)	(million STB)	(psia)	(psi)	(years)	(psi-yr)	(million STB)
1	2	3	4	5	6	7
		Water-influx calculations for $t = 0.7$ years				
0	–	2,500	–	–	–	
0.3	0.7	2,400	50	0.7	35	
0.7	**3.4**	2,300	100	0.4	40	
1.2	**9.2**				**75**	**3.4**
1.7	17.2	Water-influx calculations for $t = 1.2$ years				
2.8	44.9	2,500	–	–	–	
		2,400	50	1.2	60	
		2,300	100	0. 9	90	
		2,200	100	0.5	50	
					200	**9.2**

Table 7.4—Steady-state water-influx calculations.

7.2.1.2 Estimating C_s from Production Records. It is possible to determine the aquifer description coefficient C_s if the reservoir is operating at steady-state conditions. Reservoir voidage is equal to water volume influx at this point, and it should be assumed that the main source of pressure supply is water encroachment. Eqs. 7.4 and 7.5 can be rearranged to solve for C_s; the aquifer description coefficient is the only unknown value for this case:

$$C_s = \frac{\mathrm{d}W_e / \mathrm{d}t}{p_e - p_o}. \qquad (7.9)$$

A material balance can be developed when steady-state conditions exist:

$$\frac{\mathrm{d}W_e}{\mathrm{d}t} B_w = q_o B_o + \frac{q_g B_g}{5.61} + q_w B_w. \qquad (7.10)$$

Eq. 7.10 can be expressed in terms of cumulative production:

$$\frac{\mathrm{d}W_e}{\mathrm{d}t} B_w = \frac{\mathrm{d}N_p}{\mathrm{d}t} B_o + \left(R - R_s\right) \frac{\mathrm{d}N_p}{\mathrm{d}t} \frac{B_g}{5.61} + \frac{\mathrm{d}W_p}{\mathrm{d}t} B_w. \qquad (7.11)$$

Eq. 7.11 can be expressed in terms of the total formation volume factor, $B_t = B_o + \left(R_{si} - R_s\right) \frac{B_g}{5.61}$:

$$\frac{\mathrm{d}W_e}{\mathrm{d}t} B_w = \frac{\mathrm{d}N_p}{\mathrm{d}t} B_t + \left(R - R_{si}\right) \frac{\mathrm{d}N_p}{\mathrm{d}t} \frac{B_g}{5.61} + \frac{\mathrm{d}W_p}{\mathrm{d}t} B_w. \qquad (7.12)$$

The analysis procedure is as follows:

- Study production records and identify the presence of steady-state conditions. Production and pressure records for a particular reservoir should be plotted and examined to determine if steady-state conditions were present at some extended time during the production history. A constancy of production volumes and reservoir pressure implies the existence of a steady state.
- Produced oil, gas, and water volumes are measured at standard conditions. Eq. 7.9, 7.10, 7.11, or 7.12 can be used to convert these values to reservoir volumes. Formation volume factors can be estimated from laboratory studies or from correlations.
- Calculate the aquifer description coefficient.
- Apply the principle of superposition to calculate cumulative water influx over any particular time frame.

Exercise 7.3—Determining Cumulative Water Influx Using the Schilthuis Method. The performance history of an example reservoir is shown in **Fig. 7.4.** Analyze the data to determine if steady-state conditions have existed at any time. The properties at stabilized conditions: B_g = 0.00574 rcf/scf, B_w = 1.1 RB/STB, B_t = 1.430 RB/STB, current gas/oil ratio = R = 640 scf/STB, initial solution gas/oil ratio = R_{si} = 520 scf/STB, and N_p = 3,520 BOPD. Calculate the aquifedescription term and determine the cumulative water influx after 14 months of production history. **Table 7.5** lists the pressure history and superposition calculations for the first 14 months of production.

Solution. First, study the performance history shown in Fig. 7.4 to determine if and when steady-state conditions existed. This analysis indicates that the pressure and producing rates remained constant, and steady-state conditions existed during the 30- to 42-month period. During this time frame, producing rates (q_o) averaged 3,520 STB/D, average gas/oil ratio = R = 640 scf/STB, and reservoir pressure was 2,400 psia.

Convert the production to reservoir conditions. Use Eq. 7.12 because the oil fluid property is expressed in terms of the total formation volume factor (B_t).

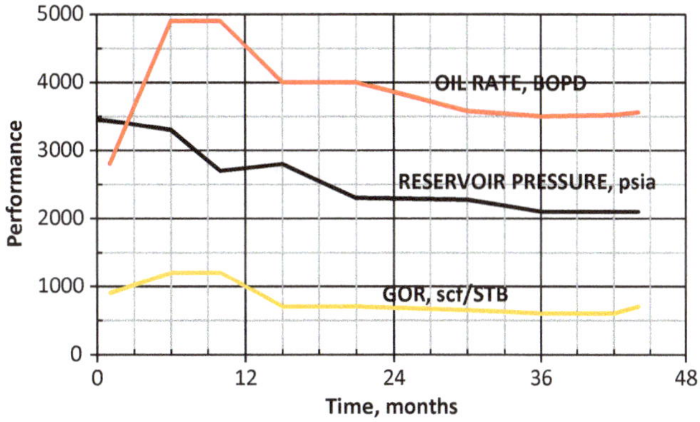

Fig. 7.4—Performance history of the example field. GOR = gas/oil ratio.

Time	Pressure (psia)	Pressure Change (psia)	Effective Time (months)
0	3,456	–	–
1	3,440	8	14
6	3,302	77	13
10	2,700	370	8
14	2,832	235	4

Table 7.5—Performance history of the example reservoir.

$$\frac{\mathrm{d}W_e}{\mathrm{d}t}B_w = \frac{\mathrm{d}N_p}{\mathrm{d}t}B_t + \left(R - R_{si}\right)\frac{\mathrm{d}N_p}{\mathrm{d}t}\frac{B_g}{5.61} + \frac{\mathrm{d}W_p}{\mathrm{d}t}B_w;$$

$$\frac{\Delta W_e}{\mathrm{d}t}(1.1) = 3,520\left[1.43 + \frac{(640-520)(0.00574)}{5.61}\right] + 0 = 4,969 \text{ STB/D}.$$

Notice that the oil producing rate is 3,520 BOPD but the reservoir voidage rate is 4,969 RB of total fluids per day. The production of free gas represents a significant reservoir volume.

Next, calculate the aquifer description coefficient using Eq. 7.9:

$$C_s = \frac{\mathrm{d}W_e/\mathrm{d}t}{p_e - p_o} = \frac{4,969}{(3,456 - 2,400)} = 4.71 \text{ STB/D-psi}.$$

The producing rate is converted to monthly time units because the history is expressed in months:

$$C_s = 4.71 \text{ STB/D-psi} \times 30.4 \text{ days/month} = 143.26 \text{ STB/month-psi}.$$

The aquifer description term can be viewed with some confidence because the performance history was used in the calculations and there was little variation in performance after the initial production.

Next, calculate the cumulative water influx. The principles of superposition and effective time are applied to calculate the volume of water influx by the end of the first 14 months. Steady-state conditions are assumed to exist during this period. The principle of superposition is used to calculate the Δpt_e values for each of the individual layers; those values are given in Table 7.5. Eq. 7.5 is applied to determine cumulative water influx during the 14-month period:

$$W_{e(t-t_j)} = C_s\int_0^t (p_e - p_o)\mathrm{d}t = C_s\int_0^t \Delta\, p\mathrm{d}t_e;$$

$$W_e = C_s\int_0^t \Delta\, p\mathrm{d}t_e = 143.26 \text{ STB/month-psi } \left[(8)(14)+(77)(13)+\right.$$

$$\left.(370)(8)+(235)(4)\right] = 0.7182 \text{ million BW.}$$

An analysis of the performance history presented in Fig. 7.4 indicates that exact steady-state conditions did not exist over the entire 14-month period. However, the ability to calculate a good description of the aquifer from performance records probably outweighs the uncertainties of using a steady-state equation when the reservoir is actually under transient conditions. A reasonable check would be to apply the most-likely aquifer parameters using Eq. 7.3 to calculate the aquifer description term and compare it with the value developed from the analysis of performance history.

7.2.2 van Everdingen and Hurst Solution. van Everdingen and Hurst (1949) solved the diffusivity equation for constant pressure and constant rate conditions, which in dimensionless terms is written as

$$\frac{\partial^2 p_d}{\partial^2 r_d} + \frac{1}{r_d}\frac{\partial p_d}{\partial r_d} = \frac{\partial p_d}{\partial t_d}. \quad (7.13)$$

The boundary conditions for this particular case, of flowing water from the aquifer to the reservoir, can be defined as

- Outer boundary (limited aquifer)
  - No flow (closed aquifer)
  - Constant pressure (aquifer recharge at outcrop)
- Outer boundary (infinite aquifer)
- Inner boundary
  - Constant pressure at the OWC
  - Constant rate
- Initial condition
  - Uniform pressure in the system

The constant pressure solutions were provided in Chapter 4, Well and Reservoir Models, for the diffusivity equation in terms of dimensionless variables.

Eq. 4.18 was given as $q_D = \dfrac{141.2qB\mu}{kh(p_i - p_{wf})}$. It is expressed here as $q_D = \dfrac{141.2q\mu B_w}{kh(p_e - p_o)}$ as a function of $(p_e - p_o) = \nabla p$ = pressure drop at the OWC boundary in psi, where p_e is the initial reservoir pressure (pressure at the original OWC) in psi and p_o is the current reservoir pressure (pressure at the OWC) in psi.

Eq. 4.19 was given as $Q_{pD} = q_D t_D$, and Eq. 4.20 was given as $Q_w = q_w t$.

Eq. 4.21, given as $r_{eD} = \frac{r_e}{r_o}$, compares the radius of the outermost boundary r_e to that of the innermost boundary, which for this case is the reservoir radius r_o, which is the distance to the OWC. Eq. 4.22 presented the dimensionless-time/real-time relationship. The time units are days: $t_D = \dfrac{0.00633kt}{\phi\mu c_t r_w^2}$.

The cumulative water influx can be obtained by integrating Eq. 7.2:

$$W_e = \int_0^t q\mathrm{d}t. \qquad (7.14)$$

van Everdingen and Hurst (1949) solved the diffusivity equation (Eq. 7.13) for an infinite-acting (transient/unsteady-state) aquifer system and presented the result in a dimensionless form for the constant terminal pressure solution. This unsteady-state flowing condition, which is also called transient, is defined as the period of time during which the boundary has no effect on the pressure behavior, and as a result, the system is considered mathematically infinite. Eq. 7.15 expresses the van Everdingen and Hurst dimensionless water influx:

$$Q_{pD} = \int_0^{t_D} q_D \mathrm{d}t_D. \qquad (7.15)$$

Dimensionless water influx Q_{pD} is also referred to as W_D in the literature.

The results of multiplying by $\mathrm{d}t/\mathrm{d}t_D$ and setting up the integral are shown as

$$\int_0^{t_D} q_D \mathrm{d}t_D \frac{\mathrm{d}t}{\mathrm{d}t_D} = \frac{\mu}{2\pi kh(p_e - p_o)} \int_0^t q\mathrm{d}t. \qquad (7.16)$$

Substituting the Q_{pD} and W_e definitions into Eq. 7.16 results in

$$Q_{pD} \frac{\mu\phi c_t r_o^2}{k} = \frac{\mu}{2\pi kh(p_e - p_o)} W_e. \qquad (7.17)$$

Using Eq. 4.16 for system compressibility and the isothermal compressibility definition and assuming a radial-shaped aquifer, an expression for W_e can be obtained, as shown by Eq. 7.18a.

$c_t = S_o c_o + S_g c_g + S_w c_w + c_f$, where $S_o = 0; S_g = 0; c_t = S_w c_w + c_f$; and $S_w = 1.0$.

$$W_e = 2\pi\phi h c_t r_o^2 (p_e - p_o) Q_{pD}. \quad (7.18a)$$

Eq. 7.18b provides the expression for W_e in oilfield units:

$$W_e = 1.119\phi h c_t r_o^2 (p_e - p_o) Q_{pD}. \quad (7.18b)$$

Eq. 7.18b (van Everdingen and Hurst 1949) assumes that water is encroaching in a radial form and from the outer infinite boundary to the inner reservoir. However, modifications must be made because aquifer geometries can be partially cylindrical; a constant for that condition was defined as C_v, which is also referred to in the literature as B.

$$C_v = \frac{1.119\phi h c_t r_o^2}{B_w}\left(\frac{\theta}{360}\right), \text{ STB/psi}. \quad (7.19)$$

$$W_e = C_v \sum_0^t Q_{pD}(p_e - p_o) = C_v \sum_0^t Q_{pD}\Delta p, \text{ STB}. \quad (7.20)$$

Eqs. 7.18a through 7.20 are in oilfield units, where k is the aquifer permeability (md), t is time (days), ϕ is porosity (fraction), μ is the aquifer-water viscosity (cp), c_t is the total aquifer compressibility ($c_f + c_w$) (1/psi), r_e is the external radius of the aquifer (ft), r_o is the external radius of the reservoir (ft), Dp is the pressure drop ($p_i - p$) (psia), and W_e is the cumulative water influx (bbl).

The pressure performance history can be expressed as a series of small, constant pressure increments by applying the principle of superposition to solve a water-influx problem.

The analysis procedure is as follows:

- Calculate the van Everdingen and Hurst (1949) correlation values:

  $$t_D = \frac{0.00633kt}{\phi\mu c_t r_o^2}, \text{ where time is in days and } r_{eD} = \frac{r_e}{r_o},$$

  where r_o is the external radius of the reservoir (ft).
- Apply the principle of superposition to convert a smooth pressure history to a layered format.
- Determine the dimensionless water-influx volume for each pressure drawdown layer.

Exercise 7.4—Using the Infinite-Acting (Transient/Unsteady-State) Solution To Calculate Water Influx. Apply the transient solution to calculate water influx when there is a 10-psi pressure drop occurring over a 90-day period. **Table 7.6** lists pertinent field data.

Solution. First, calculate the aquifer description term:

$$C_v = \frac{1.119\phi h c_t r_o^2}{B_w}\left(\frac{\theta}{360}\right) = \frac{1.119(0.12)(30)(6\times10^{-6})(2{,}000^2)}{1.03}\left(\frac{95}{360}\right) = 24.8 \text{ STB/psi}.$$

k_w = 100 md	h = 30 ft	r_e = 1,000,000 ft	B_w = 1.03 RB/STB	$c_t = 6 \times 10^{-6}$ psi^{-1}
μ_w = 0.6 cp	θ = 95°	r_o = 2,000 ft	Δp = 10 psi	ϕ = 12%

Table 7.6—Data for example problem.

Dimensionless Water Influx and Dimensionless Pressure for Infinite Radial Aquifers

t_p	Q_t	p_D	t_p	Q_t	t_p	Q_t	t_p	Q_t
1.0×10^{-2}	0.112	0.112	1.5×10^{3}	4.136×10^{2}	1.5×10^{7}	1.828×10^{6}	1.5×10^{11}	1.17×10^{10}
5.0×10^{-2}	0.278	0.229	2.0×10^{3}	5.315×10^{2}	2.0×10^{7}	2.398×10^{6}	2.0×10^{11}	1.55×10^{10}
1.0×10^{-1}	0.404	0.315	2.5×10^{3}	6.466×10^{2}	2.5×10^{7}	2.961×10^{6}	2.5×10^{11}	1.92×10^{10}
1.5×10^{-1}	0.520	0.376	3.0×10^{3}	7.590×10^{2}	3.0×10^{7}	3.517×10^{6}	3.0×10^{11}	2.29×10^{10}
2.0×10^{-1}	0.606	0.424	4.0×10^{3}	9.757×10^{4}	4.0×10^{7}	4.610×10^{6}	4.0×10^{11}	3.02×10^{10}
1.0×10^{1}	7.417	1.651	2.5×10^{5}	4.066×10^{4}	2.5×10^{9}	2.328×10^{8}		
1.5×10^{1}	9.965	1.829	3.0×10^{5}	4.817×10^{4}	3.0×10^{9}	2.771×10^{8}		
2.0×10^{1}	1.229×10^{1}	1.960	4.0×10^{5}	6.267×10^{4}	4.0×10^{9}	3.645×10^{8}		
2.5×10^{1}	1.455×10^{1}	2.067	5.0×10^{5}	7.699×10^{4}	5.0×10^{4}	4.510×10^{8}		
3.0×10^{1}	1.681×10^{1}	2.147	6.0×10^{5}	9.113×10^{4}	6.0×10^{9}	5.368×10^{8}		
4.0×10^{1}	2.088×10^{1}	2.282	7.0×10^{5}	10.51×10^{5}	7.0×10^{9}	6.220×10^{8}		

Table 7.7—Example table, developed by van Everdingen and Hurst (1949).

Next, calculate the real time and dimensionless time relationship:

$$t_D = \frac{0.00633kt}{\phi \mu c_t r_o^2} = \frac{0.00633(100)t}{0.12(0.6)(6 \times 10^{-6})(2{,}000^2)} = 0.366t.$$

Thus, for $t = 90$ days, $t_D = 32.9$, and $r_{eD} = r_e/r_o = (1 \times 10^6)/(2 \times 10^3) = 500$.

Using **Table 7.7** (developed by van Everdingen and Hurst 1949) and interpolating between the values provided the following: $t_D = 30.0$ and Q_{pD} $(Q_t) = 16.81$ and $t_D = 40.0$ and Q_{pD} $(Q_t) = 20.88$; at $t_D = 32.9$, a value of Q_{pD} $(Q_t) = 18.03$ is calculated.

$t_D = 30.0$	Q_{pD} $(Q_t) = 16.81$
$t_D = 32.9$	Q_{pD} $(Q_t) = x = 18.03$
$t_D = 40.0$	Q_{pD} $(Q_t) = 20.88$

Table 7.8—Determining the value of Q_{pD} (Q_t) at $t_D = 32.9$.

Table 7.8 summarizes the interpolation process for calculating Q_{pD} (Q_t).

Then use Eq. 7.20 to calculate the cumulative water influx:

$$W_e = C_v \sum_0^t Q_{pD} \nu p = (24.8 \text{ STB/psi})(18.03)(10 \text{ psi}) = 4{,}464 \text{ STB}.$$

The amount of water influx is estimated to be approximately 4,500 bbl of water during the 90-day period.

Exercise 7.5—Estimating Cumulative Water Influx Using the van Everdingen and Hurst Solution. Estimate the cumulative water influx for the San Miguel Field through the 1.2-year period. A preliminary analysis indicates transient conditions exist. The pressure history and aquifer reservoir rock and fluid properties were given in Table 7.3, with values obtained assuming a steady state and assuming $c_w = 3 \times 10^{-6}$ psi^{-1}:

Solution. First, calculate the total compressibility and the van Everdingen and Hurst (1949) aquifer description term:

$$c_t = c_f + c_w = 6.7\times10^{-5}\ \text{psi}^{-1} + 3\times10^{-6}\ \text{psi}^{-1} = 7.0\times10^{-5}\ \text{psi}^{-1}.$$

$$C_v = \frac{1.119\phi h c_t r_o^2}{B_w}\left(\frac{\theta}{360}\right) = \frac{1.119(0.27)(120)(7\times10^{-5})(1,800^2)}{1.0}\left(\frac{360}{360}\right) = 8,223\ \text{STB/psi}.$$

Next, calculate the real and dimensionless relationships:

$$t_D = \frac{0.00633kt}{\phi\mu c_t r_o^2} = \frac{0.00633(212)t}{0.27(0.8)(7\times10^{-5})(1,800^2)} = 0.0274t \text{ when } t \text{ is in days,}$$

$$r_{eD} = \frac{r_e}{r_o} = \frac{10,800}{1,800} = 6.$$

Follow the same process of dividing the pressure history into a series of steps as discussed in Section 7.2.1. The results of the calculation procedure assuming steady-state flow for the $t = 0.7$- and $t = 1.2$-years time periods are illustrated in Table 7.3 for $\bar{p}$ (psia) and Δp (psi):

$$\bar{p}_1 = \frac{p_i + p_1}{2} = \frac{2,500 + 2,400}{2} = 2,450\ \text{psi},$$

$$\bar{p}_j = \frac{p_{j-1} + p_j}{2} = p_2 = \frac{p_1 + p_2}{2} = \frac{2,400 + 2,300}{2} = 2,350\ \text{psi},$$

$$\Delta p_0 = \bar{p}_i - \bar{p}_1 = 2,500 - 2,450 = 50\ \text{psi},$$

$$\Delta p_1 = \bar{p}_1 - \bar{p}_2 = 2,450 - 2,350 = 100\ \text{psi},$$

$$\Delta p_2 = \bar{p}_2 - \bar{p}_3 = 2,350 - 2,250 = 100\ \text{psi}.$$

Calculating t_d results in

- When $t = 0.3$ years, $t_D = (0.0274)(365)(0.3) = 3.0$.
- When $t = 0.7$ years, $t_D = (0.0274)(365)(0.7) = 7.0$.
- When $t = 1.2$ years, $t_D = (0.0274)(365)(1.2) = 12.0$.

Then, calculate the dimensionless water-influx Q_{pD} values for $r_{eD} = 6$ at the corresponding dimensionless times (t_D values) that were just calculated.
The values of Q_{pD} $(Q_t) = W_D$ for each t_D value are obtained from **Table 7.9 or Fig. 7.5:**

- Q_{pD} $(Q_t) = W_D$ at $t_D = 3.0$ results in 3.09.
- Q_{pD} $(Q_t) = W_D$ at $t_D = 7.0$ results in 5.749.
- Q_{pD} $(Q_t) = W_D$ at $t_D = 12.0$ results in 8.436.

Dimensionless Water Influx and Dimensionless Pressure for Infinite Radial Aquifers

t_p	Q_t	p_D	t_p	Q_t	t_p	Q_t	t_p	Q_t
1.0×10^{-2}	0.112	0.112	1.5×10^{3}	4.136×10^{2}	1.5×10^{7}	1.828×10^{6}	1.5×10^{11}	1.17×10^{10}
5.0×10^{-2}	0.278	0.229	2.0×10^{3}	5.315×10^{2}	2.0×10^{7}	2.398×10^{6}	2.0×10^{11}	1.55×10^{10}
1.0×10^{-1}	0.404	0.315	2.5×10^{3}	6.466×10^{2}	2.5×10^{7}	2.961×10^{6}	2.5×10^{11}	1.92×10^{10}
1.5×10^{-1}	0.520	0.376	3.0×10^{3}	7.590×10^{2}	3.0×10^{7}	3.517×10^{6}	3.0×10^{11}	2.29×10^{10}
2.0×10^{-1}	0.606	0.424	4.5×10^{3}	9.757×10^{2}	4.0×10^{7}	4.610×10^{6}	4.0×10^{11}	3.02×10^{10}
2.0	2.442	1.020	4.0×10^{4}	7.634×10^{3}	4.0×10^{-8}	4.071×10^{7}		
2.5	2.838	1.101	5.0×10^{4}	9.342×10^{3}	5.0×10^{-8}	5.032×10^{7}		
3.0	3.209	1.169	6.0×10^{4}	11.03×10^{4}	6.0×10^{-8}	5.984×10^{7}		
4.0	3.897	1.275	7.0×10^{4}	12.69×10^{4}	7.0×10^{-8}	6.928×10^{7}		
5.0	4.541	1.362	8.0×10^{4}	14.33×10^{4}	8.0×10^{-8}	7.865×10^{7}		
6.0	5.148	1.436	9.0×10^{4}	15.95×10^{4}	9.0×10^{-8}	8.797×10^{7}		
7.0	5.749	1.500	1.0×10^{5}	17.56×10^{4}	1.0×10^{-9}	9.725×10^{7}		
8.0	6.314	1.556	1.5×10^{5}	2.538×10^{4}	1.5×10^{-9}	1.429×10^{8}		
9.0	6.861	1.604	2.0×10^{5}	3.308×10^{4}	2.0×10^{-9}	1.880×10^{8}		
1.0×10^{1}	7.417	1.651	2.5×10^{5}	4.066×10^{4}	2.5×10^{-9}	2.328×10^{8}		

Table 7.9— Portion of Table I developed by van Everdingen and Hurst (1949).

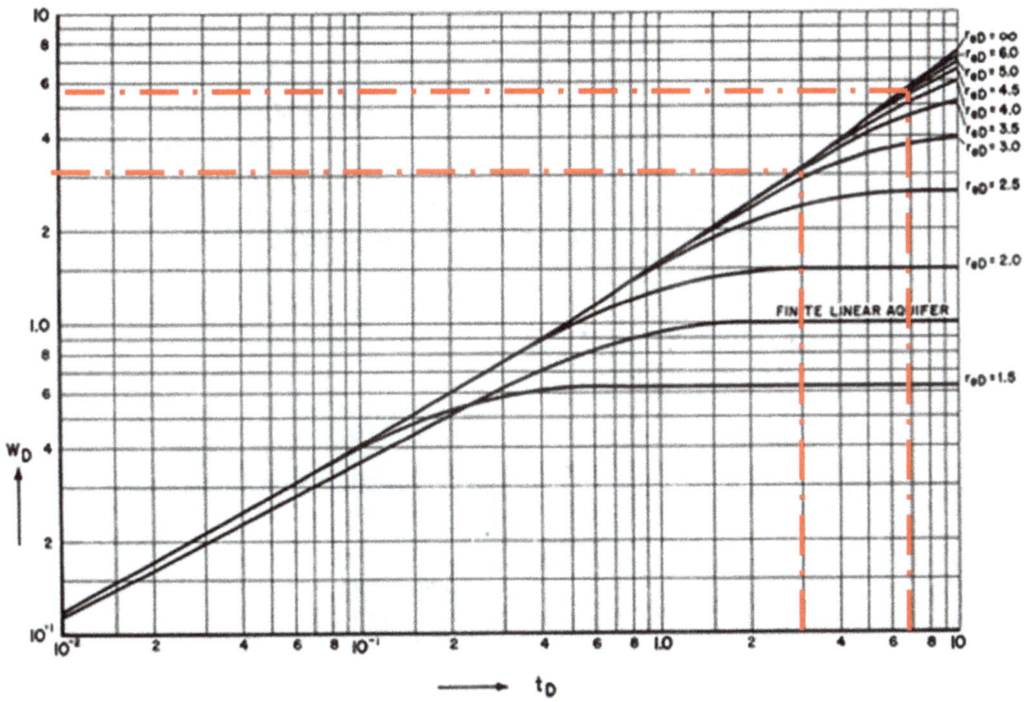

Fig. 7.5—Graphical form of van Everdingen and Hurst (1949) unsteady-state dimensionless water influx (after Todd 2021).

Follow the van Everdingen and Hurst (1949) solution procedure:

- Calculate the water influx (W_e) during the first period ($t = t_1 = 0.3$ years) using the data provided in **Table 7.10.** Note that the only pressure drop acting during that first period is υ_{p0}. Therefore, use the dimensionless cumulative water influx W_D corresponding to a dimensionless time t_D calculated to a real time $t = t_1 = 0.3$ years.

$$W_{e_1} = C_v \Delta p_0 W_D \left(t_{D_1 \text{ period}}\right) = (8{,}223)(50)(3.09) = 1.270 \text{ million RB}.$$

The time/pressure history is shown in **Fig. 7.6.**

- Follow the same procedure to calculate the cumulative water influx (W_e) during the second period ($t = t_2 = 0.7$ years). At the end of this second period,
  - Δp_0, which has now lasted for two periods = 1 year
  - Δp_1, which has lasted for only one period = 0.7 years

Time (years)	t_D	Pressure (psia)	Average Pressure (psia)	Δp (psi)	Δp (psi)	$W_D = Q_{pD}$	W_e (STB)
			Water-influx calculations for $t = 0.7$ years				
0	–	2,500	–	–	–	–	–
0.3	3	2,400	2,450	Δp_0	50	3.09	1,270,453.5
0.7	**7**	2,300	2,350	Δp_1	100	5.749	**4,904,608.4**
1.2	**12**	2,200	–	Δp_2	–	8.436	–
1.7	–		Water-influx calculations for $t = 1.2$ years				
2.8	–	2,500	–	–	–	–	–
		2,400	2,450	Δp_0	50	3.09	–
		2,300	2,350	Δp_1	100	5.749	
		2,200	2,250	Δp_2	100	8.436	**10,736,771**

Table 7.10—Calculations using van Everdingen and Hurst (1949) unsteady-state water influx.

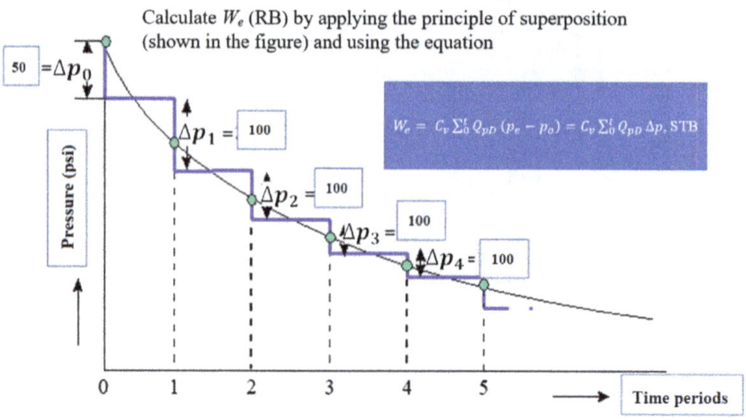

Fig. 7.6—The application of the superposition principle to the van Everdingen and Hurst (1949) equation under unsteady-state aquifer influx.

The cumulative water influx at the end of the second period (W_{e_2}), which equals 1 year, would be

$$W_{e_2} = C_v\left[\Delta p_0 W_D\left(t_{D_2 \text{ periods}}\right) + \Delta p_1 W_D\left(t_{D_1 \text{ period}}\right)\right] =$$
$$8,223\left[(50)(5.749) + (100)(3.09)\right] = 4.905 \text{ million RB.}$$

Follow the procedure again to calculate the cumulative water influx (W_e) during the third period ($t = t_3 = 1.2$ years). At the end of the third period,

- Δp_0, which has now lasted for three periods = 1.2 years
- Δp_1, which has now lasted for two periods = 1.0 year
- Δp_2, which has lasted only for one period = 0.3 years

The cumulative water influx at the end of the third period (W_{e_3}) would be

$$W_{e_3} = C_v\left[\Delta p_0 W_D\left(t_{D_3 \text{ periods}}\right) + \Delta p_1 W_D\left(t_{D_2 \text{ periods}}\right) + \Delta p_2 W_D\left(t_{D_1 \text{ period}}\right)\right].$$

Applying the principle of superposition and substituting the corresponding numerical values,

$$W_{e_3} = 8,223\left[(50)(8.436) + (100)(5.749) + (100)(3.09)\right].$$

$$W_e(t = 1.2 \text{ years}) = C_v \sum_0^t Q_{pD}\Delta p = W_{e_3} = 10.737 \text{ million RB.}$$

Table 7.11 shows for comparison the calculated values between steady-state water-influx calculations and van Everdingen and Hurst (1949) unsteady-state water influx, which considers infinite-acting aquifer behavior.

7.2.3 Carter and Tracy Method. The step function must be applied to a smooth pressure decline to apply the constant pressure approaches of Schilthuis (1936) and van Everdingen and Hurst (1949). Carter and Tracy (1960) developed an alternative method for calculating cumulative water influx by applying an approximation technique to the van Everdingen and Hurst (1949) constant rate solution to the diffusivity equation for finite-acting conditions. The solution procedure to apply the Carter and Tracy method is straightforward because the principle of superposition is omitted from the calculation process.

Time (years)	W_e Steady State (Table 7.4) (million STB)	W_e Unsteady State (million STB)	Difference (%)
0	–	–	–
0.3	0.7	1.3	46%
0.7	3.4	4.9	31%
1.2	9.2	10.7	14%

Table 7.11—Comparison between steady-state water-influx calculations and van Everdingen and Hurst (1949) unsteady-state water influx.

Application of the van Everdingen and Hurst (1949) constant rate solution to calculate cumulative water influx requires that the smooth rate/time curve be divided into a series of instantaneous constant rate cases.

Assumptions

- Works for finite and infinite aquifers.
- Radial flow.
- Production history is expressed as a series of incremental constant rate steps instead of the van Everdingen and Hurst (1949) constant pressure solution.

The following definitions, which were developed in Chapter 4, are valid for the constant pressure case. Integration is not required because the problem is treated as a series of constant rate steps.

Recall the constant pressure solution (Eq. 4.18), which was also given in Section 7.2.2: $q_D = \dfrac{141.2 q\mu B_w}{kh(p_e - p_o)}$.

Cumulative water influx is defined as

$$W_e = qt . \qquad (7.21)$$

Dimensionless water influx was defined in Chapter 4 as $Q_{pD} = q_D t_D$.

Substituting the $t_D = \dfrac{0.00633kt}{\phi\mu c_t r_o^2}$ relationship into Eq. 7.21 results in

$$W_e = qt = \frac{\phi\mu c_t r_o^2 q t_D}{0.00633k}. \qquad (7.22)$$

Eq. 7.22 defines the rate term applicable to a particular timestep. This term contains the reservoir parameters included in the dimensionless time definition.

$q*$ is defined as the pseudo flow rate:

$$q^* = \frac{\phi\mu c_t r_o^2 q}{0.00633k}. \qquad (7.23)$$

Substituting Eq. 7.22 into the dimensionless water-influx equation from Chapter 4 results in an expression for cumulative water influx referenced to rate and dimensionless time:

$$W_e = q^* t_D. \qquad (7.24)$$

Eq. 7.24 can be expanded into a series of constant rate cases:

$$W_e(t_{D3}) = q_0^*(t_{D1} - t_{D0}) + q_1^*(t_{D2} - t_{D1}) + q_2^*(t_{D3} - t_{D2}) + q_3^*(t_{D4} - t_{D3}). \qquad (7.25)$$

Substituting Eq. 7.25 into the $t_D = \dfrac{0.00633kt}{\phi\mu c_t r_o^2}$ relationship results in an equation expressed in summation form:

$$W_e\left(t_{Dj}\right)=\sum_{n=0}^{j-1}q_n^*\left[t_{D(n+1)}-t_{Dn}\right]. \qquad (7.26)$$

Eq. 7.26 can be divided into two components:

$$W_e\left(t_{Dj}\right)=\sum_{n=0}^{j-1}q_n^*\left[t_{D(n+1)}-t_{Dn}\right]+\sum_{n=i}^{j-1}q_n^*\left[t_{D(n+1)}-t_{Dn}\right]. \qquad (7.27)$$

Eq. 7.27 represents the cumulative water influx from t_{D0} to t_{Di} + the cumulative water influx from t_{Di} to t_{Dj}. This equation can be transformed to

$$W_e\left(t_{Dj}\right)=W_e t_{Di}+\sum_{n=i}^{j-1}q_n^*\left[t_{D(n+1)}-t_{Dn}\right], \qquad (7.28)$$

where the cumulative influx from t_{D0} to t_{Di} equals

$$W_e t_{Di}, \qquad (7.29)$$

and the cumulative influx summed as a series of constant rate steps from t_{Di} to t_{Dj} equals

$$\sum_{n=i}^{j-1}q_n^*\left[t_{D(n+1)}-t_{Dn}\right]. \qquad (7.30)$$

The development of Eq. 7.30 was critical because the pressure-loss/water-influx response for the first iteration is a known value (zero). The cumulative water influx occurring during only the i to j interval must be calculated. The sum of the two parts is then included as the first part of Eq. 7.28 to calculate water influx at the next timestep.

If $i = j - 1$, then

$$W_e\left(t_{Dj}\right)=W_e t_{D(j-1)}+q_{j-1}^*\left[t_{Dj}-t_{D(j-1)}\right], \qquad (7.31)$$

which can be expressed as the convolution integral,

$$W_e\left[t_{D(j-1)}\right]=C_v\int_0^{t_{Dj}}\Delta p Q_{pD}\left[t_{D(j+1)}-\tau\right]\mathrm{d}\tau, \qquad (7.32)$$

where τ is the dummy variable of integration.

Combining Eqs. 7.31 and 7.32, applying Laplace transforms, solving for and inserting the normal counting term n, and including the step function to account for superposition result in

$$W_{en}=W_{e(n-1)}+\left[t_{Dn}-t_{D(n-1)}\right]\left[\frac{C_v\Delta p_n-W_{e(n-1)}\,p'_{Dn}}{p_{Dn}-t_{D(n-1)}\,p'_{Dn}}\right]. \qquad (7.33)$$

p'_{Dn} is the slope of the p_D/t_D curve whose relationships are shown in Table 7.7 from van Everdingen and Hurst (1949). **Fig. 7.7** shows an example slope of the line calculation.

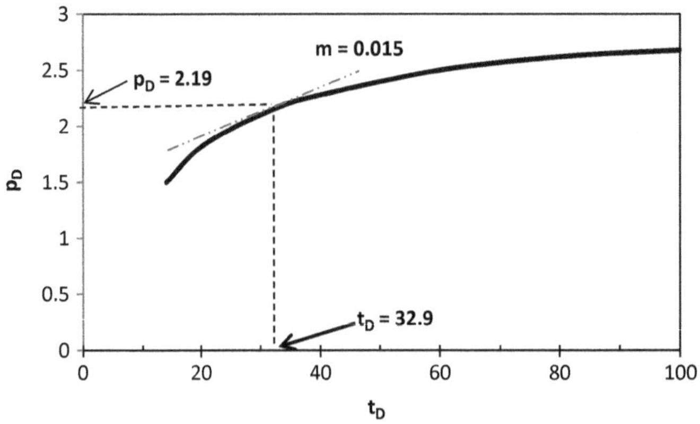

Fig. 7.7—An example slope of the line plot for the Carter and Tracy (1960) calculation procedure.

Observations

- The Carter and Tracy (1960) method is an approximation technique and is not rigorously correct. However, experience has shown the technique works equally well for most infinite- and finite-acting aquifers.
- In the Carter and Tracy (1960) analysis method, cumulative pressure drops represent the superposition method of average pressures.
- Water influx at initial time is equal to zero. Therefore, there is no problem with having to evaluate the derivative at initial time when the slope of the line is steepest.

The analysis procedure is as follows:

- Select time/pressure relations over the production history.
- Calculate the equivalent dimensionless time (t_D) for each selected real time.
- Use the constant rate table (a portion is shown in Table 7.7) from van Everdingen and Hurst (1949) to select the appropriate p_D value.
- Using the equivalent dimensionless time t_D and the appropriate p_D value, build a graph and draw a smooth curve of the p_D vs. t_D points and calculate p'_{Dn} for each selected time.
- Use Eq. 7.33 from Carter and Tracy (1960) to calculate the cumulative water influx for each progressive timestep.

Exercise 7.6—Calculating Cumulative Water Influx Using the Carter and Tracy Method. An aquifer undergoes a 10-psi pressure drop for 90 days. Calculate the cumulative water influx using the Carter and Tracy method. Use the data presented in Exercise 7.4 and Table 7.6.

Solution. Recall the calculated values of C_v, t_D, and r_{eD} from Exercise 7.4:

$$t_D = \frac{0.00633kt}{\phi\mu c_t r_o^2} = \frac{0.00633(100)t}{0.12(0.6)(6\times10^{-6})(2{,}000^2)} = 0.366t.$$

For $t = 90$ days, $t_D = 32.9$, and $r_{eD} = r_e/r_o = (1\times10^6)/(2\times10^3) = 500$.

$$C_v = \frac{1.119\phi h c_t r_o^2}{B_w}\left(\frac{\theta}{360}\right) = \frac{1.119(0.12)(30)(6\times10^{-6})(2{,}000^2)}{1.03}\left(\frac{95}{360}\right) = 24.8 \text{ STB/psi}.$$

Interpolate to find the p_D value at $t_D = 32.9$. Note in **Table 7.12** that the upper and lower bounds of the t_D and p_D columns are $t_D = 30$ and $t_D = 40$. **Table 7.13** represents the interpolation process for this particular problem.

Dimensionless water influx and dimensionless pressure for infinite radial aquifers

t_D	Q_t	p_D	t_D	Q_t	t_D	Q_t	t_D	Q_t
1.0×10^{-2}	0.112	0.112	1.5×10^{3}	4.136×10^{2}	1.5×10^{7}	1.828×10^{6}	1.5×10^{11}	1.17×10^{10}
5.0×10^{-2}	0.278	0.229	2.0×10^{3}	5.315×10^{2}	2.0×10^{7}	2.398×10^{6}	2.0×10^{11}	1.55×10^{10}
1.0×10^{-1}	0.404	0.315	2.5×10^{3}	6.466×10^{2}	2.5×10^{7}	2.961×10^{6}	2.5×10^{11}	1.92×10^{10}
1.5×10^{-1}	0.520	0.376	3.0×10^{3}	7.590×10^{2}	3.0×10^{7}	3.517×10^{6}	3.0×10^{11}	2.29×10^{10}
2.0×10^{-1}	0.606	0.424	4.0×10^{3}	9.757×10^{2}	4.0×10^{7}	4.610×10^{6}	4.0×10^{11}	3.02×10^{10}
1.0×10^{1}	7.417	1.651	2.5×10^{5}	4.066×10^{4}	2.5×10^{9}	2.328×10^{8}		
1.5×10^{1}	9.965	1.829	3.0×10^{5}	4.817×10^{4}	3.0×10^{9}	2.771×10^{8}		
2.0×10^{1}	1.229×10^{1}	1.960	4.0×10^{5}	6.267×10^{4}	4.0×10^{9}	3.645×10^{8}		
2.5×10^{1}	1.455×10^{1}	2.067	5.0×10^{5}	7.699×10^{4}	5.0×10^{4}	4.510×10^{8}		
3.0×10^{1}	1.681×10^{1}	2.147	6.0×10^{5}	9.113×10^{4}	6.0×10^{9}	5.368×10^{8}		
4.0×10^{1}	2.088×10^{1}	2.282	7.0×10^{5}	10.51×10^{5}	7.0×10^{9}	6.220×10^{8}		

Table 7.12—Example table, developed by van Everdingen and Hurst (1949) and used in the Carter and Tracy method.

Using the values specified in Table 7.12 and interpolating at $t_D = 32.9$,

$$\frac{(32.9-30)}{(40-30)} = \frac{(x-2.147)}{(2.282-2.147)};\ p_D = 2.186 \text{ at } t_D = 32.9.$$

t_D	p_D
30	2.147
32.9	?
40	2.282

Table 7.13—Interpolated values to determine dimensionless pressure p_D.

From the values given in Table 7.12, the slope can be determined, and it is shown graphically in Fig. 7.7.

$$p'_D \text{ at } t_{Dj} = \frac{2.282-2.147}{40-30} = 0.0135.$$

Calculating cumulative water influx at 90 days,

$$W_{en} = W_{e(n-1)} + \left[t_{Dn} - t_{D(n-1)}\right]\left[\frac{C_v \Delta p_n - W_{e(n-1)}\, p'_{Dn}}{p_{Dn} - t_{D(n-1)}\, p'_{Dn}}\right];$$

$$W_e = 0 + (32.9-0)\left[\frac{(24.8\times 10-0)}{(2.186-0)}\right] = 3{,}733 \text{ STB}.$$

The water-influx value at zero time is zero at the first timestep. Thereafter, there is no difficulty in evaluating the derivative of the first data point.

Exercise 7.7—Calculating Cumulative Water Influx Using the Carter and Tracy Method. Apply the Carter and Tracy method to calculate cumulative water influx in the San Miguel Field at 2.8 years. The aquifer reservoir rock and fluid properties were given in Table 7.3. The time, pressure, and calculation results are shown in **Table 7.14.** Reservoir data were developed in Exercise 7.5:

$$c_t = c_f + c_w = 6.7\times 10^{-5} \text{ psi}^{-1} + 3\times 10^{-6} \text{ psi}^{-1} = 7.0\times 10^{-5} \text{ psi}^{-1}.$$

Time (years)	p (psi)	t_{Dn}	Δp_n (psi)	p_{Dn}	p'_{Dn}	W_e (million BW)
0	2,500	0	–	–	–	–
0.3	2,400	3	100	1.167	0.128	2.1
0.7	2,300	7	200	1.511	0.067	6.7
1.2	2,200	12	300	1.815	0.058	14.1
1.7	2,100	17	400	2.103	0.058	22.9
2.8	1,800	28	700	2.732	0.057	50.7

Table 7.14—Pressure, time, and results of calculations, Carter and Tracy method.

$$C_v = \frac{1.119\phi h c_t r_o^2}{B_w}\left(\frac{\theta}{360}\right) = \frac{1.119(0.27)(120)(7\times10^{-5})(1,800^2)}{1.0}\left(\frac{360}{360}\right) = 8,223 \text{ STB/psi.}$$

Then, the real and dimensionless relationships were calculated:

$$t_D = \frac{0.00633kt}{\phi\mu c_t r_o^2} = \frac{0.00633(212)t}{0.27(0.8)(7\times10^{-5})(1,800^2)} = 0.0274t \text{ when } t \text{ is in days.}$$

Because t in Table 7.14 is in years, $t_D = 0.0274 \times 365 = 10t$.

$$r_{eD} = \frac{r_e}{r_o} = \frac{10,800}{1,800} = 6.$$

Solution. Use the Carter and Tracy solution to calculate water influx for the first two timesteps of the example problem. Note the absence of superposition.

$$W_{en} = W_{e(n-1)} + (t_{Dn} - t_{D(n-1)})\left[\frac{C_v \Delta p_n - W_{e(n-1)}\, p'_{Dn}}{p_{Dn} - t_{D(n-1)} \times p'_{Dn}}\right] =$$

$$0 + (3.0 - 0)\left[\frac{8,223\times100 - 0\times0.128}{1.167 - 0\times0.128}\right] = 2.1 \text{ Million BW.}$$

$$W_{e2} = 2,114,000 + (7.0 - 3.0)\left[\frac{8,223\times200 - 2,114,000\times0.067}{1.511 - 3.0\times0.067}\right] = 6.70 \text{ million BW.}$$

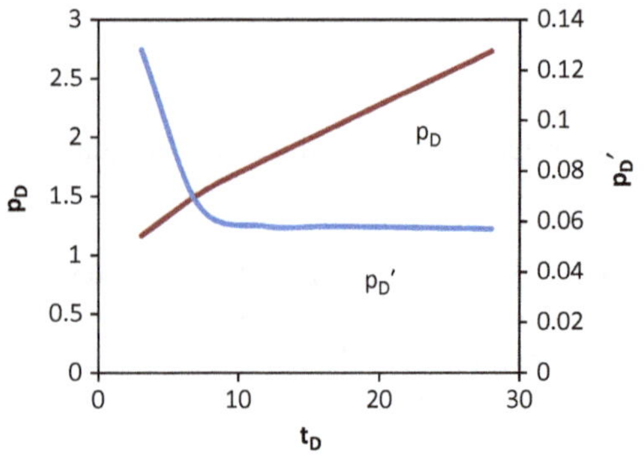

Fig. 7.8—Resulting plots of changing p_D and p'_{Dn} values for the calculations shown in Table 7.14.

The p_D/t_D plot for this problem is shown in **Fig. 7.8.**

7.2.4 Fetkovich Method. Fetkovich (1971) combined the material-balance (Chapter 5) and PI equations (Chapter 6) to develop a method for calculating water influx from small to medium-sized aquifers. The PI models describe the fluid flow within the aquifer and into a reservoir, while a material balance of water inflow is maintained at the water contact. Irregularities occur when the method is applied to very small aquifers because of the effect of the shape factor $\left(\frac{\theta}{360}\right)$, which depends on the value of the contact angle between the aquifer and the reservoir. In addition, because steady-state or pseudosteady-state conditions are assumed to exist, the technique does not apply to very large aquifers.

The Fetkovich method is applicable to closed, finite aquifers when $3 < r_{eD} < 10$. This method does not require superposition and can be applied to any reservoir/aquifer geometry. The equations are expressed in terms of real time, and transient flow is ignored.

Assumptions

- Works best for closed, finite aquifers when $3 < r_{eD} < 10$.
- The PI models flow within the aquifer.
- A material balance is maintained to calculate water outflow from the aquifer.
- Total compressibility effects cause water movement.

The aquifer PI equation is

$$q_w = \frac{dW_e}{dt} = J\left(\bar{p}_a - p_o\right), \quad (7.34)$$

where J is the aquifer PI (bbl/D-psi) and $\bar{p}_a$ is the average aquifer pressure (psi) at any timestep.

The theoretical maximum water volume that can be expelled from the aquifer W_{ei} can be expressed as

$$W_{ei} = W_i c_t p_i, \quad (7.35)$$

where W_i is the volume of water initially in place (STB), c_t is the aquifer compressibility (psi^{-1}) ($c_t = c_f + c_w$), and p_i is the initial pressure (psi).

A material balance calculates the water-influx volume as a function of aquifer compressibility and the pressure drop between an initial and any later condition:

$$W_e = W_i c_t \left(p_i - \bar{p}_a\right). \quad (7.36)$$

Combining Eqs. 7.34 and 7.35 with Eq. 7.36 results in

$$\bar{p}_a = p_i\left(1 - \frac{W_e}{W_{ei}}\right). \quad (7.37)$$

This can be rearranged to

$$\bar{p}_{a(n-1)} = p_i\left[1 - \frac{\sum_{j=1}^{n-1} \Delta W_{ej}}{W_{ei}}\right]. \quad (7.38)$$

Eq. 7.38 establishes a relationship between the decline of aquifer pressure and the expelled water volume caused by a pressure drop in the reservoir. Combining Eqs. 7.34 and 7.37 and performing mathematical manipulation result in

$$W_e = \frac{W_{ei}}{p_i}\left(p_i - p_o\right)\left(1 - e^{-Jp_i t/W_{ei}}\right). \quad (7.39)$$

Eq. 7.39 establishes a causal relationship between an instantaneous pressure decline and the resulting volume of water expelled from the aquifer. However, changing inner boundary conditions precludes the use of the equation. Fetkovich (1971) showed a different form of Eq. 7.39, which can replace the need to include the superposition principle to account for these changing conditions. Eq. 7.40 calculates incremental water influx occurring during some time span n (timestep Δt_n):

$$\Delta W_{en} = \frac{W_{ei}}{p_i}\left[\overline{p}_{a(n-1)} - p_{on}\right]\left(1 - e^{-Jp_i\Delta t_n/W_{ei}}\right), \quad (7.40)$$

where p_{on} is the pressure at the OWC at any timestep n, measured at the end of the previous timestep, and Δt_n is the time interval over which the pressure decline occurred. W_{ei}, p_i, and J are constant values calculated from reservoir characteristics.

Eq. 7.40 can be used in a stepwise manner to closely approximate the van Everdingen and Hurst (1949) solution for pseudosteady-state water influx into a finite aquifer.

At first glance, Eq. 7.40 appears formidable, and the majority of the terms are constants throughout the entire calculation. However, Eq. 7.40 can be used in a compact form when these coefficients are determined at the beginning of the calculation procedure.

The PI is a function of aquifer geometry and flow conditions. Equations expressed in field units for linear- and radial-flow conditions are listed in **Table 7.15.**

	Radial Aquifers	Linear Aquifers
Pseudosteady State	$\frac{0.00708kh(\theta/360°)}{\mu\left[\ln(r_e/r_o)-0.75\right]}$	$\frac{0.00338khw}{\mu L}$
Steady State	$\frac{0.00708kh(\theta/360°)}{\mu\left[\ln(r_e/r_o)\right]}$	$\frac{0.00113khw}{\mu L}$

Table 7.15—PI equations.

The analysis procedure is as follows:

- Calculate the aquifer properties, area (acres), and water volume (RB).
- Set up a pressure vs. time table.
- Calculate W_{ei}, p_i, J, and Δt_n, and reduce Eq. 7.40 to a more-compact form.
- Use Eq. 7.40 to calculate the cumulative water influx at each timestep. Only two terms are treated as variables; the remaining terms are constants particular to the field problem and need to be solved only once.

$$\Delta W_{en} = \frac{W_{ei}}{p_i}\left[\overline{p}_{a(n-1)} - p_{on}\right]\left(1 - e^{-Jp_i\Delta t_n/W_{ei}}\right),$$

where $\overline{p}_{a(n-1)}$ is the average aquifer static pressure measured at a previous timestep,

$$\overline{p}_{a(n-1)} = p_i\left(1 - \frac{\sum_{j=1}^{n-1}\Delta W_{ej}}{W_{ei}}\right),$$

and Δt_n is the time interval between two measurements.

- Calculate the average aquifer pressure and enter this value in Eq. 7.40 to calculate the cumulative water influx W_e.

Exercise 7.8—Calculating Cumulative Water Influx Using the Fetkovich Method. Calculate the cumulative water influx at the end of 2.8 years for the San Miguel Field. Recall that $r_{eD} = 6$ and $t_D = 10t$ for this field. Assume a radial aquifer with pseudosteady-state producing conditions. Use the dimensionless time definition to determine when the van Everdingen and Hurst solution (Tables 7.7 and 7.12 and **Fig. 7.9**) changes from transient flow to boundary-dominated flow.

A review of Tables 7.7 and 7.12 and Fig. 7.9 by van Everdingen and Hurst (1949) indicates that the first data point for the $r_{eD} = 6$ value begins at $t_D \approx 7$, which can be observed from Fig. 7.9. Therefore, the break to a pseudosteady state occurs at $t = 0.7$ years.

In **Table 7.16,** Columns 1 and 2 list the pressure history, while Columns 3 through 7 list the results of the calculations.

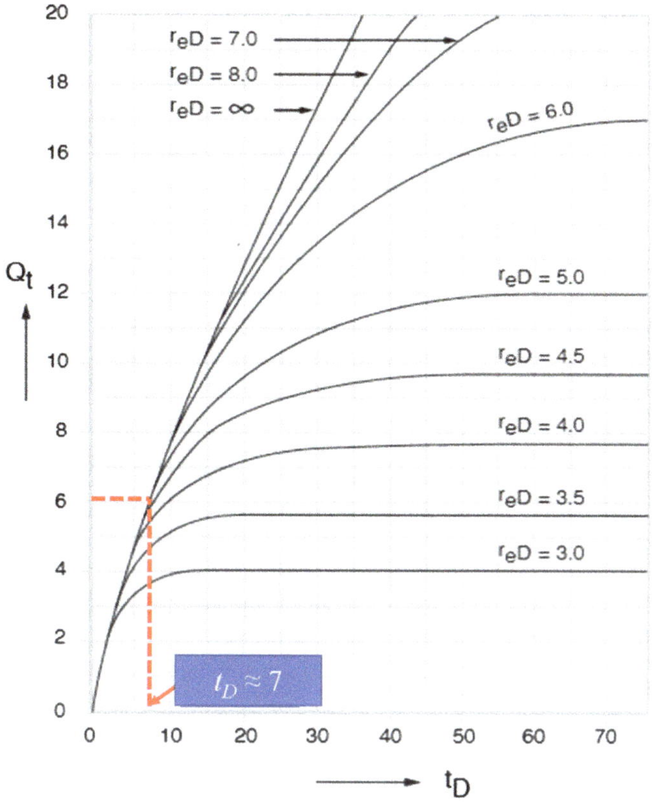

Fig. 7.9—Graphical form of van Everdingen and Hurst (1949) unsteady-state dimensionless water influx (after Todd 2021).

Solution. Calculate the aquifer properties. Assuming circular areas, the contact angle $\left(\dfrac{\theta}{360}\right) = \left(\dfrac{360}{360}\right) = 1.$

$$\text{Aquifer area} = A_a = A_t - A_o = \pi(10{,}800)^2 - \pi(1{,}800)^2 = 8{,}175 \text{ acres.}$$

k_w = 212 md			h = 120 ft	r_e = 10,800 ft	B_w = 1.0 RB/STB	$c_t = 7 \times 10^{-5}$ psi^{-1}
μ_w = 0.8 cp			θ = 140°	r_o = 1,800 ft	Δp = 10 psi	ϕ = 27%
1	2	3	4	5	6	7
Time (years)	t_D	p_{on} (psia)	$\bar{p}_{a(n-1)} - p_{on}$ (psia)	ΔW_{en} (million BW)	ΣW_e (million BW)	$\bar{p}_{an}$ (psia)
0		2,500	–	–	–	–
0.3	3	2,400	100	2.6	**2.6**	2,485
0.7	7	2,300	185	5.2	**7.8**	2,448
1.2	12	2,200	248	8.6	16.14	2,390
1.7	17	2,100	290	10.0	26.14	2,319
2.8	28	1,800	519	33.8	59.9	–

Table 7.16—Aquifer properties and a list of calculations.

W_i = aquifer volume = 7,758 bbl/acre-ft(8,175 acres)(120 ft)(0.27) = 2.054 × 10⁹ RB.

Determine when the production history enters the pseudosteady state:

$$t_D = \frac{0.00633kt}{\phi\mu c_t r_o^2} = \frac{(0.00633)(212)t}{(0.27)(0.8)\left(7\times10^{-5}\right)\left(1,800\right)^2} =$$
$$0.0274t \ (t \text{ is expressed in days}) = 10t \ (t \text{ is expressed in years})$$

It can be estimated from Fig. 7.9 that pseudosteady-state flow begins at approximately $t_D \approx 7$, which in real time is 0.7 years.

Next, calculate the maximum-possible water-influx volume W_{ei}:

$$W_{ei} = W_i c_t p_i = \left(2.054\times10^9\right)\left(7\times10^{-5}\right)\left(2,500\right) = 359,450,000 \text{ RB}.$$

Calculate the aquifer PI. Use the data provided in Table 7.16.

$$J = \frac{0.00708kh(\theta/360°)}{\mu\left[\ln\left(r_e/r_w\right)-0.75\right]} = \frac{(0.00708)(212)(120)}{0.8\left[\ln(10,800/1,800)\right]-0.75} = 263.55 \text{ RB/D-psi}.$$

Calculate $\left(1-e^{-Jp_it/W_{ei}}\right)$:

$$\left(1-e^{-Jp_i\Delta t_n/W_{ei}}\right) = 1-e^{-263.55(2,500)(\Delta t_n)(365)/(359,450,000)} = 1-e^{(-0.669\Delta t_n)}.$$

Calculate the $\frac{W_{ei}}{p_i}$ term:

$$\frac{359,450,000}{2,500} = 143,780 \text{ RB/psia}.$$

Reduce Eq. 7.40, $vW_{en} = \frac{W_{ei}}{p_i}\left[\bar{p}_{a(n-1)} - p_{on}\right]\left(1-e^{-Jp_ivt_n/W_{ei}}\right)$, to

$$\Delta W_{en} = 143,780\left[\bar{p}_{a(n-1)} - p_{on}\right]\left[1-e^{(-0.669\Delta t_n)}\right].$$

Sample calculations for $t = 0.3$ years and $t = 0.7$ years are provided here. Other calculated values are summarized in Table 7.16:

$$\text{At } t = 0.3 \text{ years}, \left[\bar{p}_{a(n-1)} - p_{on}\right] = 2,500 - 2,400 = 100 \text{ psi},$$

$$\Delta W_e = (143,780)(100)\left[1-e^{(-0.669\times0.3)}\right] = 2.6. \text{ million BW}.$$

Applying Eq. 7.38,

$$\overline{p}_{a(n-1)} = p_i\left(1-\frac{\sum_{j=1}^{n-1}\Delta W_{ej}}{W_{ei}}\right) = 2,500\left(1-\frac{2.175}{359.5}\right) = 2,485 \text{ psia.}$$

At $t = 0.7$ years, $\left[\overline{p}_{a(n-1)} - p_{on}\right] = 2,485 - 2,300 = 185$ psi,

$$\Delta W_e = (143,780)(185)\left[1-e^{(-0.549\times 0.4)}\right] = 5.2. \text{ million BW.}$$

$$\Delta W_e = 2.6 + 5.2 = 7.8 \text{ million RB,}$$

$$\overline{p}_{a(n-1)} = p_i\left(1-\frac{\sum_{j=1}^{n-1}\Delta W_{ej}}{W_{ei}}\right) = 2,500\left(1-\frac{7.42}{359.5}\right) = 2,448 \text{ psia.}$$

Values for the cumulative water influx at the end of 2.8 years for the San Miguel Field are shn in Table 7.16. $W_e = 59.9$ million bbl of water.

7.2.5 Comparison of Methods. The following observations can be made:

- The Schilthuis and van Everdingen and Hurst methods include the step function to segment the smoothly declining pressure curve.
- Carter and Tracy (1960) and Fetkovich (1971) circumvented the need for superposition with approximation techniques.
- Schilthuis (1936) assumes steady-state water influx, while van Everdingen and Hurst (1949) and Carter and Tracy (1960) are appropriate for both transient and pseudosteady-state flow conditions.

The cumulative water-influx values calculated by means of the van Everdingen and Hurst, Carter and Tracy, and Fetkovich methods are virtually the same. However, the values calculated with the steady-state Schilthuis method are less than the norm at early time but diverge to more than the norm at later time. The inability of the Schilthuis method to model water influx is probably the result of an erratic production history causing transients during a significant portion of the reservoir history. The only explanation for the errant Fetkovich solution is that transient flow conditions were not considered in the development of the method. The results of the calculations for the four methods are shown in **Fig. 7.10.**

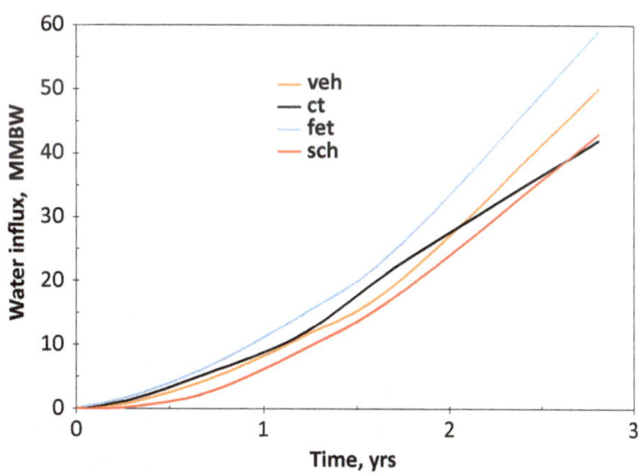

Fig. 7.10—Comparing the van Everdingen and Hurst (veh), Schilthuis (sch), Carter and Tracy (ct), and Fetkovich (fet) water-influx calculation techniques.

Schilthuis method

- Only applicable when steady-state conditions exist
- Superposition required

van Everdingen and Hurst method

- Applicable to infinite and boundary-dominated conditions
- Analytically correct but requires superposition
- Applies constant pressure solutions to the diffusivity equation

Carter and Tracy method

- Designed for boundary-dominated conditions but appears to work equally well for the infinite case
- Based on the van Everdingen and Hurst (1949) constant rate solutions for the diffusivity equation
- An approximation technique with no need for superposition

Fetkovich method

- Only applicable to steady-state or pseudosteady-state flow conditions
- Works best for $3 < r_{eD} < 10$
- An approximation technique not requiring superposition

7.3 Summary

Production from a reservoir often changes the fluid saturation distribution, and replacement by means of water encroachment from a contiguous aquifer often results. Water influx might or might not provide an energy source during reservoir depletion. Calculating these fluid replacement effects involves considerable uncertainty because aquifer rock properties and size are some of the greatest unknowns in the study of oil and gas depletion.

Idealized geometric models provide the starting point for all water-influx studies. Some type of "best-fit" statistical study is included in the calculation process because information concerning aquifer character is generally unknown. The statistical side of the procedure entails entering various combinations of aquifer properties and determining a "most likely" estimate (which is the aquifer coefficient).

The following developments were discussed:

- Applying the steady-state Darcy equation to calculate the aquifer coefficient obviates the need for a companion statistical treatment to determine aquifer properties. Unfortunately, the occurrence of unsteady-state conditions often requires the inclusion of step functions to divide the smooth pressure decline into a series of instantaneous steps.
- The van Everdingen and Hurst constant rate solution to the diffusivity equation provides a basis for calculating water influx. The Carter and Tracy method applies the van Everdingen and Hurst constant pressure solution to the diffusivity equation. The Carter and Tracy solution avoids the need to include the step function in the solution procedure.
- Fetkovich (1971) applied the constant pressure flow rate definition to determine a water-influx procedure whose solution may be calculated on a spreadsheet. The solution is limited to finite reservoirs.

7.4 Questions

- Why is it important to have knowledge regarding the water-influx effect on an oil reservoir? (Answer: It is important to have some idea of expected reservoir pressure behavior because recovery is a function of the reservoir pressure reaction to oil, gas, and water production. These calculations could be particularly useful coupled with a material-balance study to calculate the maximum efficient rate for a partial waterdrive reservoir.)
- The inner limit of the geometry of aquifer models differs from that of well flow models in terms of what significant characteristic? (Answer: The inner limit of an aquifer model is the OWC, while the inner limit of a well flow equation is wellbore radius. The aquifer-model characteristic must be predetermined, and this value is seldom known with certainty.)
- Why is the superposition method introduced into a solution procedure when studying declining reservoir pressure caused by oil and gas production? (Answer: The method divides a smoothly declining pressure into a series of finite instantaneous pressure drops.)
- Why is the Fetkovich method much easier to apply than the van Everdingen and Hurst approach? [Answer: The application of the complex dimensionless solutions from van Everdingen and Hurst (1949) and superposition are not required. Fetkovich (1971) can be applied using a simple spreadsheet approach.]

7.5 Nomenclature

A	=	area, acres
A_a	=	aquifer area, acres
A_o	=	reservoir area, acres
A_t	=	total area, acres
B	=	formation volume factor, RB/STB
B_g	=	gas formation volume factor, rcf/scf
B_o	=	oil formation volume factor, RB/STB
B_{oi}	=	oil formation volume factor at initial pressure and temperature conditions, RB/STB
B_t	=	total formation volume factor, RB/STB
B_w	=	water formation volume factor, RB/STB
c_f	=	formation compressibility, psi^{-1}
c_g	=	gas compressibility, psi^{-1}
c_o	=	oil compressibility, psi^{-1}
c_t	=	total compressibility, psi^{-1}
c_v (or B)	=	aquifer description term, STB/psi
c_w	=	water compressibility, psi^{-1}
C_s	=	Schilthuis steady-state aquifer description coefficient, STB/D-psi
h	=	height or thickness, ft
i	=	subindex
j	=	subindex
J	=	aquifer productivity index, bbl/D-psi or RB/D-psi
k	=	permeability, md
k_w	=	effective permeability to water, md
L	=	length, ft
m	=	slope of a curve
n	=	counting term
N_p	=	cumulative oil production, BOPD

p	=	pressure, psia or psi
$\bar{p}$	=	average pressure, psi or psia
$\bar{p}_a$	=	average aquifer pressure, psi or psia
p_{an}	=	aquifer pressure at any timestep n, psi or psia
$\bar{p}_a$	=	average aquifer pressure at any timestep n, psi or psia
$\bar{p}_{a(n-1)}$	=	average aquifer static pressure measured at a previous timestep $n-1$, psi or psia
p_d	=	pressure, dimensionless
p_D	=	pressure, dimensionless
p'_D	=	pressure derivative, dimensionless
p_{Dn}	=	pressure at any timestep n, dimensionless
p'_{Dn}	=	slope of p_D/t_D curve
p_e	=	pressure at aquifer limit, psi or psia
p_i	=	initial pressure, psi or psia
$\bar{p}_i$	=	average initial pressure, psi or psia
p_j	=	reservoir pressure at any timestep j, psi or psia
$\bar{p}_j$	=	average reservoir pressure at any timestep j, psi or psia
p_o	=	pressure at OWC, psi or psia
p_{on}	=	pressure at OWC at any timestep n, psi or psia
p_{res}	=	reservoir pressure, psi or psia
p_{wf}	=	flowing bottomhole pressure, psi or psia
p_1, p_2, p_3, p_4	=	pressures at timesteps 1, 2, 3, and 4, respectively, psi or psia
$\bar{p}_1, \bar{p}_2, \bar{p}_3, \bar{p}_4$	=	average pressures at timesteps 1, 2, 3, and 4, respectively, psi or psia
q	=	flow rate, STB/D

q^*	=	pseudo flow rate, STB/Dq_D = flow rate, dimensionless
q_g	=	gas flow rate, scf/D or MMcf/D
q^*_{j-1}	=	pseudo flow rate at timestep $j-1$, STB/D
q^*_n	=	pseudo flow rate during a certain time n, STB/D
q_o	=	oil flow rate, STB/D
Q_{pD} (Q_t) or W_D	=	water influx, dimensionless
Q_t	=	van Everdingen and Hurst (1949) water influx, dimensionless
q_w	=	water flow rate, STB/D
Q_w	=	cumulative water, STB
r	=	radius, ft
r_a	=	aquifer radius, ft
r_d	=	radius, dimensionless
r_e	=	distance from center of circle to aquifer limit or external radius of aquifer, ft
r_{eD}	=	external radius, dimensionless
r_o	=	distance from center of circle to edge of reservoir at OWC or external radius of reservoir, ft
r_t	=	distance from center of wellbore to reservoir boundary or at any reference plane at a certain time t, ft
r_w	=	wellbore radius, ft
R	=	gas/oil ratio, scf/STB
R_s	=	amount of gas soluble in oil, gas/oil solubility, scf/STB
R_{si}	=	amount of gas soluble in oil at initial conditions of pressure and temperature, gas/oil solubility, scf/STB
S_g	=	gas saturation, percent
S_o	=	oil saturation, percent
S_w	=	water saturation, percent
t	=	time, days, months, or years
t_d	=	time, dimensionless
t_D	=	time, dimensionless
t_{Di}	=	dimensionless time at period specified by subindex i, dimensionless
t_{Dj}	=	time at period specified by subindex j, dimensionless
t_{Dn}	=	time at certain time n, dimensionless
t_e	=	effective time, days, months, or years
t_j	=	time at period specified by subindex j, days, months, or years
t_1, t_2, t_3, t_4	=	times 1, 2, 3, and 4, respectively, days, months, or years
w	=	width, ft
W_D or Q_{pD} (Q_t)	=	water influx, dimensionless
W_e	=	cumulative water influx, STB, BW, RB, or bbl
W_{ei}	=	maximum water volume, RB
W_{ej}	=	cumulative water influx at period specified by subindex j, STB, BW, RB, or bbl
W_{en}	=	cumulative water influx at a certain time n, STB, BW, RB, or bbl
$W_{e(t-tj)}$	=	water influx between a period of time $(t-t_j)$, STB, BW, RB, or bbl
W_i	=	volume of water initially in place, STB or RB
W_p	=	volume of water produced, STB
Δp	=	pressure drop $(p_i - p)$, psi or psia
Δ_{Pn}	=	pressure deference during a certain time n, psi or psia
Δ_{P0}	=	pressure drop during first time period, psi or psia
Δt_n	=	time interval over which pressure decline occurred, days, months, or years
θ	=	angle, degrees
μ	=	viscosity, cp
μ_w	=	water viscosity, cp
τ	=	dummy variable of integration
ϕ	=	porosity, percent or fraction

7.6 Problems

Exercise 7.9—Calculating Cumulative Water Influx Using the van Everdingen and Hurst Method. Use the van Everdingen and Hurst method to calculate cumulative water influx at 1,100 days for the Centennial Reservoir. The performance history and aquifer characteristics are provided in **Table 7.17.**

Solution. First, calculate the aquifer description term:

$$C_v = \frac{1.119\phi h c_t r_o^2}{B_w}\left(\frac{\theta}{360°}\right) = \frac{1.119(0.12)(30)(6\times10^{-6})(2{,}000^2)}{1.03}\left(\frac{95°}{360°}\right) = 24.8 \text{ STB/psi}.$$

$k_w = 100$ md, $B_w = 1.03$ RB/STB, $\theta = 95°$, $h = 30$ ft, $\mu_w = 0.6$ cp, $\phi = 12\%$, $r_{eD} = 500$, $r_o = 2{,}000$ ft, $c_w = 3 \times 10^{-6}$ psi^{-1}, $c_f = 3 \times 10^{-6}$ psi^{-1}

1	2	3	4	5	6
Time (days)	Pressure (psia)	Average Pressure (psia)	Pressure Change (psia)	Effective Time (days)	Δpt_e (1,000 psi-D)
0	5,000	5,000	–	–	–
300	4,700	4,850	150	1,100	165.0
600	4,500	4,600	250	800	200.0
1,000	4,750	4,625	−25	500	−12.5
1,200	4,300	4,525	100	100	10.0
					Total: 362.5

Table 7.17—Aquifer characteristics and calculation results.

Pressure Change (psia)	Pressure Change (psia)	t_e	t_D	$t_D = 0.366t$	W_D
ΔP_0	150	1,100	$t_{D4\text{periods}}$	403.0	135.7
ΔP_1.	250	800	$t_{D3\text{periods}}$	293.0	103.5
ΔP_2	−25	500	$t_{D2\text{periods}}$	183.0	70.0
ΔP_3	100	100	$t_{D1\text{periods}}$	36.6	19.4

Table 7.18—Cumulative water influx using the van Everdingen and Hurst method, results of calculations.

Then, calculate the dimensionless time/real-time relation:

$$t_D = \frac{0.00633kt}{\phi\mu_w c_t r_o^2} = \frac{(0.00633)(100)t}{(0.12)(0.6)(6\times10^{-6})(2{,}000^2)} = 0.366t;$$

$$r_{eD} = r_e / r_o;\ r_e = r_{eD} \times r_o = 500 \times 2{,}000 = 1{,}000{,}000 \text{ ft}.$$

Then, calculate the cumulative water influx using the data provided in **Tables 7.17 and 7.18:**

$$W_{e_3} = C_v\left[\Delta p_0 W_D\left(t_{D_4\text{ periods}}\right) + \Delta p_1 W_D\left(t_{D_3\text{ periods}}\right) + \Delta p_2 W_D\left(t_{D_2\text{ periods}}\right) + \Delta p_3 W_D\left(t_{D_1\text{ period}}\right)\right];$$

$$W_e = 24.8\left[(150)(135.7) + (250)(103.5) + (-25)(70.0) + (100)(19.4)\right] = 1.151 \text{ million STB}.$$

Using the van Everdingen and Hurst method is best because it is able to handle the transient pressure effect. The pressure buildup at the $t = 1{,}000$-days timestep obviously introduced transient conditions into the aquifer.

Exercise 7.10—Calculating Cumulative Water Influx Using the Fetkovich Method. Use the Fetkovitch method to calculate water influx at the end of 2.3 years. The aquifer characteristics are provided in **Table 7.19.**

Assume the aquifer radius is seven times greater than the radius of the oil reservoir and that pseudosteady-state conditions prevail. **Table 7.20** shows the pressure performance history, which was monitored over 2.3 years.

h = 27 ft	S_w = 19%	B_o = 1.3 RB/STB	$c_w = 3 \times 10^{-6}$ psi^{-1}	μ_w = 0.7 cp
ϕ = 30%	k = 100 md	Original oil-in-place (OOIP): 6.0 million STB	$c_f = 4 \times 10^{-6}$ psi^{-1}	θ = 180°

Table 7.19—Aquifer characteristics.

Time (years)	Pressure (psia)
0	3,675
1	3,400
2	3,220
2.3	3,180

Table 7.20—Time and pressure history.

Solution. First, calculate the area of the oil reservoir using $OOIP = 7,758\ \dfrac{\phi hA(1-S_w)}{B_{oi}}$:

(6,000,000 STB)(1.3 RB/STB) = (7,758 bbl/acre-ft)(0.30)(27 ft)(A)(1 – 0.19); A = 153 acres.

Calculate the radius of the semicircle for the oil reservoir using the just-calculated reservoir area:

$$A = \frac{\pi r^2}{2} = 153 \text{ acres},\ r_o = 2{,}060 \text{ ft}.$$

Next, calculate the aquifer area. If $r_{eD} = 7$, then $r_e = r_a$ = 14,420 ft.

Recall that $r_t = r_a + r_o$ and $A_t = A_a + A_o$. For concentric circular areas and using the conversion factor 43,560 ft²/acre:

$$A_t = \frac{(14{,}420^2)\pi}{2(43{,}560 \text{ ft}^2/\text{acre})} = 7{,}497 \text{ acres}.$$

$$A_a = A_t - A_o = 7{,}497 \text{ acres} - 153 \text{ acres} = 7{,}344 \text{ acres}.$$

Next, calculate the maximum water influx:

$$W_{ei} = W_i c_t p_i = \frac{(7\times10^{-6})\left(\dfrac{180°}{360°}\right)(3.141)(14{,}420^2 - 2{,}060^2)(27)(0.3)(3{,}675)}{5.615} = 11{,}871{,}463 \text{ RB}.$$

Calculate the pseudosteady-state radial PI:

$$J = \frac{0.00708kh\left(\dfrac{\theta}{360°}\right)}{\mu_w(\ln r_e/r_o - 0.75)} = \frac{(0.00708)(100)(27)\left(\dfrac{180}{360}\right)}{0.7\left[\ln\left(\dfrac{14{,}420}{2{,}060}\right) - 0.75\right]} = 11.42 \text{ RB/D-psi}.$$

Calculate the terms:

$$(1-e^{-Jp_i\Delta t_n/W_{ei}}) = 1-e^{(-11.42)(3,675)(\Delta t_n)(365)/11,871,463} = 1-e^{-1.29\Delta t_n},$$

$$\frac{W_{ei}}{p_i} = \frac{11,871,463}{3,675} = 3,230 \text{ RB/psi}.$$

Reduce Eq. 7.40, $\Delta W_{en} = \frac{W_{ei}}{p_i}\left[\bar{p}_{a(n-1)} - p_{on}\right]\left(1 - e^{-Jp_i\Delta t_n/W_{ei}}\right)$, to

$$\Delta W_{en} = 3,230\left[\bar{p}_{a(n-1)} - p_{on}\right]\left(1 - e^{-1.29\Delta t_n}\right).$$

At $t = 1$ year, $\left[\bar{p}_{a(n-1)} - p_{on}\right] = 3,675 - 3,400 = 275$ psi,

$$\Delta W_{en} = 3,230\left[\bar{p}_{a(n-1)} - p_{on}\right]\left(1 - e^{-1.29\Delta t_n}\right) = 3,230\left[\bar{p}_{a(n-1)} - p_{on}\right]\left(1 - e^{-1.29\times 1}\right),$$

$$\Delta W_{en} = (3,230)(275)(1 - 0.275) = 0.643 \text{ million STB}.$$

At $t = 2$ years with $\Delta t_n = 1$:

$$\Delta W_{en} = 3,230\left[\bar{p}_{a(n-1)} - p_{on}\right]\left(1 - e^{-1.29\Delta t_n}\right) = 3,230\left[\bar{p}_{a(n-1)} - p_{on}\right]\left(1 - e^{-1.29\times 1}\right),$$

$$\Delta W_{en} = (3,230)(257)(1 - 0.275) = 0.601 \text{ million STB}.$$

Table 7.21 summarizes the calculated values of incremental water influx during the specified periods and the cumulative water influx after 2 years of pressure and production history.

Time (years)	p_{on} (psia)	$\bar{p}_{a(n-1)} - p_{on}$ (psia)	ΔW_{en} (million STB)	ΣW_{en} (million STB)	p_{an} (psia)
0	3,675	–	–	–	–
1.0	3,400	275	0.643	0.643	3,477
2.0	3,220	257	0.601	1.244	3,290

Table 7.21—Cumulative water influx using the Fetkovich method, results of calculations.

7.7 References

- Carter, R. D. and Tracy, G. W. 1960. An Improved Method for Calculating Water Influx. In *Transactions of the Society of Petroleum Engineers*, Vol. 219, SPE-1626-G, 415–417. Richardson, Texas: Society of Petroleum Engineers. https://doi.org/10.2118/1626-G.
- Craft, B. C. and Hawkins, M. F. 1959. *Applied Petroleum Reservoir Engineering*. Englewood Cliffs, New Jersey: Prentice-Hall.
- Fetkovich, M. J. 1971. A Simplified Approach to Water Influx Calculations—Finite Aquifer Systems. *J Pet Technol* **23** (7): 814–828. SPE-2603-PA. https://doi.org/10.2118/2603-PA.
- Schilthuis, R. J. 1936. Active Oil and Reservoir Energy. In *Transactions of the Society of Petroleum Engineers*, Vol. 118, SPE-936033-G, 33–52. Richardson, Texas: Society of Petroleum Engineers. https://doi.org/10.2118/936033-G.
- Todd, A. C. 2021. Chap 17—Water Influx. Course notes, Heriot-Watt University Institute of Petroleum Engineering, Edinburgh, Scotland.
- van Everdingen, A. F. and Hurst, W. 1949. The Application of the Laplace Transformation to Flow Problems in Reservoirs. *J Pet Technol* **1** (12): 305–324. SPE-949305-G. https://doi.org/10.2118/949305-G.

Chapter 8

Displacement Concepts

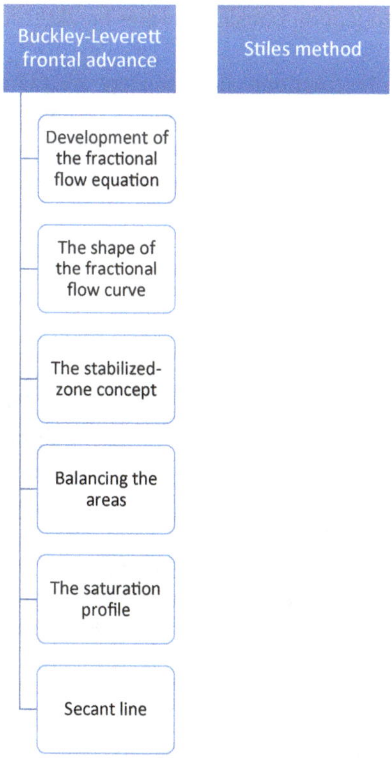

Material-balance studies are based on reservoir production-pressure responses. Mobility ratio characteristics and displacement efficiencies are not included in this zero-dimensional model. For instance, an expanding gas cap physically displaces oil ahead of the invading front. The material-balance method accounts for the changing reservoir volumes as a function of pressure loss, but the effects of the changing relative permeability and viscosity values of the displacing and displaced phases are ignored. The fractional flow concept, presented in this chapter, accounts for these constantly changing relative permeability values.

The calculation procedure results in a graph showing the fraction of the displacing phase flowing in the reservoir as a function of changing saturation. The graph can be used to illustrate how changing reservoir conditions affect the relationship between the displacing and the displaced phases. Solutions are concerned primarily with the efficiency of oil displacement by means of water in a linear system when predicting waterflood performance. Modifications to the 1D fractional flow equation have been proposed that consider the effects of gravity segregation (Dietz 1953; Dake 1978). However, the use of reservoir numerical simulation (finite-difference and streamline models) has been widely substituted in those particular cases, and it can account for the multidimensional effects of high heterogeneity in rock and fluid properties, structural complexities, and dipping reservoirs (Coats et al. 1967; Morse and Whiting 1970; Datta-Gupta and King 2007).

8.1 Buckley-Leverett Frontal Advance

The fractional flow equation is a significant component of the Buckley-Leverett frontal advance theory, in which changes in the relative volumes of fluid moving through a system are calculated from relative permeability curves, which are converted to a fractional flow curve.

This rate/time prediction technique can be divided into

- Time and saturation characteristics in a reservoir at the time of breakthrough
- Expected oil and water production rates after water breakthrough until flood-out

The assumptions underlying the fractional flow concept are (1) a linear flow system, but it can also be applied to radial flow; (2) that the driving force across the system is the sum of an induced pressure differential and the effect of gravity; (3) that the pressure differential induces positive flow in an upward manner, while gravity acts as a downward and negative force; (4) that the vertical and areal sweep efficiencies are 100%; (5) that the fluids are incompressible and immiscible and flow in a steady-state manner; and (6) that relative permeability curves are representative of the system.

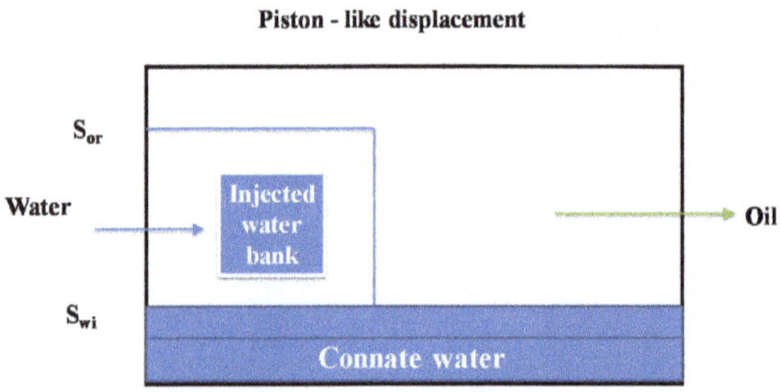

Fig. 8.1—Buckley-Leverett frontal advance theory. The ideal displacement of oil would lead to complete residual saturation by means of the displacing water phase. Piston-like displacement is a highly idealized model.

Fig. 8.1 shows water moving through a volume of the reservoir. In front of the flood front, the oil saturation is

$$S_o = 1 - S_{wi}. \quad (8.1)$$

Behind the flood front, the oil and water saturations are, respectively,

$$S_o = S_{or} \text{ and}$$

$$S_w = 1 - S_{or}. \quad (8.2)$$

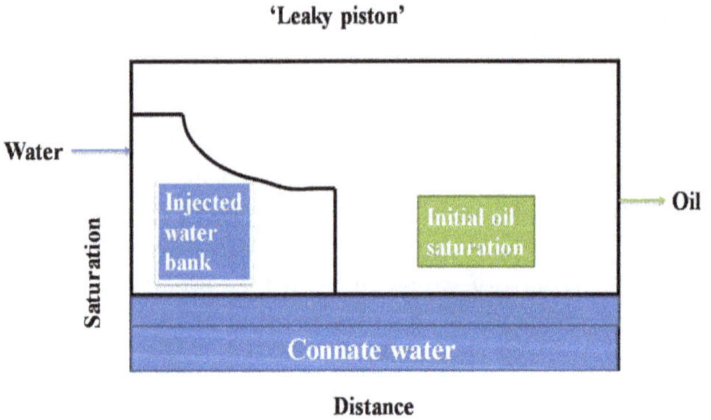

Fig. 8.2—In this model, water displaces the oil in such a manner that the majority of the oil is displaced ahead of the front. The remaining oil is swept out by the water moving behind the flood front according to relative permeability relations.

Waterflooding displaces oil by pushing it to the producer. In an ideal setting, it would be a piston-like displacement.

A more-realistic model of the displacement process is shown in **Fig. 8.2,** where the oil saturation decreases behind the flood front until the oil saturation reaches a minimum value, which is the residual oil saturation (S_{or}). Fig. 8.2 represents a reservoir above the bubblepoint, in what is called leaky piston-like displacement.

If the waterflood is to be initiated under depletion conditions below the bubblepoint, a free-gas saturation exists in the pore spaces. Reservoir-pressure reduction can cause gas-cap expansion and secondary gas-cap creation, and water injection can prevent or reverse these effects. As the reservoir pressure increases by means of water injection to a value greater than the bubblepoint, gas will go back into solution with the oil. Waterflood response will not occur until this condition, known as fill-up, has occurred. The more depleted the reservoir, the longer the time to fill-up, and therefore, the longer the time to waterflood response. **Fig. 8.3** shows a schematic for this possible and most-common case.

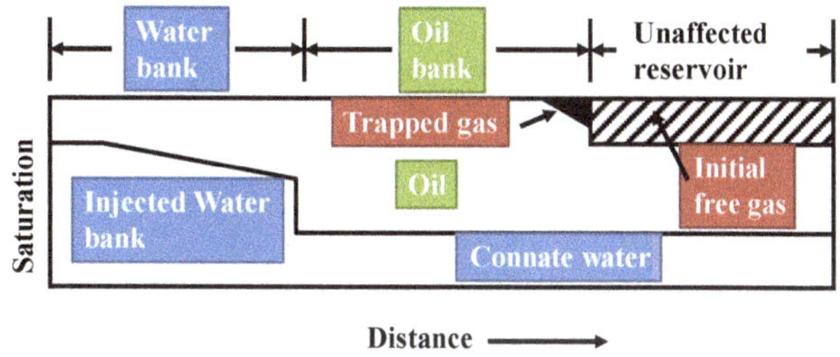

Fig. 8.3—Water displaces oil, and a free-gas saturation exists in the pore spaces at the beginning of the project when the waterflood is initiated below the bubblepoint pressure of the oil.

8.1.1 Development of the Fractional Flow Equation. The fractional flow equation is a model used to determine the water fraction of the total fluid flow at a particular location and time in a linear reservoir waterflood.

This model provides insight into the immiscible waterflood displacement process and the relative effects of different rock, fluid, and operational properties on the displacement efficiency.

The location and time for a fractional flow value are obtained by determining the saturation history for that location.

The total flow through the system is the sum of the oil and water flow rates:

$$q_t = q_o + q_w. \quad \text{(8.3)}$$

A similar set of equations can be determined for gas/water or gas/oil systems.

The fraction of water flowing in the system can be defined as

$$f_w = \frac{q_w}{q_o + q_w} = \frac{q_w}{q_t}. \quad \text{(8.4)}$$

The Darcy equations defining the linear flow of water and oil in a porous medium are shown in Eqs. 8.5 and 8.6. The potential function is the sum of the pressure drop and the gravity effects measured across the system.

$$q_w = -\frac{0.001127 k_w A}{\mu_w}\left(\frac{\mathrm{d}p_w}{\mathrm{d}L} + 0.433\gamma_w \sin\theta\right), \quad \text{(8.5)}$$

$$q_o = -\frac{0.001127 k_o A}{\mu_o}\left(\frac{\mathrm{d}p_o}{\mathrm{d}L} + 0.433\gamma_o \sin\theta\right), \quad \text{(8.6)}$$

where A is the cross-sectional area measured perpendicular to the direction of flow (ft^2); θ is the angle of dip of the bed (degrees); L is the length of the flow system (ft); q_w and q_o are the flow rates for the water and oil phases, respectively (RB/D); k_o is the effective permeability to oil (md); k_w is the effective permeability to water (md); μ_o is the oil viscosity (cp); μ_w is the water viscosity (cp); and γ_o and γ_w are the specific gravities of oil and water, respectively, when referenced to water (fraction).

Capillary pressure is defined as

$$p_c = p_o - p_w. \quad (8.7)$$

A capillary pressure gradient exists across the system, and therefore, Eq. 8.7 can be rewritten in terms of a differential:

$$\frac{dp_c}{dL} = \frac{dp_o}{dL} - \frac{dp_w}{dL}. \quad (8.8)$$

Combining Eqs. 8.5, 8.6, and 8.8 results in

$$\frac{dp_c}{dL} = -\frac{q_o \mu_o}{0.001127 k_o A} + \frac{q_w \mu_w}{0.001127 k_w A} + 0.433(\gamma_w - \gamma_o)\sin\theta. \quad (8.9)$$

Dividing Eq. 8.9 by q_t and adding $\dfrac{\mu_o}{0.001127 k_o A}$ and then rearranging to solve Eq. 8.4 for f_w result in the generalized fractional flow equation, Eq. 8.10. This frontal advance equation is used to determine the movement of these fluid fronts (Dake 1978).

$$f_w = \frac{1 + \dfrac{1.127 \times 10^{-3} A}{q_t} \dfrac{k_o}{\mu_o} \left(\dfrac{\partial p_c}{\partial L} - 0.433 \Delta\gamma \sin\alpha \right)}{1 + \dfrac{\mu_w}{\mu_o} \dfrac{k_o}{k_w}}, \quad (8.10)$$

where q_t is the total flow rate ($q_o + q_w$) (RB/D), q_o is the oil flow rate (RB/D), q_w is the water flow rate (RB/D), k_o is the effective permeability to oil (md), k_w is the effective permeability to water (md), L is the length of the flow system (ft), A is the cross-sectional area available to flow (ft^2), μ_o is the oil viscosity (cp), μ_w is the water viscosity (cp), p_c is the capillary pressure (psia), γ is the specific gravity of the fluids (fraction), and α is the angle of inclination, positive updip (degrees).

The capillary pressure effects are represented by the capillary pressure number, which is defined as N_{pc}. (It is usually ignored.)

$$N_{pc} = \frac{1.127 \times 10^{-3} A}{(q_o + q_w)} \frac{k_o}{\mu_o} \frac{\partial p_c}{\partial L}. \quad (8.11)$$

It is difficult, if not impossible, to evaluate the actual magnitude of capillary forces; therefore, they are typically omitted from Eq. 8.11.

Gravity effects are represented by the gravity number, which is defined as N_g:

$$N_g = \frac{1.127 \times 10^{-3} A}{(q_t)} \frac{k_o}{\mu_o} 0.433 \Delta\gamma \sin\alpha. \quad (8.12)$$

The fractional flow equation for a horizontal reservoir is given by

$$f_{wh} = \frac{1}{1 + \dfrac{\mu_w}{\mu_o} \dfrac{k_{ro}}{k_{rw}}}. \quad (8.13)$$

Eq. 8.10 defines the fraction of water flowing in an oil/water system in terms of capillary pressure, gravity, and relative permeability effects. Eq. 8.10 reduces to Eq. 8.13 when flow is in a horizontal plane because gravity and capillary pressure effects can be ignored. Eq. 8.13 is evaluated for a location in the reservoir at a point in time, which explicitly defines a water saturation. Using the relative permeabilities as a function of water saturation data provides the values of k_{ro} and k_{rw}, which are entered into Eq. 8.13. Oil and water viscosities at the average reservoir pressure are obtained from laboratory data or correlations.

By combining the defined Eqs. 8.10 through 8.13, a compacted Eq. 8.14 is obtained:

$$f_w = f_{wh} \times \left(1 + N_{pc} - N_g\right). \quad (8.14)$$

Eq. 8.10 can be expressed when only capillary pressure effects are ignored. The constant gravity terms are collected and defined as a coefficient C', which is shown in Eq. 8.15:

$$C' = \frac{0.000488A\left[(\Delta\gamma)(\sin\theta)\right]}{\mu_o q_t}. \quad (8.15)$$

Substituting in Eq. 8.10 results in Eq. 8.16a:

$$f_w = \frac{1 - C'k_o}{1 + \dfrac{k_o\mu_w}{k_w\mu_o}} = \frac{1 - C'kk_{ro}}{1 + \dfrac{k_{ro}\mu_w}{k_{rw}\mu_o}}. \quad (8.16a)$$

Dividing the equation into the fractional flow and gravity components results in Eq. 8.16b:

$$f_w = f_{wh} - f_{w(g)} = \frac{1}{1 + \dfrac{k_o\mu_w}{k_w\mu_o}} - \frac{C''k_{ro}}{1 + \dfrac{k_o\mu_w}{k_w\mu_o}}. \quad (8.16b)$$

Substituting $k_{ro} = k_o/k$ and $k_{rw} = k_w/k$ results in Eq. 8.16c:

$$f_w = \frac{1}{1 + \dfrac{k_o\mu_w}{k_w\mu_o}} - \frac{C''k_{ro}}{1 + \dfrac{k_o\mu_w}{k_w\mu_o}} = \frac{1}{1 + \dfrac{k_{ro}\mu_w}{k_{rw}\mu_o}} - \frac{C''k_{ro}}{1 + \dfrac{k_{ro}\mu_w}{k_{rw}\mu_o}}, \quad (8.16c)$$

where

$$C'' = C'k = \frac{0.000488kA\left[(\Delta\gamma)(\sin\theta)\right]}{\mu_o q_t}. \quad (8.16d)$$

In summary, the forces operating during a waterflood are

- Viscous force
  - Resulting from pressure gradients imposed
  - Mainly controls horizontal fluid movement
  - Related to viscosity
- Gravity force
  - Resulting from fluid density difference
  - Controls gravity segregation and vertical movement

- Capillary force
  - Active at interface between immiscible fluids
  - Generally weak force in fluid displacement
  - Has a significant effect in establishing residual saturations

An indicator of the importance of gravity forces in the displacement process ignoring capillary forces is known as the viscous-gravity number and is expressed by Eq. 8.17:

$$N_{gv} = \frac{k_v k_{rw} \Delta\rho g \cos\alpha A}{887.2 q \mu_w} \frac{L}{h}. \quad (8.17)$$

Eq. 8.17 uses field units, and α is the dip angle from the horizontal plane. The higher the dip, the lower the viscous-gravity number N_{gv}. Willhite (1986) defines the dimensionless viscous-gravity number as

- For viscous-dominated flow, $N_{gv} < 0.1$.
- For gravity-dominated flow, $N_{gv} > 10$.
- For transitional flow, $0.1 < N_{gv} < 10$.

An indicator of the importance of capillary force in the displacement process ignoring gravity forces is known as the capillary-gravity number N_{cv}. Some authors instead use a dimensionless Péclet number, which is defined inversely with similar variables (Rapoport and Leas 1953).

$$N_{cv} = \frac{k_v A L \Delta p_c}{887.2 q \mu h^2}. \quad (8.18)$$

Better floods occur with a high N_{cv} (small Péclet number). For an N_{cv} greater than 10, a waterflood would show the following characteristics:

- A uniform flood front
- Sustained peak oil rate
- Late water breakthrough
- Rapid watering out of wells after breakthrough
- Small post-breakthrough oil recovery

Example 8.1—Calculating the Fraction of Water. Contemplate the waterflooding of a rectangular-shaped reservoir, with the water-injection well located at one end of the rectangle and the producing well at the opposing end. The angle of dip of the beds is 5°. The cross-sectional area of the reservoir is 1,000,000 ft². Rock and fluid properties are available in **Table 8.1.**

The expected water-injection rate is 10,000 B/D. Calculate the fraction of water flowing in the reservoir when (1) flow is horizontal, (2) flow is updip, and (3) flow is downdip.

Solution. Use Eq. 8.13 to describe the horizontal flow effect:

μ_w = 0.7 cp	k_w = 20 md	ρ_w = 65 lbm/ft³	γ_w = 65/62.4 = 1.04
μ_o = 1.0 cp	k_o = 100 md	ρ_o = 50 lbm/ft³	γ_o = 50/62.4 = 0.80

Table 8.1—Reservoir and fluid properties, fractional flow.

$$f_{wh} = \frac{1}{1 + \frac{\mu_w}{\mu_o} \frac{k_{ro}}{k_{rw}}},$$

and expressing it as a function of effective permeabilities,

$k_o = k_{ro}k$ and $k_w = k_{rw}k$,

$$f_{wh} = \frac{1}{1 + \frac{k_o \mu_w}{k_w \mu_o}} = \frac{1}{1 + \left(\frac{100}{20}\right)\left(\frac{0.7}{1.0}\right)} = 0.222.$$

Gravity effects are calculated with Eq. 8.12:

$$N_g = \frac{1.127 \times 10^{-3} A}{(q_t)} \frac{k_o}{\mu_o} 0.433(\Delta\gamma)(\sin\alpha) = \frac{(0.001127)(1 \times 10^6)(100)(0.433)(1.04 - 0.80)(0.0872)}{(1.0)(10{,}000)}$$

$$= 0.102.$$

Therefore,

$$f_{w(\text{horizontal})} = f_{wh} = 0.222.$$

$$f_{w(\text{updip})} = f_{wh}(1 - N_g) = 0.222(1 - 0.102) = 0.1993.$$

$$f_{w(\text{downdip})} = f_{wh}(1 + N_g) = 0.222(1 + 0.102) = 0.2447.$$

Injecting water from a downdip location [$f_{w(\text{updip})}$] in a waterflood promotes the working of oil/water gravity segregation in opposition to the imposed pressure differential. Gravity segregation promotes a more-efficient process for this case because less water is produced in conjunction with 1 bbl of oil.

On the other hand, injecting water from an updip location [$f_{w(\text{downdip})}$] increases the ability of water to flow and increases the fraction of water flowing per unit volume of oil. The water flow rate increases when injecting in the downdip direction. The combination of gravity effects and water injection is the reason peripheral waterfloods are so efficient. A peripheral waterflood also combines water encroaching from a downdip location and gravity segregation to herd oil in an updip direction, where it is eventually corralled at the structural high. Less water is processed during the oil-recovery process.

Fig. 8.4 shows water moving through a volume of the reservoir and the effect of dip angle on the fractional flow curve. When the water displaces oil updip (i.e., the injection well is located downdip), a more-efficient performance is obtained.

8.1.2 The Shape of the Fractional Flow Curve. The fractional flow equation is typically expressed as a function of saturation. **Fig. 8.5** (Latil 1980) compares the fractions of water flowing in a water-displacing-oil system as a function of oil viscosity. A decreasing oil viscosity effectively increases the mobility ratio.

The bounds of each curve are the initial water saturation and the residual oil saturation. The location and steepness of the fractional flow curve define the efficiency of the displacement process. Fig. 8.5 indicates that the water saturation for a particular f_w increases as a function of declining oil viscosity. Oil recovery is equivalent to an increased water saturation. The extent of the tail at the end of each curve is linked to increasing oil viscosity. The absence of a tail for the $\mu_o = 0.1$ cp curve reflects the reduced volume of processed water in this case, the most efficient of the three studies. Residual oil saturations are also lower in this case.

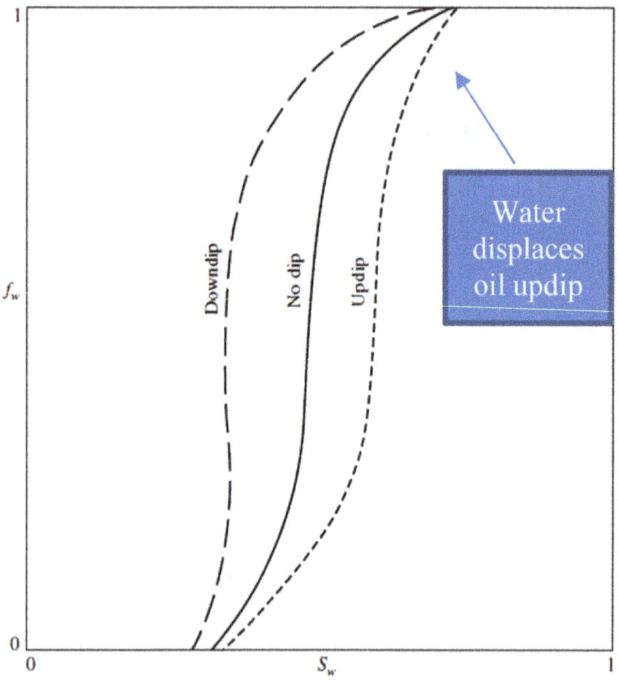

Fig. 8.4—Buckley-Leverett frontal advance theory. Effect of dip on fractional flow curve response.

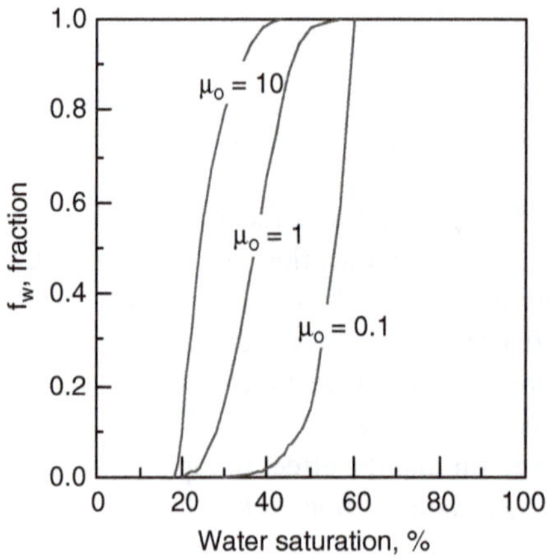

Fig. 8.5—The effect of oil viscosity on the fractional flow plot (after Latil 1980). Recovery is dramatically enhanced for the lower-viscosity oils.

A steep curve for the μ_o = 10 cp sample indicates a marginal decrease in the oil saturation value after water breakthrough to the producing well. Water-handling costs will be reduced in this instance because the majority of the movable oil is recovered by the time the fraction of water flowing in the system increases to its maximum value. However, the residual oil saturation for the system could be quite high, and large volumes of potentially recoverable oil would be left in the reservoir at the time of economic abandonment. Enhanced oil recovery processes strive to shift the fractional flow curve to the right to increase recovery.

The fractional flow equation is a model used to predict the immiscible waterflood displacement process and the relative effects of different rock and fluid properties on the displacement efficiency. **Fig. 8.6** shows the resulting curve when the fractional flow equation is used to determine the water saturation in a linear reservoir waterflood at a particular time after injection has been started and at a certain distance from the injection point. The equation will be developed applying a volumetric balance in a finite element of the reservoir.

Fig. 8.7 shows what occurs when a volumetric balance is applied in a finite element of a linear system.

The volume of water V_w in the volume element is given by

$$W = V_w = A\phi\Delta x S_w. \quad \text{(8.19)}$$

The rate of water accumulation within the differential element is given by

$$\frac{\partial W}{\partial t} = \frac{\partial(A\phi\Delta x S_w)}{\partial t} = A\phi\Delta x \frac{\partial S_w}{\partial t}. \quad \text{(8.20a)}$$

In field units,

$$\frac{dW}{dt} = \frac{\phi A dx}{5.615}\left(\frac{dS_w}{dt}\right)_x, \text{ B/D}, \quad \text{(8.20b)}$$

where A is the cross-sectional area of the waterflood (ft²), $\frac{\phi A dx}{5.615}$ is the pore volume of the differential area (bbl), and $\left(\frac{dS_w}{dt}\right)_x$ is the change in the saturation front as a function of time.

The fractional flow equation (Eq. 8.4), $f_w = \dfrac{q_w}{q_o + q_w} = \dfrac{q_w}{q_t}$, can be rewritten as a function of the water flow rate:

$$q_w = f_w q_t. \qquad (8.21)$$

The rate of increase of mass equals the mass flow rate minus the mass flow rate in the volume element, or

$$\frac{\partial W}{\partial t} = \left(q_w\right)_x - \left[\left(q_w\right) + \left(\frac{\partial q_w}{\partial t}\right)\Delta x\right], \qquad (8.22)$$

and

$$\frac{\mathrm{d}W}{\mathrm{d}t} = \left[f_w - \left(f_w - \mathrm{d}f_w\right)\right]q_t = q_t \mathrm{d}f_w. \qquad (8.23)$$

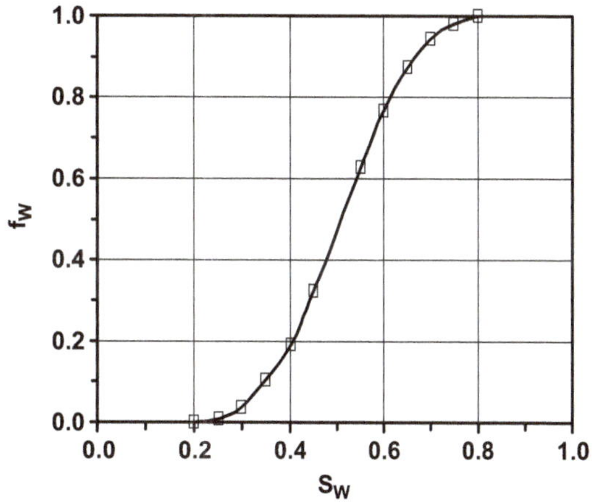

Fig. 8.6—The fractional flow equation is used to determine the movement of the water/oil fluid fronts.

Combining Eqs. 8.20b and 8.23 results in

$$\frac{\mathrm{d}S_w}{\mathrm{d}t} = \frac{5.615 q_t}{\phi A}\frac{\mathrm{d}f_w}{\mathrm{d}x}. \qquad (8.24)$$

The derivation has shown that the rate of change of the water saturation front is a function of reservoir volume and the change in the fraction of water flowing through the element.

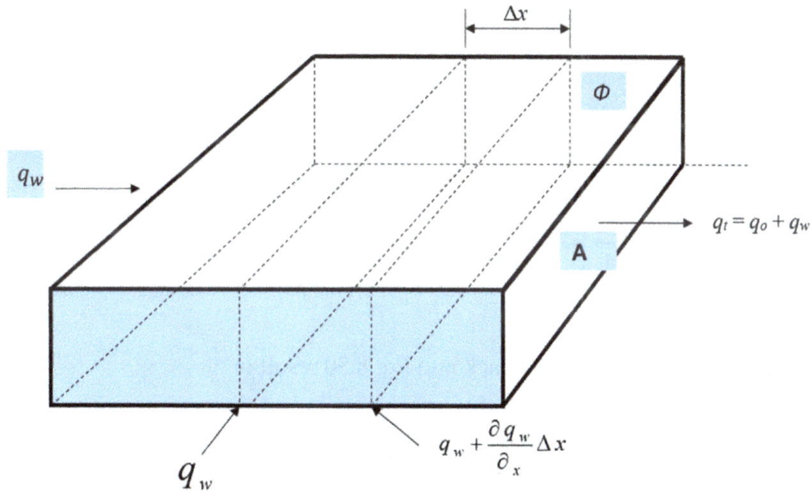

Fig. 8.7—The fractional flow equation is used to determine the movement of water displacing oil. Fig. 8.7 assumes a linear flood and the application of a mass balance rate through a volume element.

Applying limits to Eq. 8.24 relates the change in water saturation at any point x along the system to the fraction of water flowing at that time (t), and the water saturation change as a function of time can be expressed as

$$\left(\frac{dS_w}{dt}\right)_x = \left(\frac{5.615q_t}{\phi A}\frac{df_w}{dx}\right)_t \quad (8.25)$$

Because $S_w = f(x,t)$, the differential of the water saturation is

$$dS_w = \left(\frac{\partial S_w}{\partial x}\right)_t dx + \left(\frac{\partial S_w}{\partial t}\right)_x dt. \quad (8.26)$$

We are interested in determining the movement of a plane of constant water saturation S_w, which implies that $(dS_w) = 0$ at S_w = constant.

$$0 = \left(\frac{\partial S_w}{\partial x}\right)_{t,S_w} dx + \left(\frac{\partial S_w}{\partial t}\right)_{x,S_w} dt, \quad (8.27a)$$

and

$$\left(\frac{\partial S_w}{\partial t}\right)_{x,S_w} = -\left(\frac{dx}{dt}\right)_{S_w}\left(\frac{\partial S_w}{\partial x}\right)_{t,S_w} \quad (8.27b)$$

Combining Eqs. 8.25, 8.27a, and 8.27b results in

$$-\frac{5.615q_t}{A\phi}\left(\frac{\partial f_w}{\partial x}\right)_{t,S_w} = -\left(\frac{dx}{dt}\right)_{S_w}\left(\frac{\partial S_w}{\partial x}\right)_{t,S_w} \quad (8.28)$$

$$\frac{dx}{dt} = \frac{5.615q_t}{A\phi}\left(\frac{\partial f_w}{\partial t}\right)_{t,S_w}\left(\frac{\partial x}{\partial x}\right)_{t,S_w} \text{ or } \frac{dx}{dt} = \frac{5.615q_t}{A\phi}\left(\frac{\partial f_w}{\partial t}\right)_{t,S_w} \quad (8.29)$$

$$\frac{dx}{dt} = \frac{\frac{5.615q_t}{A\phi}\left(\frac{\partial f_w}{\partial t}\right)_{t,S_w}}{\left(\frac{\partial S_w}{\partial t}\right)_{t,S_w}} \text{ or } \frac{dx}{dt} = \frac{5.615q_t}{A\phi}\left(\frac{\partial f_w}{\partial t}\right)_{t,S_w}\left(\frac{\partial x}{\partial S_w}\right)_{t,S_w} \quad (8.30)$$

Separating variables and substituting back into Eq. 8.30 result in

$$\left(\frac{dx}{dt}\right)_{S_w} = \frac{5.615q_t}{A\phi}\left(\frac{\partial f_w}{\partial S_w}\right)_{S_w}, \text{ ft/D.} \quad (8.31)$$

For a constant rate of water injection q_w and a constant total produced fluids flow rate q_t, the velocity $\left(\frac{dx}{dt}\right)_{S_w}$ of a plane of constant water saturation S_w is directly proportional to the derivative of the fractional flow equation evaluated at that particular saturation.

The rate of advance of any water saturation value can be calculated by evaluating the slope of the fractional flow curve.

Integrating over the distance x that the saturation plane could travel results in

$$x = \frac{5.615 q_t}{\phi A}\left(\frac{df_w}{dx}\right). \qquad (8.32)$$

One can now evaluate the saturation plane at any point in the reservoir by tracking the change in the fraction of water flowing at that point.

Eq. 8.31 can be integrated for the total time because the different water saturation planes of the injected water began to reach defined positions in the waterflooded area.

$$dx = \frac{5.615 q_t}{\phi A}\left(\frac{df_w}{dS_w}\right)dt. \qquad (8.33)$$

$$x_{Sw} = \frac{5.615}{\phi A}\frac{df_w}{dS_w}\int_0^t \underbrace{q_t \, dt}_{W_i = q_w t} \text{ and } x_{Sw} = \frac{5.615 W_i}{\phi A}\left(\frac{df_w}{dS_w}\right)_{Sw}, \qquad (8.34)$$

where W_i is the cumulative water injected. It is assumed as an initial condition that at $t = 0$, $W_i = 0$ and x_{Sw} is the distance to the positions of the different water saturation planes.

8.1.3 The Stabilized-Zone Concept. The saturation equation (Eq. 8.34) shows that the distance x that a saturation front S_w has moved in a porous medium can be evaluated at any time t. Eq. 8.34 implies that by knowing the slope of the fractional flow curve, the distance x can be calculated at a given time. The positions of different water saturation planes can be plotted by using Eq. 8.34 and determining the slope of the fractional flow curve for that particular saturation value. However, there is a mathematical difficulty in applying this technique because there is a point of inflection in the fractional flow curve, and a plot of f_w vs. S_w (**Fig. 8.8a**) indicates that three slopes or saturation values would result if the fractional flow curve were used in this S-shaped form.

The presence of three slopes in the equation implies the need to make simplifying assumptions. In addition, the inflection point can be observed on the plot of $\frac{df_w}{dS_w}$ vs. S_w (Fig. 8.8b).

8.1.4 Balancing the Areas. A problem arises because capillary and gravity effects were ignored in the derivation of the fractional flow equation. The profile shown in Fig. 8.8a is physically impossible

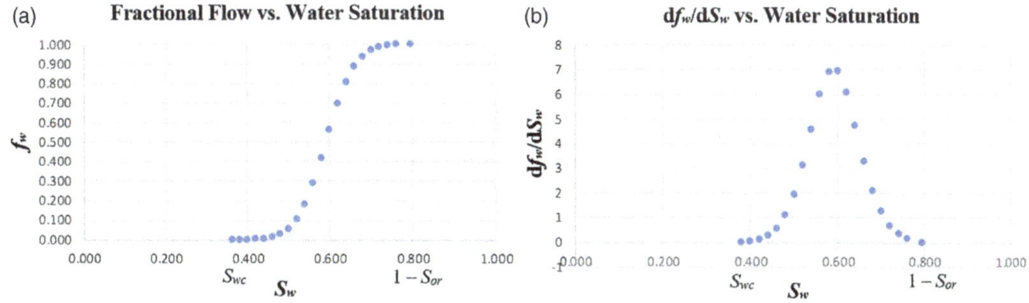

Fig. 8.8—(a) A point of inflection is observed in the fractional flow curve; (b) a graph of df_w/dS_w vs. S_w will have a maximum point.

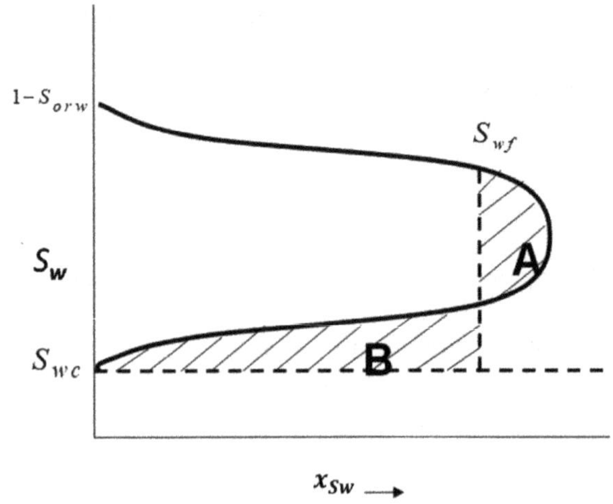

Fig. 8.9—Resulting water saturation distribution in the displacement path, showing the positions of different water saturation planes x_{Sw}.

because it indicates that multiple water saturations can coexist at a given point in the reservoir. The triple-value saturation curve could be forced to represent a true saturation profile when the areas are balanced.

A complex derivation by Welge (1952) shows that the assumption is analytically correct, and the determination of the vertical dashed line shown in **Fig. 8.9** is required, in such a way that shaded areas A and B are equal. The dashed line represents the shock front of saturation discontinuity S_{wf}.

In reality, the saturation distribution within an oil reservoir can be divided into

- An initial water saturation ($S_{wi} = S_{wc}$)
- A residual oil saturation (S_{or})
- Movable oil saturation (S_{mov}) regions

8.1.5 The Saturation Profile. Welge (1952) demonstrated that the average water saturation behind the front can be obtained by integrating the saturation distribution over the distance from the injection point to the front. This is shown graphically in **Fig. 8.10** by the symbol $\bar{\bar{S}}_w$. The situation depicted is at a fixed time t, before water breakthrough in the producing well, and it corresponds to an amount of water injection W_i. At this time, the maximum water saturation ($S_w = 1 - S_{or}$) has moved some distance, and its velocity is proportional to the slope of the fractional flow curve evaluated for the maximum saturation ($S_w = 1 - S_{or}$). The flood-front saturation S_{wf} is located at a position x_2, measured from the injection point.

The displacement process occurs in two stages:

- An initial stage: Constant rate and a zero water/oil ratio (WOR) lasts until breakthrough of the constant saturation front to the producing well.
- A subordinate stage: This stage is characterized by a constantly increasing water saturation in the reservoir.

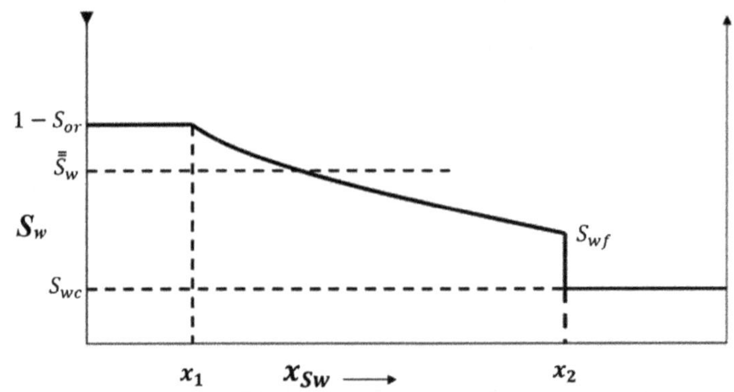

Fig. 8.10—Water saturation distribution as a function of distance x_{Sw}, before breakthrough in the producing well (after Dake 1978).

Laboratory studies have shown that the stabilized zone remains constant. Therefore, the df_w/dS_w slope must remain constant. In reality, the displacement process occurs in two stages, and not three, as predicted by the S-shaped curve of the fractional flow equation.

8.1.6 Secant Line. The fundamentals of analytic geometry indicate that the slope of a secant (tangent) line intersecting a curve at two or more points represents the average slope of the curve. Recall, the secant angle is equal to 1/cosine.

Therefore, the df_w/dS_w slope at the time of water breakthrough is represented by a straight line drawn tangent to the fractional flow curve.

The fractional flow curve has now been divided into stabilized and variable saturation zones. **Fig. 8.11** shows the relation of the secant (tangent) line to the fractional flow curve and water saturations as a function of the fraction of water flowing (f_w).

Several important pieces of information can be derived from the fractional flow curve. By drawing a straight line tangent to the fractional flow curve, starting at $f_w = 0$ and $S_w = S_{wi}$,

- At the tangent point, the corresponding S_w is the water saturation at the flood front.
- The corresponding f_{wf} is the fraction of water flowing at the flood front.
- The water saturation value where the tangent line intersects $f_w = 1.0$ is the average water saturation in the reservoir at breakthrough.

The displacement efficiency E_D at breakthrough is calculated as

$$E_D = \frac{\bar{S}_{wBT} - S_{wi}}{1 - S_{wi}}. \quad (8.35)$$

Note, this is applicable to a single-layer system.

In Fig. 8.11, note the locations of

- The secant (tangent) line
- Initial water saturation (S_{wi}) and residual oil saturation (S_{or}) values
- Water saturation at the flood front (S_{wf})
- Average water saturation behind the flood front at breakthrough in the system ($\bar{S}_{wBT}$)

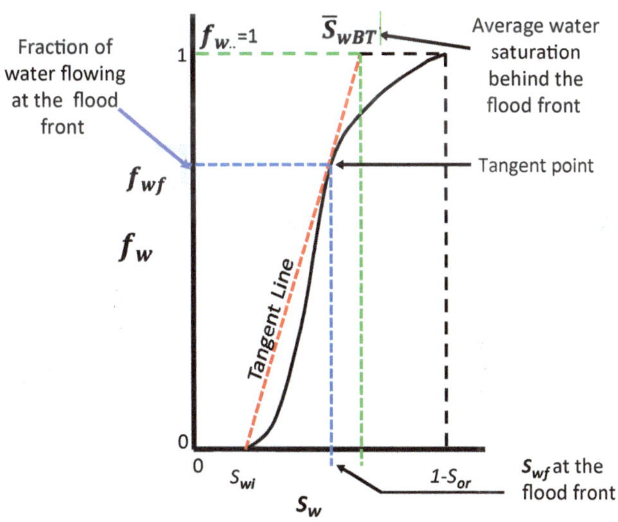

Fig. 8.11—Fractional flow curve showing the secant (tangent) line and the different water saturation values obtained from its analysis.

The displacement concept represented by Fig. 8.11 can be divided into a stabilized and a variable saturation zone:

- The first point of tangency defines the fraction of water flowing (f_{wf}) at the time of breakthrough and the water saturation (S_{wf}) at the flood front.
- Extending the straight line to the f_w = 1 curve defines the average water saturation ($\bar{S}_{wBT}$) behind the flood front at breakthrough.
- A straight line initiated at the f_w axis and drawn to a point of tangency on the fractional flow curve will result in the water saturation (S_{wf}) measured at the front.

A waterflood project feasibility analysis implies the performance of some tasks that require the determination of reservoir performance/cumulative recovery, the producing WOR, and the cumulative water injected. Knowledge of reservoir volume is required to calculate the required volume of water to be injected, as shown by Eq. 8.36.

Volume of water injected = (reservoir volume) (Δ water saturation):

$$W_i = xA\phi\left(\bar{S}_w - S_{wi}\right). \qquad (8.36)$$

To determine $\bar{S}_w$, recall Eq. 8.34, $x_{Sw} = \dfrac{5.615W_i}{\phi A}\left(\dfrac{df_w}{dS_w}\right)_{S_w}$.

Setting Eqs. 8.34 and 8.36 equal to each other results in an expression showing the change in the water saturation as a function of the slope of the secant line:

$$\left(\bar{S}_w - S_{wi}\right) = \frac{1}{\dfrac{df_w}{dS_w}} \text{ or } \frac{df_w}{dS_w} = \frac{1}{\left(\bar{S}_w - S_{wi}\right)}. \qquad (8.37)$$

To calculate the slope of the secant line,

$$m = \frac{df_w}{dS_w} = \frac{y_2 - y_1}{x_2 - x_1} = \frac{1-0}{\left(\bar{S}_w - S_{wi}\right)}. \qquad (8.38)$$

To calculate the oil volume produced at any average water saturation value $\bar{S}_w$ behind the front,

$$N_p = \frac{V_p\left(\bar{S}_w - S_{wi}\right)}{B_o}, \text{ STB}. \qquad (8.39a)$$

Applying Eq. 8.39a to determine the oil volume produced at the average water saturation value $\bar{S}_{wBT}$ at the front breakthrough is then given by Eq. 8.39b:

$$N_p = \frac{V_p\left(\bar{S}_{wBT} - S_{wi}\right)}{B_o}, \text{ STB}. \qquad (8.39b)$$

To calculate the cumulative water injected at any time, rearrange Eq. 8.32 to

$$W_i = q_t t = \frac{xA\phi}{\dfrac{df_w}{dS_w}}, \text{ RB}. \qquad (8.40)$$

Because pore volume (V_p) = $xA\phi$, W_i = (pore volume displaced) (inverse slope of secant line):

$$W_i = q_t t = \frac{1}{\frac{df_w}{dS_w}}\left(V_p\right). \quad (8.41)$$

To calculate the producing WOR at any average water saturation $\bar{S}_w$,

$$WOR = \frac{q_w}{q_o} = \frac{f_w}{f_o} = \left(\frac{f_w}{1-f_w}\right)\frac{B_o}{B_w}. \quad (8.42)$$

Analysis procedure

- Calculate and plot the f_w vs. S_w curve from relative permeability data.
- Calculate the reservoir pore volume and original oil-in-place (OOIP).
- Draw a tangent line to represent times at and after breakthrough.
- Calculate
  - Cumulative water injected.
  - Time to attain breakthrough.
  - Cumulative recovery.
  - Producing WOR.

Example 8.2—Predicting Performance. A reservoir is being studied as a potential candidate for a waterflood project. Predict the performance of the project at the time of breakthrough and at WOR = 2, and calculate cumulative oil recovery, the percent of oil recovered, the time required to achieve breakthrough, and the volume of water injected.

Reservoir data are available in **Table 8.2** (assume fluid properties are constant at the specified values determined at initial conditions), and **Fig. 8.12** shows the reservoir configuration as an areal view of the static model and approximated using the geometric shape of a rectangle.

Fig. 8.13 shows the relative permeability curves from laboratory tests for the reservoir being studied.

The producing interval is fairly uniform, and well locations appear to afford a reasonable sweep efficiency.

Production/injection facilities: Natural water influx appears absent, and water-injection facilities are capable of processing and injecting 5,000 BWPD.

Perform the following calculations:

Part 1: Predict reservoir performance at the time of breakthrough.

1. Calculate cumulative oil recovery.
2. Calculate percent oil recovered.
3. Calculate the time required to achieve breakthrough.
4. Calculate the volume of water injected.

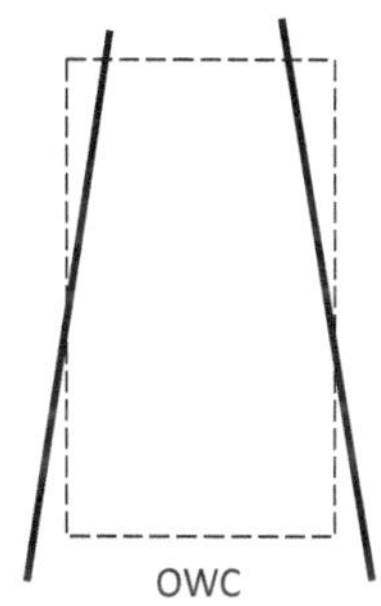

Fig. 8.12—Areal view of the reservoir, approximated using the geometric shape of a rectangle and showing the oil/water contact (OWC) at the south and the boundaries to the east and west.

Size	h = net pay = 40 ft	x = length = 5,000 ft	w = width = 500 ft	dip = 2.5°
Fluids	B_{oi} = 1.5 RB/STB, B_{wi} = 1.0 RB/STB	μ_o = 2.0 cp	μ_w = 1.0 cp	γ_w = 1.1
Rock	ϕ = 25%	S_{wi} = 20%	k = 893 md	γ_o = 0.866

Table 8.2—Reservoir data for Example 8.2.

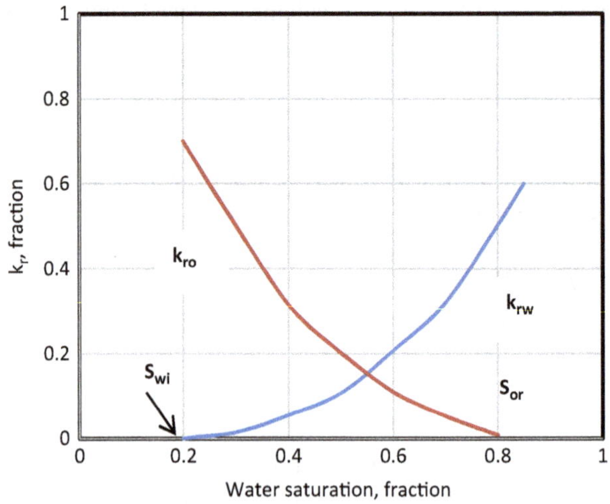

Fig. 8.13—Relative permeability curves, Example 8.2.

Part 2: Predict performance for WOR = 2.

1. Calculate cumulative oil recovery.
2. Calculate percent oil recovered.
3. Calculate the volume of water injected.

Solution. Part 1: Predict reservoir performance at the time of breakthrough.

First, construct the fractional flow curve. Use Eqs. 8.15, 8.16a, and 8.16d (assume the capillary pressure term equals zero):

$$f_w = \frac{1 - C'k_o}{1 + \dfrac{k_o \mu_w}{k_w \mu_o}} = \frac{1 - C'kk_{ro}}{1 + \dfrac{k_{ro} \mu_w}{k_{rw} \mu_o}} \text{ and } C' = \frac{0.000488A(\Delta\gamma)(\sin\theta)}{\mu_o q_t},$$

$$C'k = \frac{0.000488Ak(\Delta\gamma)(\sin\theta)}{\mu_o q_t} = \frac{0.000488 \times 40 \times 500 \times 893(0.23 \times 0.04)}{2.0 \times 500} = 0.08,$$

$$f_w = \frac{1 - 0.08k_{ro}}{1 + \dfrac{k_{ro} \mu_w}{k_{rw} \mu_o}}.$$

Table 8.3 shows the fractional flow calculations using the relative permeability vs. S_w data (Columns 1–3) read from Fig. 8.13. The fractional flow vs. S_w curve **(Fig. 8.14)** is built using data from Table 8.3 (Columns 1 and 7).

Using the constructed fractional flow curve, draw two lines of tangency, at breakthrough (red dashed line) and at WOR = 2 (blue dashed line), as seen in Fig. 8.14.

Values read from Fig. 8.14 (red dashed line) are summarized as follows:

- Fractional flow at the flood front: $f_{wf} = 0.78$
- Average water saturation at breakthrough behind the flood front: $\overline{S}_{wBT} = 0.71$
- Water saturation at the flood front: $S_{wf} = 0.60$

1	2	3	4	5	6	7
S_w	k_{rw}	k_{ro}	$k_o\mu_w/k_w\mu_o$	$1 - 0.08k_{ro}$	$1 + k_o\mu_w/k_w\mu_o$	f_w
0.20	0	0.700	–	0.930	–	0.000
0.30	0.015	0.500	16.667	0.960	17.667	0.054
0.40	0.058	0.315	2.716	0.975	3.716	0.262
0.50	0.107	0.202	0.944	0.979	1.944	0.506
0.60	0.207	0.110	0.272	0.991	1.266	0.783
0.70	0.320	0.053	0.083	0.996	1.083	0.920
0.80	0.503	0.008	0.008	0.999	1.008	0.991
0.85	0.600	0.000	–	1.000	1.000	1.000

Table 8.3—Fractional flow calculations, Example 8.2.

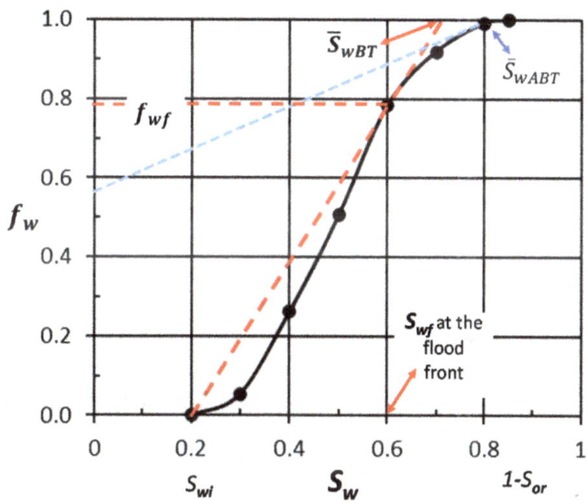

Fig. 8.14—Fractional flow curve, Example 8.2.

Next, calculate pore volume and OOIP:

Area $(A) = w \times h$,

$$\text{Pore volume } (V_p) = xA\phi = x \times w \times h \times \phi = \frac{(5{,}000 \text{ ft})(500 \text{ ft})(40 \text{ ft})(0.25)}{5.615 \text{ ft}^3 / \text{bbl}} = 4{,}452{,}360 \text{ RB},$$

$$\text{Pore volume } (V_p) \text{ expressed in STB} = \frac{xA\phi}{B_{oi}} = \frac{4{,}452{,}360 \text{ RB}}{1.5 \text{ RB / STB}} = 2{,}968{,}240 \text{ STB},$$

$$OOIP = \frac{(V_p)(1 - S_{wi})}{B_{oi}} = \frac{(4{,}452{,}360 \text{ RB})(1 - 0.2)}{1.5 \text{ RB / STB}} = 2{,}374{,}592 \text{ STB}.$$

The secant (tangent) line, which is the red dashed line in Fig. 8.14, indicates that the average water saturation behind the front at breakthrough is $\bar{S}_{wBT} = 71\%$.

1. Calculate cumulative oil recovery. The oil recovered to change the water saturation S_w from S_{wi} to $\bar{S}_{wBT}$ is calculated using Eq. 8.39b, and because it is assumed that fluid properties are constant at the specified values at initial conditions, $B_o = B_{oi}$:

$$N_p = \frac{V_p\left(\bar{S}_{wBT} - S_{wi}\right)}{B_o} = \frac{(4{,}452{,}360)\left(0.71-0.20\right)}{1.5} = 1{,}211{,}042 \text{ STB.}$$

2. Calculate percent oil recovered:

$$\text{Fraction of oil recovered} = \frac{N_p}{OOIP} = \frac{1{,}211{,}042}{2{,}374{,}592} = 0.51,$$

$$\text{Percent recovery} = \left(\frac{N_p}{OOIP}\right)\times 100 = \left(\frac{1{,}211{,}042}{2{,}374{,}592}\right)\times 100 = 51\%.$$

3. Calculate the time required to achieve breakthrough. Injection volumes must equal oil production volumes until breakthrough. For incompressible displacement, the oil recovery is equal to the volume of water injected, and there is no water production during this phase.

$$\text{Because } q_w = 5{,}000 \text{ BWPD, } q_o = \frac{5{,}000 \text{ RB / D}}{1.5 \text{ RB / STB}} = 3{,}333 \text{ BOPD.}$$

$$t = \frac{1{,}211{,}042 \text{ STB}}{3{,}333 \text{ BOPD}} = 363.4 \text{ days or approximately 1 year.}$$

4. Calculate the volume of water injected:

Volume of water injected = 363.4 days × 5,000 BWPD = 1,816,744 BW.
Part 2: Predict performance for WOR = 2.
First, construct a tangent line on the fractional flow curve for the WOR = 2 case.
Calculate the f_w equivalent to WOR = 2 using Eq. 8.42:

$$WOR = \left(\frac{f_w}{1-f_w}\right)\frac{B_o}{B_w};\ 2 = \left(\frac{f_w}{1-f_w}\right)\frac{1.5}{1};\ f_w = 0.57.$$

Draw a tangent line (blue dashed line), as seen in Fig. 8.14 beginning at $f_w = 0.57$. The intersection point with the x-axis gives a value of the average water saturation behind the flood front after breakthrough, $\bar{S}_{wABT}$, which is 0.80. The slope is determined by reading in Fig. 8.14 the corresponding values at any pair f_w and S_w. For instance, at $f_w = 1$, the corresponding $S_w = 0.8$, and for $f_w = 0.57$, it reads $S_w = 0.0$. Thus,

$$\frac{df_w}{dS_w} = \frac{1.0-0.57}{0.8-0.0} = 0.54.$$

Calculated and determined values from the tangent line (blue dashed line) in Fig. 8.14 are summarized as follows:

- Fractional flow at WOR = 2: $f_w = 0.57$
- Average water saturation after breakthrough behind the flood front: $\bar{S}_{wABT} = 0.80$
- Water saturation at the flood front: $S_{wf} = 0.60$

1. Calculate cumulative oil recovery:

$$N_p = OOIP\left(\bar{S}_{wABT} - S_{wi}\right) = 2{,}374{,}592\left(0.80 - 0.20\right) = 1{,}424{,}755 \text{ STB}.$$

2. Calculate percent oil recovered:

$$\text{Oil recovered (\%)} = \frac{N_p}{OOIP} = \frac{1{,}424{,}755}{2{,}374{,}592} = 60\%.$$

3. Calculate volume of water injected:

$$W_i = \frac{1}{\frac{df_w}{dS_w}}(V_p) = \frac{2{,}968{,}240 \text{ STB}}{0.54} = 4{,}397{,}393 \text{ STB},$$ or approximately 4.4 million STB of water injected.

Example 8.2 illustrates how to estimate oil recovery and the growth of water production as a function of changing relative permeabilities using the Buckley-Leverett frontal advance solution and the calculated fractional flow curve.

8.2 Stiles Method

The Stiles method (Stiles 1949) accounts for piston-like displacement of the varying flood-front positions in a multilayered oil reservoir undergoing water encroachment. It presents the fraction of water flowing f_w as a function of the total rate ($q_o + q_w$). Flood-front advancement is a function of layer permeability and the vertical distribution of the layers.

Fig. 8.15 shows the displacement of oil by means of water where the sands K1 and K3 have flooded out, while K2, K4, and K5—which are lower-permeability layers—are still undergoing displacing water pushing oil to the producing well in a piston-like manner.

Assumptions

- The velocity of the advancing linear water front is proportional to the permeability of each idealized layer and is the same as the velocity of oil.
- There is no vertical flow or crossflow between zones, and there is piston-like immiscible displacement, which assumes a mobility ratio equal to unity.
- Water front breakthrough occurs in a sequence that starts in the layer with the highest permeability. At water front breakthrough and after it, all higher-permeability beds are only flowing water.

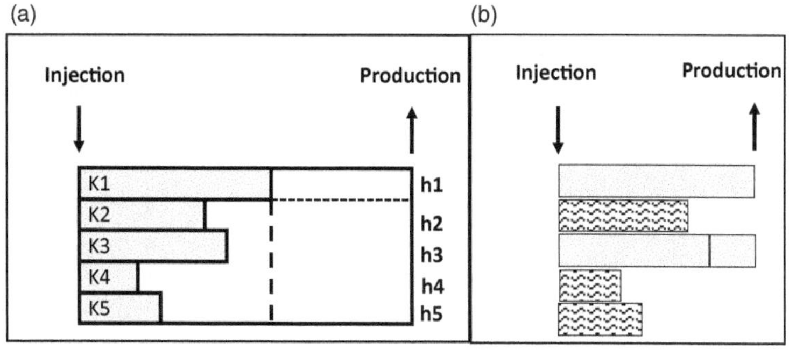

Fig. 8.15—(a) Displacement of oil by means of water in a stratified reservoir before breakthrough. (b) Stratified reservoir during waterflooding; breakthrough has occurred in Layers 1 and 3.

- There is no provision for an established gas saturation.
- All beds have similar values of porosity, relative permeability, and oil saturation changes, and are arranged sequentially in decreasing order, with the maximum *kh* layer at the top.
- Fractional recovery of the total movable oil *R* is equal to k_k/k_j.

Stiles (1949) defined terms for forecasting production and oil and gas recoveries as functions of time in a layered reservoir, in addition to estimating values of fill-up and WOR as functions of time. The capacity *C* of each individual layer can be expressed as a function of successive layers flooding out as a series expansion and is shown in Eq. 8.43:

$$C_j = (h_i)(k_i) + \sum hk_{i-1}, \quad (8.43)$$

where C_j is the water capacity of the last flooded-out layer. The series expands sequentially from the most- to the least-permeable layer.

The change in water capacity *C* is a function of a flood-out of decreasing-permeability layers. Recovery can be defined as

$$R = (1 - C). \quad (8.44)$$

Oil recovery from all beds is the sum of the flooded-out layers and the fractional portions of the layers still undergoing oil displacement. Thus, *R* can be calculated as a function of the flooded-out layers, as expressed by Eq. 8.45:

$$R = \frac{h_j k_j + (C_t - C_j)}{h_t k}. \quad (8.45)$$

Fig. 8.16 shows the relationship between recovery *R* and water capacity *C* in a layered system.

The water capacity distribution curve in Fig. 8.16 shows the accumulated capacity as a function of accumulated thickness, starting with the highest permeability. Capacities and thicknesses are expressed as fractions of the total capacity and the total thickness, respectively.

Production measurements are taken at surface conditions while the water is being injected to displace oil in the reservoir being studied.

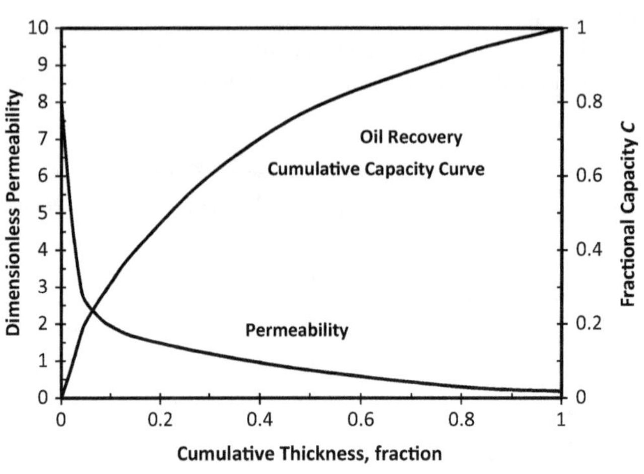

Fig. 8.16—Stiles method in a stratified reservoir during waterflooding.

The mobility ratio measured at reservoir conditions is calculated as

$$M = \frac{k_{rw}\mu_o}{k_{ro}\mu_w}. \quad (8.46)$$

The mobility ratio measured at surface conditions is calculated as

$$M' = \frac{k_{rw}\mu_o}{k_{ro}\mu_w} \times \frac{B_o}{B_w}. \quad (8.47)$$

The fraction of water flowing was given as $f_w = \frac{q_w}{q_o + q_w}$ in Eq. 8.4. Applying Darcy's law and expressing f_w as measured at surface conditions,

$$f_{w(\text{surface})} = \frac{M'C_j}{M'C_j + \left(C_t - C_j\right)}. \quad (8.48)$$

Cumulative production is calculated as

$$N_p = f_{w(\text{surface})} \times \left(\text{reserves}\right). \quad (8.49)$$

Oil production is given as

$$q_o = \frac{(\text{injection rate})\left[1 - f_{w(\text{res})}\right]}{B_o}. \quad (8.50)$$

The fraction of water flowing in the reservoir is calculated as

$$f_{w(\text{res})} = \frac{MC_j}{MC_j + \left(C_t - C_j\right)}. \quad (8.51)$$

As given in Eq. 8.42, $WOR = \left(\frac{f_w}{1 - f_w}\right)\frac{B_o}{B_w}$.

The maximum recoverable oil (including residual oil saturation) is calculated using Eq. 8.52:

$$\text{Recoverable oil} = \left(1 - S_{wi} - S_{or}\right) \times OOIP. \quad (8.52)$$

The remaining recoverable oil (measured at current reservoir conditions) is calculated using Eq. 8.53a and Eq. 8.53b:

$$\text{Remaining recoverable oil} = (\text{recoverable oil} - \text{primary recovery}) \times \frac{B_o}{B_{oi}}. \quad (8.53a)$$

The calculated remaining recoverable oil is, in general, adjusted to account for fingering effects, and a value between 70 and 85% is used as the estimated sweep efficiency, as expressed by Eq. 8.53b:

$$\text{Remaining recoverable oil} = \left(\text{recoverable oil} - \text{primary recovery}\right) \times \frac{B_o}{B_{oi}} \times \text{sweep efficiency} \quad (8.53b)$$

Eq. 8.54 is used to calculate the fill-up volume, and Eq. 8.55 is used to calculate the fill-up time:

$$\text{Fill-up volume (including gas saturation)} = \left[7{,}758\ \text{bbl/acre-ft (isopach thickness volume)(acre-ft)}\right](\phi)\left(1 - S_{wi} - S_{or} - S_g\right). \quad \text{(8.54)}$$

$$\text{Fill-up time} = \text{fill-up volume/water-injection rate.} \quad \text{(8.55)}$$

Example 8.3—Predicting Performance. Use the data provided in **Table 8.4** to predict performance from a 29-ft layered producing interval. Assume an 85% sweep efficiency.

Determine the fill-up volume and fill-up time, and predict future oil recovery as a function of time.

Solution. First, calculate the mobility ratio measured at surface and reservoir conditions:

$$M' = \frac{k_{rw}\mu_o}{k_{ro}\mu_w} \times \frac{B_o}{B_w} = \frac{0.2 \times 4.34}{0.8 \times 0.82} \times \frac{1.073}{1.0} = 1.42.$$

$$M = \frac{k_{rw}\mu_o}{k_{ro}\ \mu_w} = \frac{0.2 \times 4.34}{0.8 \times 0.82} = 1.32.$$

Second, calculate OOIP and reserves using Eq. 1.3:

$$OOIP = \frac{\left[7{,}758\ \text{bbl/acre-ft (isopach thickness volume)(acre-ft)}\right](\phi)\left(1 - S_{wi}\right)}{B_{oi}\ \text{(RB/STB)}}$$

$$= \frac{(7{,}758\ \text{bbl/acre-ft})(100\ \text{acre-ft})(0.19)\left(1 - 0.24\right)}{1.215\ \text{RB/STB}} = 92{,}200\ \text{STB}.$$

The maximum recoverable oil (including residual oil saturation) is calculated using Eq. 8.52:

$$\text{Recoverable oil} = \left(1 - S_{wi} - S_{or}\right) \times OOIP = \left(1 - 0.24 - 0.21\right) \times 92{,}200 = 50{,}710\ \text{STB}.$$

The remaining recoverable oil (measured at current reservoir conditions) is calculated using Eqs. 8.53a and 8.53b:

$$\text{Remaining recoverable oil} = \text{(recoverable oil} - \text{primary recovery)} \times$$

$$\frac{B_o}{B_{oi}} = \left(50{,}710 - 11{,}150\right)\frac{1.073\ \text{RB/STB}}{1.215\ \text{RB/STB}} = 34{,}940\ \text{STB}.$$

Remaining recoverable oil (assuming 85% sweep efficiency) = remaining recoverable oil × sweep efficiency = (34,940)(0.85) = 29,699 STB.

Reservoir volume = 100 acre-ft	ϕ = 19%	Water-injection rate = 100 BWPD	k_{rw} = 0.2
B_{oi} = 1.215 RB/STB	S_{wi} = 24%	Primary recovery = 11,150 STB	k_{ro} = 0.8
B_o = 1.073 RB/STB	S_{or} = 21%		μ_w = 0.82
B_w = 1.0 RB/STB	S_g = 17%		μ_o = 4.34

Table 8.4—Reservoir and performance data for Example 8.3 (after Craft and Hawkins 1959).

Calculate the fill-up volume using Eq. 8.54 and the fill-up time using Eq. 8.55:

$$\text{Fill-up volume (including gas saturation)} = \left[7{,}758 \text{ bbl/acre} - \text{ft (isopach thickness volume)(acre} - \text{ft)}\right]$$
$$(\phi)\left(1 - S_{wi} - S_{or} - S_g\right) = (7{,}758 \text{ bbl/acre} - \text{ft})(100 \text{ acre} - \text{ft})(0.19)\left(1.0 - 0.24 - 0.21 - 0.17\right) = 56{,}015 \text{ RB}.$$

Fill-up time = fill-up volume/water-injection rate = 56,015/100 =560 days.

Table 8.5 illustrates a calculation setup to predict the producing history for Example 8.3. Calculated values specified in Table 8.5 include daily and cumulative oil production and WOR as a function of time.

1	2	3	4	5	6	7	8	9	10
h (ft)	*k* (md)	C_j (md-ft)	*R* (fraction)	$f_{w(surface)}$ (fraction)	N_p (STB)	$f_{w(res)}$ (fraction)	q_o (STB/D)	*t* (days)	WOR (BW/BO)
0	reservoir	reservoir	reservoir	surface	surface	reservoir	surface		surface
1	776	776	0.226	0.204	6,698	0.192	75.3	89	0.3
2	454	1,230	0.361	0.312	10,722	0.297	65.5	129	0.5
3	349	1,579	0.449	0.391	13,331	0.374	58.4	152	0.7
4	308	1,887	0.495	0.457	14,697	0.439	52.3	161	0.9
5	295	2,182	0.511	0.517	15,354	0.499	46.7	171	1.1
6	282	2,464	0.526	0.573	15,627	0.555	41.5	179	1.4
7	273	2,737	0.537	0.624	15,939	0.607	36.6	189	1.8
8	262	2,999	0.549	0.672	16,307	0.656	32.1	226	2.2
9	228	3,227	0.590	0.713	17,518	0.697	28.2	291	2.7
10	187	3,414	0.651	0.745	19,337	0.731	25.1	309	3.1
11	178	3,592	0.667	0.775	19,797	0.762	22.2	350	3.7
12	161	3,753	0.697	0.801	20,698	0.789	19.6	355	4.3
13	159	3,912	0.700	0.827	20,804	0.816	17.1	388	5.1
14	148	4,060	0.719	0.850	21,361	0.841	14.8	466	6.1
15	127	4,187	0.758	0.870	22,522	0.862	12.9	558	7.2
16	109	4,296	0.798	0.887	23,705	0.879	11.3	713	8.4
17	88	4,384	0.857	0.900	25,451	0.893	9.9	722	9.7
18	87	4,471	0.860	0.913	25,544	0.907	8.7	722	11.3
19	87	4,558	0.860	0.926	25,544	0.921	7.4	829	13.4
20	77	4,635	0.887	0.937	26,334	0.933	6.3	908	16.1
21	71	4,706	0.903	0.948	26,829	0.944	5.2	1,056	19.4
22	62	4,768	0.929	0.957	27,601	0.953	4.3	1,137	23.7
23	58	4,826	0.941	0.965	27,951	0.962	3.5	1,230	29.5
24	54	4,880	0.952	0.973	28,277	0.971	2.7	1,338	38.1
25	50	4,930	0.962	0.980	28,572	0.978	2.0	1,431	51.8
26	47	4,977	0.968	0.986	28,762	0.985	1.4	1,431	77.4
27	47	5,024	0.968	0.993	28,762	0.992	0.7	1,967	150.1
28	35	5,059	0.981	0.998	29,143	0.998	0.2	4,462	481.8
29	16	5,075	1.000	1.000	29,699	1.000	0.0	–	–

Table 8.5—Calculating production as a function of layer depletion, Example 8.3.

Note that the layers are sorted as a function of permeability, not as observed in the reservoir. Each layer is 1 ft thick for this case.

For example calculations for the 5-ft level (identified in Table 8.5 in red)

- Columns 1 and 2 show input thickness and permeability data.
- Column 3 gives the cumulative capacity (coverage), which is the increasing capacity open to water production as each layer floods out. Measurement at reservoir conditions.

$$C_j = (h_i)(k_i) + \sum hk_{i-1},$$

where subscript j refers to the cumulative capacity of the corresponding Layer i, subscript i refers to the properties of the corresponding layer number, and subscript $i - 1$ refers to the properties corresponding to the previous layer.

The cumulative capacity of Layer 5 = C_5 = (1)(295) + (1)(1,887) = 2,182 md-ft.

The eventual capacity of the flooded-out system is calculated and shown in Table 8.5 in Column 3: Cumulative capacity of Layer 29 = C_{29}= (1)(16) + (1)(5,059) = 5,075 md-ft.

Thus, C_t = 5,075 md-ft.

- In Column 4, recovery R is calculated as a function of the flooded-out layers. Measurements are at reservoir conditions.

$$R = \frac{h_j k_j + (C_t - C_j)}{h_t k_j} = \frac{5 \times 295 + (5{,}075 - 2{,}182)}{29 \times 295} = 0.511.$$

- In Column 5, the mobility properties are included to convert the coverage and recovery calculations to water production measured at the surface:

$$f_{w(\text{surface})} = \frac{M'C_j}{M'C_j + (C_t - C_j)} = \frac{1.42 \times 2{,}182}{1.42 \times 2{,}182 + (5{,}075 - 2{,}182)} = 0.517.$$

- Column 6 calculates cumulative production by layer. Measured at stock-tank conditions.

$$N_p = f_{w(\text{surface})} \times \text{reserves} = f_{w(\text{surface})} \times \text{remaining recoverable oil}$$

$$(\text{assuming 85\% sweep efficiency}) = 0.517 \times 29{,}699 = 15{,}354 \text{ STB}.$$

- In Column 7, the mobility value is applied to calculate the fraction of water flowing f_w in a reservoir:

$$f_{w(\text{res})} = \frac{MC_j}{MC_j + (C_t - C_j)} = \frac{1.32 \times 2{,}182}{1.32 \times 2{,}182 + (5{,}075 - 2{,}182)} = 0.499.$$

- Column 8 determines the oil flow rate q_o. Recall, the remaining secondary recovery potential is 29,699 STB. The daily oil flow rate is calculated at stock-tank conditions.

$$q_o = \frac{(\text{injection rate})\left[1 - f_{w(\text{res})}\right]}{B_o} = \frac{100 \text{ BWPD}(1 - 0.499)}{1.073 \text{ RB/STB}} = 46.7 \text{ BOPD}.$$

- Column 9 presents the cumulative days.
- In Column 10, WOR is determined as a function of time using Eq. 8.42 and refers to fractional flow at surface conditions:

$$WOR = \left(\frac{f_w}{1 - f_w} \right) \frac{B_o}{B_w},$$

$$WOR = \left[\frac{f_{w(\text{surface})}}{1 - f_{w(\text{surface})}} \right] \frac{B_o}{B_w} = \left(\frac{0.517}{1 - 0.517} \right) \left(\frac{1.073}{1.0} \right) = 1.1 \text{ BW/BO}.$$

Fig. 8.17 was plotted to present a forecast of the future production rate and water production.

8.3 Summary

The secondary recovery process injects water to replace and displace oil toward a producing well. Displacement does not occur on a one-to-one basis because of varying and relative permeability relationships. Oil and water interactions are nearly impossible to model because of these generally unknown subsurface variations. This chapter reviews variable water/oil flow in waterflood cases.

Reservoir models are idealized to fit some norm, which seldom applies to an actual reservoir case. However, studying these models can provide insight into the interactions occurring in the subsurface.

The fractional flow equation models the simultaneous oil and water flow in a porous medium and is a significant component of the Buckley-Leverett frontal advance theory. Gravity effects and dynamically changing oil and water saturation values are accounted for. The outcome of these studies predicts future WOR changes as a function of the injected water volume.

The Buckley-Leverett frontal advance theory studies the effects of saturation changes and relative permeability on the waterflood process. The theory divides the waterflooding process into an initial stage of piston-like flow and a subordinate stage after water breakthrough. The producing stream shifts from all oil to an oil-and-water combination after water breakthrough to the producing well. Oil and water relative permeability relations predominate after breakthrough, and water production slowly increases after this point. Because the model assumes constant rock properties, the fractional flow curve remains constant. A more-thorough discussion applying this concept to field case history studies can be found in Poston et al. (2019).

The idealized effects of rock layering on waterflooding results were studied by Stiles (1949). The Stiles method accounts for piston-like displacement effects in a layered reservoir without any effects from crossflow between layers. Crossflow between adjacent layers is well-documented in oilfield operations. The Stiles application is seldom applied, but it can be useful as a water-handling predictor.

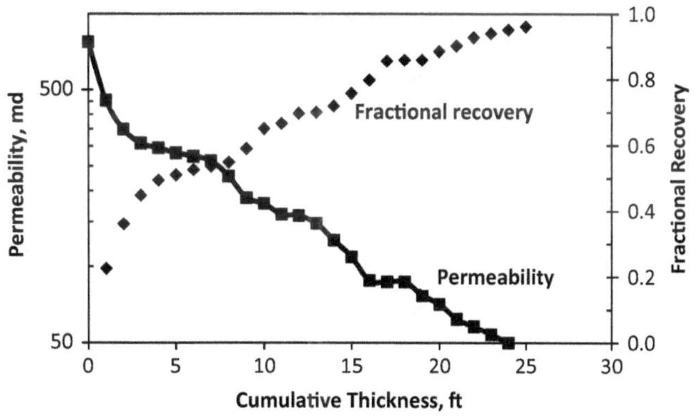

Fig. 8.17—Stiles method in a stratified reservoir during waterflooding. Permeability and recovery are presented as functions of cumulative thickness. Note how the abrupt loss in permeability reduces recovery efficiency.

8.4 Questions

- The Buckley-Leverett concept of fluid flow in a reservoir accounts for the effects of multiphase flow. Where are the produced volumes measured? (Answer: Produced volumes are measured at the separators, which measure the flow rate of each phase separately at standard conditions. However, the study is conducted at reservoir conditions.)
- Are the effects of capillarity and gravity accounted for when using the Buckley-Leverett approach? (Answer: Gravity effects are included in the fractional flow curve, but capillarity is ignored.)
- What is the primary disadvantage of applying the Stiles method to study waterflooding a layered reservoir? (Answer: Assumptions used in the Stiles method imply that there is no communication between sands. The potential for crossflow between perforations or layers is totally discounted.)
- Should the models developed in this chapter be viewed in a quantitative or qualitative sense? (Answer: They should be viewed as qualitative in nature. The models present idealized geometric representation and idealized reservoir conditions.)

8.5 Nomenclature

A = cross-sectional area, ft^2
B_o = oil formation volume factor, RB/STB
B_{oi} = oil formation volume factor at initial reservoir conditions of pressure and temperature, RB/STB
B_w = water formation volume factor, RB/STB
B_{wi} = water formation volume factor at initial reservoir conditions of pressure and temperature, RB/STB
C = water capacity, md-ft
C' = coefficient for constant gravity terms
C'' = constant including gravity and effective permeability terms
C_j = water capacity of flooded-out layer, md-ft
C_t = water capacity of final flooded-out layer, md-ft
E_D = displacement efficiency, fraction or percent
f_o = fraction of oil flowing in system
f_w = fraction of water flowing in system
f_{wbt} = fraction of water flowing at time of breakthrough
$f_{w(\text{downdip})}$ = fraction of water flowing in system when injecting water from updip location in waterflood
f_{wf} = fraction of water flowing at flood front
$f_{w(g)}$ = fraction of water flowing in system when injecting water accounting for gravity effects in waterflood
f_{wh} = fractional flow in a horizontal reservoir
$f_{w(\text{horizontal})}$ = fraction of water flowing in system when injecting water horizontally in waterflood
$f_{w(\text{res})}$ = fraction of water flowing in reservoir
$f_{w(\text{surface})}$ = fraction of water measured at surface conditions
$f_{w(\text{updip})}$ = fraction of water flowing in system when injecting water from downdip location in waterflood
g = gravity
h = height or thickness, ft
h_i = height or thickness of Layer i, ft
h_j = height or thickness of Layer j, ft
h_t = total height or thickness, ft
k = permeability, md
k_i = permeability of Layer i, md
k_j = permeability of Layer j, md
k_k = permeability of Layer k, md
k_o = effective permeability to oil, md
k_r = relative permeability, fraction
k_{ro} = relative permeability to oil, md
k_{rw} = relative permeability to water, md
k_v = vertical permeability, md
k_w = effective permeability to water, md
L = length of flow system, ft
m = slope of secant line
M = mobility ratio measured at reservoir conditions
M' = mobility ratio measured at surface conditions
N_{cv} = capillary-gravity number
N_g = gravity term
N_{gv} = viscous-gravity number
N_p = produced oil volume, STB
N_{pc} = capillary pressure term
p_c = capillary pressure, psia
p_o = oil pressure, psia
p_w = water pressure, psia
q = flow rate, RB/D, STB/D, or BOPD
q_o = oil flow rate, RB/D, STB/D, or BOPD
q_t = total flow rate, RB/D, STB/D, or BOPD
q_w = water flow rate, RB/D, STB/D, or BWPD
R = oil recovery, fraction
S_g = gas saturation, fraction or percent
S_{mov} = movable oil saturation
S_o = oil saturation, fraction or percent
S_{or} = residual oil saturation, fraction or percent
S_{org} = residual oil saturation from oil/gas relative permeability measurements, fraction or percent
S_{orw} = residual oil saturation from oil/water relative permeability measurements, fraction or percent
S_w = water saturation, fraction or percent
$\overline{S}_w$ = average water saturation in system, fraction or percent

$\overline{\overline{S}}_w$ = water saturation as function of distance x_{Sw} before breakthrough, fraction or percent
$\overline{S}_{wABT}$ = average water saturation after breakthrough, fraction or percent
$\overline{S}_{wBT}$ = average water saturation at breakthrough, fraction or percent
S_{wc} = connate-water saturation, fraction or percent
S_{wf} = saturation discontinuity or water saturation at flood front, fraction or percent
S_{wi} = initial water saturation, fraction or percent
t = time, days
V_p = pore volume, RB or STB
V_w or W = volume of water, RB, STB, or BW
w = reservoir width, ft
W or V_w = volume of water, RB, STB, or BW
W_i = cumulative water injected, RB, STB, or BW
x = horizontal length, ft
α = angle of inclination from the horizontal plane, positive updip, degrees
γ = specific gravity of the fluids, fraction
γ_o = specific gravity of oil when referenced to water, fraction
γ_w = specific gravity of water when referenced to water, fraction
$\Delta\rho$ = density change, lbm/ft^3
θ = angle of dip of bed, degrees
μ = viscosity, cp
μ_o = oil viscosity, cp
μ_w = water viscosity, cp
ρ_o = oil density, lbm/ft^3
ρ_w = water density, lbm/ft^3
ϕ = porosity, percent

8.6 Problems

Exercise 8.4—Using the Buckley-Leverett Frontal Advance Theory. Apply the frontal advance concepts to interpret the producing characteristics for a waterdrive reservoir using the data provided in **Table 8.6.**

The relative permeability curves for Exercise 8.4 are provided in **Fig. 8.18.**

Apply the relative permeability data to construct a fractional flow curve (f_w vs. S_w) for an edge-waterdrive reservoir. Be sure to include the gravity effects.

What is the mobility ratio? In addition, determine f_w and S_w at the flood front and $\overline{S}_w$ at breakthrough, and calculate the oil displaced, cumulative water injected, and elapsed time to achieve breakthrough.

Solution. Recall Eqs. 8.15 through 8.16c, the equations used for fractional flow when capillary effects are ignored and water is moving updip. The C' coefficient accounts for gravity effects. Remember, the fractional flow equation accounts for changes in effective permeability. Include absolute relative permeability when relative permeability is applied in the equation.

$$f_w = \frac{1 - C'k_o}{1 + \dfrac{k_o\mu_w}{k_w\mu_o}} = \frac{1 - C'kk_{ro}}{1 + \dfrac{k_{ro}\mu_w}{k_{rw}\mu_o}},$$

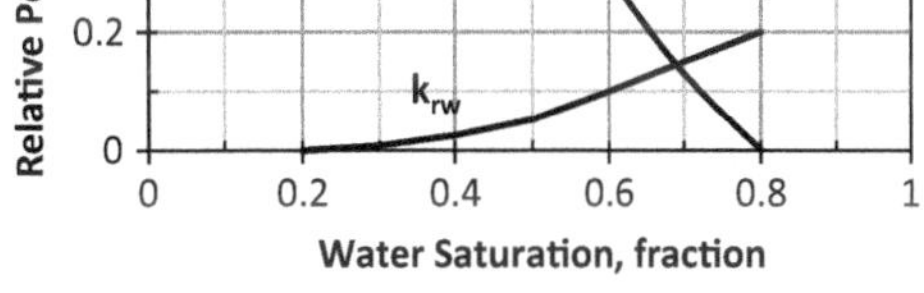

Fig. 8.18—Relative permeability data.

m_w = 0.5 cp	m_o = 10.0 cp	g_w = 1.07	g_o = 0.6
B_{oi} = 1.3 RB/STB	B_{wi} = 1.0 RB/STB	S_{wi} = 20%	q_t = 10,000 STB/D
k = 493 md	Dip = 17.5°, sin 17.5° = 0.3	S_{or} = 20%	
Rectangle-shaped area = A = 520,000 ft^2, L = 1,650 ft	xA = rock volume = 8.58 × 10^8 ft^3	ϕ = 20%	Pore volume (V_p) = $xA\phi$ = 1.72 × 10^8 ft^3

Table 8.6—Field data.

where $k_o = kk_{ro}$ and C' is given in Eq. 8.15,

$$C' = \frac{0.000488A[(\Delta\gamma)(\sin\theta)]}{\mu_o q_t}.$$

Using Eqs. 8.16a through 8.16d:

$$f_w = f_{wh} - f_{w(g)} = \frac{1}{1+\frac{k_o\mu_w}{k_w\mu_o}} - \frac{C''k_{ro}}{1+\frac{k_o\mu_w}{k_w\mu_o}} = \frac{1}{1+\frac{k_{ro}\mu_w}{k_{rw}\mu_o}} - \frac{C''k_{ro}}{1+\frac{k_{ro}\mu_w}{k_{rw}\mu_o}},$$

calculated as

$$C'' = \frac{0.000488kA\left[(\Delta\gamma)(\sin\theta)\right]}{\mu_o q_t} = \frac{0.000488\times 493\text{ md}\times 520{,}000\text{ ft}^2\left[(0.47)\times(\sin 17.5)\right]}{10\times 10{,}000} = 0.176.$$

$$f_{wh} = \frac{1}{1+\frac{k_o\mu_w}{k_w\mu_o}} = \frac{1}{1+0.05\frac{k_o}{k_w}} \text{ and } f_{w(g)} = \frac{0.176k_{ro}}{1+0.05\frac{k_o}{k_w}}.$$

$$f_w = \frac{1-C''k_{ro}}{1+\frac{k_o\mu_w}{k_w\mu_o}} = \frac{1-0.176k_{ro}}{1+0.05\frac{k_o}{k_w}}.$$

Then, calculate the fractional flow equation (as shown in **Table 8.7**) and plot the results, as seen in **Fig. 8.19.**

Next, calculate the mobility ratio using Eq. 8.46. At $S_w = 0.2$, $k_{ro} = 0.9$. At $S_w = 0.8$, $k_{rw} = 0.2$. Recall, $\mu_o = 10$ cp and $\mu_w = 0.5$ cp.

$$M = \frac{k_{rw}\mu_o}{k_{ro}\mu_w} = \frac{0.2\times 10}{0.9\times 0.5} = 2.5.$$

Fig. 8.20 shows the effect of gravity on the fraction of water expected to flow in the reservoir. The gravity effect, although marginal, does reduce the water volume flowing in the system.

S_w	k_{ro}	k_{rw}	$k_{ro}/k_{rw} = k_o/k_w$	f_{wh}	$f_{w(g)}$	f_w
0.2	0.90	0				
0.3	0.81	0.008	101.25	0.17	0.024	0.141
0.4	0.68	0.025	27.20	0.42	0.051	0.373
0.5	0.52	0.052	10.00	0.67	0.061	0.606
0.6	0.30	0.100	3.00	0.87	0.046	0.824
0.7	0.13	0.150	0.87	0.96	0.022	0.937
0.8	0	0.200	0.00	1.00	0.000	1.000

Table 8.7—Results of fractional flow calculations.

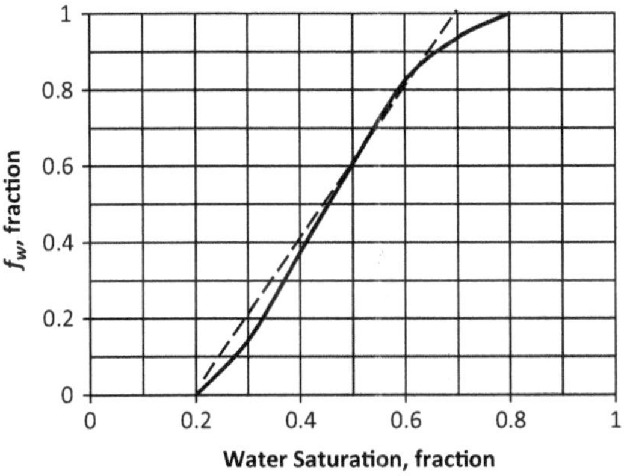

Fig. 8.19—Fractional flow results.

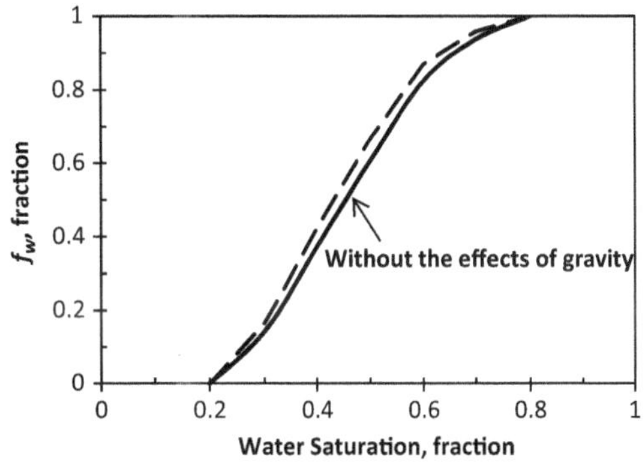

Fig. 8.20—Considering gravity effects on fractional flow.

Next, determine f_w and S_w at the flood front and $\overline{S}_w$ at breakthrough by drawing the secant (tangent) line to the curve, as shown in **Fig. 8.21.**

At breakthrough,

$$f_w = 0.70,\ S_w = 0.54,\ \text{and}\ \overline{S}_w = 0.68;\ m = \frac{df_w}{dS_w} = \frac{y_2 - y_1}{x_2 - x_1} = \frac{1-0}{\left(\overline{S}_w - S_{wi}\right)} = \frac{1-0}{0.68-0.2} = 2.08.$$

$$\text{Pore volume } (V_p) = xA\phi = \frac{1.72\times10^8 \text{ ft}^3}{5.615 \text{ ft}^3/\text{bbl}} = 30.561 \text{ million RB}.$$

$$\text{Pore volume } (V_p) \text{ expressed in STB} = \frac{xA\phi}{B_{oi}} = \frac{30.561 \text{ million RB}}{1.3 \text{ RB/STB}} = 23{,}508{,}462 \text{ STB}.$$

$$OOIP = \frac{\left(V_p\right)\left(1-S_{wi}\right)}{B_{oi}} = \frac{\left(30{,}561{,}000 \text{ RB}\right)\left(1-0.2\right)}{1.3 \text{ RB/STB}} = 18{,}806{,}769 \text{ STB}.$$

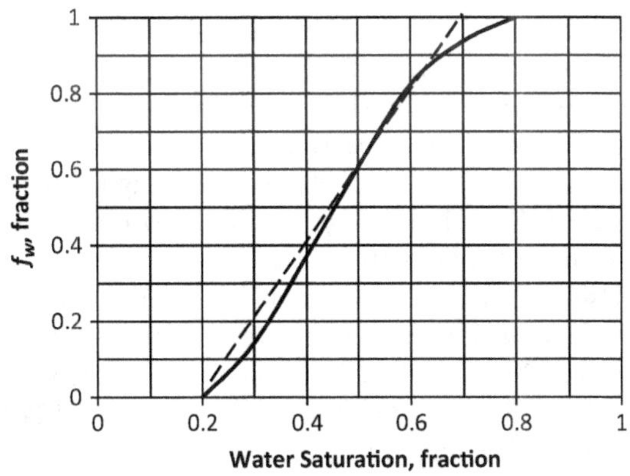

Fig. 8.21—Fractional flow curve. The secant (tangent) shows the average saturation at breakthrough is 68% and the initial water saturation is 20%.

At breakthrough,

$$N_p = \frac{V_p\left(\bar{S}_w - S_{wi}\right)}{B_o} = \frac{30.561 \text{ million RB}\left(0.68-0.20\right)}{1.3 \text{ RB / STB}} = 11.284 \text{ million STB.}$$

$$W_i = q_t t = \frac{xA\phi}{\frac{df_w}{dS_w}} = \frac{23{,}508{,}462 \text{ STB}}{2.08} = 11{,}302{,}145 \text{ STB.}$$

Estimated time to achieve breakthrough

$$= \frac{11{,}302{,}145 \text{ STB}}{10{,}000 \text{ STB / D}} = 1{,}130 \text{ days } = 3.09 \text{ years} = 37.15 \text{ months.}$$

Then, calculate the producing WOR at breakthrough when $f_w = 0.70$:

$$WOR = \frac{q_w}{q_o} = \frac{f_w}{f_o} = \left(\frac{f_w}{1-f_w}\right)\frac{B_o}{B_w} = \left(\frac{0.7}{1-0.7}\right)\frac{1.3}{1.0} = 3 \text{ BW/BO.}$$

8.7 References

- Coats, K. H., Nielsen, R. L., Terhune, M. H. et al. 1967. Simulation of Three-Dimensional, Two-Phase Flow in Oil and Gas Reservoirs. *SPE J.* **7** (4): 377–388. SPE-1961-PA. https://doi.org/10.2118/1961-PA
- Craft, B. C. and Hawkins, M. F. 1959. Applied Petroleum Reservoir Engineering. Englewood Cliffs, New Jersey: Prentice-Hall.
- Dake, L.P. 1978. *Fundamentals of Reservoir Engineering*. Amsterdam: Developments in Petroleum Science 8, Elsevier Science.
- Datta-Gupta, A. and King, M. J. 2007. Streamline Simulation: Theory and Practice, Vol. 11. Richardson, Texas: Textbook Series, Society of Petroleum Engineers.
- Dietz, D. N. 1953. A Theoretical Approach to the Problem of Encroaching and By-Passing Edge Water. *Proc.*, Akad. van Wetenschappen, Amsterdam.
- Latil, M. 1980. *Enhanced Oil Recovery*, 232. Houston, Texas: Gulf Publishing Co.
- Morse, R. A. and Whiting, R. L. 1970. A Numerical Model Study of Gravitational Effects and Production Rate on Solution Gas Drive Performance of Oil Reservoirs. *J Pet Technol* **22** (5): 625–636. SPE-2696-PA. https://doi.org/10.2118/2696-PA.
- Poston, S. W., Laprea-Bigott, M., and Poe, B. D., Jr. 2019. *Analysis of Oil and Gas Production Performance*. Richardson, Texas: Society of Petroleum Engineers.
- Rapoport, L. A. and Leas, W. J. 1953. *Properties of Linear Waterfloods*. *J Pet Technol* **5** (5): 139–148. SPE-213-G. https://doi.org/10.2118/213-G.
- Stiles, W. E. 1949. Use of Permeability Distribution in Water Flood Calculations. *J Pet Technol* **1** (1): 9–13. SPE-949009-G. https://doi.org/10.2118/949009-G.
- Welge, H. J. 1952. A Simplified Method for Computing Oil Recovery by Gas or Water Drive. *J Pet Technol* **4** (4): 91–98. SPE-124-G. https://doi.org/10.2118/124-G.
- Willhite, G. P. 1986. *Waterflooding*, Vol. 3, 153. Richardson, Texas: Textbook Series, Society of Petroleum Engineers.

Part 2: Introduction to Waterflood Design

In the second part of the book, Waterflood Design, the reader is introduced to the subject of waterflooding.

One must recognize there is a difference between the primary depletion and waterflood displacement processes, and these concepts are reviewed in Chapter 9, Waterflooding Process. Primary depletion assumes that subsurface drive mechanisms supply energy to drive the oil to the surface. Waterflooding, on the other hand, involves the injection of water through a wellbore to physically displace the reservoir oil toward a producing well. Producing-well perforations must be hydraulically connected to a water-injection well for the process to occur. The injection of water affects recovery either by helping to maintain reservoir pressure or by actively displacing reservoir oil to a producing well. Pressure maintenance by means of water injection helps to partly replace the reservoir fluid volume loss caused by production. Fluid replacement reduces the pressure decline rate, which in turn affords a greater reservoir production rate.

Chapter 10, Waterflood Components, reviews important aspects of a successful waterflood project. Clastic sediment fields are typically developed using appraisal drilling to test for reservoir boundaries. Completions are typically performed in the best pay zone. Because carbonates are generally very heterogeneous and less amenable to interwell correlations, they are often developed in a pattern configuration because so many pay zones appear to be discontinuous. Perforations are shot on net pay without regard to interwell continuity.

Shales are generally present in a dispersed form or as discrete layers. The presence of either can affect performance during a waterflood. Gravity segregation in thick sand layers during reservoir depletion reduces the thickness of the oil column, allowing the underlying water column to preferentially flow toward a producing well as a result of relative permeability effects.

The presence of intervening shale layers divides the oil column into a series of productive layers, which inhibits the gravity segregation effect. A more-efficient piston-like displacement process occurs. Dispersed shale particles lower matrix permeability, and in some cases, the clay minerals react to the introduction of injected water.

Monitoring and reacting to changes in the oil, water, and gas production rates displayed by a waterflood are some of the primary ways to ensure a successful project. Changes in the relative fluid volumes typically signal fluctuations within the system. These unforeseen changes typically represent an unanticipated issue concerning the integrity of the flow system. This is why it is important to relate the internal structure of the reservoir to current and proposed well locations. Structure and isopach maps and cross-section and fence diagrams coupled with production information form the basis for these studies. Map updating and comparing the character changes of an intervening period can be readily performed when set up on a computer.

Chapter 9

Waterflooding Process

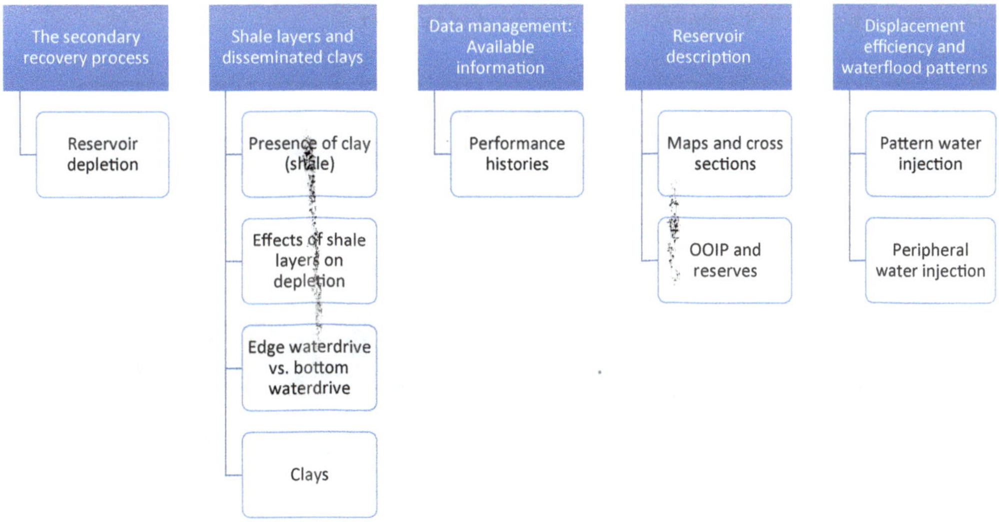

9.1 The Secondary Recovery Process

9.1.1 Reservoir Depletion. Oil recovery is a function of the effectiveness of fluid and rock expansion, gas-cap expansion, and water influx. A pressure decline occurs when reservoir fluid withdrawal from a reservoir is greater than the volume replacement by rock, gas, and water expansion. A loss of reservoir pressure also reduces production rates and, ultimately, oil recovery. Gravity segregation provides a more subtle influence on ultimate recovery. Capillary pressure is more of a hindrance than a help.

Water or gas can be injected into the reservoir to replace at least a portion of the fluid lost, to augment pressure and increase ultimate recovery. Both methods have positive and negative aspects:

- Because gravity segregation rearranges fluids according to their relative densities, the gas-injection process tends to push oil outward to more-downdip areas while injected water pushes oil upward. **Figs. 9.1 and 9.2** show the two processes. The water-injection process pushing oil updip into a more confined space would require fewer wells to deplete the reservoir. On the other hand, increasing the well drainage area reduces sweep efficiency for the gas-injection process.

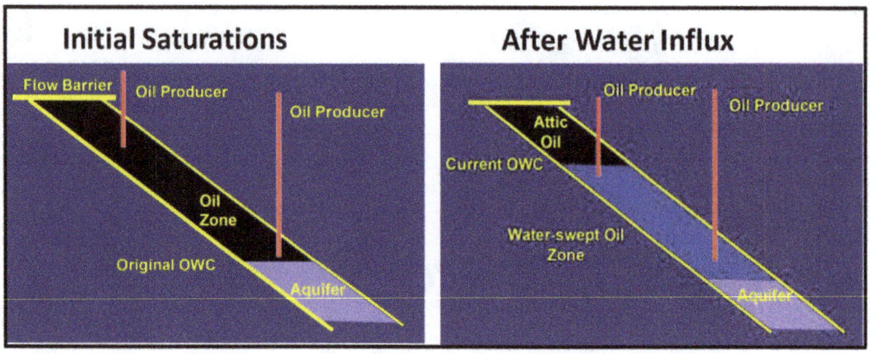

Fig. 9.1—Edge waterdrive. Compare the oil/water contact (OWC) levels in the two images. In the right-side image, the OWC level is continually pushed up, decreasing the size of the original oil zone, showing that the oil producer highest in the structure has been invaded by water

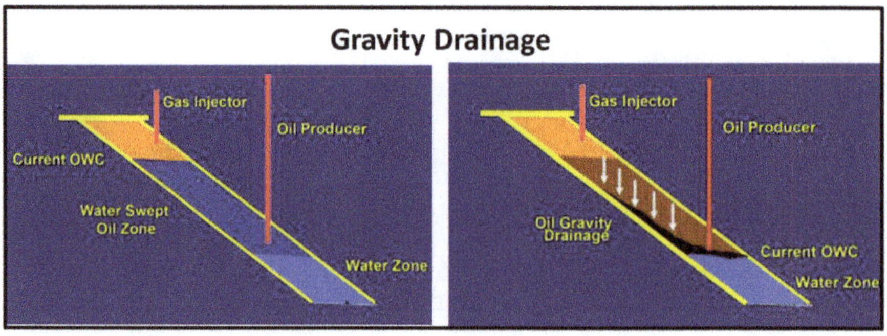

Fig. 9.2—Gas injection. Note the effect of gravity on oil recovery. OWC = oil/water contact.

- The capital expenditure and maintenance costs for installing and maintaining a high-pressure gas-injection station are greater than those for an equivalent water-injection project. The operating expenses for maintaining compressors are generally high.
- Wellbore impairment caused during the gas-injection process is not an important consideration, but it is a constant problem in terms of maintaining adequate water-injection rates. Gas is relatively inert when introduced to reservoir rocks. Injection water can react with clay particles, which can be a significant problem.
- Increased hydrostatic head occurs after water breakthrough, and installing artificial lift to offset decreased flowing pressure can be a major expense.

Fig. 9.1 illustrates aquifer-water influx into an oil field. Encroaching water moves the denser water upward to efficiently trap the less-dense oil into a more-confined area.

The technical team involved in planning an improved recovery project should be able to present to management an estimate of remaining reservoir or field potential. Can ultimate recovery be increased by means of installing a secondary recovery project? If the expense of installing an alternative recovery method is not justified by an increased value, one has done his/her job. A gas- or water-injection project should potentially be installed to supply additional energy.

Gas injection requires a gas source, and gas is nonreactive when introduced to reservoir rock. Fig. 9.2 shows a schematic of gas injection in a tilted reservoir where gravity drainage is dominant.

However, higher maintenance costs, decreased well drainage volumes, and a possible loss of injected gas (which has an inherent value) to the reservoir can result in significant expense. On the other hand, source water is typically available at moderate expense. However, treating costs can be significant, and reservoir rocks often react negatively to the introduction of a reactive water. Water injection to augment reservoir energy is typically preferred because of the minimal cost of water, less-significant environmental concerns, and moderate operating costs.

Water injection can be divided into two essentially different processes:

- Pressure maintenance: A production-rate decline occurs when the offtake rate is greater than the reservoir replacement because of the effect of rock and fluid compressibility. A continued loss of reservoir pressure causes a reduced well rate and increases oil viscosity and residual oil saturation. Supplementing reservoir pressure with injected water helps maintain the maximum efficient rate (MER), which is defined as the maximum-permissible production rate when there is a marginal loss in reservoir pressure. Actual displacement of oil is not the objective of water injection; rather, reservoir fluid replacement is. Gravity segregation separates the two immiscible fluids as a function of density and viscosity differences and permeability.

  Excessive gas production accelerates the depletion of reservoir energy. In addition, introducing the gas phase into an oil wellstream reduces the ability of the well to produce oil at full potential. Producing gas in excess of solution gas during a waterflood means additional water-injection horsepower is required to supplement reservoir pressure and push oil toward a well pressure sink. It is best to shut in high gas/oil ratio (HGOR) wells. Pressure maintenance is prevalent in large reservoirs containing thick, high-permeability producing intervals where wells can be produced at high rates without harming the reservoir.
- Displacement: Water injection displaces otherwise immovable oil toward a pressure sink. In this case, water displaces oil by means of an advancing interface. Displacement efficiency becomes paramount. Pressure-depletion conditions introduce a pressure differential between the well perforations (p_{wf}) and the outer drainage limit (p_e). The induced pressure differential encourages oil to migrate toward the wellbore. The waterflood displacement process introduces a different flow system to the depletion process. Injecting water obviates fluid and rock expansion effects. Water immiscibly displaces oil by physically pushing into the oil-saturated pore spaces and increasing the water saturation value.

Fig. 9.3 illustrates the difference in producing conditions between the pressure-depletion and water-injection processes in a layered heterogeneous reservoir.

Look at the effect of reservoir continuity between a production well (A in Fig. 9.3) and an injection well (B in Fig. 9.3) in a discontinuously layered oil field. Interwell communication exists for Layers 1, 4, and 5, while Layers 2 and 3 are not affected by the injection process.

Oil depletion is affected in different ways. Oil flows from the outer limits of the drainage pattern toward the perforated intervals for solution-gas drive and rock and fluid expansion during primary recovery. However,

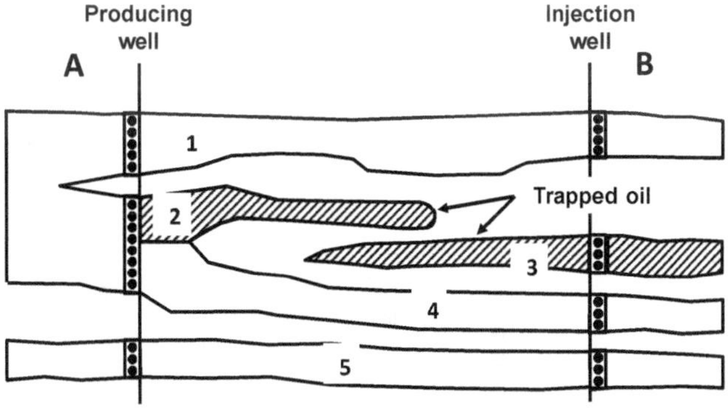

Fig. 9.3—Differences in ultimate recovery in a layered heterogeneous reservoir. Recognizing interwell connectivity is of great importance in waterflooding. Note the importance of well completion and placement.

water injected into Well B does not affect the oil in Layers 2 and 3, and Well A will continue to produce by means of primary depletion as long as the reservoir pressure is sufficient. In fact, oil could be forced back into these low-pressure layers (2 and 3) if water injection increases the current reservoir pressure. This illustrates the differences in estimating oil recovery between the pressure-depleting case and the waterflood case. However, as a rule of thumb, the secondary recovery quantity will typically approximate that achieved by means of primary recovery.

9.2 Shale Layers and Disseminated Clays

9.2.1 Presence of Clay (Shale). Shale is abundantly encountered in the geologic record. It is a fine-grained clay material formed in place or transported to and deposited in a very-low-energy environment. **Fig. 9.4** (Berg 1986) relates grain size to the erosion, tranportation, and deposition sequences as a function of velocity. The movement and deposition of larger-sized grains becomes increasingly dependent on current velocity, while a shale-sized particle could still be transported at very-low-energy conditions. Thick sections of shale commonly occur in low-energy environments such as lakes, ocean basins, and shelf sedimentary environments. Cyclic environments with alternating high- and low-energy conditions often contain extensive layers of very-low-permeability shale that serve as baffles to the vertical migration of fluids located in the sand streaks.

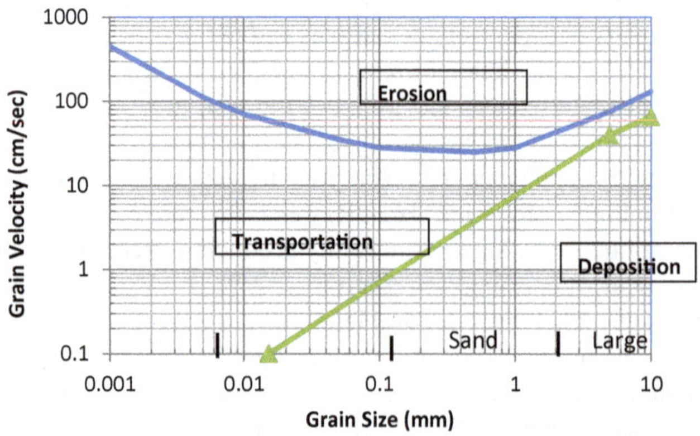

Fig. 9.4—Current velocity and grain-size effects on the erosion, transportation, and deposition envelopes. A minor decline in transport energy can cause the deposition of larger-grained material such as coarse sand, while fine-grained material movement can occur over a wide range of conditions. After Berg (1986).

9.2.2 Effects of Shale Layers on Depletion. The presence of a single or of multiple pay sands within a producing zone can materially affect water-production history. **Fig. 9.5** illustrates that completing in the oil portion of a homogeneous sand (Case A) does not prevent water from migrating upward from the water leg. The oil/water contact will gradually rise until viscous forces cause the water to cone upward in defiance of gravity.

The upward movement of water for Case B is hindered by the presence of a shale barrier. Each sand member depletes as a function of the encroaching water front.

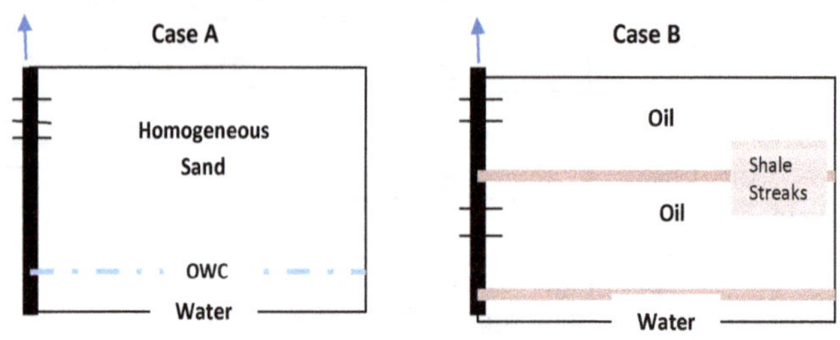

Fig. 9.5—Effect of shale layers on vertical fluid movement during depletion. OWC = oil/water contact.

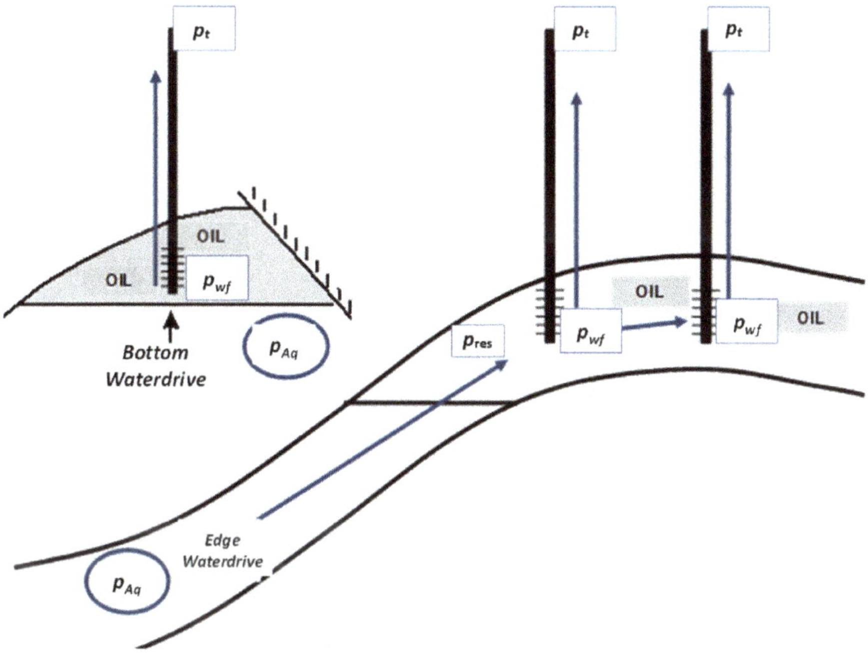

Fig. 9.6—Edge waterdrive and bottom waterdrive affect vertical fluid movement.

9.2.3 Edge Waterdrive vs. Bottom Waterdrive. There are two types of waterdrive mechanisms **(Fig. 9.6):**

- Edge waterdrive is a function of viscous forces (i.e., oil viscosity and permeability).
- Bottom waterdrive is a function of the interplay between gravity and viscous forces.

Edge waterdrive is the preferred of the two because of the presence of a piston-like displacement process, while bottom waterdrive involves simultaneous oil and water production.

In conclusion, shale layers of minor thicknesses can profoundly affect well performance history.

9.2.4 Clays. Clay particles are generally alkaline calcite materials derived from a marine environment. They generally result from the dissolution of carbonate materials that are transported and subsequently deposited as very-fine-grained sediments in thick to thin beds or disseminated among silt or sand deposits. These fine-grained bodies can form barriers to the vertical segregation of oil, gas, and water. Alternatively, clay minerals can be formed by the degradation of clay and organic materials during the movement of groundwater.

Clays comprise a variety of minerals and compounds, some of which can be significantly affected by the introduction of foreign waters. Clays can be transported in by an external force (allogenic) or formed in place (authigenic) through the decomposition of material. Injecting fresh or brackish water into the subsurface can cause swelling and plugging of the pores, particularly when the reservoir rock contains a significant reactive clay volume. Therefore, the type and quantity of clay material present in a formation can significantly affect performance.

Laboratory tests to determine the likelihood of clay swelling, movement of formation fines, and scaling for the particular reservoir rock and fluid should be conducted during a feasibility study if problems are anticipated. There is no point attempting the project if water cannot be injected into the ground in reasonable quantities. Laboratory measurements of clay swelling should be made before the waterflood project is initiated if this problem is expected to occur. **Fig. 9.7** compares types of clays and the pore structures where bridging might be present.

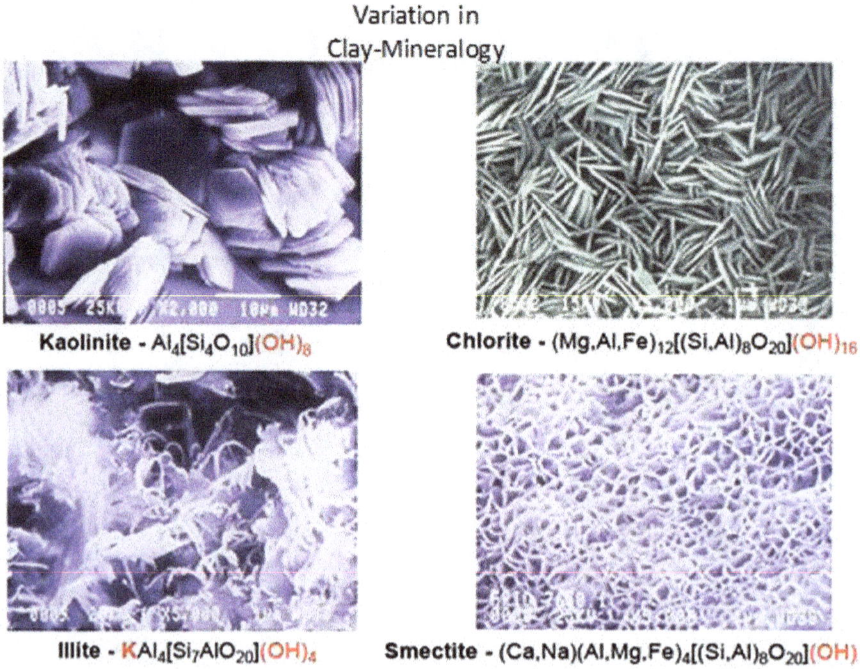

Fig. 9.7—Various clay types, showing the differences in pore spaces. Courtesy of Mamdouh A. Shebl.

The different types of clay shown in Fig. 9.7 have specific characteristics, which can be described as follows:

- Kaolinite and chlorite. These smaller-sized materials are plugging or partially plugging larger-sized pores, and they are relatively impervious to degradation. There are few to no interlayer adsorption sites, which helps make kaolinite and chlorite stable.
- Illite and smectite. These smaller-sized grains are scattered along pore linings. These very-fine-grained materials are extremely reactive to the introduction of foreign liquids. They show large and expandable interlayer surface areas that can accommodate many water molecules, making them expandable and very unstable "reactive" clays.

Table 9.1 provides a general description of the types of accumulations and significant variations of the problematic clay minerals.

Kaolinite	Discrete particles, least likely to swell in the presence of fresh water, thus has a moderate effect on permeability
	Acid material derived from a freshwater environment
	Caused by a breakdown of potassium feldspars or mica
Illite	Is pore bridging, can cause a significant permeability reduction
	Alkaline material derived from a marine environment
	Smectite is converted to illite
	Caused by weathering of potassium feldspars as a result of increasing temperature and ion exchange
Chlorite	Lines pore spaces, with a significant reduction in permeability
	Alkaline material derived from brackish water
	Caused by weathering of volcanic material and iron/magnesium materials

Smectite	Alkaline material derived from a brackish to marine environment
	Has very large and expandable interlayer surface area and is a very unstable clay that can absorb water in the interlayer space
	Caused by a breakdown of plagioclase material and mafic rocks rich in calcium and sodium; some members of this group are montmorillonite, bentonite, nontronite
Montmorillonite	Derived from authigenic smectite and is the most likely to swell of all the clay minerals
	Will immediately rehydrate and swell if fresh water is introduced into the system

Table 9.1—Reactivity of clays.

9.3 Data Management—Available Information

All field, well, production, and subsurface maps should be included in any reservoir characterization and analysis. Their creation involves professionals from different disciplines and is directed toward understanding the geological and depositional environment, in addition to the reservoir structure, rock and fluid properties, and reservoir energy sources. The entire process of reservoir characterization, static-model construction, and updating the model with dynamic data is highly iterative and requires constant interaction between geologists, geophysicists, and reservoir engineers. The quality of the information can range from very high to very low. Data quality generally depends on the operator and the age of the field. In any case, one works with what one has.

Data collection, data management, and quality checks are important components of waterflood reservoir management, but they can potentially be time-consuming aspects of integrated reservoir waterflood studies. The data can be broadly classified into two groups: static data and dynamic data. Static data are time-invariant data such as 3D seismic, petrophysics/well logs, and conceptual geology. The dynamic data are time-varying information such as production and pressure history and time-lapse seismic.

All data need to be validated and stored in a database that is readily accessible to all the study team members. Information can be generally categorized as follows:

- Log surveys, completion logs, and log files
- Structure maps, isopach maps, cross sections, and fence diagrams
- Oil, gas, and water well production histories; flowing tubinghead pressure measurements; static pressure measurements; and production test records
- Reservoir rock and fluid property information obtained from log or laboratory analysis and correlations

9.3.1 Performance Histories.

9.3.1.1 Data Management—Production. The design, installation, and startup of a waterflood assume the expected sweep efficiency is a function of producing well locations within the flood area. To a large extent, the development plan is based on histories from the primary production records. In reality, one should visualize the actual operations of a waterflood as a constantly changing organic process. Often, a waterflood pushes reservoir fluids in different directions than expected. Recall, primary production does not require interwell communication to drive oil toward a well, but water injection requires communication between the injection and oil producing wells. The depletion process becomes more complicated. Decline curve analysis of field producing wells provides clues to flow paths within the reservoir. Monitoring production and analyzing these records is a good way to compare well histories. **Fig. 9.8** shows a semilog hyperbolic rate/time plot of a west Texas well.

However, one seldom finds a smoothly declining well history. Interruptions caused by completing an offset well, holes in the casing, or a myriad of other physical problems can result in changes in the historical production curve. **Fig. 9.9** shows a typical production history, oil rate (BOPD), water/oil ratio (BW/BO), and gas/oil ratio (GOR) for a west Texas field. The interpreted history is summarized in **Table 9.2.**

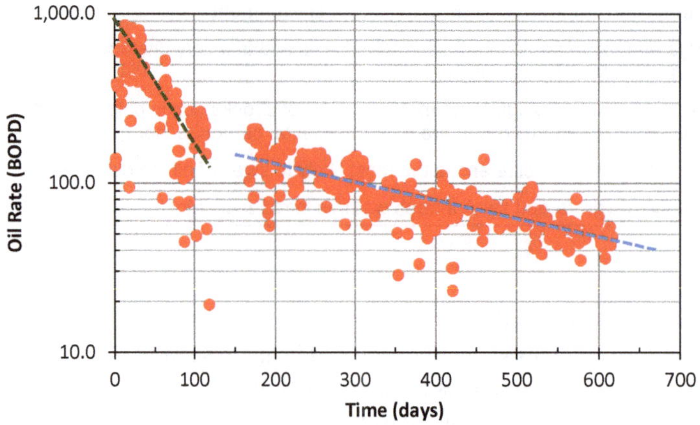

Fig. 9.8—Production history of a west Texas oil well. Changes in operating conditions divide the well history into two segments: early time of the hydraulic fracture followed by a monotonically declining flow rate that lasts until approximately 630 days. Apparently, the well was shut in during the 120- to 170-day time period. The smoothly declining line represents a time of unchanging reservoir character and implies a loss of oil volume within the confines of the reservoir.

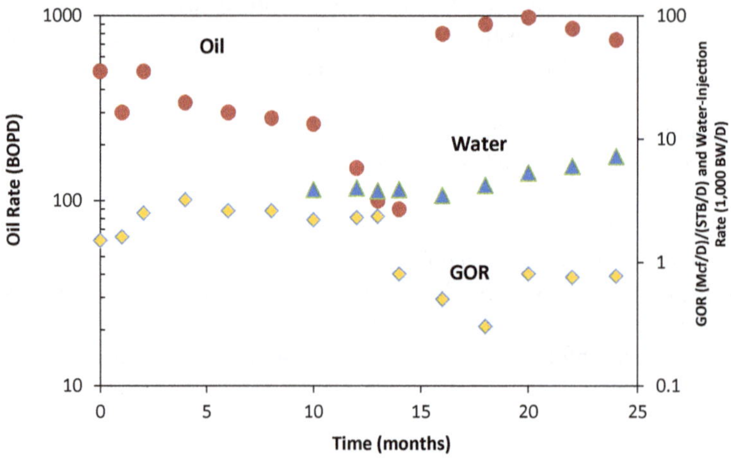

Fig. 9.9—Waterflood history divided into a series of producing segments where uniform depletion conditions are observed during the first 14 months. Estimating future performance by fitting a decline curve over the complete history would result in a qualitative answer at best.

Table 9.2 indicates that

- The producing history can be divided into four producing segments. Each new segment was initiated by a change in field operations. This behavior can also be observed in Fig. 9.9.
- Shutting in the HGOR wells materially increased the effectiveness of the waterflood.
- Water breakthrough occurred soon after water injection commenced.
- Dividing production records into segments representing specific depletion conditions furthered our knowledge of the reservoir mechanics and suggested remedial actions.

Fig. 9.10 provides an example of how knowledge of field conditions can help when interpreting a performance history. The initial review showed a good

Interval	Segment	Actions
0–10 months	Primary depletion	Producing rate exponentially declined. Pressure loss was caused by producing HGOR wells close to the gas/oil contact. GOR averaged 2,500 scf/STB.
At 10 months	Optimizing production	Shutting in HGOR wells reduced gas production from the expanding gas cap.
10–14 months	Water injection	Water injection increased the efficiency of the depletion process. The production rate increased from 500 to 900 BOPD.
14–24 months	Water breakthrough	Water production rapidly increased as a result of water cycling. Reduced GOR was a result of repressuring, and shutting in the HGOR wells increased the oil production rate.

Table 9.2—Production segments.

straight-line fit on the semilog plot, which indicated the possibility of pressure depletion. The producing section consists of two sand members of differing permeabilities; in the upper layer, $k = 83$ md, and in the bottom layer, $k = 11$ md. Analyzing Fig. 9.10, it can be observed that the history can be divided into three segments. The results of the interpretation are shown in **Table 9.3.**

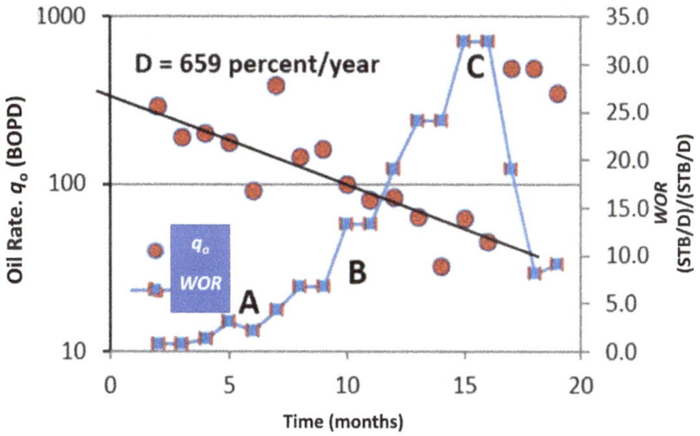

Fig. 9.10—Production history of the J12 well. Note how the well's potential could be completely misinterpreted if the operational aspects of the history are not considered and the production is divided into segments. WOR = water/oil ratio. The exponential decline (D) = 659%/year.

The initial interpretation of the oil rate shown in Fig. 9.10 indicates an apparent exponential (straight line on a semilog rate/time plot) until Month 15, when a dramatic rate increase occurred. The interpretation of performance also suggests that, in actuality, field operating conditions changed. Shutting in an adjacent water-injection well caused a complete change in the expected behavior.

Inattention to waterflood management often results in wells producing at reduced potentials because of excessive water channeling. Reducing the water-injection rate had a positive effect on the J12 well performance by significantly increasing the flow rate and reducing the water/oil ratio, as observed in Segment C in Fig. 9.10.

9.3.1.2 Data Management—Subsurface Pressures. Knowledge of the initial pressure in the reservoir is needed to determine the original hydrocarbons-in-place [original oil-in-place (OOIP) and original gas-in-place (OGIP)]. It also is used to determine if the reservoir is normally pressured, underpressured, or overpressured. A normally pressured reservoir will have a pressure that is consistent with the gradient of water at that depth. The initial pressure value is also used to calculate the formation volume factor for each phase in the reservoir. The formation volume factor is used to adjust for fluid volume differences between the reservoir and the surface that result from pressure and temperature. Knowledge of the reservoir pressure gradient allows for the determination of the fluid-contact depth, and transient well test analysis characterizes the reservoir by determining the average drainage area pressure,

Segment	Action	Remarks
A	Water breakthrough	
B	Artificial lift	The water/oil ratio continued to make a rapid ascent because the water viscosity was considerably less than that of the oil. Subsequent analysis indicated that water was being cycled from an adjacent well.
C	Adjacent injector well was shut in	The oil rate increased because less water was introduced into the system and because artificial lift had been introduced.

Table 9.3—Production segments.

permeability, near-wellbore conditions of damage or stimulation (skin effect), distance to boundaries such as faults, and reservoir continuity.

The benefits of measuring static pressures include

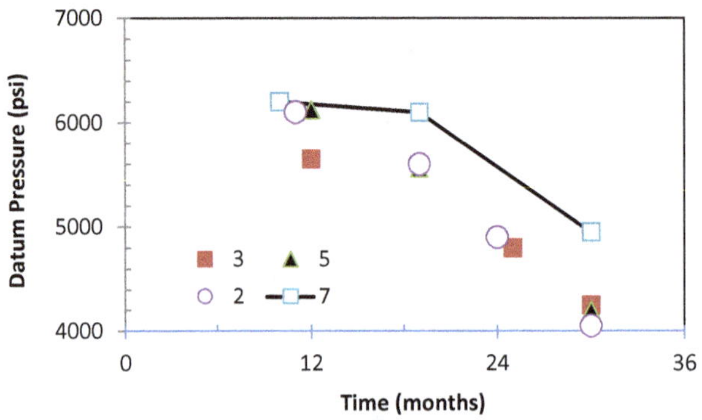

Fig. 9.11—Well connectivity for Sand "A." The No. 7 well remained at a higher reservoir pressure than its neighbors. The explanation for this behavior could be that the well was completed in a different reservoir.

- Provides an estimate of reservoir energy loss as a function of cumulative recovery
- Reflects changes in reservoir rock and fluid behavior
- Permits the use of material-balance calculations to determine OOIP and OGIP, water influx, gas-cap expansion, and compaction effects
- Allows the pressure histories of contiguous wells to be combined in a single graph **(Fig. 9.11)**, highlighting the commonality of well behavior and the extent of interwell drainage

The pressure histories of the Nos. 2, 3, and 5 wells appear quite similar, while the pressures of the No. 7 well generally remained at a higher level. Therefore, the assumption is that the No. 7 well is located outside of the "A" Reservoir and probably produces from a different sand. The historical higher static reservoir pressure could also indicate a larger drainage volume. Additional geological and engineering interpretation is required before the offsetting of the No. 7 well could be contemplated.

9.4 Reservoir Description

9.4.1 Maps and Cross Sections. Structure maps, isopach maps, logs, and producing and pressure histories should be included in the reservoir characterization to develop a complete picture of the reservoir rock architecture. A structure map relating the reservoir outline and structural relations to well completions is by far the most-important reference tool. For instance, one should ask

- Are wells located in a common reservoir? If not, is separation caused by faults or changes in lithology?
- Is the producing interval homogeneous, layered, or discontinuous? Can you visualize sand continuity with cross sections and fence diagrams?
- What is the dip of the beds? Gravity separation, although difficult to approximate, can profoundly influence displacement and production efficiency.
- Does reservoir character change areally or vertically, or both? Selective well completion intervals can influence well productivity.

Reservoirs are often divided vertically by erratically developed productive and nonproductive layers. Geologists use geological maps and cross sections interpreted from well logs and geophysics to map the continuity of the oil accumulation. Reservoir engineers commonly use average rock properties such as an average value for permeability to determine flow rates and perform reservoir production forecasting to estimate ultimate recoveries.

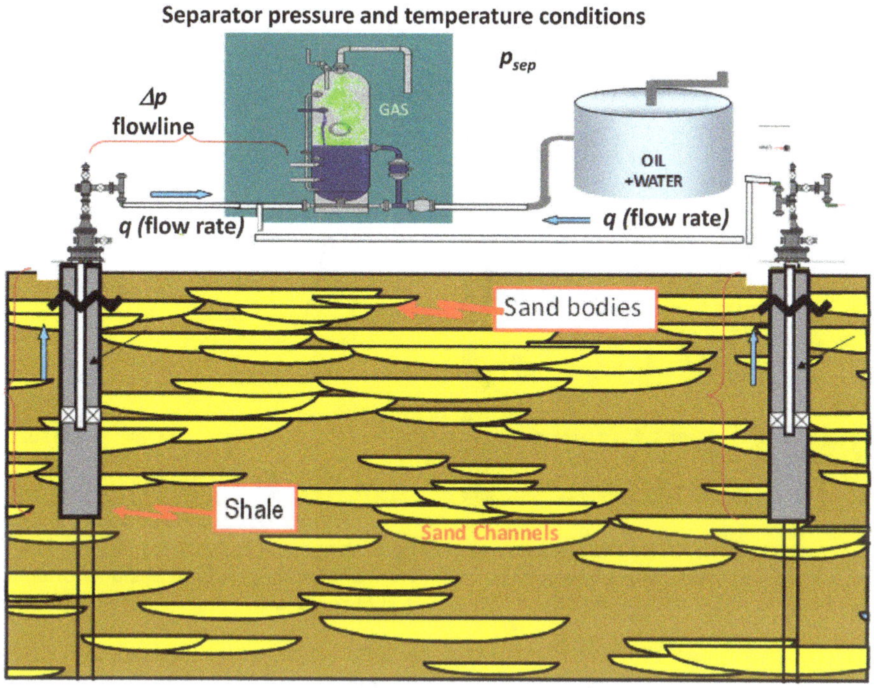

Fig. 9.12—Sand continuity within an oil field (courtesy of Schlumberger NExT). Characterization of sand distribution in wellbores is based on log interpretation and core description. The interpretation that these are channel sands is derived from a data set. Can you select areas of highest expected recovery if a waterflood were installed? Infill drilling to recover "dead-end" reserves could be useful but would depend on the associated drilling and operating costs, capital expenditures, estimated additional oil recovery, and good-quality geological and reservoir analysis.

Fig. 9.12 shows a cross section of a very heterogeneous oil field. The pay zone comprises erratically developed interbedded sand/silt/shale producing intervals. Identifying interwell flow channels by means of correlating pay zones between wells is usually very difficult for this type of field.

9.4.2 OOIP and Reserves. Original and remaining oil- and gas-in-place volumes must be identified. Remaining estimates are further divided into primary and secondary reserves. Secondary reserves are expected to be recovered through the installation of a secondary recovery project, with additional capital expenditure required.

A reserves study could comprise volumetric estimates, American Petroleum Institute (API) correlations, decline curve analysis, and material-balance calculations. If similar values are determined by means of different methods, there is an increased credibility of the estimates performed. Income from these reserves must exceed the capital and maintenance costs to justify the expense of conducting the project. One cannot spend development money without anticipating a return on investment. The results of these calculations must be plotted on maps to easily visualize these volumes at current well locations.

Characteristics to help forecast future performance can be inferred from past well histories. Even though a well might currently be producing at a low rate, perhaps at initial conditions the well showed an ability to produce at a significantly greater production rate. Both past performance histories and inflow tests provide a basis for predicting future performance. The last step of any evaluation requires that a production forecast be translated into expected cash flow estimates.

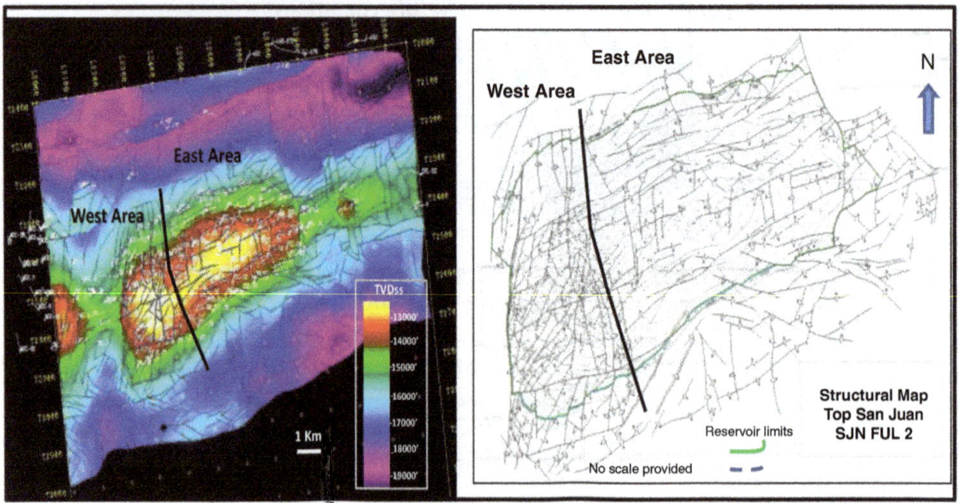

Fig. 9.13—Structure map of SJN FUL 2 Reservoir (after Pérez Amaro et al. 2020). The asymmetric anticline is 14 km long and 6 km wide with an elongated ramp. Normal and reverse fault systems oriented N70°E and normal faults oriented N40°W are characteristics of El Furrial Field.

The volume of secondary reserves that is expected to be recovered by installing a waterflood can increase incentives to pursue the project further. Comparable volumetric and material-balance study results provide a reasonable estimate of OOIP.

For forecasting production, the fraction of oil expected to be recovered should be divided into primary (pressure depletion) and secondary (waterflood) reserves. Drainage efficiency is a function of reservoir architecture and of producing-well locations, which are not always optimal. However, well and field decline curve performance studies should provide a reasonable estimate of the remaining life and ultimate recovery at an economic limit for a primary reserves estimate. Simulator studies should really help in this case, but one seldom has this luxury.

Attributing secondary reserves to a project is typically much more difficult. However, a significant decline in the oil production rate and a low recovery factor are typically hallmarks of pressure-depletion conditions. The API equation for sandstone reservoirs (Eq. 5.3 in Chapter 5, Oil Reservoirs) also provides a reliable basis for estimating secondary reserves.

Fig. 9.13 presents a structure map of El Furrial Field in Venezuela. Structure maps locate well locations in relation to their structural positions. In Fig. 9.13, the structure appears extremely faulty and compartmentalized. Note the locations of the injection wells along the periphery of the reservoir in **Fig. 9.14.** The injection wells were placed in those locations to maintain reservoir pressure and develop a more efficient displacement.

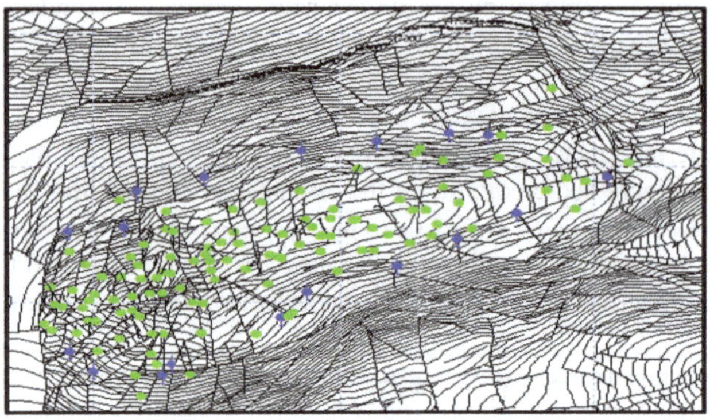

Fig. 9.14—Structure map of SJN FUL 2 Reservoir, El Furrial Field, showing the locations of injection wells. The injection wells were placed around the perimeter of the reservoir (indicated by the blue dots), creating a "peripheral" waterflood.

Injecting water into the best-developed portions of the reservoir reduces the number of injection wells required to effectively deplete the reservoir with the least amount of expense.

Constructing cross sections and fence diagrams provides a visualization of the potential flow paths. Sand continuity maps are useful when studying interwell connectivity, which is the foundation of all successful waterfloods. The degree of layering and sand discontinuity across the reservoir is the primary consideration when selecting well locations and developing a waterflood plan. A geological model must be incorporated into the engineering model. Ignoring the geological model usually results in a poor waterflood.

9.5 Displacement Efficiency and Waterflood Patterns

Geological discontinuities affect not only the efficiency of a displacement process but also the areal and vertical sweep efficiency. Therefore, reservoir displacement efficiency is a function of the vertical and areal sweep efficiency, which in turn is affected by well placement and the completion strategy in the reservoir.

Areal sweep efficiency identifies the fractionally swept and unswept areas as functions of current and expected future well locations. Vertical sweep efficiency is a function of interwell communication and dead-end productive zones, as well as permeability profiles.

9.5.1 Pattern Water Injection. In this type of injection scheme, the injection- and producing-well arrangement is repeated in a geometric manner. A pattern is particularly useful in large, flat-lying carbonates where interwell zonal communication is difficult to identify.

In a five-spot pattern, the pressure gradient between the injection and production wells is maximized to ensure the greatest-possible sweep efficiency and a rapid flood-out. However, there are a number of different injection-/producing-well patterns, which account for specific geological conditions.

A comparison of the economics of possible flooding schemes is used to determine the final selection of spacing, pattern type, and pattern orientation.

Pattern selection is often determined by well locations established during the primary field development. The cost of drilling new wells frequently dictates that existing wells be used and that few, if any, additional wells be drilled. To prevent early breakthrough caused by water channeling from injection to production wells, the line connecting adjacent injectors should be constructed parallel to the direction of maximum permeability or fracture trend, and the drilling of infill injection wells should also be considered.

Fig. 9.15 compares the traditional geometric and peripheral injection/production patterns commonly used in a waterflood.

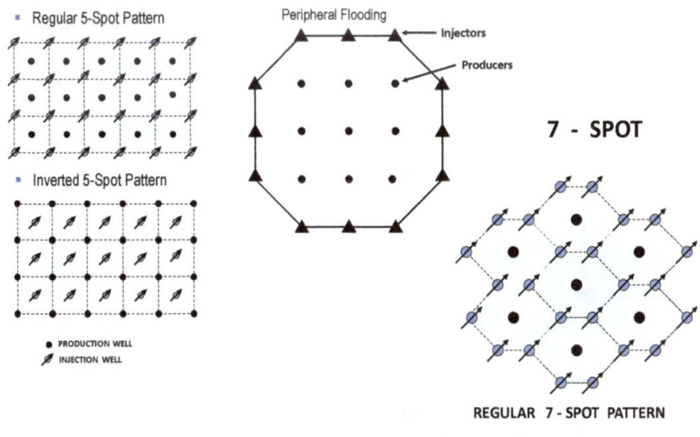

Fig. 9.15—Pattern and peripheral injection- and production-well arrangements.

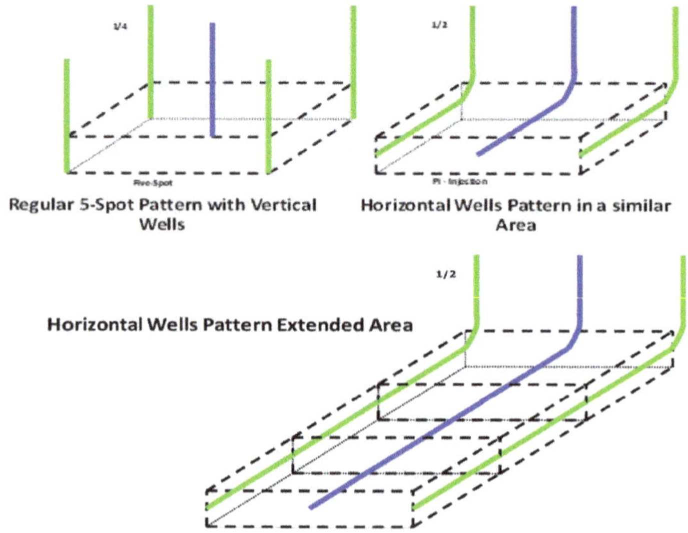

Fig. 9.16—Five-spot pattern and injection/production arrangements for horizontal wells.

Fig. 9.16 shows the geometric arrangement of injection/production patterns when drilling horizontal wells. The selection of a waterflood project development strategy will be based on economics.

9.5.2 Peripheral Water Injection. Water-injection wells are often located in the downdip periphery of moderate- to high-permeability oil reservoirs producing from correlative zones. Peripheral injection often allows the conversion of temporarily abandoned wells to water injectors and promotes the gravity segregation process.

The effect of geological discontinuities on the efficiency of a displacement process is not only defined by the remaining oil saturation but also by the areal and vertical sweep efficiency. Reservoir displacement efficiency results from the vertical and areal sweep efficiency, which is a function of well placement and the geology of the reservoir.

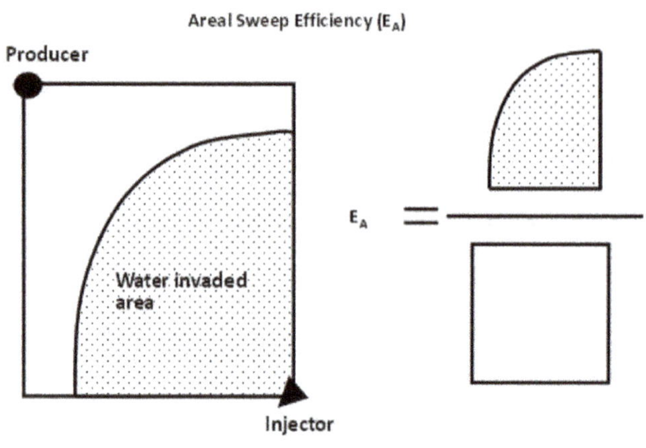

Fig. 9.17—Areal sweep efficiency E_A. It represents the fraction of the horizontal plane of the reservoir that is behind the flood front at a point in time. The mobility ratio, well spacing, and pattern geometry are important factors affecting E_A.

The areal sweep efficiency (E_A) allows a comparison of the drainage pattern swept area to the reservoir area. It is defined as the fraction of the oil recovered areally by a waterdrive and depends on the heterogeneity of the reservoir properties in the areal direction. **Fig. 9.17** graphically shows the definition of E_A.

Pattern geometry influences areal sweep efficiency. Correlations exist for determining common pattern geometries as a function of the mobility ratio, defined by Eq. 9.1. Laboratory experiments have been conducted to determine the areal sweep efficiency of various patterns for different mobility ratios. These results and correlations are discussed at length in Craig (1971). In addition, streamtube and streamline models or a 2D (*x*/*y*) fine reservoir simulation model incorporating laboratory relative permeability data could be used.

$$M = \frac{\text{mobility of water}}{\text{mobility of oil}} = \frac{\dfrac{kk_{rw}}{\mu_w}}{\dfrac{kk_{ro}}{\mu_o}} = \frac{k_{rw} \times \mu_o}{k_{ro} \times \mu_w}. \qquad (9.1)$$

In general, areal sweep efficiency and oil recovery decrease as the mobility ratio increases:

- $M = 1$: When the mobility ratio is equal to unity, the mobilities of oil and water are the same. The resistance to flow in the reservoir is identical for both phases.
- $M < 1$: A mobility ratio that is less than unity is called a favorable mobility ratio. In this scenario, oil flows more easily through the formation than water. The water moves slower than the oil, leading to higher water saturations behind the front. As a result, the water sweeps the oil toward the producer more efficiently, resulting in an improved oil recovery.
- $M > 1$: A mobility ratio that is greater than unity is called an unfavorable mobility ratio. In this scenario, water flows through the rock better than oil does. The water behind the front moves faster than the oil ahead of the front. As a result, the water does not displace the oil as efficiently as it advances toward the production well.

The vertical sweep efficiency (E_V) compares fluids displaced in the vertical direction to the net thickness of the flooded interval. It is a function of the permeability distribution and vertical discontinuities. It represents the fraction of oil recovered in the reservoir cross section by a waterdrive. **Fig. 9.18** graphically shows the definition of E_V.

Vertical (invasion) sweep efficiency is defined as the cross-sectional area contacted by injection water divided by the cross-sectional area enclosed in all layers behind the farthest waterflood front. Vertical sweep efficiency is influenced most significantly by the mobility ratio and the vertical degree of variation of permeabilities (k_v) across the reservoir section.

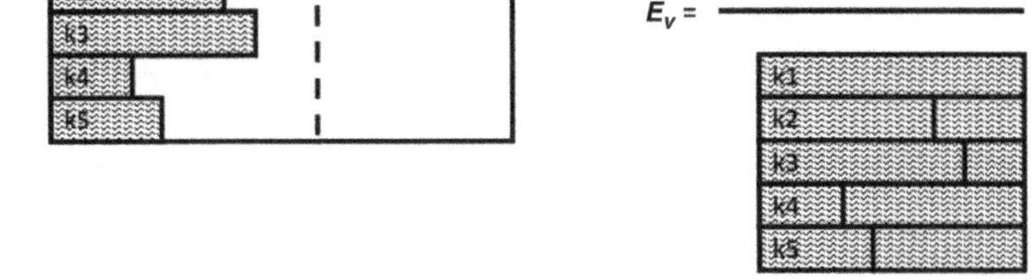

Fig. 9.18—Vertical sweep efficiency E_v, which is the fraction of the cross-sectional area contacted by injection water divided by the cross-sectional area enclosed in all layers behind the farthest waterflood front. The mobility ratio and the vertical degree of variation of permeabilities (k_v) are important factors affecting E_v.

Calculation techniques that can be used are analytical methods such as the Stiles, Craig-Geffen-Morse, and Dykstra-Parsons methods and numerical model solutions, and 2D (*y*/*z* or *x*/*z*) fine cross sections incorporating laboratory relative permeability data.

Volumetric sweep efficiency (E_D) is the product of the areal and vertical sweep efficiencies:

$$E_D = E_A \times E_V. \quad (9.2)$$

Example 9.1—Calculating Volumetric Sweep Efficiency. The areal displacement for a proposed waterflood of a sandstone reservoir is expected to be 80% as a result of the homogeneous nature of the sand body.

The reservoir is a steeply dipping fault block of the type commonly encountered around salt domes, and well placement is optimal. Optimal well locations and gravity segregation effects should result in an 80% vertical sweep efficiency. What is the overall sweep efficiency?

Solution. Use Eq. 9.2:

$$E_D = E_A \times E_V = 0.8 \times 0.8 = 64\%.$$

The waterflood is expected to displace oil from only 64% of the reservoir even though both vertical and areal sweep efficiencies appear to be at nearly optimal values. The remaining oil saturation will remain undisturbed in 36% of the reservoir.

Prediction methods share common assumptions that account for the reservoir vertical heterogeneity when forecasting reservoir performance: The reservoir is represented by a series of layers with no vertical communication (no crossflow between layers), and each layer is characterized by a thickness h, permeability k, and porosity ϕ.

Other assumptions are

- Immiscible displacement
- Linear flow
- That the distance water has traveled through each layer is proportional to the permeability of the layer
- Piston-like displacement, incompressible fluids

9.6 Summary

Oil, gas, and water production induces a pressure drop between the higher-pressured reservoir (p_{res}) and the surface. Oil production rates are a direct function of the flowing tubinghead pressure (p_t) at the surface. A loss of reservoir pressure resulting from production or increasing the hydrostatic head reduces the flow rate. Water or gas injection is often used to artificially augment reservoir pressure when the natural expansive processes are insufficient to maintain reservoir pressure.

The water-injection process is more commonly used than gas injection because of injection-fluid availability and moderate cost. Water injection can be classified as pressure maintenance or as a liquid displacement project. Pressure maintenance involves the injection of a sufficient quantity of water to maintain reservoir pressure so that the reservoir off-take rate remains stable. Typically, water injection attempts to push otherwise immovable oil toward a producing well.

Waterflooding represents a dramatically different displacement process than that observed during primary depletion. Primary depletion induces fluid flow through interconnected pores to the producing well. All flow channels connected to the completion zone contribute to the depletion process. With water injection, an energy source is used to displace injected water and hydrocarbons toward a producing well. Established flow channels must be connected between the input source and the producer. The displacement process for oil located in discontinuous sands does not occur. Knowledge of reservoir architecture and the producing character becomes of paramount importance when contemplating a waterflood.

Coupling applications of fundamental reservoir analysis concepts with commonly available geological and production information have been discussed. Field examples have been included to illustrate the use of these basic concepts.

Reservoir simulators are designed to study multiwell situations and require additional technical expertise. However, these preliminary analytical studies should serve as a first phase and should always be conducted before attempting to simulate a reservoir.

9.7 Questions

- What would be the primary concern when contemplating waterflooding a 50-ft, rather-flat-lying homogeneous 32° oil reservoir that is underlain by water? (Answer: The primary concern is the effect of gravity segregation because of underrunning of the oil leg by encroaching water. This estimation would involve knowledge of the vertical permeability and the water and oil viscosity.)
- What are the criteria for designing water-injection wells? (Answer: Injection wells should be located in areas of maximum injectivity to ensure that sufficient quantities of water get into the ground. In pressure-maintenance projects, this is likely not as important because the injection rate is a function of the reservoir off-take rate.)
- Why is it important to know the mineralogy of clay particles present in an oil reservoir? Is this knowledge only pertinent when contemplating a waterflood? (Answer: Some clays react to the introduction of fresh water. This knowledge is not just pertinent when contemplating a waterflood; it is very important to know the minerology when selecting the type of completion fluid.)
- Name the clay type that is the least and the most susceptible to the introduction of fresh water. (Answer: Kaolinite and montmorillonite, respectively.)
- What primary factors ensure a successful outcome of a water-injection project? (Answer: Integrated reservoir studies and ongoing monitoring and surveillance to react to unexpected events.)
- What are the best methods for monitoring performance? (Answer: Study pressure and production histories. Well production records tell the true story of performance.)
- Why are geologic and production maps useful tracking tools? (Answer: Computer-generated maps allow one to visualize the 3D character of a field, while comparing recent maps with historical maps allows one to note any changes in performance.)
- What are the primary factors affecting areal and vertical sweep efficiency? (Answer: For areal sweep efficiency, early well placement is typically predicated on an uncertain knowledge of the geology, and the result is uneven well placement after the completion of the primary development. Differences in sand quality that are found after additional study may show areas for additional development. For vertical sweep efficiency, section homogeneity should be studied, and the proper perforations should be selected to ensure efficient depletion.)
- Pattern injector-/producer-well locations are often proposed for developing large fields. Research has previously been conducted to estimate recovery for different well patterns. What is the main problem when applying a pattern well arrangement? (Answer: Uncertain vertical and areal reservoir heterogeneities may alter the idealized flow systems.)

9.8 Nomenclature

E_A = areal sweep efficiency, fraction or percent
E_D = volumetric sweep efficiency, fraction or percent
E_V = vertical sweep efficiency, fraction or percent
h = thickness, ft
k = permeability, md
k_{ro} = relative permeability to oil, md
k_{rw} = relative permeability to water, md
k_v = variation of permeabilities, md
M = mobility ratio
p = pressure, psia
p_{Aq} = pressure at aquifer, psia
p_e = pressure at outer drainage limit, psia
p_{res} = reservoir pressure, psia
p_{sep} = pressure at separator, psia
p_t = pressure at tubing head, psia
p_{wf} = bottomhole flowing pressure, psia
q = flow rate, STB/D
q_o = oil flow rate, STB/D or BOPD
μ_o = oil viscosity, cp or Pa·s
μ_w = water viscosity, cp or Pa·s
ϕ = porosity, percent

9.9 References

- Berg, R. R. 1986. *Reservoir Sandstones*, 479. Englewood Cliffs, New Jersey: Prentice-Hall, Inc.
- Bradley, H. B. ed. 1962. *Petroleum Engineering Handbook*. Richardson, Texas: Society of Petroleum Engineers.
- Craig, F. F., Jr. 1971. *The Reservoir Engineering Aspects of Waterflooding*, Vol. 3, Chap. 5. Richardson, Texas: Monograph Series, Society of Petroleum Engineers.
- Pérez Amaro, I., Chelhod, N. C.-Y., and Laprea-Bigott, M. 2020. Workflow for Optimization of Waterflood Performance: Field Example from El Furrial Field – East Venezuela Basin. Paper presented at the SPE Latin American and Caribbean Petroleum Engineering Conference, Bogota, Colombia, 27–31 July. SPE-199121-MS. https://doi.org/10.2118/199121-MS.

Chapter 10

Waterflood Components

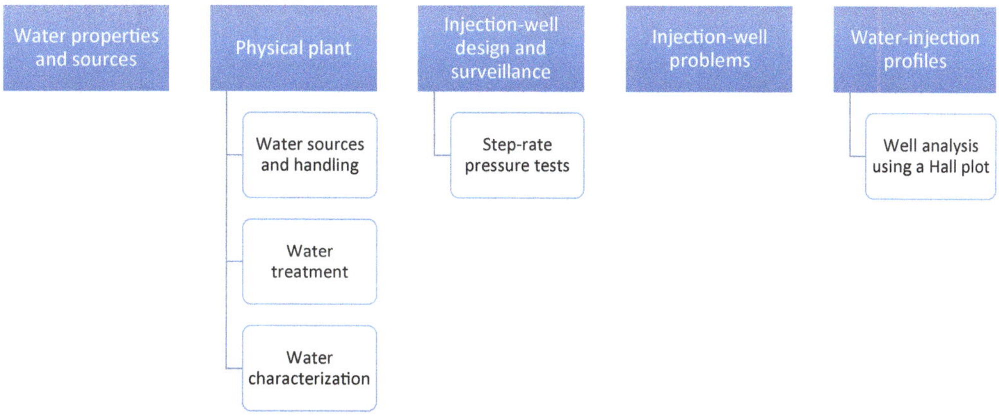

10.1 Water Properties and Sources

The successful application of a waterflood project has several components, and knowing the geochemical characteristics of the water source is essential to understanding the rock/fluid interaction phenomena. Water properties are very important components of all exploration, production, refining, and sales activities, which include assessments such as special coreflood experiments to determine fluid compatibility and formation damage, petrophysical studies, flow assurance, and analyses of scaling and corrosion problems.

Fig. 10.1 shows the water cycle process when injecting water for enhanced oil recovery. Having a comprehensive understanding of the water properties to determine the distribution of fluids and increase displacement efficiency is key to optimizing reservoir performance. Multidisciplinary studies are a fundamental part of any successful waterflood field development plan. Managing the produced and injected fluids requires the proper sampling and analysis of the chemical and physical properties of the water.

10.2 Physical Plant

The physical plant for a waterflood can be roughly divided into

- The oil, water, and gas separation system: The separator setup that is already in use is probably sufficient. However, if water cycling is expected, some expansion of separation capabilities should be contemplated. Separating oil from large water volumes could require additional tankage, particularly if the oil gravity is high.

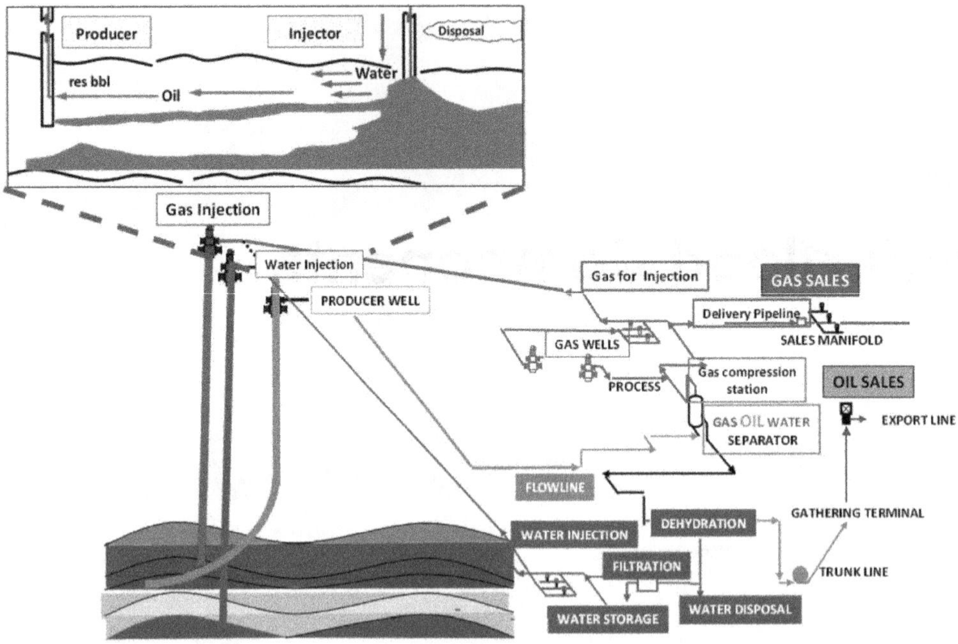

Fig. 10.1—Schematic of the water cycle process of injecting water into a reservoir to recover oil. The different components must be considered to optimize the process, the reservoir, the wells (producers and injectors), and the facilities for water separation and treatment.

- Produced- and source-water systems: These systems (surface, groundwater, saline aquifer) are commonly combined for the injection fluid. Produced water commonly contains carry-over of oil particulates, fine-grained rock particles, and organic material. An accumulation of this material close to the perforations during injection can cause plugging. It is imperative that produced water be effectively cleaned before being reinjected into the reservoir.

Source water that is obtained from rivers, lakes, or the sea must be similarly cleaned, with a particular emphasis on the injection of a biocide to inhibit growth of organic material. All water should be filtered and sanitized before injection. Chemical additives can be adjusted based on the quality of the water, such as the levels of total dissolved solids (TDS), iron, and alkalinity.

10.2.1 Water Sources and Handling. Produced water from oilfield operations is recycled for use in enhanced oil recovery projects, steam generation, agriculture, watering, and different plant processes. Often, excess water is disposed of in aquifers, the ocean, rivers, or lakes. The final use of the produced water will dictate its quality requirements.

Significant impurities found in produced water include oil, dissolved gases, dissolved solids, suspended solids, and microorganisms, and the produced water will need the appropriate treatment for these impurities before it can be reinjected or disposed of.

In oilfield operations using water injection in a reservoir, suspended oil and solids concentrations must be decreased and oxygen must be removed from the system to minimize corrosion impacts and downtime.

To manage and remediate deposits and incrustations, fouling, scaling, corrosion, material fracturing and loss, and bacteria growth (aerobic/anaerobic)—all of which occur often in surface water-processing equipment—a continuous monitoring of water parameters is required. These parameters include alkalinity, hardness, pH (a temperature-dependent measure at standard temperature and pressure), acidity, softening, TDS, turbidity, stoichiometry, and mass balance.

Waste water can include water produced with crude oil, rainwater, and washdown water. The primary constituents of waste water are calcium, magnesium, sodium bicarbonate, chloride, and sulfate, and it can also contain complex mixtures of organic chemicals such as fatty acids, aromatic hydrocarbons, and naphthenates. In addition, it can be contaminated with sulfate-reducing bacteria, which can lead to reservoir souring.

Produced water can be hazardous, and it can contain heavy metals. Radioactive added chemicals (inhibitors) can be hot and corrosive when they reach the surface, and some are complex, unstable (scaling) brines, which can have detrimental environmental effects.

10.2.2 Water Treatment. Water must be separated from the crude oil and disposed of in a manner that does not violate established environmental regulations. **Fig. 10.2** shows the general layout of a water cycling and injection system where gas/oil/water separators are the first step in surface handling.

Physical-separation methods such as settling, coalescing, flotation, and filtration are used in the primary treatments to separate suspended oil and solids. A biological treatment is applied when specifications require it for certain uses.

Produced-water treating equipment incorporates a variety of solutions and gravity separation techniques; skimmer tanks or American Petroleum Institute (API) separators are options available. The simplest form of primary treating equipment is a settling (skim) tank or vessel. This type of equipment is typically designed to provide long residence times during which coalescence and gravity separation can occur. If the outlet oil concentration is known, the dimensions of the vessel can be determined (API 1990).

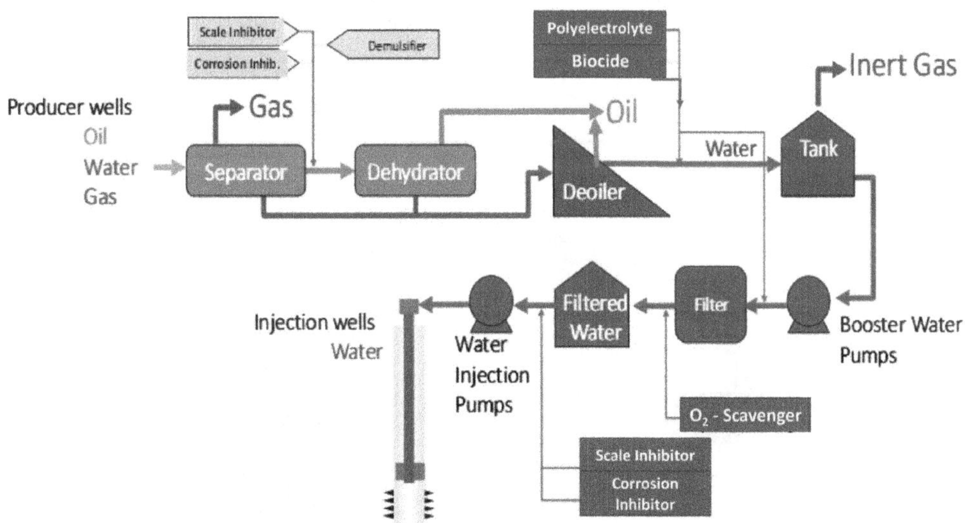

Fig. 10.2—Cleaning of seawater for injection. Note how water is cleaned and treated before injection. A breakdown in any individual step could ultimately cause injection-well plugging.

Skimmers can be either vertical or horizontal in configuration. In vertical skimmers, the oil droplets must rise upward countercurrent to the downward flow of the water. Some vertical skimmers have inlet spreaders and outlet collectors to help even the distribution of the flow. In horizontal skimmers, the oil droplets rise perpendicular to the flow of the water.

Horizontal skimmers are more efficient for water treatment because the oil droplets do not have to flow countercurrent to the water flow. However, vertical skimmers with a water outlet or a sand drain off the bottom handle sand and other solid particles better.

Fig. 10.3 shows the separation process. The wellhead connects completions to the flowline gathering. The flowline network and process facilities transport and control the flow of oil or gas from the wells to a main storage facility, processing plant, or shipping point. The phase separation is an operation in which the wellstream is passed through a vessel to separate the different phases (gas, oil, water, and sand).

Several equipment solutions have been developed by the industry for produced-water treatment, which will help reduce pollution and plugging by means of water discharge according to the required quality specification and oil-droplet size.

Separation equipment includes several options, such as plate coalescence (parallel plate interceptors or corrugated plate interceptors), enhanced coalescence, gas flotation, filtration, and hydrocyclones. In hydrocyclones (also referred to as enhanced gravity separators), centrifugal force is used to remove oil from water. "Centripetal forces cause the lighter-density [oil] droplets to move toward the low-pressure central core, where axial reverse flow occurs. The oil is removed through a small-diameter reject port located in the head of the hydrocyclone. Clean water is removed through the downstream outlet" (Stewart and Arnold 2009).

One of the most-challenging aspects of water treatment is the extraction of oil from the emulsion stage. Equipment efficiency based on the range of droplet sizes removed by deoiling is shown in **Fig. 10.4.**

In offshore areas where discharge to the sea is allowed, the governing regulatory body specifies the maximum-allowable hydrocarbon content of the water that will be discharged overboard. The range is currently 0 mg/L [1 part per million (ppm) = 1 mg/L] to 50 mg/L, depending on the specific location. The permitted associated free-oil content can vary from < 100 ppm to several thousand ppm depending on API guidelines and the efficiency of production separators.

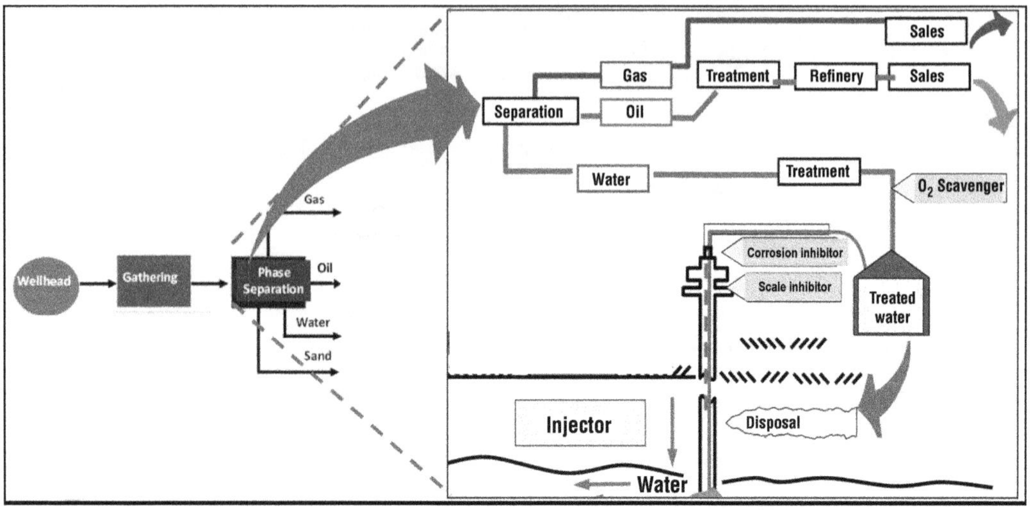

Fig. 10.3—Flow chart summarizing the steps involved in the production and separation process.

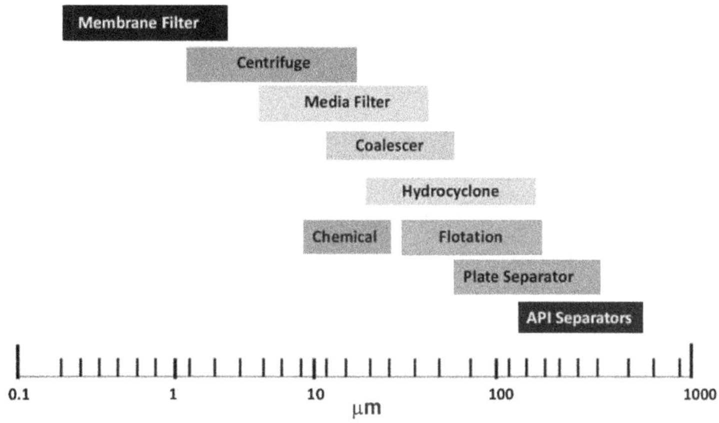

Fig. 10.4—Different types of equipment used to clean produced water by means of droplet removing (after Schlumberger NExT 2004). The horizontal axis shows the droplet sizes removed by the deoiling equipment in micrometers (also known as microns).

Table 10.1 shows the discharge regulation values for dispersed oil-in-water used by regulating agencies in different regions of the world.

Table 10.2 shows the water quality standards applied by international oil and gas companies in waterflooding projects in the North Sea.

Onshore locations also require proper water treatment because of possible salt contamination. Salinity can vary from potable (< 100 mg/L TDS) to supersaturated (> 350 000 mg/L TDS). The size of solids can range from submicrons to millimeters and should be controlled before injecting, to avoid wellbore damage resulting from plugging.

10.2.3 Water Characterization. Monitoring the physicochemical properties of production and injection water allows the identification of water-front preferential paths. The application of chemical treatment processes, on the other hand, allows planning for preventing possible scaling problems in surface facilities. Salinity and water quality values are used for the diagnosis of production problems.

Injected water mixes with connate water, and geochemical reactions can occur, changing the injected brine composition. Available physicochemical characterizations are used to build Stiff and

Region	Dispersed Oil-in-Water Discharge Regulations (mg/L)
Northeast Atlantic and Arctic oceans	< 40
North Sea	< 40
US far shore	28 mean, 42 maximum
Caspian Sea	< 20
Mediterranean Sea	10–15
Red Sea	< 15
Niger Delta	< 10
Germany offshore	0
US near shore	0

Table 10.1—Cleaning of produced water—discharge regulations for dispersed oil-in-water.

Typical Water Quality Parameters for Seawater Injection (Deepwater Suction), North Sea

Main separator => hydrocyclones (deoiling/desanding) => degassers => injection

- Suction solids < 1 mg/L
- Filtration to protect injectivity 40–80 μm
- Filtration to protect equipment < 1 mg/L

Typical Water Quality Parameters for Production Water Used for Injection

- Solids in the injection manifold < 20 mg/L
- Particle size < 4 μm
- Crude content < 200 ppm

Table 10.2—Water quality parameters of injection water, injection regulations for the North Sea.

Piper diagrams, pie charts, and Collins bar diagrams, which allow one to classify and determine the water types present in a field. The diagrams and charts are graphical representations of chemical analyses (Stiff 1951) and are used to show how the ionic composition of injected water changes over space and/or time.

Selected samples should be distributed in the area from different producing wells, and sampling from the water-injection wells in the field should also be performed. The selection should be based on those wells presenting a complete physicochemical characterization so that an ionic balance necessary to identify their representativeness of the water data can be performed. This allows one to identify when injection-water breakthrough occurs and from which formation or formations the well is producing. **Fig. 10.5** shows the different graphical options that are commonly used for representations.

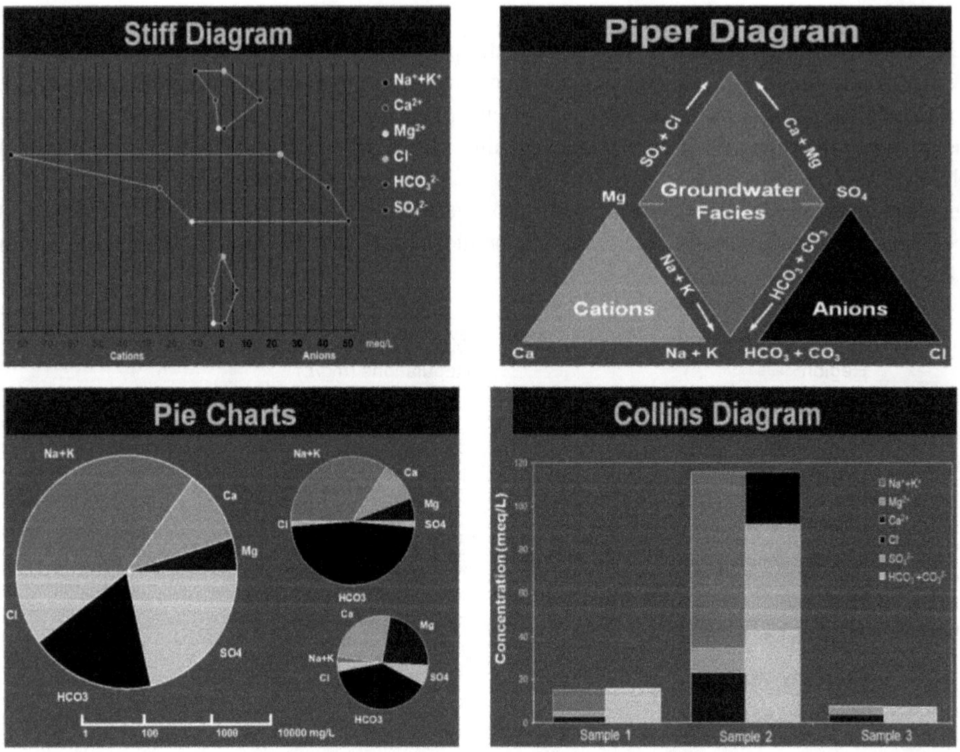

Fig. 10.5—Produced-water composition diagrams.

The chemical composition of formation water is determined by the ions present, and the differences between the concentration of each ion are what establishes that the water in the different deposits has variation in its properties. The main formation-water ion constituents that the API recommends for analysis are

- Sodium (Na^+)
- Magnesium (Mg^{++})
- Calcium (Ca^{++})
- Sulfates (SO_4=)
- Carbonates (CO_3=)
- Bicarbonates (HCO_3=)
- Chlorides (Cl^-)

Table 10.3 shows the ion concentrations calculated from an ionic balance for each one of the physicochemical characterizations found in water samples taken from wells in a waterflooded reservoir. These values are plotted in **Fig. 10.6.**

The Stiff diagram (Fig. 10.6) shows a greater concentration variability of chlorides and sodium in Well B, followed by Well A, Well D, Well C, and Well E. It can be concluded that connate water has the same geological origin, and the variability in the concentrations of anions and cations results from the existence of different minerals in the rock matrix instead of from the interaction of the rock/connate water with injected water.

Fig. 10.7 shows the Stiff diagrams for the different waters analyzed. The connate-water samples correspond to Wells A and B. It can be seen from the behavior of most of the ions that they correspond to the same type of water, where for the cations $Na^+ > Ca^{++} > Mg^{++} > Fe^{++}$ and for the anions $Cl^- > HCO_3^- > SO_4 > CO_3$. The waters are of a mixed chlorinated-sodium type and are the product of the mixing of connate and meteoric waters during the geological record.

10.3 Injection-Well Design and Surveillance

Injection tests have several objectives, which include measuring reservoir pressure, the injectivity index into the matrix, the injectivity above the fracture pressure, horizontal stress, rock strength, and fracture pressure (initiation, reopening, and propagation). In waterflood projects, it is particularly important to understand how fracturing pressure varies with time given the fact that historically, injection rates have declined and injection pressures have increased over time. Active monitoring of

Well Name	Nacl	HCO_3	Ca	CO_3	Cl	Na	Fe	Mg	SO_4	
	ppm	meq/L	meq/L	meq/L	meq/L	meq/L	meq/L	meq/L	meq/L	Na/Cl
Well A	14,371.00	38.43	4.99	0.00	232.59	266.29	0.00	0.99	6.74	1.14
Well B	34,748.63	0.00	71.86	16.78	571.73	526.84	0.01	1.97	12.88	0.92
Well C	1,642.00	20.23	1.01	1.17	28.12	48.80	0.02	0.06	1.62	1.74
Well D	1,373.67	18.68	0.24	1.10	23.49	42.36	0.02	0.07	1.31	1.80
Well E		17.31	0.60	0.00	6.20	22.61	0.01	0.00	8.10	3.65

Table 10.3—Concentrations of physically and chemically characterized ions in produced water (after Chek-yan and Pérez Amaro 2017).

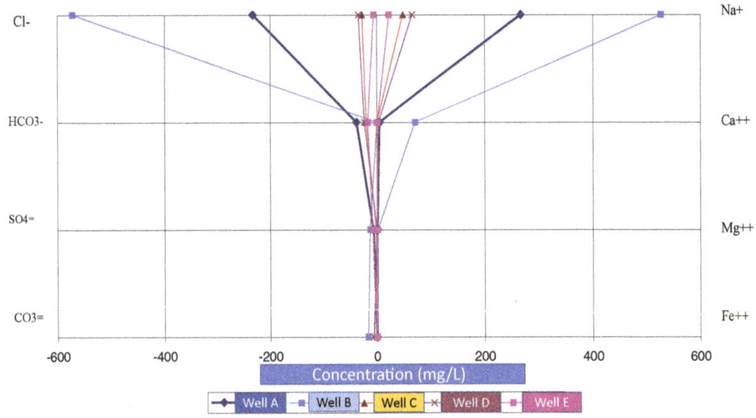

Fig. 10.6—Stiff diagram showing produced-water compositions in a sample reservoir (after Chek-yan and Pérez Amaro 2017).

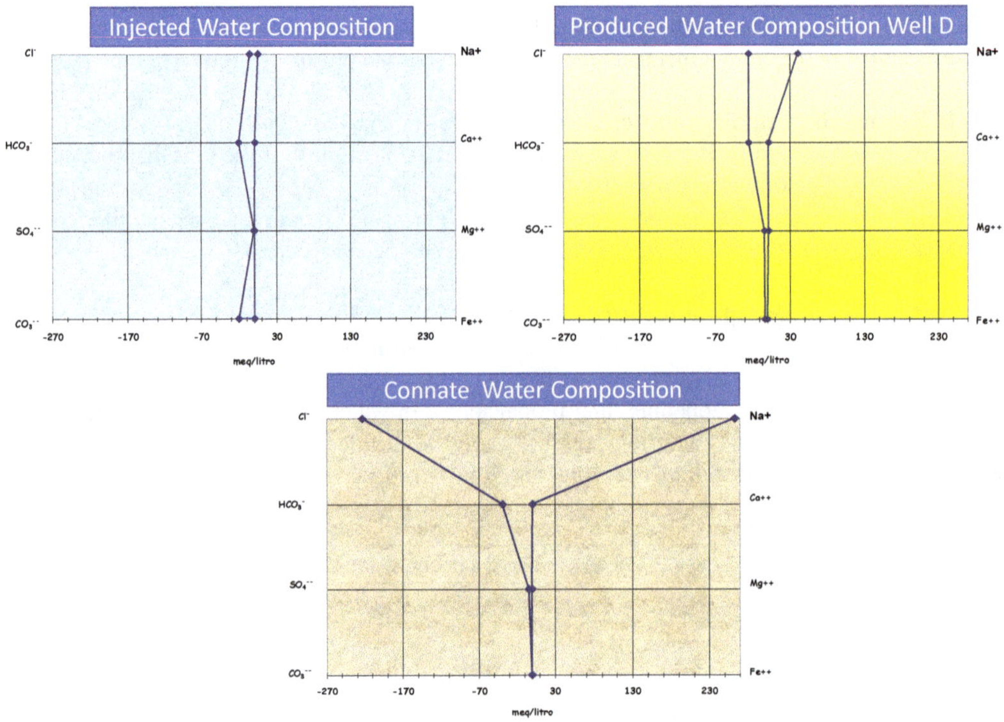

Fig. 10.7—Stiff diagrams of the injection, production, and formation waters (after Chek-yan and Pérez Amaro 2017).

injection-well capacities is a prerequisite for a successful waterflood. The sections that follow present the simplified well and reservoir testing techniques for injection wells.

10.3.1 Step-Rate Pressure Tests. For most cases, the bottomhole pressure in an injection well should remain below the fracture pressure to ensure injected fluids are properly placed into the zone or zones of interest. Fracture pressure for a water-injection well can be determined with a step-rate test. Bottomhole or surface pressures are measured as a function of increasing water-injection rates. Fracture initiation pressure occurs at the inflection point on a pressure vs. injection rate plot. **Fig. 10.8** illustrates the step-rate test for a Louisiana Gulf Coast well. Injection rate steps are designed such that all have the same duration, typically 15–20 minutes. The pressure is noted at the end of each injection rate step. Analyzing the figure, it can be seen that the extrapolated straight line at lower injection rates should pass through the origin at a zero rate.

Fig. 10.9 shows a step-rate well test conducted to determine fracture pressure at the perforations at a Timbalier Bay well. A series of injection rate vs. surface injection pressure values was recorded and plotted in Cartesian coordinates (*x*/*y*). It can be observed that two distinct straight lines can be drawn. The intersection of the two straight lines indicates a 570-psia fracture pressure measured at the surface for the reservoir.

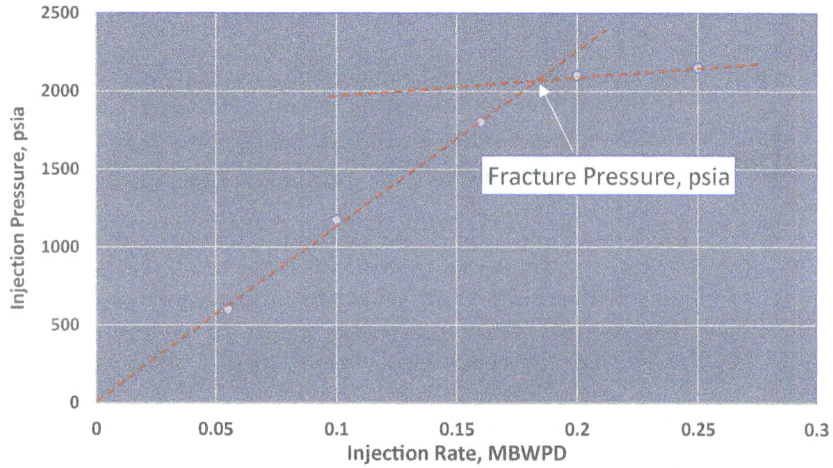

Fig. 10.8—Step-rate test example using a Louisiana Gulf Coast well.

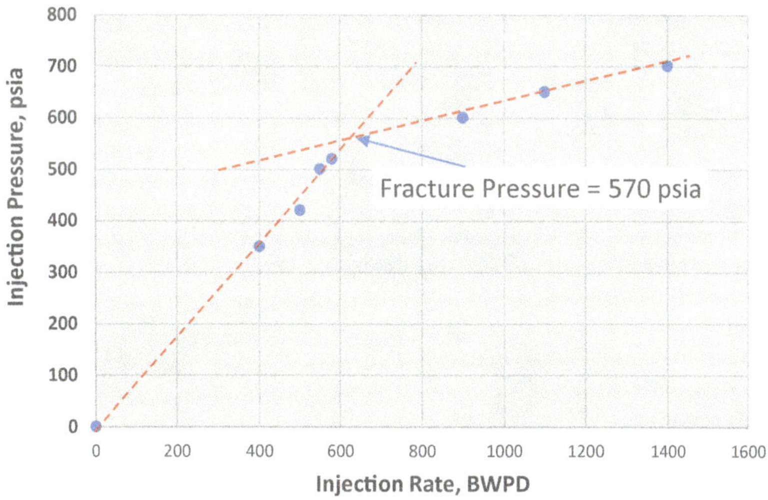

Fig. 10.9—Results of a Timbalier Bay well test.

The actual bottomhole fracture initiation pressure (BHFIP) is calculated using Eq. 10.1:

$$BHFIP = \text{test-measured surface pressure} + (\text{fluid gradient} \times \text{depth}). \quad (10.1)$$

In this case, $BHFIP$ = 570 psia + (0.465 psia/ft)(9,130 ft) = 4,816 psia.

10.4 Injection-Well Problems

The operational aspects of waterflooding are often neglected after project installation and initiation. Poor field supervision is one of the primary causes of project failure as a result of injection-well problems. Many times, the project is developed by the engineering staff but then becomes the responsibility of the field operating personnel who will need the proper training to develop the technical competencies.

Water-injection-well capabilities are expected to decline during the fill-up period as the water bank moves out into the reservoir. Additional pressure drops can be anticipated because of interference between water banks. The primary problems and concerns are a loss of injectivity resulting from plugging and a poor injection profile.

Poor water quality can originate plugging, and an injection profile can be very poor if water is being injected into only one section of the well. A poor injection profile can also result from fracturing; injecting above the fracture pressure causes low vertical sweep efficiency in vertical wells and poor areal sweep efficiency in horizontal injector wells.

Other problems that can occur include

- The water-injection well was completed in poor-quality reservoir rock.
- There is a lack of reservoir continuity. Water is injected into discontinuous sand members.
- Injection into a layered system is distributed on the basis of the product of the permeability and height (kh) of each layer. Distribution should be a function of the product of porosity and height (ϕh).
- Poor injection profiles can result from formation damage that occurred during drilling and completion, poor cement bonding, or the presence of a thief zone.
- Well impairment can result from the development of a near-wellbore skin effect caused by the formation of scale, by poor water quality, or by emulsions.

Additional problems that can occur are shown in **Fig. 10.10.**

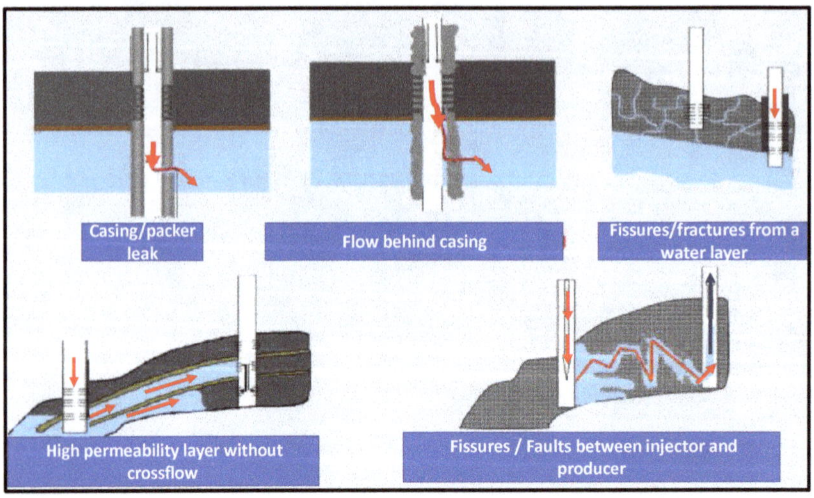

Fig. 10.10—Injector-well problems (after Arnold et al. 2004). Producer wells experience the same types of problems.

10.5 Water-Injection Profiles

Monitoring injection rates and pressures can provide information on the characteristics of the injection process if wellbore plugging is occurring.

The injection rate vs. pressure profile relationship shown in **Fig. 10.11** indicates a high degree of parallelism between two well tests, which were performed at different times during the life of the well. The sandface permeability does not appear to be affected during the water-injection process.

On the other hand, **Fig. 10.12** shows the effects of later wellbore plugging in a second well test.

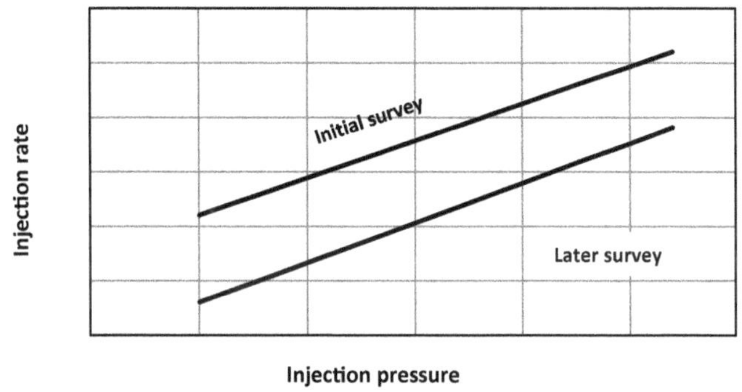

Fig. 10.11—An example showing growth of the water front but no near-wellbore plugging.

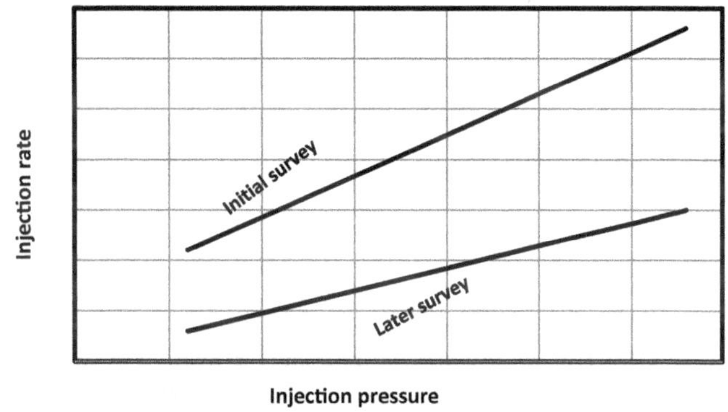

Fig. 10.12—The later well test indicates additional pressure will be required to inject water at a given rate.

10.5.1 Well Analysis Using a Hall Plot. Hall (1963) developed an analysis technique to monitor injection-well capacity in which the integral form of the Darcy equation is expressed in terms of estimated cumulative recovery. Eq. 10.2 represents cumulative water injected as a function of the total effective pressure drop measured over the same time period:

$$\left(p_e - p_{wf}\right)\times t = \frac{141.2\mu\left[\ln\left(r_e / r_w\right)+s\right]W_i}{kh}. \quad (10.2)$$

The equation of a straight line is applied to the relation, and the slope of the line represents the character of the near-wellbore transmissibility changes.

The objectives of the analysis method are to

- Evaluate the change in the injectivity of the well as a function of time
- Estimate the injection capacity (kh) and effective permeability to water
- Obtain an indication of the type of damage the injection well has suffered

The Hall plot is used to detect plugging and fracturing in injection wells. The product of the wellhead pressure and time vs. cumulative production values is plotted, and the resulting changes in the slope of the line signify positive or negative changes in injectivity.

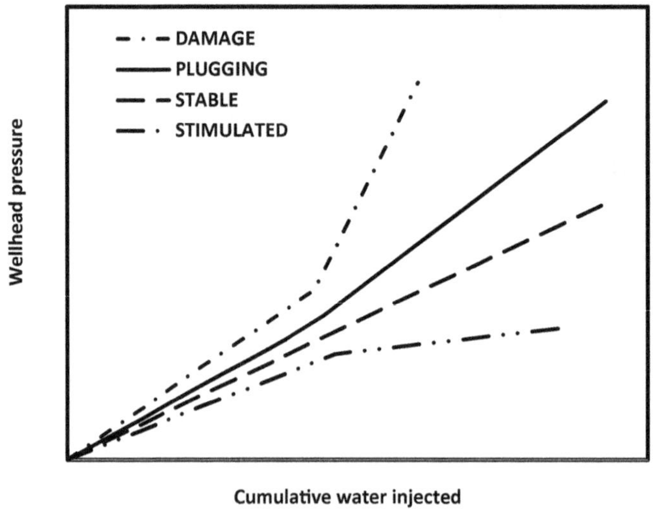

Fig. 10.13—Hall plot showing beneficial and detrimental divergences from the straight line.

Fig. 10.13 shows four different shapes representing general effects that might be observed in a waterflood.

Changes in the slope of a Hall plot indicate changes in the injection system. For instance,

- Normal operations follow a smooth upward-trending line. After fill-up, an upward line indicates stable or normal injection.
- Leftward deviation from the normal curve reflects an increasing pressure drop per unit of water injected. These effects can be caused by attaining fill-up early in the life of an injection well, and the water-zone radius will increase with time, causing the slope to curve concavely upward. Also, the leftward deviation is observed because of growth of a wellbore skin or plugging when injected water has not been sufficiently treated. An increasing slope that moves concavely upward generally indicates a positive skin or poor water quality.
- Rightward deviation from the normal curve reflects a reduced pressure effect. The deviation could possibly be caused by water channeling through fractures or joints, by a mechanical problem or holes in the casing, or by a failure of packers or a cement sheath.
- A line with a decreasing slope indicates negative skin or injection above the fracturing pressure of the formation. The injection under the latter condition can be verified by running step-rate tests. A very low slope value, as shown by the lowest line, is an indication of possible channeling or out-of-zone injection.

In **Fig. 10.14,** the Hall plot shows that poor injection performance could be the result of a water quality issue.

10.6 Summary

A waterflood consists not only of a set of rock and fluid characteristics but additional operational components that must be included in any project analysis. The chapter discusses the relative merits of obtaining water from different sources. Knowledge of the chemical compositions of the various waters involved in the selection process forms the basis of all subsequent studies. This information is a prerequisite for determining water compatibilities and cleanliness. Economics typically drives the selection process because large water volumes are required. Proper treating methods can be set up after the supply source has been selected.

The majority of water-injection projects require the installation of a water-treating plant to clean the water of particulate matter, carried-over oil particles, oxygen, and other contaminants before injection. Generally, the separation and chemical-injection components involved in this process are designed for a particular project and set of conditions. Recycled produced water is often added to the mix even though additional treating is required. The schematics of the components typically included in a treating station are discussed.

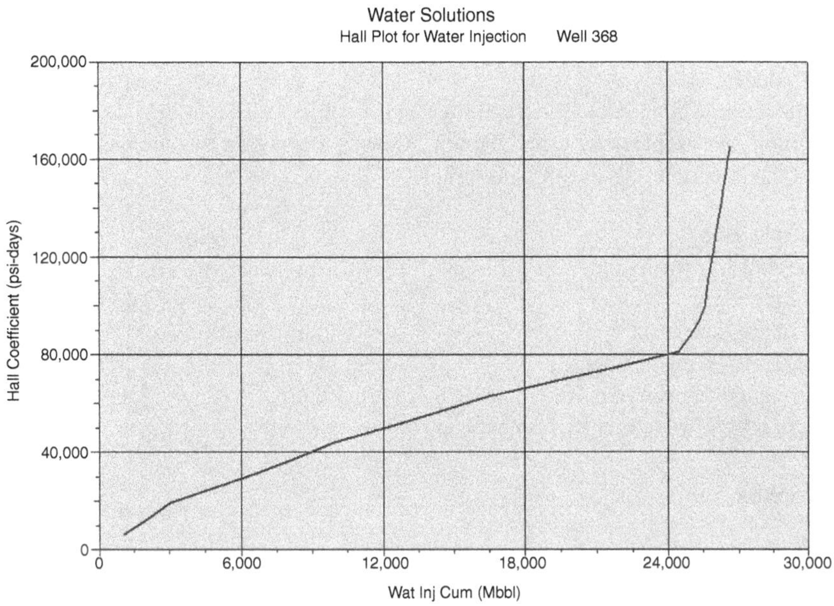

Fig. 10.14—The Hall plot shows that poor injection performance could be the result of a water quality issue (after Schlumberger NExT 2004).

Nearly all water-injection wells gradually build up a low-permeability sheath around the perforations as a function of time. Fine-grained material is typically composed of carried-over organic/inorganic material that forms a zone of reduced permeability adjacent to the wellbore. The result is a buildup of injection pressure and a loss of injection rate, both of which are detrimental to the displacement process and project economics. Water-injection wells must be constantly monitored for the onset of wellbore plugging. Exceeding the fracture pressure, on the other hand, creates zones of higher permeability, and more water can be forced out into the reservoir. Methods for monitoring these effects are discussed during the course of the chapter.

10.7 Questions

- Produced water is often reinjected to mitigate environmental concerns. Name two treatments recommended for produced water and other water sources. (Answer: Adding biocides and ensuring that suspended oil is filtered out.)
- Name at least three methods applied in an oil field to ensure the complete separation of oil from water when cleaning produced water for reinjection. (Answer: Separators, skimmers, cyclones, settling tanks, and filters.)
- What is the primary concern when using seawater as the water source for a waterflood project? (Answer: Seawater contains large amounts of organic material that must be discarded before it is injected.)
- Name diagram types that can illustrate the types and quantities of constituents present in a water sample. (Answer: Stiff and Piper diagrams, pie charts, and Collins bar diagrams.)
- Each diagram type results in a method to separate ions into two main groups. Name those groups. (Answer: Anions and cations.)

- Name the factors that can cause injection-well plugging. (Answer: Contaminated produced water, sediment carry-over, and fresh water and incompatible source water introduced into reactive shales.)
- Water-injection wells tend to plug up as a function of time. This effect can be monitored with an injection rate vs. pressure graph. How? (Answer: By noting the change and the increasing slope in the injection rate vs. pressure graph.)

10.8 Nomenclature

h = thickness, ft
k = permeability, md
p_e = effective pressure, psia
p_{wf} = bottomhole flowing pressure, psia
r_e = radius to edge of reservoir, ft
r_w = radius to edge of borehole, ft
s = skin, dimensionless
t = time, days, months, or years
W_i = cumulative water injection, RB
μ = fluid viscosity, cp
ϕ = porosity, percent

10.9 References

- Arnold, R., Burnett, D. B., Elphick, J. et al. 2004. Managing Water—From Waste to Resource. *Oilfield Review* **Summer**: 26–41.
- API. 1990. Management of Water Discharges: Design and Operation of Oil-Water Separators. API Publication 421, American Petroleum Institute, Washington, DC (February 1990).
- Chek-yan, N. and Pérez Amaro, I. M. 2017. Perfil de Producción e Inyección en el Área Este del Yacimiento SJN FUL 2 Campo El Furrial. Degree work presented to the Universidad Venezolana de los Hidrocarburos, Caracas, Venezuela.
- Hall, H. N. 1963. How to Analyze Waterflood Injection Well Performance. *World Oil* 128–130.
- Stiff, H. A., Jr. 1951. The Interpretation of Chemical Water Analysis by Means of Patterns. *J Pet Technol* **3** (10): 15–17. SPE-951376-G. https://doi.org/10.2118/951376-G.
- Stewart, M. and Arnold, K. 2009. *Emulsions and Oil Treating Equipment: Selection, Sizing and Troubleshooting*. Burlington, Massachusetts: Gulf Equipment Guides, Gulf Professional Publishing/Elsevier.

Part 3: Field Study

Chapter 11

Field Study—Analysis Workshop

11.1 Introduction

The following workshop applies commonly available geological, well, and production information to review performance efficiency. Basic reservoir analysis concepts, as well as spreadsheet formulations, are the only tools required for a multitask field analysis. Ultimately, the study provides an alternative depletion plan to maximize recovery from a candidate reservoir.

Fig. 11.1 shows a flow chart for the study. Note the progression from reservoir reconnaissance, data gathering, and in-depth evaluation to, finally, concrete conclusions and recommendations for modifying future performance. Here are a few details of note regarding the sections outlined in Fig. 11.1:

- The initial character-review effort familiarizes the reviewer with reservoir characteristics, well locations, and primary producing histories.
- The reservoir evaluation section estimates project viability after determining original oil-in-place (OOIP), primary and secondary reserves, and projected sweep efficiency.
- A project plan is assembled into a cohesive package. Future oil and water production, expected new drills and workovers, and the water-injection plan are determined.

Because oil fields differ in character and quality, the proposed work-study program may not exactly model your reservoir. However, the thinking process engendered during the discussion of applying generally available field information to study a reservoir can be used to evaluate the remaining potential for most fields.

Fig 11.2 presents a more-detailed diagram of the workflow projected for the example reservoir analysis study performed in this chapter. The effort is designed around a set of sequential tasks whose ultimate goal is to develop a complete picture of the reservoir. The primary areas of interest are listed in the left-hand column and range from reservoir boundaries to expected recoveries, and well status and requirements to predicted future production. Reserves and future production estimates provide the end points. These values are determined by more than one set of calculation procedures to develop a degree of certitude regarding the findings. Last, a development plan is produced for presentation to management. Note that decision points are included in the work profile to review the information and determine if it is useful to proceed further along the investigative path.

Section 11.4.1 contains data that provide information for the study. Refer to this data when processing and completing each particular task.

The fundamentals of the principles applied during the analysis process are discussed as the lead-in to that task. Subjects such as reservoir pressures, basic rock and fluid properties, analysis

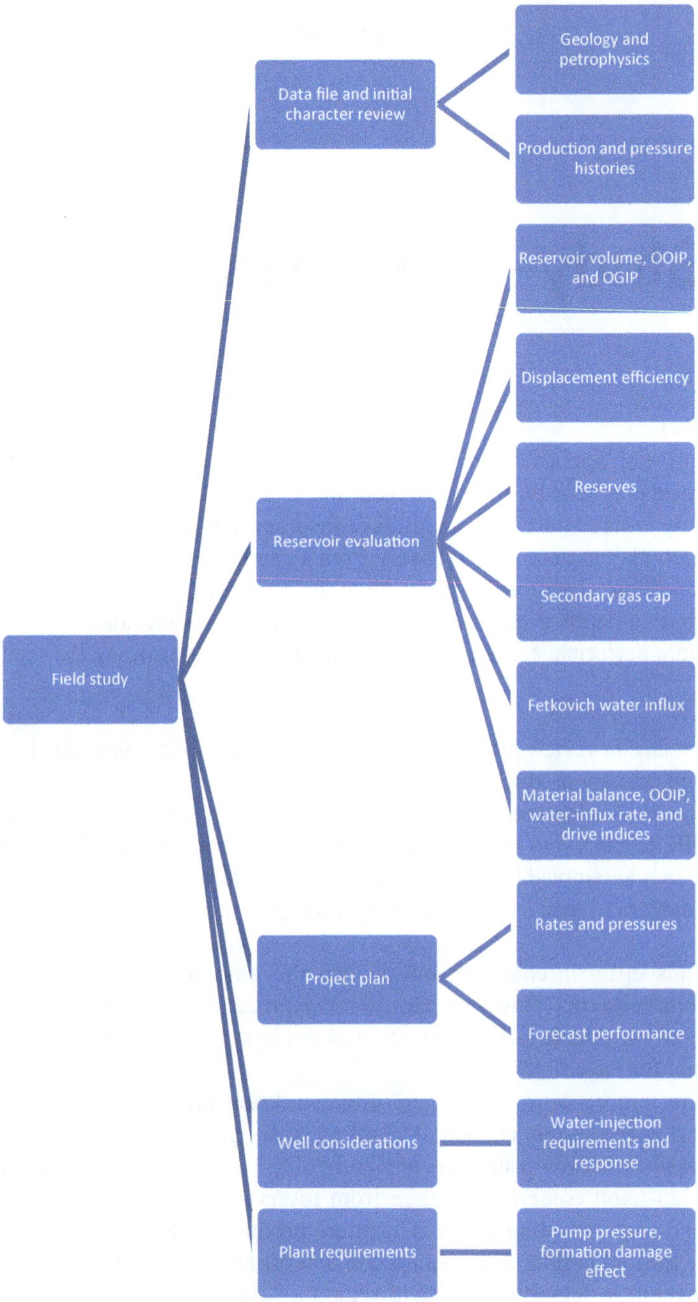

Fig. 11.1—Flow chart of analysis procedure.

of performance, material balance, Darcy flow, inflow performance, and fluid displacement are developed. The design requirements for the required injection and production facilities are included to complete the study. The primary goal of waterflooding is to displace oil with water in an efficient manner to maximize the profitable recovery of oil from a reservoir. Usually, the primary unknown is the estimation of displacement efficiency.

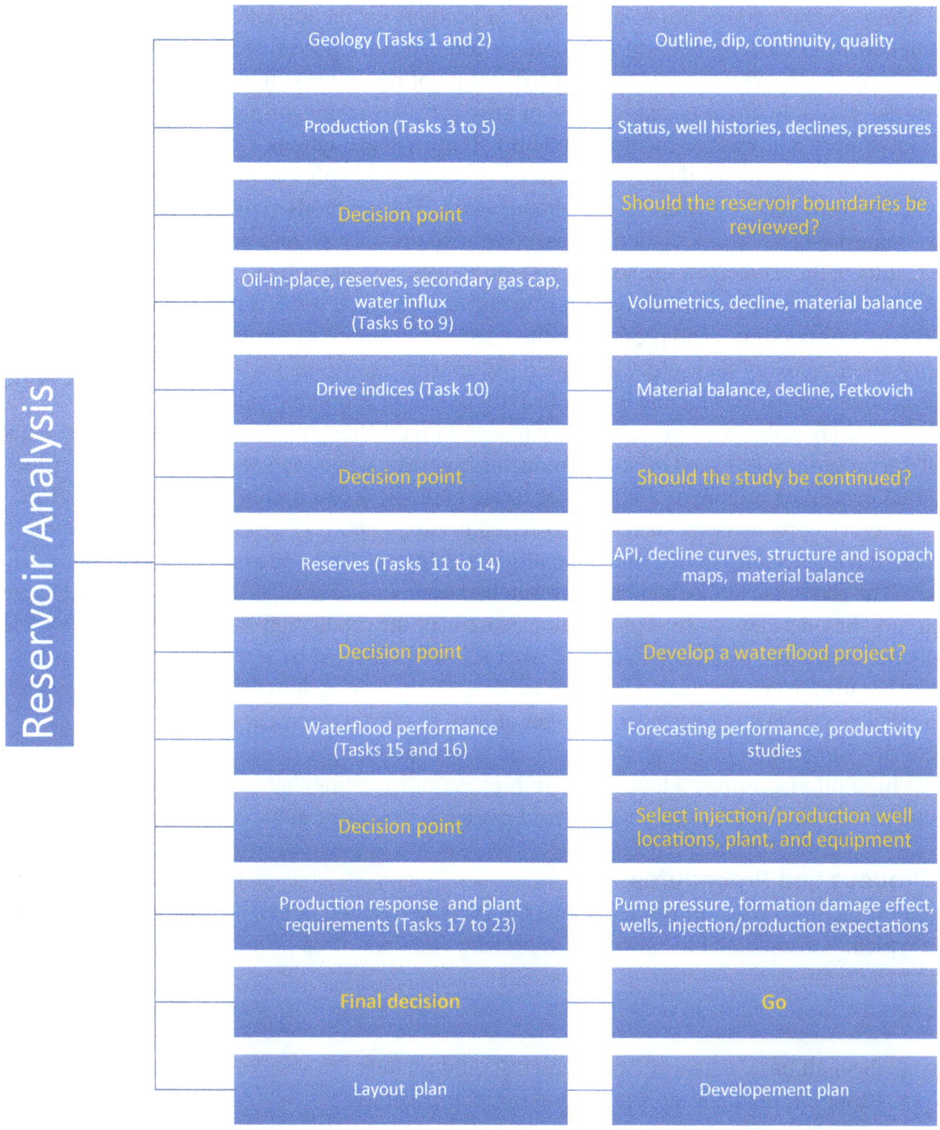

Fig. 11.2—Workflow for example reservoir analysis study.

Inserted into the evaluation procedure are decision points at which previously completed tasks are reviewed and conclusions concerning the continuation of the study are drawn. Last, as a capstone, all assumptions and conclusions should be reviewed and any additional studies proposed to further knowledge of the reservoir. A study guide starting in Section 11.5 is provided to help you complete the reservoir study, and the answers to the study are provided in Chapters 12 and 13.

11.2 Field Study Overview

This field study has been designed to show the steps typically taken during the development stages in the life of an oil field when implementing a waterflooding project. These tasks include the evaluation of geological and petrophysical information to establish the areal extent of the reservoir, net-pay thickness, porosity, and fluid saturations. Subsequent steps involve the estimation of the

OOIP using several approaches in an attempt to account for incompleteness of data. At that stage, fundamental reservoir engineering calculations are typically performed to estimate pressure and saturations in the reservoir on the basis of the results of petrophysical and core analysis and on the basis of the available direct measurements of pressure. The results of pressure/volume/temperature (PVT) laboratory analysis are incorporated into all of the calculations.

As more data become available, understanding of the reservoir should be updated by evaluating fieldwide performance and that of individual wells. The methods typically applied at this stage involve material-balance calculations, pressure transient analysis, and nodal analysis. On the basis of the results of these calculations, underperforming wells are identified, causes of poor performance are determined, and the economic benefit of working over such wells is determined.

As production continues and pressure in the reservoir declines, one must consider several methods to improve oil recovery and select the one most appropriate to the situation. The method selected should account for various factors and be heavily based on economic calculations.

This field study is designed to acquaint the reader with the thought process and the typical sequence of analysis procedures followed by an oil producing company when evaluating a field for waterflooding. The initial evaluation requires sufficiently accurate geological description and calculation of the fundamental reservoir parameters, such as porosity, saturations, and PVT properties of the reservoir fluid, which are the input values to determine the original volumes of hydrocarbons-in-place and remaining reserves. The economic attractiveness of a given hydrocarbon accumulation to be developed by means of waterflooding depends on the optimization process of the production and injection profiles and the impact on the recovery of the remaining oil reserves to be obtained by a properly designed waterdrive and pressure maintenance project. Using the proposed workflow in this chapter will assist in increasing sweep efficiency, which in one way or another affects all the development decisions.

There are limitations to the applicability of waterflooding, in particular when PVT properties indicate high oil viscosities, which result in high mobility ratios that cause fingering. While some heterogeneity is acceptable, one should avoid extensive fractures in the reservoir.

11.3 Process and Procedures

The goal of the procedure for reservoir description is to assess a field using the limited information that is typically available at the outset of a project: log data, fluid properties, production data, and pressure transient data.

The sequence of evaluation is as follows:

1. Collect all basic data available, which should include
   - Reservoir depth, temperature, and initial pressure
   - Fluid properties:
     - PVT; oil gravity (°API); viscosities of oil, water, and gas
     - Water salinity
   - Original fluid contacts
   - Well spacing and completion history, pressure well tests
   - Rock properties: Rock-quality distribution and reservoir heterogeneity
     - Porosity and permeability distribution, relative permeabilities, capillary pressure
   - Reservoir drive mechanisms
2. Obtain values of shale volume, porosity, and water saturation, and use cutoff values to determine net pay.
3. Evaluate fluid properties.
4. Calculate OOIP/original gas-in-place (OGIP), drive mechanism, and recovery factor.
5. Evaluate relative permeability data and capillary pressure data.
6. Use decline-curve analysis to estimate decline parameters, reserves, and recovery factor.
7. Generate production forecast under actual primary production strategy.

8. Analyze well tests: pressure buildup, drawdown, falloff, and injectivity-test data to determine permeability, skin and reservoir pressure, distance to heterogeneities, and boundaries.
9. Using available reservoir and fluid data, calculate decline parameters and generate forecasts for waterflooding screening studies.
10. Evaluate the benefits of placing a horizontal or highly inclined well into the reservoir. Calculate the production rate of such a well and generate a forecast.
11. Compare productivity indices of new well designs, actual wells, and stimulated wells to the base-case well.
12. Using estimated costs and forecast performances, calculate economic indicators—such as net present value (NPV), internal rate of return (ROR), payout period, and discounted profit to investment ratio of the different scenarios—and make recommendations.

The ideal data set is rarely available, and the subsurface evaluation team is challenged with significant data management tasks. The first axiom is that data must be complete and easy to access; however, more often in reality, data are contained in various locations or within various disciplines. Because of this, data must be available for sharing between team members and will require regular updates during the process and life of a given reservoir.

The overall purpose of this field study is to characterize a reservoir and to develop exploitation strategies that optimize future reservoir performance and/or meet a predetermined production level with economic objectives, such as to optimize economic parameters (e.g., NPV and ROR) and/or to meet an investment or expense schedule.

Gathering, compiling, and reviewing the data available for the selected areas and then distributing geological, seismic, drilling-history, mud-log, core, well-completion, production, engineering, and petrophysical data to all involved for discussion and quality analysis form the first step of the project.

Deliverables of the study to be provided include the following:

- A technical report with recommended management strategies and a summary of the analysis results, including key accomplishments, interim findings and recommendations, and/or problems identified
- Presentation of final results to management

11.4 Field Study Tasks

11.4.1 Data File and Initial Character Review. The actual field data included in this section are to be used in the analysis, design, and management of a waterflood. Applying the workflow process and the latest technologies in a multiple-discipline work environment will aid in reservoir characterization, drilling optimization, and exploitation of reserves and resources evaluation and will facilitate increasing production and identifying new opportunities in this reservoir by means of waterflooding.

The following tasks will guide the reviewer in assessing the reservoir by using the initially available data.

Task 1. Study maps, log profiles, and C1 sand continuity. **Fig. 11.3** presents an outline of AA Reservoir Unit, Bayou Black C1 Sand. A fault block is a portion of a field formed by sands truncating against an intrusive salt dome. An oil accumulation is typically defined by two radiating normal faults and a downdip oil/water contact (OWC). Well 1 in Fig. 11.3 was drilled near the apex of the AA fault block. The OWC lies just below Well 6 (W6). The current status of each well tested in the C1 Sand within the AA fault block is included on the map.

Refer to **Table 11.1** for the well performance history condensed from a study of well files. Completed wells are generally gravel packed because sand production is a problem in this area.

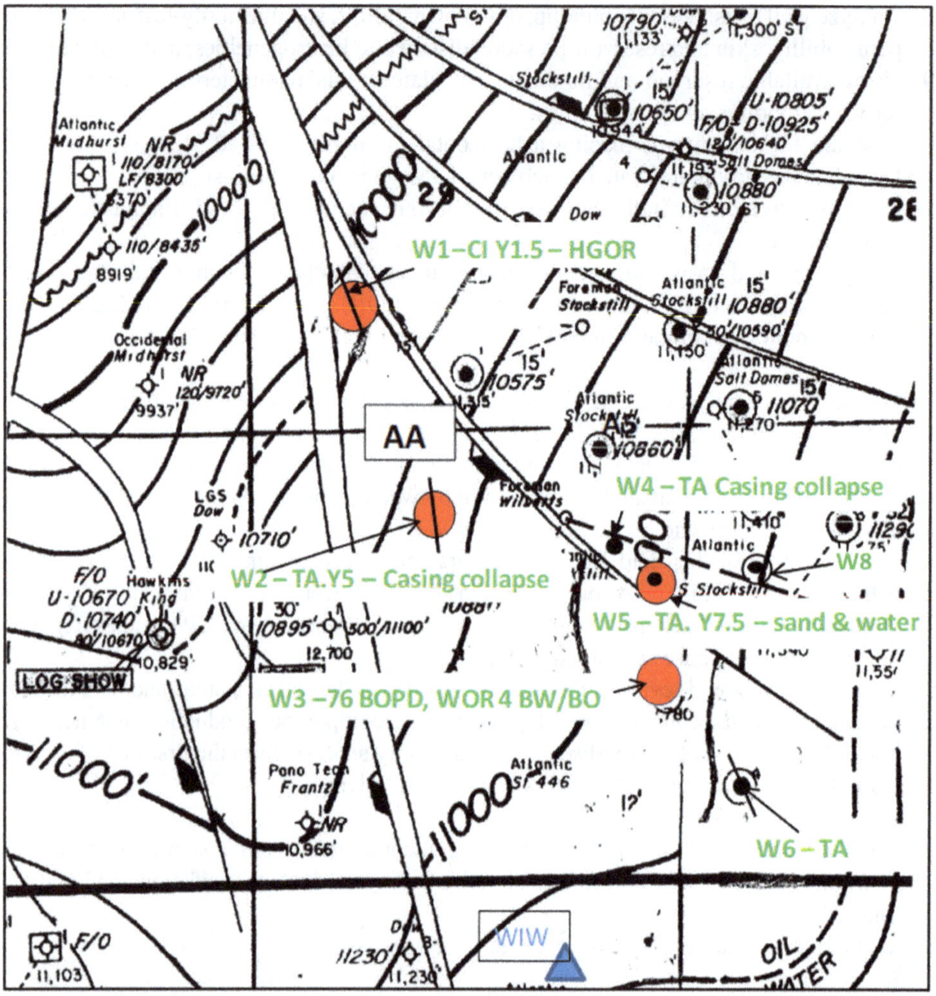

Fig. 11.3—Structure map, Bayou Black C1 Sand, AA Reservoir; average dip ≈ 4° to 10°. Scale: 1 in. = 2,000 ft. WIW = water-injection well.

The southeast limiting faults die out downdip. There is some question whether Well 8 should be included in the AA Unit. Determining the extent of the fault block will be one of the objectives of building a geological static reservoir model.

Next, study the geological and petrophysical aspects using well logs, a cross-sectional well correlation, and an oil isopach map.

The log shown in **Fig. 11.4** indicates the C1 Sand is divided into two producing zones of widely differing permeability. Segments A, B, C, D, and E in Fig. 11.4 correlate with the rock properties shown in **Table 11.2.**

The cross-sectional well correlation **(Fig. 11.5)** indicates that a loss of section occurs in the uppermost sand member (net oil = 20 ft) for the uppermost well, Well 1. Should the pay intervals be treated as continuous or as separate entities throughout the AA reservoir?

Core analysis (Table 11.2) indicates there is a three- to fourfold difference in permeability between the upper and lower sand lobes. Would gravity segregation affect production from both upper and lower sand members?

Year	Well	Status
1	1	Initial flow rate (q_{oi}) = 685 STB/D; gas/oil ratio (GOR) = 685 scf/STB; basic sediment and water (BS&W) = 0%
	2	q_{oi} = 590 STB/D; GOR = 600 scf/STB; BS&W = 0%
1.5	1	Shut-in (CI); went to high GOR (HGOR)
2	3	q_{oi} = 620 STB/D; GOR = 520 scf/STB; BS&W = 0%
3	4	Temporarily abandoned (TA)—drilled and completed (D&C) well; tested but casing collapsed during completion
	5	D&C well; q_{oi} = 395 STB/D; GOR = 420 scf/STB; BS&W = 0%
5	2	TA—collapsed casing—sand and water
	8	D&C in C1 Sand—in AA Reservoir?
7	6	D&C in deeper horizon; tested mainly water in C1 Sand; good candidate to be recompleted and become a water injector; currently TA
7.5	5	TA—sand and water
8	3	Producing q_o = 76 BOPD; GOR = 350 scf/STB; water/oil ratio (WOR) = 6 BWPD/BOPD

Table 11.1—Well performance history and status.

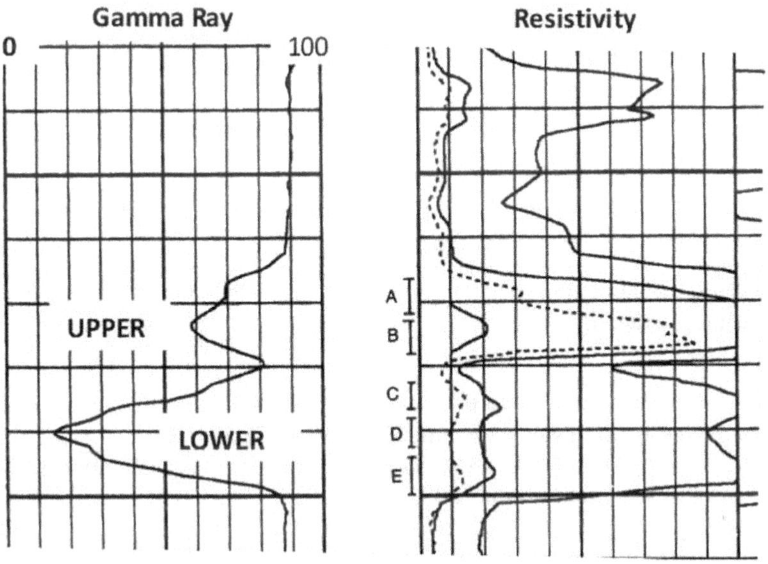

Fig. 11.4—Log section. Do the two members of the C1 Sand maintain similar character throughout the AA Unit? Do you think the upper sand member will have production rates similar to those of the lower member?

Zone	Depth (ft)	h (ft)	ϕ (%)	S_{wi} (%)	$\bar{k}$ (md)
A	10,066–70	4	23	45	31
B	10,072–78	6	27	39	75
C	10,082–89	7	29	26	92
D	10,089–95	6	33	27	300
E	10,095–99	4	29	36	220
Σ		27			

Table 11.2—Rock properties determined from log and core information on Well 3.

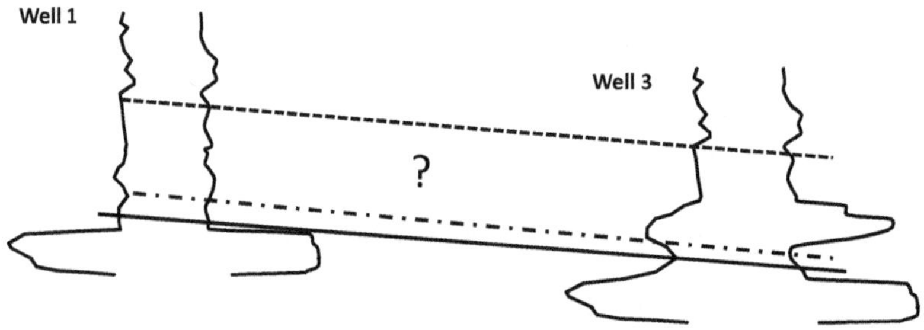

Fig. 11.5—Cross section of a correlation between Well 3 and Well 1 exhibiting sand continuity between downdip (Well 3) and updip (Well 1) wells for the C1 Sand located in the AA Reservoir. Reduced resistivity indicates a pronounced loss of porosity in the upper sand body in Well 1. Where does one select the porosity pinchout line?

1	2	3	4	5	6	7	8	9	10
Zone	Depth (ft)	h (ft)	ϕ (%)	$h \times \phi$ (Column 3 × Column 4)	ϕ (%)	S_{wi} (%)	$h \times S_{wi}$ (Column 3 × Column 7) (%)	$\overline{S}_{wi}$ (%)	$\overline{k}$ (md)
A	10,066–70	4	23	92		45	180		31
B	10,072–78	6	27	162		39	234		75
C	10,082–89	7	29	203		26	182		92
D	10,089–95	6	33	198		27	162		300
E	10,095–99	4	29	116		36	144		220
Σ		27		771	28.56			33.4	

Table 11.3—Average porosity, permeability, and initial water saturation properties determined from log and core information on Well 3.

Table 11.3 relates each zone to its vertical location within the C1 producing interval. Columns 6, 9, and 10 show the results from calculating the average rock properties.

Note the following calculations for a weighted average porosity, water saturation, and permeability:

$$\overline{\phi} = \frac{\Sigma h_n \times \phi_n}{h_t} = 28.56,$$

$$\overline{S}_{wi} = \frac{\Sigma h_n \times S_{wi}}{h_t} = 33.4\%,$$

$$\overline{k} = \frac{\Sigma h_n \times k_n}{h_t} = 144.4.$$

Rock properties were derived from log analysis except for permeability ($\overline{k}$), which was developed from core analysis studies.

An isopach map has been constructed and is shown in **Fig. 11.6.**

Task 2. Study log profiles. Study Figs. 11.4 and 11.5 to determine reservoir and flow-channel character, and study the petrophysical aspects obtained from core data (Table 11.2).

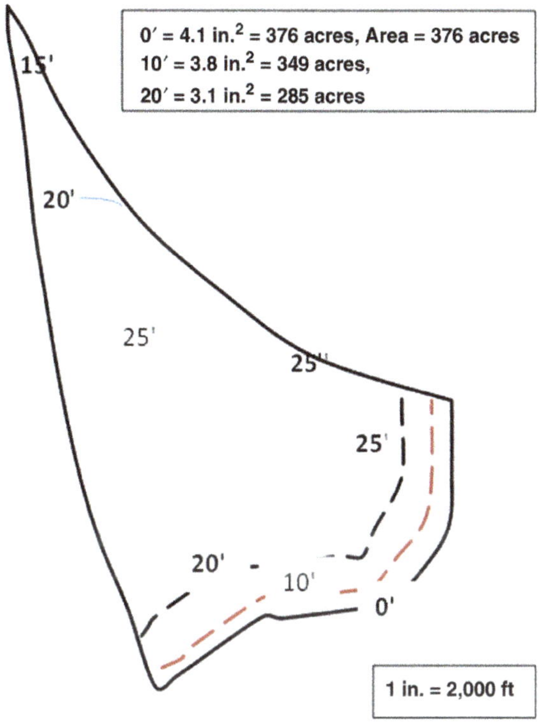

Fig. 11.6—An oil isopach map outlining the mapped original oil accumulation for the C1 Sand/AA Reservoir. Within Fig. 11.6, the apostrophe sign indicates feet.

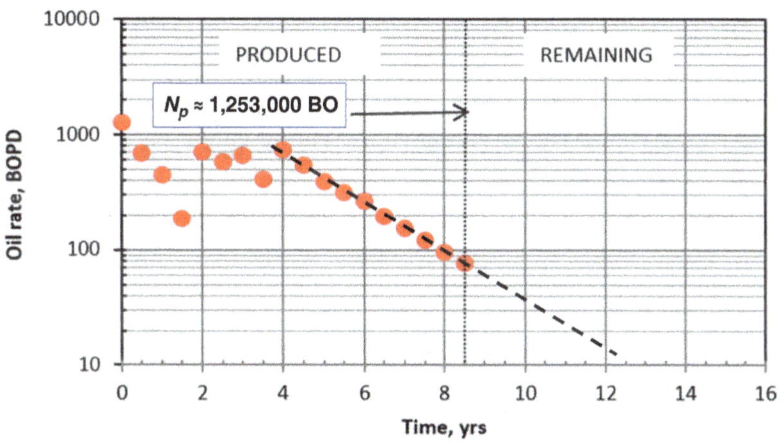

Fig. 11.7—Reservoir performance history. Cumulative production was calculated by summing individual well histories in a spreadsheet.

Task 3. Next, study the reservoir history **(Fig. 11.7),** well histories **(Figs. 11.8a through 11.8d),** and Table 11.1 to analyze reservoir and general well performances.

The reservoir performance history is depicted in Fig. 11.7, which shows a straight line on a semilog plot of flow rate vs. time, indicating exponential decline.

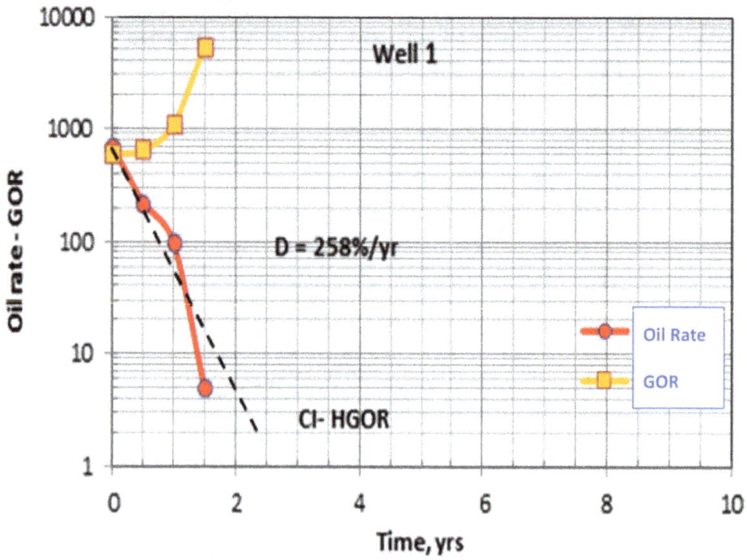

Fig. 11.8a—Well performance history, Well 1. The well was closed in (CI) after 1.5 years of production as a result of excessive gas production. Does this knowledge provide a clue regarding the probable depletion characteristics? Refer to the structure map in Fig. 11.3.

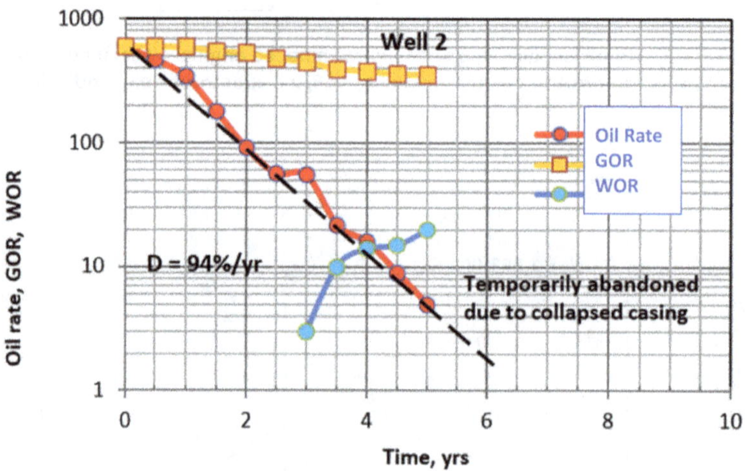

Fig. 11.8b—Well performance history, Well 2. The well suffered mechanical problems and was temporarily abandoned in Year 5. It produced excessive sand and water near the end.

The available producing histories, for four wells completed in the AA Unit, are presented in **Figs. 11.8a** through **11.8d.** Well production histories are shown with attendant decline rates. The reservoir histories display an exponential decline. Do the shapes of the performance-decline curves signify reservoir depletion? How many production segments do the depletion histories display? Pay close attention to what these histories are trying to tell you. This knowledge will also help when future workovers and recompletions are planned.

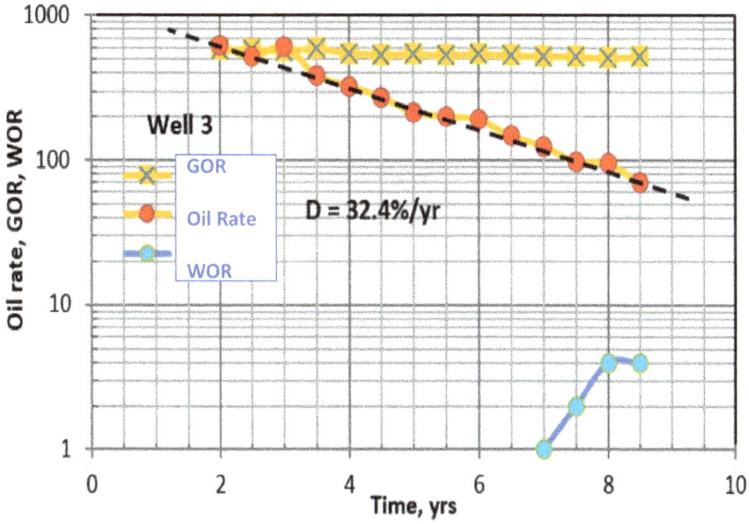

Fig. 11.8c—Well performance history, Well 3. The well is cutting small volumes of water but is still `producing under gas lift.

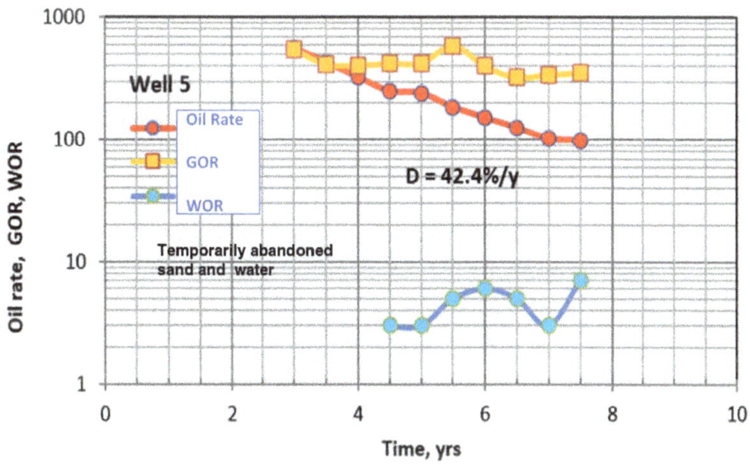

Fig. 11.8d—Well performance history, Well 5. In Year 7, the oil rate ≈ 100 STB/D, GOR ≈ 360 scf/STB, and WOR ≈ 5 BWPD/BOPD.

Points to consider when evaluating the well performance histories in Figs. 11.8a through 11.8d:

- How do the producing characters of the downdip wells differ from the character of the most updip well, Well 1?
- The reservoir is currently producing q_o = 76 BOPD, q_g = 28.9 Mscf/D of gas, and q_w = 456 BWPD (WOR = 6).

Study the profiles in **Table 11.4,** which summarizes well decline curve characteristics. Table 11.4 shows the values calculated on the basis of the analysis of Figs. 8.11a through 11.8d.

1	2	3	4	5	6
Well	q_i (BOPD)	q_{last} (BOPD)	Arps' continuous decline rate D (%/yr), $D = \ln(q_1/q_2)/t$	$N_{p(current)}$ (bbl)	$N_{p(primary)}$ (bbl)
1	685	5	258	96,202	96,202
2	590	5	93.6	227,154	227,154
3	620	71	32.4	612,840	670,429
5	395	54	42.4	316,812	316,812
				Total: 1,253,008	Total: 1,310,597

Table 11.4—Exponential decline curve characteristics for the producing wells.

In Table 11.4, initial and current production rates and decline rates for each well are listed in Columns 2, 3, and 4; extrapolation (exponential decline) of best-fit straight lines to the economic limit results in the estimated total recovery values shown in Column 5; and Column 6 presents expected primary production from the AA Reservoir when production declines to the 5 BOPD economic limit. The difference between Columns 5 and 6 is the remaining primary production. Here are some questions to consider during evaluation:

- What can one say about the expectations of future performance from the AA Reservoir?
- All wells exhibited an exponential decline. What type of depletion does that indicate?

Task 4. Study general reservoir and field information.

Table 11.5 presents general field information derived from various sources. These data are provided to complete some of the ensuing tasks. Note that the left-hand column categorizes the information.

Volume	Reservoir volume = 6,795 acre-ft	A = 376 acres	
Fluid properties	B_{oi} = 1.44 RB/STB B_{obp} = 1.45 RB/STB B_o =1.373 RB/STB	R_{si} = 600 scf/STB R_s = 350 scf/STB in Year 8.5	B_w = 1.0 RB/STB
Saturations	S_{or} = 25%	S_{rg} = 20%	
Relative permeabilities	k_{rw} = 0.25	k_{ro} = 0.75	
Viscosity	μ_o = 2.0 cp	μ_w = 1.0 cp	μ_{obp} = 2.0 cp
Pressure	p_i = 5,750 psia, measured from well test on Well 1 at time of discovery p_{bp} = 5,500 psia p_{last} = 2,300 psia in Year 8.2 as measured in Well 3	p_a = 1,000 psia	$p_{wf(current)}$ = 2,000 psia
Production	EL_{res} = 50 BOPD	*ELwell = 5 BOPD*	N_p = 1,253,000 BO q_o = 76 BOPD, q_g = 28.9 Mscf/D q_w = 456 BWPD (WOR = 6)
Field	Datum depth = 10,800 ft Reservoir temperature = 232°F	r_w = 0.25 ft	γ_{ob} = 1.0 psi/ft γ_{sw} = 0.465 psi/ft

Table 11.5—Reservoir and field information. Original conditions of reservoir pressure and temperature indicate that an original gas cap is not present.

Time (year)	Well 1 (psia)	Well 2 (psia)	Well 3 (psia)	Well 4 (psia)	Well 8 (psia)
0.25	5,326	5,590	5,334	5,500	
0.75					
1.25					5,333
2.25				4,562	
2.75	4,035				
3.25		3,615		3,000	5,122
3.75			3,051		
4.2		2,650			
5.5	2,510				4,955
6.05			2,410	2,380	
8		2,410			4,800
8.2	2,350		2,300		

Table 11.6—Measured static-BHP history, corrected to datum (10,800 ft).

Task 5. Analyze bottomhole pressures (BHPs) to study well connectivity. **Table 11.6** presents static-BHP-data information derived from well tests. These data are provided to understand well connectivity and pressure depletion as a function of time. Note that the left-hand column categorizes the information by year of production.

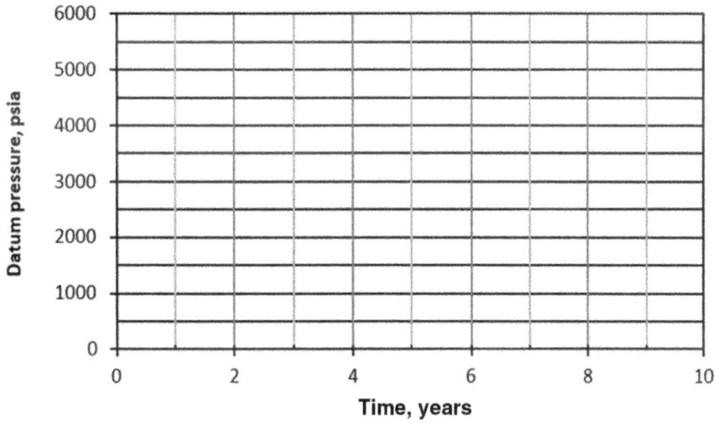

Fig. 11.9—Use the graph to plot historical datum pressures.

Plot the values indicated in the example graph **(Fig. 11.9)** to provide the results that define reservoir boundaries. Estimate initial pressure (Year 0) and reservoir pressure at Year 8 from the pressure study. This information will later be used to find fluid and reservoir properties.

Decision point*—Define reservoir boundaries.*

11.4.2 Reservoir Evaluation. The tasks in this section should be performed to further evaluate the characteristics of the reservoir. The results from these tasks will later serve as a foundation for the project plan. **Table 11.7** shows the reservoir

	Pressure (psia)	B_o (RB/STB)	B_g (RB/scf)	R_s (scf/STB)
Year 1	5,500	1.440		600
Year 8.5	2,380	1.373	0.00129	350

Table 11.7—Fluid properties and average reservoir pressure at the beginning and at the end of history (Year 8.5).

1	2	3	4	5	6	7
Time (year)	p_o (psia)	Δp (psi)	W_e, Cumulative Water Influx (bbl of water influx in the time period)	W_e, Water Influx per Day (BWPD)	ΣW_e	$p_{a(n-1)}$
0	5,765					
1	5,465	532	193,140	529	193,140	5,498
2	4,915	515	375,573	1,030	568,713	4,980
3	4,015	796	621,195	1,701	1,189,907	4,122
4	2,790	1,077	687,143	1,866	1,877,050	3,183
5	2,600	152	198,752	562	2,075,803	2,900
6	2,400	174	150,867	333	2,226,670	2,736
7	2,400	−3	16,781	169	2,243,451	2,651
8	2,380	17	14,473	151	2,258,193	2,575

Table 11.8—Results of water-influx calculations, the Fetkovich method.

pressure and average fluid information listed as a function of the initial values at Year 1 and final history at Year 8.5.

Task 6. Calculate reservoir volume, OOIP, and OGIP from isopach map and rock properties. One must know if there are sufficient hydrocarbons remaining when primary and secondary reserves are quantified.

Task 7. Study performance records and review exponential decline principles to determine remaining primary reserves. A study of well histories provides clues regarding which wells produced at initial—and later—conditions.

Task 8. Study the structure map and production records to determine whether a secondary gas cap was formed.

Task 9. Apply the Fetkovich water-influx method (see Chapter 7) to calculate historical water encroachment into the AA Reservoir.

The results are shown in **Table 11.8.**

Plot the values given in Column 5 vs. the values given in Column 1 using the example graph **(Fig. 11.10).**

Task 10. Apply the material-balance study results shown in **Table 11.9** to determine the OOIP, water-influx rate, and drive indices for the AA Reservoir. Drive indices should help to reinforce previous conclusions about reservoir mechanics. Compare the material-balance OOIP calculation with the previous volumetric answer (specified in Task 6).

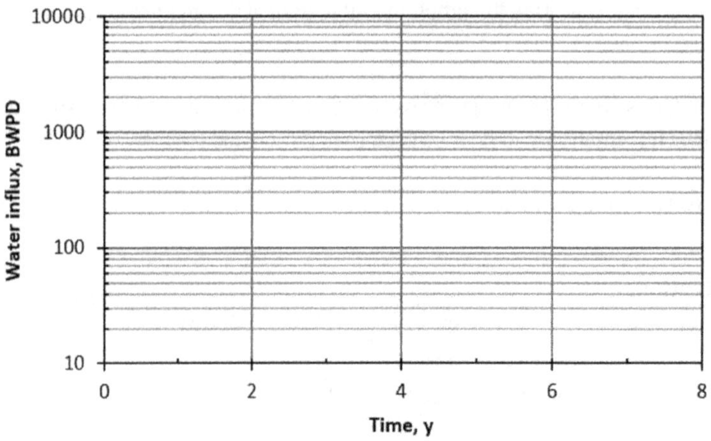

Fig. 11.10—Use the graph to plot the results of the Fetkovich water-influx calculations using the Fetkovich model of water-encroachment history. Note the reservoir behavior: Does the model seem to fit the reservoir conditions?

Time (year)	0	1	2	3	4	5	6	7	8
N_a (million bbl of oil)		4.6	4.2	4.9	5.0	6.15	6.3	6.25	6.2
Water influx (thousand bbl of water)		193	375	621	680	204	119	62	55
Solution-gas-drive index		0.12	0.15	0.23	0.40	0.55	0.9	0.9	0.9
Waterdrive index		0.38	0.82	0.67	0.50	0.40	0	0.05	0

Table 11.9—Results of material-balance study.

Production, average pressure, and equivalent fluid properties were included in the material-balance equation, which has been written in various forms (refer to Chapter 5), to calculate the following:

- Apparent oil-in-place (N_a)
- Water influx (W_e) over the 8-year history
- Relative influences of drive mechanisms on reservoir behavior

***Decision point**—Couple the volumetric and material-balance results to determine a realistic estimate for OOIP and OGIP.*

Task 11. Calculate the maximum-oil recoverable using the difference-in-saturation equation (Eq. 5.1 in the Oil Reservoirs chapter). This answer provides the upper limiting values when estimating reserves, and it is a function of the residual oil saturation.

Task 12. Apply the American Petroleum Institute (API) Subcommittee on Recovery Efficiency (1967) equations (Eqs. 5.2 and 5.3 in the Oil Reservoirs chapter) to calculate reserves for waterdrive and solution-gas-drive cases.

Task 13. Estimate the displacement efficiency if a waterflood were installed.

Task 14. Calculate primary and secondary reserves. The study of primary and secondary reserves estimates is the most important. Ultimately, the production forecast of these reserves translates into future cash flow.

Structure and isopach maps, residual oil saturations, and the API correlations, in addition to the material-balance calculations, are used to estimate the reserves.

***Decision point**—Calculate the reserves.*

11.4.3 Project Plan. The tasks presented in this section will serve to form the project plan, which encompasses the rates and pressures, performance forecast, and plant requirements.

Task 15. Couple the core analysis record with the Stiles calculations to predict future performance. The results of the Stiles method study are shown in **Table 11.10,** predicting the WOR as a function of time and cumulative oil production.

Task 16. Evaluate your findings: Do you believe a waterflood is feasible? Have you performed sufficient studies to form an opinion on the reservoir to present to management a plan for additional development?

Questions one should ask include

- Should a peripheral or a patterned waterflood be installed?
- Will the layered sand aid or hurt recovery?
- Are any new drills or workovers needed?

***Decision point**—Review the result of all previous tasks to determine whether you deem it worthwhile to continue with selecting injection/production well locations, determining equipment needs, and predicting performance.*

1	2	3	4	5	6	8	9	10	11
h (ft)	k (md)	C_j (md-ft)	R	f_w	q_o (BOPD)	N_p (STB)	t (years)	Δt (years)	WOR
1	300	300	0.48	0.071	676	70,440	0.29	0.290	0.1
2	300	600	0.48	0.143	624	141,758	0.62	0.337	0.2
3	300	900	0.48	0.216	571	213,971	1.03	0.405	0.4
4	300	1,200	0.48	0.290	517	287,095	1.52	0.495	0.6
5	300	1,500	0.48	0.365	462	361,149	2.14	0.618	0.8
6	300	1,800	0.48	0.441	407	436,149	2.94	0.795	1.1
7	220	2,020	0.58	0.497	366	491,762	3.68	0.745	1.4
8	220	2,240	0.58	0.554	325	547,901	4.62	0.942	1.7
9	220	2,460	0.58	0.611	283	604,574	5.85	1.231	2.2
10	220	2,680	0.58	0.669	241	661,788	7.53	1.675	2.8
11	92	2,772	0.86	0.694	223	685,877	8.42	0.895	3.1
12	92	2,864	0.86	0.718	205	710,062	9.48	1.054	3.5
13	92	2,956	0.86	0.743	187	734,344	10.74	1.260	4.0
14	92	3,048	0.86	0.767	169	758,725	12.27	1.532	4.5
15	92	3,140	0.86	0.792	151	783,203	14.17	1.904	5.2
16	92	3,232	0.86	0.817	133	807,781	16.60	2.431	6.1
17	92	3,324	0.86	0.842	115	832,458	19.81	3.210	7.3
18	75	3,399	0.91	0.862	100	852,649	23.31	3.492	8.6
19	75	3,474	0.91	0.883	85	872,908	28.03	4.728	10.3
20	75	3,549	0.91	0.903	70	893,233	34.79	6.760	12.8
21	75	3,624	0.91	0.924	55	913,627	45.26	10.461	16.7
22	75	3,699	0.91	0.945	40	934,088	63.60	18.346	23.5
23	75	3,774	0.91	0.966	25	954,617	104.14	40.538	38.4
24	31	3,805	1.00	0.974	19	963,123	139.99	35.854	51.7
25	31	3,836	1.00	0.983	13	971,640	211.70	71.707	78.2
26	31	3,867	1.00	0.991	6	980,169	426.82	215.122	157.6
27	31	3,898	1.00	1.000	0	988,710			

Table 11.10—Spreadsheet results for the Stiles calculation.

11.4.4 Well Considerations. Water-injection well requirements are important aspects to be considered in the design and operation of waterflood projects. Determining injection pressures is key because water-injection wells must have an injectivity capacity that is limited by the maximum formation pressure to avoid fracturing. Formation damage is the result of a permeability reduction near the wellbore, and it is represented by a positive skin factor and estimated by performing well tests. These possible flow restrictions must be considered, and awareness of those parameters will assist in the proper well design.

Task 17. Calculate water-injection requirements. Apply the material-balance equation (see Chapter 5) to compare current reservoir production with the expected water-injection requirements.

Task 18. Estimate the production response to water injection. Review production records and apply the Klins and Clark method discussed in Chapter 6, Section 6.5.3, to determine the effect of reservoir pressure on the production rate.

A well test was conducted on Well 3, and the Fetkovich backpressure equation (Eq. 6.14 in the Inflow Performance chapter) was developed from a log-log flowing pressure vs. rate plot. This equation was further refined to reflect J' and n values for 5,500-psia-bubblepoint conditions for the Klins and Clark adjustment analysis method; the results are shown in **Table 11.11.**

$$q = 0.043\left(5{,}500^2 - p_{wf}^2\right)^{0.72}$$

Table 11.11—Equation developed using the Klins and Clark method using results from a well test conducted on Well 3.

Task 19. Summarize well status. Use **Table 11.12** to summarize information regarding the wells located in the proposed waterflood project. At least two workovers and one new drill should be assigned to the work schedule.

11.4.5 Plant Requirements. The following studies should be viewed as approximations of the expected operating conditions.

Task 20. Apply the pressure information shown in Table 11.6 to calculate the gradients and expected hydrostatic, injection, and formation fracture pressures.

Task 21. Calculate the pressure rating for the pump pressure gradient.

Task 22. Calculate the expected pressure response resulting from an expansion of the flood front. Apply and rearrange the Darcy equation (see Chapter 4, Section 4.3) to determine if pressure increases out to an estimated drainage radius of 2,000 ft.

Task 23. Estimate the effects of formation damage on the water-injection rate.

Decision point*—Evaluate your results and create a basic outline of a development plan.*

Well No.	Status	Remarks

Table 11.12—Use the table to create a summary of the wells located in the proposed waterflood project.

11.5 Study Guide

The study guide reviews the tasks presented in the first part of the chapter to help the reader evaluate the records of a mature oil reservoir. The chapter outlines the general thinking process required to completely initiate a field or reservoir review. Using geologic and surface data, knowledge of well facilities, well histories, and well-testing records, the reader can review and relate the theoretical knowledge discussed in the first two parts of the book to an actual field situation. Capstones for such a study develop an understanding of oil-in-place, reserves, and oil and water producing rates.

Additionally, the logical reduction of available field information should serve as a prelude to more-sophisticated reservoir simulation studies where flow systems must be developed in greater detail. This process has been proven to be very successful in developing the competencies needed by the engineer or geoscientist who will be working on an asset team for his/her company.

11.5.1 Task 1—Study Maps, Log Profiles, and C1 Sand Continuity. To begin to understand reservoir behavior, study maps and production records. Some questions to ask during these studies include the following:

- What has happened to the producing rates over time? Determine the maximum well producing rate. This value will be of interest when predicting future performance.
- Are there observable mechanical problems?
- Compare the uppermost well history to that of the lower-structure wells. What are the differences?
- Is there evidence of significant water production?

11.5.2 Task 2—Study Log Profiles. Use these profiles to determine reservoir and flow-channel character and study the petrophysical aspects. Closely review the cross section and sand continuity shown in Fig. 11.5 to estimate reservoir continuity.

11.5.3 Tasks 3, 4, and 5—Study Well History Profiles and General Reservoir and Field Information, and Plot a Corrected Static BHP Record. After studying the well history profiles in Table 11.4 and the general reservoir and field information in Table 11.5, use the values from Table 11.6 to plot a corrected static BHP record in Fig. 11.9. Remember, wells located in a common reservoir often display similar pressures. Are all of the wells connected?

Draw an average pressure line through the pressure record to use later in the evaluation. Measured pressures often display a fair amount of scatter. Using Fig. 11.9, draw a smooth line through the plotted data points. Reservoir pressures measured at specific time intervals are particularly important when applying the material-balance techniques. Estimate initial pressure (p_i) and current reservoir pressure (p_{res}) at Year 8.2.

11.5.4 Task 6—Calculate Reservoir Volume, OOIP, and OGIP. Use the values from Table 11.2 to calculate OOIP and OGIP. Calculated total average porosity and water saturation values are given in Columns 6 and 9 in Table 11.3; **Table 11.13** summarizes the calculations for determining original hydrocarbons-in-place.

Calculate pore volume and OOIP by using equations from Chapter 1:

$$\text{Pore volume } (V_p) = \text{reservoir volume} \times \phi,$$

$$\text{Pore volume } (V_p) = 7{,}758 \text{ bbl/acre-ft} \times \text{reservoir volume (acre-ft)} \times \text{porosity, RB}$$

1	2	3	4	5	6	7	8	9
Zone	Depth (ft)	h (ft)	θ (%)	Average Porosity (%), $\bar{\phi} = \frac{\Sigma h_n \times \phi_n}{h_T}$	S_{wi} (%)	Average Initial Water Saturation (%), $\bar{S}_{wi} = \frac{\Sigma h_n \times S_{wi}}{h_T}$	$h\phi(1-S_{wi})/B_{oi}$ (STBO/acre)	$\bar{k}$ (md)
A	10,066–70	4	23		45			31
B	10,072–78	6	27		39			75
C	10,082–89	7	29		26			92
D	10,089–95	6	33		27			300
E	10,095–99	4	29		36			220
Σ		27		28.56		33.4		

Table 11.13—Rock and fluid properties for estimating original reservoir volumes of oil- and gas-in-place.

A pore volume map (ϕh) is calculated using Eq. 1.4:

$$OOIP = 7{,}758 \text{ bbl/acre-ft } \frac{[\text{isopach } \phi h \text{ volume (acre} - \text{ft)}](1 - S_{wi})}{B_{oi} \text{ (RB / STB)}}, \text{ STBO.}$$

Hydrocarbon pore volume (HCPV) is calculated using Eqs. 1.5, 1.7a, and 1.7b:

$$\frac{[\phi h(1 - S_{wi})]}{B_{oi}}, \text{ ft-STB/RB.}$$

$$OOIP = 7{,}758 \text{ bbl/acre-ft} \frac{[\phi(1 - S_{wi})]}{B_{oi}}, \text{ STBO/acre-ft,}$$

$$OOIP = 7{,}758 \text{ bbl/acre-ft } \frac{[\phi(1 - S_{wi})(HCPV)]}{B_{oi}}, \text{ STBO.}$$

In gas reservoirs, the gas-in-place is calculated using Eqs. 1.10a and 1.10b:

$$OGIP = \frac{43{,}560 \text{ ft}^3 \text{ / acre} - \text{ft}(\phi)(1 - S_{wi}) p_{\text{res}} T_{sc}}{z_{\text{res}} p_{sc} T_{\text{res}} 10^6}, \text{ MMscf/acre-ft,}$$

$OGIP$ = reservoir volume (acre-ft) × MMscf/acre-ft, MMscf.

In oil reservoirs at initial conditions at or above the bubblepoint pressure, OGIP is calculated using Eq. 1.10c:

$OGIP$ = $OOIP$ (STBO) × R_{si} (scf/STBO),

where

$$OOIP = 7{,}758 \text{ bbl/acre-ft } \frac{[\text{isopach } \phi h \text{ volume (acre} - \text{ft)}](1 - S_{wi})}{B_{oi} \text{ (RB / STB)}}, \text{ STBO.}$$

Now calculate OOIP and OGIP (R_{si} = 600 scf/STB in Table 11.4):

$OOIP$ =

$OGIP = N \times R_{si} =$

11.5.5 Task 7—Study Performance Records and Decline Curves To Predict Remaining Primary Recovery. Recall that the straight line seen on a log rate vs. time plot identifies the reservoir as pressure depleting and is classed as an exponential curve. Extending this straight line from the last measured rate to the economic limit determines the remaining production, if no additional work is performed on the wells.

The exponential equations used to calculate cumulative oil production between the current rate and the economic limit flow rate are

$$N_p = \frac{q_1 - q_2}{D} \text{ and } D = \frac{\ln \frac{q_1}{q_2}}{t}.$$

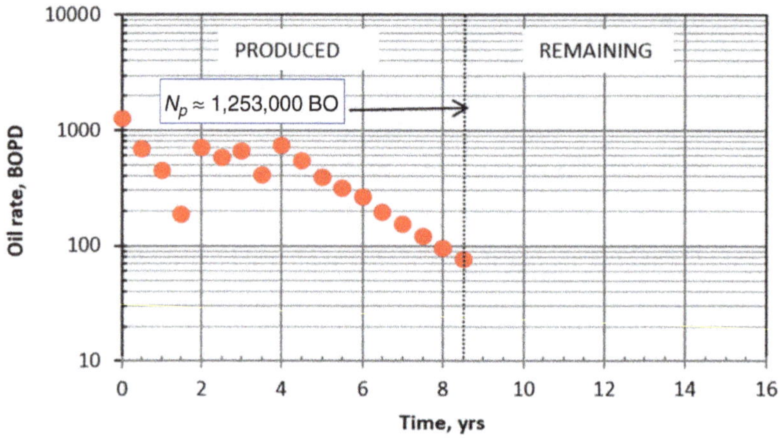

Fig. 11.11—Producing history of the C1 Sand, AA Reservoir; interpreted primary reserves from the current well arrangement. Into how many segments can the producing history be divided?

Fig. 11.11 shows the 8.5-year producing history of the AA Reservoir. Credited cumulative production is 1,253,008 bbl of oil. Figs. 11.8a through 11.8d illustrate the histories of Wells 1, 2, 3, and 5, which have produced from the C1 Sand, AA Reservoir, and Fig. 11.7 shows the reservoir performance history, which has been modified and is presented in Fig. 11.11.

Calculate the remaining primary production out to the point at which the economic limit = 5 BOPD. What seems to be the primary cause for loss of wells? Coupling Table 11.1 with the well-decline figures (Figs. 11.8a through 11.8d) would be a great help here.

11.5.6 Task 8—Determine Whether a Secondary Gas Cap Has Formed. Study the structure map (Fig. 11.3) and well production histories and records (Table 11.1) to determine whether a secondary gas cap has formed.

From the pressure plot (Task 5, Fig. 11.9), it can be seen that apparent pressure-depletion conditions occurred with little increase in the producing GOR. Where can free gas migrate in a high-permeability sand?

11.5.7 Task 9—Apply the Fetkovich Method To Calculate Water-Influx History. Table 11.8 presents the results of applying the Fetkovich model (see Chapter 7) to the study of the water-influx history for the C1 Sand.

Fetkovich (1971) developed a relationship between the decline of the average pressure in an aquifer and the water volume expelled as pressure declines. The method establishes a relationship between the cumulative water-influx volume measured at any time period and the difference between the average pressure in the aquifer ($\bar{p}_a$) and the pressure at the OWC (p_o).

Interpret the results from the Fetkovich model, which works best for closed, finite aquifers; the productivity index models aquifer flow, and the material balance calculates water outflow from the aquifer.

The results of the calculations are listed in Table 11.8. Use Fig. 11.10 to plot the results and interpret their meaning, and analyze the water-influx effect.

11.5.8 Task 10—Apply Material-Balance Equation Technique. To interpret OOIP, water influx, and the effect of drive mechanisms on performance, apply the material-balance-equation technique (see Chapter 5). The material balance compares fluid-produced volumes with changes in the volume of materials originally present in the reservoir as a function of pressure history. Historical production, pressure, and fluid properties must be known to apply this method.

Plotting and interpreting the results shown in Table 11.9 will help determine OOIP, water influx as a function of time (W_e), and the relative effects of solution-gas-drive (SGDI) and waterdrive (WDI) indices on performance. An original gas cap has been deemed absent, but there is evidence of a secondary gas cap whose relative effect is represented by the corresponding gas-cap-drive index (GCDI).

The study procedure is as follows:

1. Develop the calculations listed in Table 11.9. Plot the apparent oil-in-place (N_a) vs. time data in **Fig. 11.12** and interpret the results to determine OOIP.
2. Assume zero water influx to calculate and interpret OOIP as a function of cumulative production. Back extrapolate to $t = 0$.
3. Rewrite the material-balance equation (Eq. 5.12a in the Oil Reservoirs chapter) to include the apparent OOIP (N_a) to calculate and interpret water influx as a function of time.
4. Calculate and plot in **Fig. 11.13** the water-influx (W_e) volumes listed in Table 11.9 as a function of time.
5. Compare these results to those calculated using the Fetkovich method in Task 9 (see Section 11.5.7).

Apply the results from Table 11.9 and use **Fig. 11.14** to plot the calculated drive indices shown as a function of time for the AA Unit. Is there a fundamental change in water-influx effectiveness with producing life? Are there producing segments? Compare to the Fetkovich calculations.

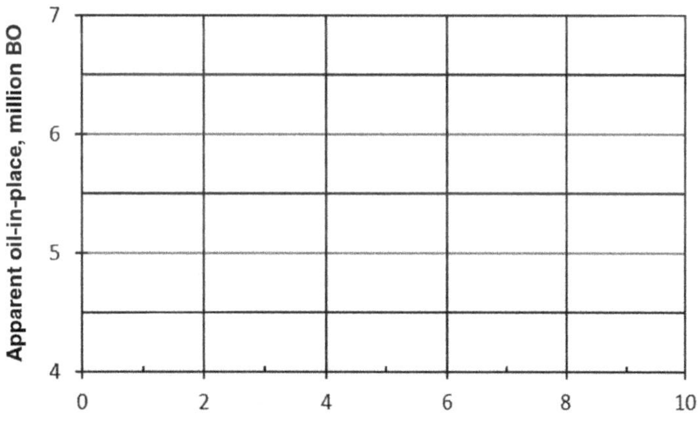

Fig. 11.12—Plot your material-balance OOIP estimates. Can you interpret changes in the performance history from your figure?

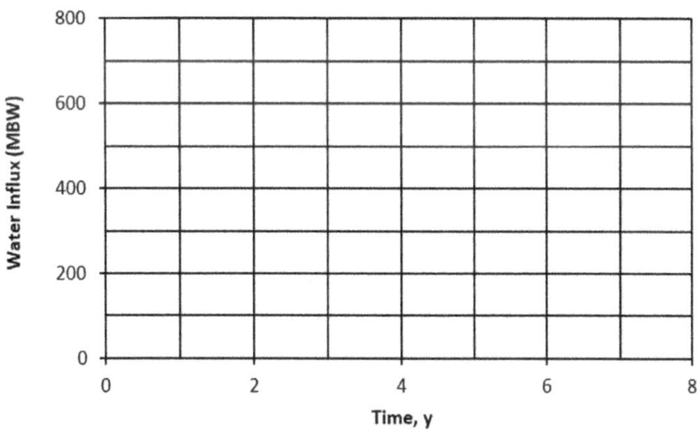

Fig. 11.13—Use the graph to plot the material balance water-influx calculations.

11.5.9 Tasks 11 and 12—Estimate Reserves. The following development is probably the most important of the entire discussion: For the reservoir of interest, what do you estimate the amount of reserves to be, and how much do you think can be produced if we were to install a waterflood?

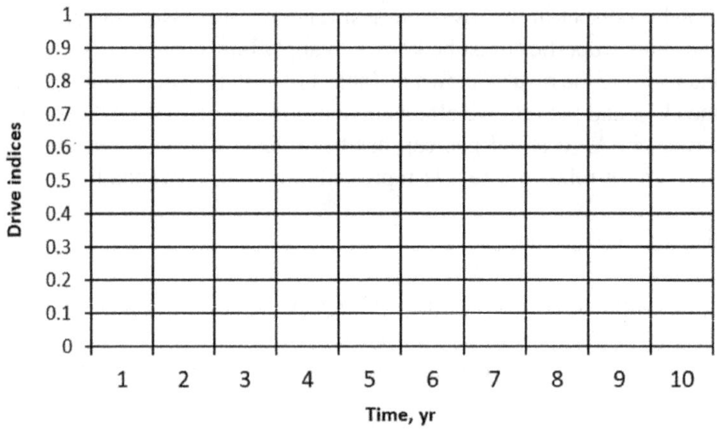

Fig. 11.14—Use the graph to plot and interpret drive index effects.

Initially, apply the change-in-oil-saturation (Eq. 5.1 in the Oil Reservoirs chapter) and the API Subcommittee on Recovery Efficiency (1967) equations (Eqs. 5.2 and 5.3) to develop an understanding of potential reserves boundaries.

Task 11—Recovery as a Function of Residual Oil Saturation. Calculate the change in oil saturation (ΔS_o) between the initial and residual conditions to determine the maximum volume of oil that could be produced on a unit basis. Tables 11.2 and 11.5 provide pertinent information.

$$N_p = (7{,}758 \text{ bbl/acre-ft})(\phi)[\frac{(1-S_{wi})}{B_{oi}} - \frac{S_{or}}{B_o}], \text{ STB/acre-ft.}$$

Task 12—API Subcommittee on Recovery Efficiency (1967) Equations. Apply Eqs. 5.2 and 5.3 from the Oil Reservoirs chapter to calculate the pressure depletion (solution gas drive, RE_{SGD}) and waterflood (waterdrive, RE_{WD}) recovery factors.

$$RE_{SGD} = 41.8\left\{\left[\frac{\phi(1-S_{wi})}{B_{obp}}\right]^{0.1611}\left(\frac{k}{\mu_{obp}}\right)^{0.0979}\left(S_{wi}\right)^{0.3722}\left(\frac{p_{bp}}{p_a}\right)^{0.1741}\right\}, \%$$

$$RE_{SGD} =$$

$$RE_{WD} = 54.9\left\{\left[\frac{\phi(1-S_{wi})}{B_{oi}}\right]^{0.0422}\left(\frac{k\mu_w}{\mu_{oi}}\right)^{0.077}\left(S_{wi}\right)^{-0.19}\left(\frac{p_i}{p_a}\right)^{-0.2159}\right\}, \%$$

$$RE_{WD} =$$

The abandonment pressure for the pressure-depletion case is 2,000 psig, while the abandonment pressure for the waterflood case is 3,000 psig. Both estimates were made from nodal analysis of the flow system. Data for these calculations are shown in Tables 11.2 and 11.5.

11.5.10 Task 13—Estimate Displacement Efficiency. Study the structure map, type logs, and cross section (well correlations) to estimate areal and vertical sweep if we were to install a waterflood project. Remember, the flow system will change with the installation of water injection.

Assume the temporarily abandoned downdip Well 6 is recompleted to become a water injector. Recall that the upper member of the C1 Sand is discontinuous and trending updip. Should we drill the location shown on the structure map (in Fig. 11.3, shown as WIW) to improve areal sweep?

***Vertical Sweep Efficiency* (*E_V*)**—Review the log profiles in Figs. 11.4 and 11.5 and in Table 11.2. Divide the high-permeability layer located in the lower-lobe thickness by total thickness when estimating waterflood reserves.

***Areal Sweep Efficiency* (*E_A*)**—The well locations coupled with the sand quality and reservoir dip seem to indicate a reasonably good horizontal sweep.

Volumetric sweep efficiency (E_D) is the product of the areal and vertical sweep efficiencies:

Displacement efficiency $E_D = E_A \times E_V$.

11.5.11 Task 14—Calculate Remaining Primary and Secondary Recovery Reserves. Estimate the following parameters, which are considered to be key performance indicators:

- Recoverable oil (STB)
- Recovery factor (%)
- Maximum recoverable oil
- Waterdrive index
- Solution-gas-drive index
- Oil production to 8.5 years (STB)
- Remaining primary recovery (decline-curve study) (STB)
- Total oil primary production (STB)
- Remaining movable oil (STB)
- Displacement efficiency (%)
- Waterflood reserves—assuming same well outlets

11.5.12 Task 15—Apply the Stiles Calculations To Forecast Performance. The Stiles method (see Chapter 8) assumes that the producing interval can be divided into a series of discrete, noncommunicating layers. The only variables in this idealized method are permeability and thickness.

The calculation procedure sums water and oil production as a sequential breakthrough of the flood front as a function of layer permeability.

The following equations are involved in the calculation procedure:

- $C_j = (h_i)(k_i) + \sum hk_{i-1}$ (Eq. 8.43), where C defines the capacity of the water flowing as a function of successive layers flooding out as a series expansion, C_j is the water capacity of the flooded-out layer (the series expands sequentially from the most- to the least-permeable layer), and C_t is the water capacity of the last and final flooded-out layer.

Recovery is the amount of oil remaining:

- $R = (1 - C)$ (Eq. 8.44).

Oil recovery from all beds is the sum of the flooded-out layers plus the fractional portions of the layers still undergoing oil displacement.

- R = oil remaining to be produced; $R = \dfrac{h_j k_j + (C_t - C_j)}{h_t k}$ (Eq. 8.45).

Remember, production measurements are at surface conditions while we are studying what is occurring in the reservoir:

- Mobility measured at reservoir conditions: $M = \dfrac{k_{rw}\mu_o}{k_{ro}\ \mu_w}$ (Eq. 8.46).
- Mobility measured at surface conditions: $M' = \dfrac{k_{rw}\mu_o}{k_{ro}\ \mu_w}\dfrac{B_o}{B_w}$ (Eq. 8.47).
- Fraction of water flowing in a reservoir: $f_w = \dfrac{q_w}{q_w + q_o}$ (Eq. 8.4).
- Fraction of water flowing at surface conditions: $f_{w(\text{surface})} = \dfrac{M'C_j}{M'C_j + \left(C_t - C_j\right)}$.
- Cumulative production: $N_p = f_{w(\text{surface})} \times \left(\text{reserves}\right)$ (Eq. 8.49).
- Oil production: $q_o = \dfrac{(\text{injection rate})\left[1 - f_{w(\text{res})}\right]}{B_o}$ (Eq. 8.50).
- Fraction of water flowing in the reservoir: $f_{w(\text{res})} = \dfrac{MC_j}{MC_j + \left(C_t - C_j\right)}$.
- $WOR = \left(\dfrac{f_w}{1 - f_w}\right)\left(\dfrac{B_o}{B_w}\right)$ (Eq. 8.42).

The Stiles waterflood problem was divided into an upper low-permeability layer and a lower high-permeability layer, with 1-ft intervals. Refer to the type log shown in Fig. 11.4.

- Upper layer weighted-average permeability: (4 ft × 31 md + 6 ft × 75 md) ÷10 ft = 57 md.
- Lower layer weighted-average permeability: (7 ft × 92 md + 6 ft × 300 + 4 ft × 220 md) ÷ 17 ft = 196 md.

A spreadsheet was set up, and the results of the required calculations are shown in Table 11.10. **Fig. 11.15** shows the amended results of the effort.

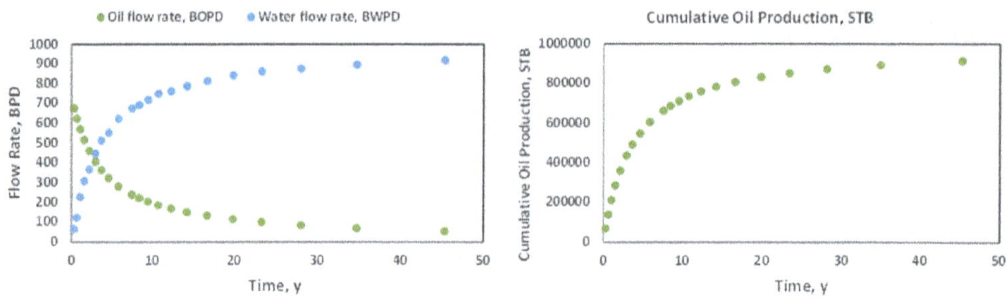

Fig. 11.15—Results of the use of the Stiles method to predict future performance for the C1 Sand, AA Reservoir. The left side shows oil and water flow rates vs. time; observe the steady increase in the water flow rate in the first 5 years. On the right, cumulative oil production is shown as a function of time, showing that approximately 80% of reserves are expected to be produced in the first 10 years.

11.5.13 Task 16—Evaluate Findings and Determine Waterflood Project Feasibility.

- Have your efforts up until now indicated there could be an economic waterflood potential for this fault block?
- What necessary part of the analysis has not been discussed?

11.5.14 Task 17—Calculate Water-Injection Requirements. Apply the material-balance equation (see Chapter 5) to calculate the volume of injected water required to offset the near-peak reservoir production rate. Assume the water-injection rate must, at a minimum, offset the reservoir depletion rate.

Assume the fluid properties listed in Tables 11.5 and 11.7 are for the Year 8 pressure level. The reservoir has produced N_p = 830,680 BO, and Rs = solution GOR = 350 scf/STB near peak production.

Use the concept of material balance and Eq. 5.4 in the Oil Reservoirs chapter to calculate the water-injection requirements on a daily basis. Total voidage replacement refers to the replacement of the volumes of oil, gas, and water produced from the reservoir with the injected fluids, which can be calculated on an instantaneous basis. The voidage replacement ratio (VRR) is the ratio of reservoir barrels of injected fluid to reservoir barrels of produced fluid. Mathematically (for water injection), VRR can be expressed as

VRR = injected reservoir volumes/underground withdrawal,

Underground withdrawal = oil produced + free gas produced + water produced,

where the oil produced is q_oB_o, the free gas produced is $\left(q_g - q_oR_s\right)B_g = q_o\left(GOR - R_s\right)B_g$, and the water produced is q_wB_w. Thus,

VRR = water injected/[oil produced (q_oB_o) + free gas produced $\left(q_g - q_oR_s\right)B_g$
+ water produced q_wB_w)].

How much water should be injected? Just replacing the produced volume will not measurably move unresponsive oil toward a producing well.

11.5.15 Task 18—Estimate the Production Response to Water Injection. Review production records and apply the Klins and Clark method discussed in Chapter 6, solving the inflow performance equation listed in Table 11.11 to determine the effect of reservoir pressure on flow rate. Determine the effect of reservoir pressure on the flow rate.

Assume water injection increases reservoir pressure from the current $\bar{p}$ = 2,380 psia to $\bar{p}$ = 3,000 psia, and $p_{wf(\text{current})}$ = 2,000 psia.

Note that the equation has already been normalized to bubblepoint conditions at $\bar{p}$ = 5,500 psia. Therefore, the first task is to change the equation to reflect $\bar{p}$ = 3,000 psia and calculate the rate when p_{wf} = 2,000 psia.

The second task adjusts the equation to reflect $\bar{p}$ = 2,380-psia reservoir pressure to calculate the rate when p_{wf} = 2,000 psia. Compare the two rate values to see how overinjection could increase the production rate.

11.5.16 Task 19—Summarize the Status of the Wells. Review your work in **Table 11.12,** where you were asked to summarize information regarding the wells located in the proposed waterflood project.

Recommend possible workovers and the locations of new drills. Refer to the map shown in Fig. 11.3, in which the only new well is an additional water-injection well drilled in the southwest corner of the reservoir. (The location is indicated by a blue triangle and marked as WIW.)

11.5.17 Task 20—Calculate Expected Hydrostatic Pressures. The actual bottomhole fracture initiation pressure is calculated using Eq. 10.1 in the Waterflood Components chapter when data from a step-rate pressure test are available.

Pore pressure (p_p) is the pressure of the fluid contained in the pore spaces of sediments or other rocks. For example, pore pressure increases with true vertical depth: At 10,000 ft, the typical p_p = 4,650 psia, which gives a gradient of 0.465 psi/ft.

In a "normal" situation, the pore pressure will be equal to the hydrostatic pressure, and the pressure gradient for this is the normal gradient of water density (1 g/cm^3 = 0.465 psi/ft), while seawater is typically 1.02 g/cm^3, or 0.442 psi/ft, because the density of the formation water varies with the salinity.

The overburden gradient is the total weight of overlying rocks and water, which apply stress to the rock layer at a vertical depth of interest:

Overburden = effective vertical stress + pore pressure,

where overburden is S_v and the effective vertical stress is σ_v. Thus,

$$S_v = \sigma_v + p_p.$$

The effective vertical stress at any particular depth is determined as

Effective stress = overburden gradient – pore pressure (from normal compaction at that depth).

For example, with an overburden of 9,000 psia, if there is a pore pressure of 5,000 psia, the rock matrix will undergo only 4,000 psia of stress.

For this study, assume the overburden gradient = 1.0 psi/ft, and recall that current reservoir pressure = 2,380 psi. γ_{sw} = 0.465 psi/ft.

Therefore, to estimate fracture pressure at current reservoir pressure and injection mid-interval depth D (ft), Eaton's formula can be used:

$$p_{\text{frac}} = \{NOB[v/(1-v)] + p_{PV}\} \times D, \quad (11.1)$$

where p_{frac} is the fracture pressure (psi), NOB is the net overburden pressure gradient (psi/ft), p_{PV} is the pore pressure gradient (psi/ft), v is Poisson's ratio, and D is the injection mid-interval depth (ft).

$$NOB = \text{overburden gradient} - \text{pore pressure gradient}. \quad (11.2)$$

$$\text{Poisson's ratio} = v = -\varepsilon_p / \varepsilon_\alpha. \quad (11.3)$$

Poisson's ratio is a measure of the lateral expansion to the longitudinal contraction. Elastic properties can be measured using sonic and density logs or by testing the core in the laboratory.

Typical values of Poisson's ratio are 0.27 for limestone and 0.33 for sandstone. (Our case study is a sandstone.)

To calculate the pressures to be considered in the pumping surface facilities, use the following assumed values:

- Overburden gradient = 1.0 psi/ft.
- Depth = 10,800 ft.
- Reservoir pressure = 2,380 psia.
- $\gamma_{sw} = 0.465$ psi/ft.
- $p_{PV} = 0.22037$ psi/ft = pressure/depth.
- $NOB = 0.77963$ psi/ft.
- Fracture pressure gradient = 0.604 psi/ft.

11.5.18 Task 21—Calculate Pressure Rating for Pump Pressure Gradient. This value equals the initial injection pressure, which was calculated in Task 20.

11.5.19 Task 22—Calculate the Expected Pressure Response Resulting from an Expansion of the Flood Front. Apply and rearrange a form of the Darcy equation shown as Eq. 4.2 (see the Well and Reservoir Models chapter) to determine whether pressure increases out to an estimated drainage radius of 2,000 ft.

The Darcy's law form using rock and fluid properties for water injection, assuming a water-injection rate i_w of 1,000 STBW/D, a drainage radius r_e of 2,000 ft, a wellbore radius r_w of 0.25 ft, and water viscosity μ_w of 1 cp:

$$q = i_w,\ \mu = \mu_w,\ B = B_w, \text{ and } k = k_w = \bar{k}.$$

Table 11.14 shows the average permeability calculations ($\bar{k}$) for the proposed injection interval.

$$p_{wf} = \frac{i_w \mu_w \ln\left(\frac{r_e}{r_w}\right)}{0.00708 h k_w} + p_e = \frac{i_w \mu_w \ln\left(\frac{r_e}{r_w}\right)}{0.00708 h \bar{k}} + p_e.$$

Production and facilities engineers possibly should be brought into the discussion on evaluating and ordering the proper pump rating: How would the buildup of formation damage affect the injection rate?

11.5.20 Task 23—Determine the Effects of Formation Damage on the Water-Injection Rate. Assume formation damage extends 3 ft. Apply the Craft and Hawkins (1959) equation (Eq. 6.17 in the Inflow Performance chapter) or the Hawkins (1956) equation (Eq. 4.8 in the Well and Reservoir Models chapter) for formation damage to calculate skin *s* as a function of permeability loss:

1	2	3	4	5	6
Zone	Depth (ft)	*h* (ft)	*k* (md)	*h*×*k* (Column 3 × Column 4)	$\bar{k}$ (md)
C	10,082–89	7	92	644	
D	10,089–95	6	300	1,800	
E	10,095–99	4	220	880	
Σ		17		3,324	195.5

Table 11.14—Rock permeability properties for estimating the average permeability at the water-injection interval.

$$s = \left(\frac{k - k_s}{k_s}\right) \ln \frac{r_s}{r_w},$$

where $k = k_f = k_{res}$ is the reservoir permeability in the non-altered zone (virgin zone); $k_s = k_d$ is the reservoir permeability in the altered zone (damaged zone); and r_s is the radius of the altered zone (damaged zone).

Example Calculation. Set up a spreadsheet and, assuming the damaged zone extends 3 ft, estimate the bottomhole injection pressure by sequentially assuming a permeability reduction from the original average permeability. The first effort applies the Darcy equation for series flow (Eq. 2.9 in Chapter 2, Reservoir Rock Properties; Eq. 4.13 in Chapter 4, Well and Reservoir Models):

$$\bar{k} = \frac{\ln(r_e / r_w)}{\frac{\ln(r_d / r_w)}{k_d} + \frac{\ln(r_e / r_d)}{k_{res}}}.$$

Again, use the rearranged Darcy equation from Task 22 to calculate the required BHP to maintain the minimum water-injection rate (Task 17):

$$p_{wf} = \frac{i_w \mu_w \ln\left(\frac{r_e}{r_w}\right)}{0.00708 h\bar{k}} + p_e.$$

Use **Fig. 11.16** to relate the effect of formation damage or skin on injection pressure.

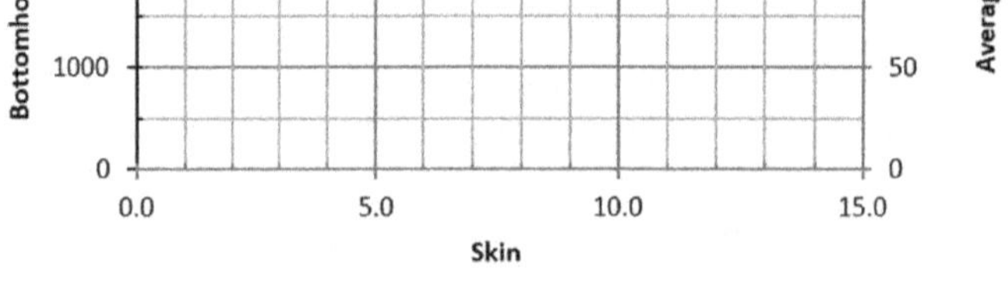

Fig. 11.16—Use the graph to plot the effect of a 3-ft skin buildup on injection pressure and average permeability.

11.5.21 Additional Considerations. The field development plan under a waterflood project should also consider additional aspects that will have an important influence when performing the calculations of the estimated-economics key performance indicators. Some of the following uncertainties should be analyzed:

- Are additional wells required if a waterflood project were installed? If so, where?
- Are well recompletions needed? Which ones? Is any special completion equipment required?
- What should the recommended water-injection rate be? How many wells should be used?
- The project will use produced field water. What type of cleaning equipment should be included in the project equipment?
- What will be the life of the project?

11.6 Workshop Summary

In this chapter, the reader performs an actual analysis workshop to apply most of the principles outlined in the first two parts of the book to an actual field situation. (The results of the analysis are provided in Chapter 12.) The tasks presented here illustrate the importance of applying basic analysis principles and field data when evaluating field potential.

Structure maps, cross sections for well correlations, and production pressure data are commonly available for interpretation. The reader is led through a series of tasks covering subsurface-to-surface subjects to evaluate the waterflood potential and equipment needs. Different principles to study the same characteristics (e.g., OOIP) are applied so that the quality of the final conclusions can be relied on. The projection of physical requirements and the prediction of future performance extend the formulation to the cash flow and economic analysis studies boundary, which are not being considered in this book.

A set of sequential tasks leads the reader to the ultimate goal, which is the development of a complete picture of historical and predicted reservoir behavior. Design requirements for the required injection and production facilities are addressed to complete the study.

Completing the study process acquaints the reader with the thinking process involved when using commonly available information to develop an understanding of reservoir potential. Emphasis is placed on comparing conclusions derived from different investigative approaches, and a similarity of values lends credibility to the final conclusions.

11.7 Nomenclature

A	=	area, acres
B	=	formation volume factor, RB/STB
B_g	=	gas formation volume factor, RB/scf
B_o	=	oil formation volume factor, RB/STB
B_{obp}	=	oil formation volume factor at bubblepoint conditions, RB/STB
B_{oi}	=	initial oil formation volume factor, RB/STB
B_w	=	water formation volume factor, RB/STB
C	=	water capacity, md-ft
C_j	=	water capacity of flooded-out layer, md-ft
C_l	=	water capacity of last and final flooded-out layer, md-ft
D	=	Arps' continuous decline rate, %/yr
D	=	injection mid-interval depth (Eaton's formula), ft
E_A	=	areal sweep efficiency, fraction or percent
E_D	=	volumetric sweep efficiency or displacement efficiency, fraction or percent
EL_{res}	=	reservoir economic limit, BOPD
EL_{well}	=	well economic limit, BOPD
E_V	=	vertical sweep efficiency, fraction or percent
f_w	=	fraction of water flowing in system
$f_{w(res)}$	=	fraction of water flowing in reservoir
$f_{w(surface)}$	=	fraction of water flowing measured at surface conditions
h	=	height or thickness, ft
h_i	=	height or thickness of Layer *i*, ft
h_j	=	height or thickness of Layer *j*, ft
h_n	=	height or thickness of Layer *n*, ft
h_t	=	total height or thickness, ft
i_w	=	water-injection rate, STBW/D
J'	=	productivity index intercept, BOPD/psi
k	=	permeability, md (or darcies when using API correlations)
$\overline{k}$	=	average permeability, md
k_d or k_s	=	permeability of damaged region, md
k_f or k_{res}	=	reservoir permeability in non-altered zone (virgin zone), md
k_i	=	permeability of Layer *i*, md
k_{i-1}	=	permeability of Layer *i*–1, md
k_j	=	permeability of Layer *j*, md
k_n	=	permeability of Layer *n*, md
k_{res} or k_f	=	reservoir permeability in non-altered zone (virgin zone), md
k_{ro}	=	relative permeability to oil, md
k_{rw}	=	relative permeability to water, md
k_s or k_d	=	permeability of damaged region, md
k_w	=	effective permeability to water, md
M	=	mobility ratio measured at reservoir conditions
M'	=	mobility ratio measured at surface conditions
n	=	turbulence, also defined in the Fetkovich method as an exponent term, dimensionless
N	=	original oil-in-place, STBO
N_a	=	apparent oil-in-place, bbl
N_p	=	production rate or cumulative oil production, bbl, BO, STB, or STB/acre-ft
$N_{p(current)}$	=	current production rate, bbl
$N_{p(primary)}$	=	initial production rate, bbl
$\overline{p}$	=	average pressure, psia
p_a	=	abandonment pressure (API correlations), psig
$\overline{p}_a$	=	average pressure in aquifer, psia

$p_{a(n-1)}$ = aquifer static pressure measured at a previous timestep $n-1$, psia
p_{bp} = bubblepoint pressure, psia or psig
p_e = pressure at drainage radius, psia or psig
p_{frac} = fracture pressure, psi
p_i = initial pressure, psia or psig
p_{last} = final pressure, psia
p_o = pressure at oil/water contact, psia
p_p = pore pressure, psia
p_{PV} = pore pressure gradient, psi/ft
p_{res} = reservoir pressure, psia
p_{sc} = pressure at standard conditions (sometimes called at base conditions) (in this book, p_{sc} = 14.65 psia), psia
p_{wf} = bottomhole flowing pressure, psia
$p_{wf(current)}$ = current bottomhole flowing pressure, psia
q = flow rate, STB/D, BOPD, or B/D
q_g = gas flow rate, Mscf/D
q_i = initial flow rate, BOPD
q_{last} = final flow rate, BOPD
q_o = oil flow rate, BOPD or STB/D
q_{oi} = initial oil flow rate, STB/D
q_w = water flow rate, BWPD
q_1 and q_2 = flow rate at a certain time, RB/D or STB/D
R = fractional recovery of the total movable oil (Stiles), fraction
r_d = radius of damaged zone, ft
r_e = drainage radius, ft
RE_{SGD} = solution-gas-drive reservoir recovery, %
RE_{WD} = waterdrive reservoir recovery, %
r_s = radius of altered zone (damaged zone)
R_s = solution gas/oil ratio, scf/STB
R_{si} = initial solution gas/oil ratio, scf/STB
r_w = wellbore radius, ft
s = skin, dimensionless
S_{oi} = initial oil saturation, fraction or %
S_{or} = residual oil saturation, fraction or %
S_{rg} = residual gas saturation, fraction or %
S_v = overburden, psia
S_{wi} = initial water saturation, fraction or %
$\overline{S}_{wi}$ = average initial water saturation, fraction or %
t = time, days, months, or years
T_{res} = reservoir temperature, °**F**
T_{sc} = temperature at standard conditions (sometimes called at base conditions) (in this book, T_{sc} = 60°F or 520°R), °F or °R
V_p = pore volume, RB
W_e = cumulative water influx, bbl or BWPD
z_{res} = compressibility factor (also called the gas deviation factor, supercompressibility, or z-factor at reservoir conditions)
γ_{ob} = gradient of oil density, psi/ft
γ_{sw} = gradient of water density, psi/ft
Δp = change in pressure, psi
ΔS_o = change in oil saturation, fraction or percent
Δt = change in time, years
ε_α = longitudinal contraction (Poisson's ratio)
ε_p = lateral expansion (Poisson's ratio)
μ = viscosity, cp
μ_o = oil viscosity, cp
μ_{obp} = oil bubblepoint viscosity, cp
μ_{oi} = initial oil viscosity, cp
μ_w = water viscosity, cp
ν = Poisson's ratio
σ_v = effective vertical stress, psia
ϕ = porosity, fraction or %
$\overline{\phi}$ = caverage porosity, fraction or %
ϕ_n = porosity of Layer n, fraction or %

Chapter 12

Results of Analysis

12.1 Tasks 1 and 2—Study Maps, Log Profiles, and C1 Sand Continuity

The conclusions from this study are as follows:

- The C1 producing interval is comprised of two sand members separated by a thin shale.
- The bottommost sand member (highest permeability) appears continuous throughout the AA Reservoir (see Fig. 11.5 in Chapter 11, Field Study—Analysis Workshop), while the upper (much-lower permeability) sand member transitions to a shale trending-up structure.
- Producing wells are optimally located, except those in the southwest portion of the AA Reservoir.
- Production records indicated sand production and collapsed casing were problems in some wells.
- The well production rate was greater than 600 STB/D per well at initial conditions.

12.2 Tasks 3 and 4—Study Production Records, Well History Profiles, and General Reservoir and Field Information

The conclusions from this study are as follows:

- There was a significant reservoir pressure loss caused by production.
- The most updip well (Well 1) rapidly went to a high gas/oil ratio (HGOR), indicating the formation of a secondary gas cap; after that measurement, Well 1 was shut in.
- From the information provided in Table 11.5 in Chapter 11, it can be seen that reservoir pressure declined significantly:

p_i = 5,750 psia, p_{last} = 2,300 psia in Year 8.2 as measured in Well 3.

12.3 Task 5—Analyze Bottomhole Pressures

The conclusions from this study are as follows:

- Pressure and production records reflect a pressure-depletion history. Gravity segregation appears prevalent between the updip and downdip wells.
- You were asked in Chapter 11 to use Fig. 11.9 to plot the static pressure behavior (see Table 11.6) from well-test measurements as a function of time. The result is given in **Fig. 12.1,** which indicates there is a lack of connectivity between Well 8 and all other wells in the field (1, 2, 3, and 4).
- Remember, wells located in a common reservoir should display similar pressures.

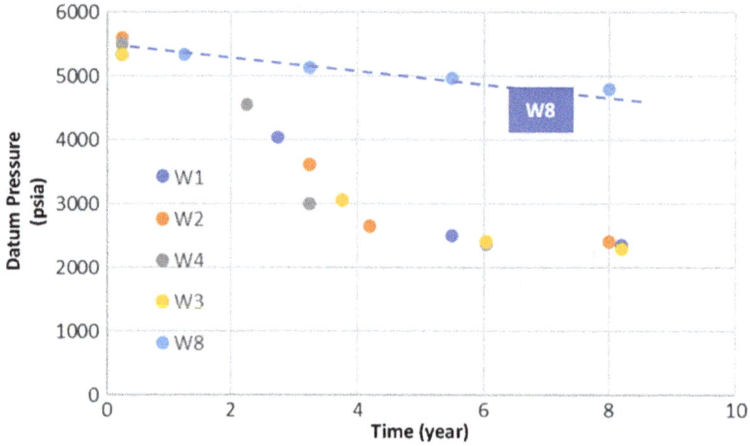

Fig. 12.1—Plot of historical datum pressures, interpreted bottomhole pressure vs. time.

- As can be observed from Fig. 12.1, not all wells included in the original interpretation are hydraulically connected.
- It can be seen from the pressure histories and Fig. 12.1 that there is an outlier (Well 8).
- The average pressure line drawn through the data indicates Well 8 is not located in the AA Reservoir.

Decision point interpretation: The pressure analysis allows for the interpretation and assists in the definition of reservoir boundaries delineation, indicating that the minor fault must be sealing.

12.4 Task 6—Calculate Reservoir Volume, Original Oil-in-Place, and Original Gas-in-Place

The data used to calculate the original oil-in-place (OOIP) and the original gas-in-place (OGIP) is taken from Tables 11.2 and 11.5 in Chapter 11 and is summarized in **Table 12.1,** where rock properties were derived from log analysis, except for permeability ($\overline{k}$), which was developed from core analysis studies.

Fig. 11.6 in Chapter 11 provides the following data:

- h (isopach intervals) = 10 ft.
- Reservoir area (0 ft) = 376 acres.

1	2	3	4	5	6	7	8	9	10
Zone	Depth (ft)	h (ft)	ϕ (%)	$h \times \phi$ (Column 3 × Column 4)	ϕ (%)	S_{wi} (%)	$h \times S_{wi}$ (Column 3 × Column 7) (%)	$\overline{S}_{wi}$ (%)	$\overline{k}$ (md)
A	X066–70	4	23	92		45	180		31
B	X,072–78	6	27	162		39	234		75
C	X,082–89	7	29	203		26	182		92
D	X,089–95	6	33	198		27	162		300
E	X,095–99	4	29	116		36	144		220
Σ		27		771	28.56			33.4	

Table 12.1—Rock properties for estimating reservoir volume, OOIP, and OGIP.

- Reservoir area (10 ft) = 349 acres.
- Reservoir area (20 ft) = 285 acres.

Using these data and the data provided in Tables 11.2 and 11.5 in Chapter 11, the following can be calculated:

- Reservoir volume:

$$V=\frac{h}{2}(A_0+2A_1+2A_2+2A_{n-1}\cdots A_n)=\frac{10}{2}[376+2(349)+285]=6{,}795\text{ acre-ft.}$$

- OOIP:

$$OOIP=\frac{7{,}758\text{ bbl / acre}-\text{ft}\left(0.2856\right)\left(1-0.334\right)}{1.44\text{ RB / STB}}=1{,}023\text{ STB/acre-ft.}$$

$$\text{Total } OOIP=1{,}023\frac{\text{STB}}{\text{acre}-\text{ft}}\left(6{,}795\text{ acre}-\text{ft}\right)=6{,}951{,}285\text{ STB.}$$

- OGIP:

$$OGIP = N \times R_{si} = 6{,}951{,}285\text{ STB} \times 600\text{ scf/STB} = 4.171\text{ billion scf.}$$

The upper low-permeability zone contains more than half of the OOIP.

12.5 Task 7—Study Performance Records and Decline Curves To Predict Remaining Primary Recovery

Predict primary recovery by studying the performance decline curve found in Fig. 11.11 in Chapter 11.

Fig. 12.2 shows the reservoir performance exponential decline behavior. The extrapolation of the straight line provides an estimate of the remaining primary reserves, indicating that approximately 55,500 bbl of oil (BO) will be recovered under the actual production conditions.

Extending the straight line in Fig. 12.2 from the last measured rate to the economic limit (which is equal to 5 BOPD) determines the remaining production under the actual production strategy, which assumes the reservoir will not have any well interventions or workovers.

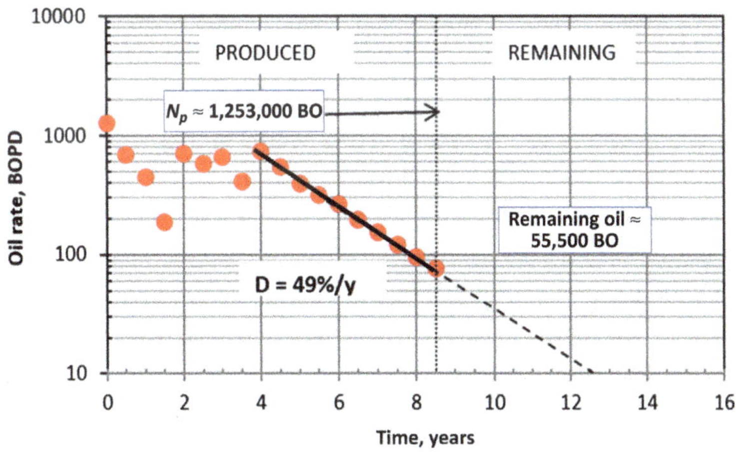

Fig. 12.2—Reservoir performance exponential decline. The estimate of remaining primary reserves is approximately 55,500 BO.

- $D = \frac{In\frac{q_1}{q_2}}{t}$. From the straight line in Fig. 12.2 (traced through the red dots, q_1 = 700 BOPD in Year 4 and q_2 = 100 BOPD in Year 8),

  D = [ln (700/100)]/4 = 48.65, or approximately 49%/yr (0.49 fraction/yr).

- Cumulative oil production = 1,253,008 STB.
- $N_p = \frac{q_1 - q_2}{D}$. From the extrapolated straight line in Fig. 12.2 (using the dashed line, $q_1 \approx 79$ BOPD, which differs by only 3 BOPD from the reported value of 76 BOPD in Tables 11.1 and 11.5 in the Field Study—Analysis Workshop chapter, and q_2 = 5 BOPD at the economic limit),

  N_p = (79 BOPD – 5 BOPD) × 365 D/yr 0.4865/yr = 55,519 STB.

  N_p = (76 BOPD – 5 BOPD) × 365 D/yr 0.4865/yr = 53,268 STB.

These estimated calculations show a small difference of approximately 4% when using values read from the plotted points in Fig. 12.2.

- Expected primary reservoir production:

  Expected primary reservoir production = 1,253,008 + 55,519 = 1,308,527 STB.

  Expected primary reservoir production = 1,253,008 + 53,268 = 1,306,276 STB.

These calculations show a small difference of approximately 0.17% in the expected recovery when using values read from the plotted points in Fig. 12.2. The calculated expected primary reservoir production differs from the value given in Table 11.4 in Chapter 11 (total estimated primary production = 1,310,597) by only 0.15%.

12.6 Task 8—Determine Whether a Secondary Gas Cap Has Formed

A secondary gas cap has most likely formed: Updip Well 1 did gas out early in reservoir life.

12.7 Task 9—Apply the Fetkovich Method To Calculate Water Influx

Reservoir performance seems to have been affected by water influx from a very small aquifer. Water influx exerted a profound influence on performance early in reservoir life but tapered off after exhausting its water-expansive capabilities.

Table 11.8 in Chapter 11 presents the results of applying the Fetkovich model to the study of the water-influx history. The Fetkovich model works best for closed, finite aquifers; the productivity index approximates aquifer flow; and a material balance calculates water outflow from the aquifer.

You were asked in Chapter 11 to use Fig. 11.10 to plot the data provided in Table 11.8, which helps to analyze the water-influx effect on reservoir performance. **Fig. 12.3** shows the result and indicates that water influx was significant for the first 3 years. The effect tapered off and has been dormant since Year 6.

12.8 Task 10—Apply the Material-Balance Equation Technique

The material-balance equation (see Chapter 5) allows one to estimate OOIP, as well as water-influx and drive-indices histories. You were asked in Chapter 11 to use Fig. 11.12 to plot your material-balance OOIP estimates using the calculations in Table 11.9. **Fig. 12.4** shows the result.

Fig. 12.4 seems to indicate two depletion histories, which transition between Year 4 and Year 5, in the AA Reservoir. The transition from waterdrive to depletion drive mirrors the results of the previous Fetkovich calculations (Task 9).

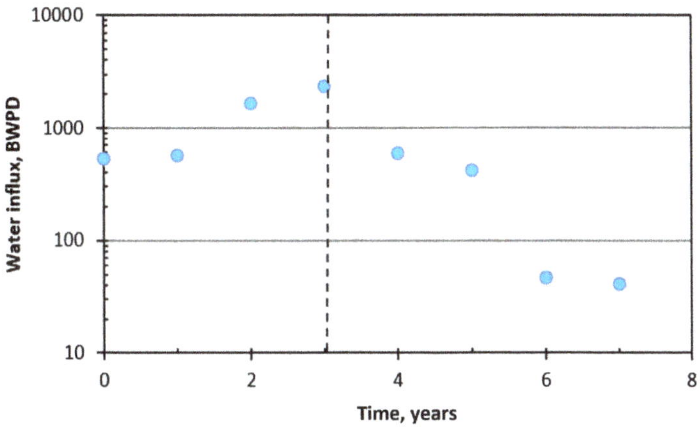

Fig. 12.3—Fetkovich model of water encroachment history. Note the reservoir behavior—does the model seem to fit reservoir conditions?

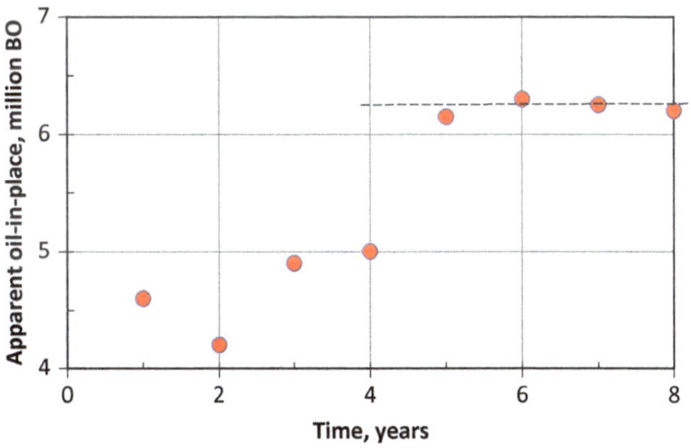

Fig. 12.4—The material-balance study on the C1 Sand.

A best-fit line during the Year 5 to Year 8 period seems to indicate OOIP = 6.2 million STB. This observation compares with the volumetric calculations performed in Task 6 (OOIP = 6,951,285 STB), with a difference of approximately 10%.

A rearranged form of the material-balance equation (Eq. 5.12a in Chapter 5, Oil Reservoirs) calculates water influx as a function of time. The results of this study are shown in Table 11.9 in Chapter 11, and you were asked to use Fig. 11.13 to plot the data. **Fig. 12.5** shows the resulting interpretation.

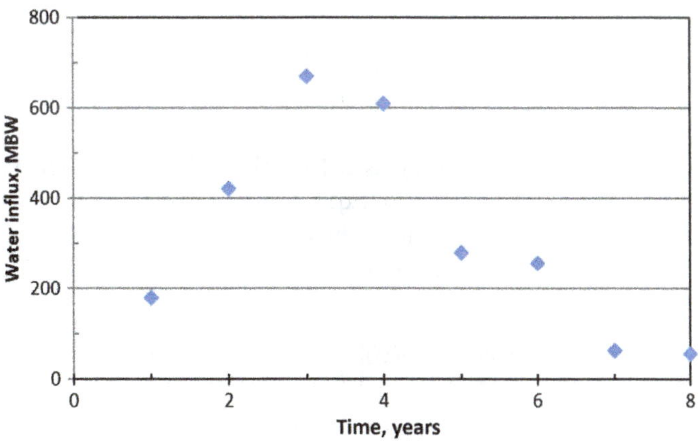

Fig. 12.5—Material-balance water-influx calculations. The graph reinforces the interpretation of the Fetkovich water-influx study, as well as the interpretation of the OOIP.

Water influx seems to have occurred in two phases: early in the producing life followed by much less late in life. These results roughly concur with the

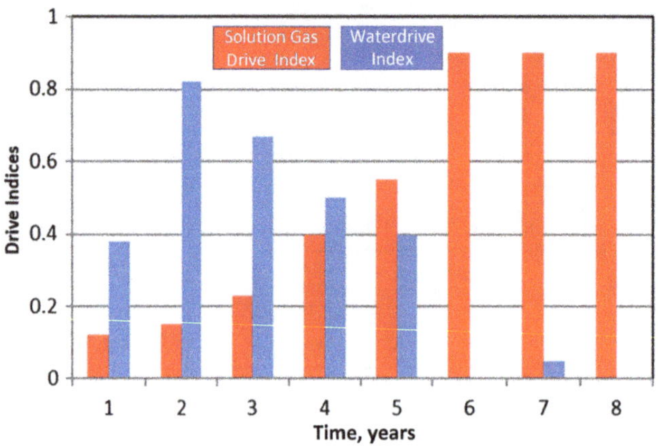

Fig. 12.6—Drive-index history.

previous Fetkovich study (Fig. 12.3) and show the early-time replacement of reservoir production by encroaching water.

You were asked in Chapter 11 to use Fig. 11.14 to plot the calculated drive indices. The resulting **Fig. 12.6** shows that less-efficient solution-gas drive increases as the influence of water influx declines, as indicated by the waterdrive index. Recall that an original gas cap was assumed to be absent because the initial reservoir pressure was greater than the bubblepoint pressure, as indicated in Table 11.5.

Drive-index calculations indicate a two-stage depletion history, with an early water encroachment transitioning to pressure depletion.

Fig. 12.6 shows that solution gas is the predominant drive mechanism. All the analysis of reservoir performance seems to indicate the effect of a rather small aquifer on the producing history. Note that the sum of the solution-gas drive and the water influx does not equal unity. This fact indicates there is some uncertainty in the calculations because of the data used, such as production data allocation, pressure data measurements, fluid property estimates, and rock property estimates from logs and cores.

Decision point interpretation: The volumetric and material-balance estimates differ by approximately 10%. The zero-dimensional material-balance and Fetkovich dynamic models account for rock and fluid properties and pressure and production data measured during the production history. To reduce the uncertainty, more-realistic estimates of OOIP and OGIP based on the areal and volumetric calculations should be considered, by reviewing the input values from logs and core data used in the maps.

12.9 Tasks 11 and 12—Estimate Reserves

The volumetric value of OOIP ≈ 6.95 million STB tests the measured extent of the oil saturation while the material-balance results, OOIP = 6.2 million STB, represent the interaction of all connected reservoir fluids. A recommended practice is to assume an OOIP average value for the C1 Sand, AA Reservoir = 6.6 million STB and possibly go back and review the volumetric calculations.

In this section, the ultimate recovery is estimated using different methods to arrive at a "most-probable" value. Changes in residual oil saturation, the American Petroleum Institute (API) correlation equations for pressure depletion, and the waterdrive reserves are considered in these decisions.

12.9.1 Task 11—Recovery as a Function of Residual Oil Saturation. Apply the concept of a change in oil saturation (ΔS_o) (Eq. 5.1 in the Oil Reservoirs chapter) to calculate the maximum volume of oil that can be produced by means of waterflood displacement. (Tables 11.2, 11.5, and 11.7 in Chapter 11 list pertinent information.)

$$N_p = (7{,}758\ \text{bbl/acre-ft})(\phi)\left(\frac{S_{oi}}{B_{oi}} - \frac{S_{or}}{B_o}\right) = (7{,}758\ \text{bbl/acre-ft})(0.2856)\left(\frac{0.666}{1.44} - \frac{0.25}{1.373}\right)$$

$$= 621\ \text{STB/acre-ft}.$$

The maximum oil that could be recovered on a reservoirwide basis would be

(621 bbl/acre-ft)(6,795 acre-ft) = 4,219,695 STB, or approximately 4.2 million STB.

As calculated in Task 6, $OOIP$ = 1,023 STB/acre-ft.

Recovery factor $= \dfrac{N_p}{OOIP} = \dfrac{621}{1{,}023} = 0.61$, or 61% when 100% sweep efficiency is assumed.

This value should be treated as the uppermost reserves limit. If it is accurate, one should review the geology.

12.9.2 Task 12—API Subcommittee on Recovery Efficiency (1967) Equations. Apply Eqs. 5.2 and 5.3 from Chapter 5 to calculate the pressure depletion and waterflood recovery factor.

Fluid and rock properties are provided in Tables 11.2, 11.5, and 11.7 in Chapter 11.

$$RE_{SGD} = 41.8\left\{\left[\frac{\phi(1-S_{wi})}{B_{obp}}\right]^{0.1611}\left(\frac{k}{\mu_{obp}}\right)^{0.0979}\left(S_{wi}\right)^{0.3722}\left(\frac{p_{bp}}{p_a}\right)^{0.1741}\right\} = 41.8$$

$$\left\{\left[\frac{0.2856(1-0.334)}{1.45}\right]^{0.1611}\left(\frac{0.144}{2}\right)^{0.0979}\left(0.334\right)^{0.3722}\left(\frac{5{,}500}{2{,}000}\right)^{0.1741}\right\} = 18.5\%$$

= 1,023 STB/acre-ft × 0.185 = 189 STB/acre-ft.

$$RE_{WD} = 54.9\left\{\left[\frac{\phi(1-S_{wi})}{B_{oi}}\right]^{0.0422}\left(\frac{k\mu_w}{\mu_{oi}}\right)^{0.077}\left(S_{wi}\right)^{-0.19}\left(\frac{p_i}{p_a}\right)^{-0.2159}\right\} = 54.9$$

$$\left\{\left[\frac{0.2856(1-0.334)}{1.44}\right]^{0.0422}\left(\frac{0.144\times 1}{2}\right)^{0.077}\left(0.334\right)^{-0.19}\left(\frac{5{,}750}{3{,}000}\right)^{-0.2159}\right\} = 44.1\%$$

= 1,023 STB/acre-ft × 0.441= 451 STB/acre-ft.

Based on the values calculated in this task, it can be observed that estimated waterflood recovery factors could be between 2 to 3 times the estimated primary recovery. The estimated values can vary significantly (up to 1.5 times, between 40 to 60%), depending on the method used and the assumptions regarding the estimated input values of pressure data and rock and fluid properties.

12.10 Task 13—Estimate Displacement Efficiency

Assume only the high-permeability members of the C1 Sand (D and E intervals) are the most productive in the proposed waterflood; therefore, $h_{(D+E)}$ = 10 ft and h_t = 27 ft.

Vertical sweep efficiency: E_V = 10/27 = 37%.

Review the proposed map for the injection/production arrangement if a waterflood were installed. The estimated areal sweep efficiency (E_A) is approximately 80%.

Therefore, the displacement efficiency is

$E_D = E_A \times E_V$ = 80% × 37% = 30%.

12.11 Task 14—Calculate Remaining Primary and Secondary Recovery Reserves

12.11.1 Recovery.

Decline-Curve Analysis—Primary.

- Production to Year 8.5 = 1,253,008 STB.
- Remaining production = 55,519 STB.
- Total primary = 1,253,008 + 55,519 = 1,308,527 STB. This value is obtained by using the best fit of the exponential decline shown in Fig. 12.2, which is based on the consolidated total reservoir production data. The total primary production value, which can be approximated to 1.310 million STB and was obtained by analyzing the best fit of the exponential decline for each well, is shown in Table 11.4 in Chapter 11 and indicates a very small difference of 0.15%.
- Percent of total primary = 1,308,527/6,600,000 = 0.198, or 19.8% of the OOIP will be recovered.
- Remaining oil = 6.6 million STB – 1.253 million STB = 5.347 million STB.

Recovery Estimates—On a Per-Unit Basis.

- RE_{SGD} (API) = 18.5%.
- RE_{WD} (API) = 44.1%.

These are estimates developed from correlations and should be viewed as boundaries. These values should be compared to the previous estimate to arrive at a most-likely answer. Remember, you should be constantly reviewing your perceptions about reservoir character as you proceed with the study.

Primary Recovery.

- 1,308,527 STB, or 19.8%, of OOIP reserves are predicted if no additional well work is contemplated. This value is very close to the value calculated by the API Subcommittee on Recovery Efficiency (1967) correlation method of 18.5%, particularly when compared to the current well placement. Assume the API waterflood recovery factor is also a valid approximation.

12.11.2 Waterflood Reserves.

- Waterflood reserves = (0.443)(6,600,000) = 2,923,800 STB.
- Primary recovery =1,308,527 STB.
- Remaining waterflood potential ≈ 1,615,270 STB.
- Calculated waterflood potential (sweep efficiency ≈ 30%) = 484,581 STB (approximately 500,000 STB).

Notes.

- Nonuniform sweep with primary recovery.
- Excess sand and water production caused early casing collapse. Well life was cut short. Well problems could be endemic.
- Only lower sand potential was included in this analysis, and it should be viewed as a conservative estimate.

Decision point interpretation: We have now reached a point in our investigation where we need to evaluate the viability of continuing with the study. The use of the API correlations indicates that a potential additional 500,000 bbl of oil could be produced by implementing a waterflood project, considering the sweep efficiency of 30%.

The decision to implement a waterflood project should be based on an analysis of economic key performance indicators. Economics should be applied to determine if the project is viable.

12.12 Task 15—Apply the Stiles Calculations To Forecast Performance

The results of these calculations are shown in Table 11.10 and Fig. 11.15 in Chapter 11. The same results are also plotted in **Fig. 12.7** and reveal the following:

- The oil production rate declines to 100 STB/D in approximately 23.3 years, and from Fig. 12.7, it can be seen that an estimated cumulative production of 850,000 STB would have been produced.
- Water production never exceeds 900 BWPD, well within current field capabilities.
- The Stiles model presents idealized conditions. In acuality, there should be a crossflow effect, and because of gravity segregation and the fortuitous well placement, the sweep efficiency should approach 100%.

12.13 Task 16—Determine Waterflood Project Feasibility

The calculated 484,581 STB of reserves and the results of the Stiles calculations seem to indicate the benefit of installing a waterflood in the Cl Sand, AA Reservoir.

Decision point interpretation: A detailed review of the results of all previous tasks indicates that it is worthwhile to continue with selecting injection/production well locations, determining equipment needs, and predicting performance.

The remaining tasks involve the design of additional facilities and proposed well work if a go-ahead signal is given.

12.14 Task 17—Calculate Water-Injection Requirements

Apply material balance to determine the water-injection rate. How much equivalent water should be injected to compensate the current off-take rate? Convert the current production rate measured at surface conditions to an equivalent underground withdrawal rate. Remember, all produced fluids are measured at surface conditions. Table 11.1 in Chapter 11 indicates that the reservoir currently produces at q_o = 76 STB/D, WOR (q_w) = 6 BWPD/BOPD, and q_g = 28.9 Mscf/D (GOR = gas/oil ratio = 350 scf/STB). Well 3 is the only active well in the reservoir at the end of the production history. Tables 11.5 and 11.7 list fluid properties.

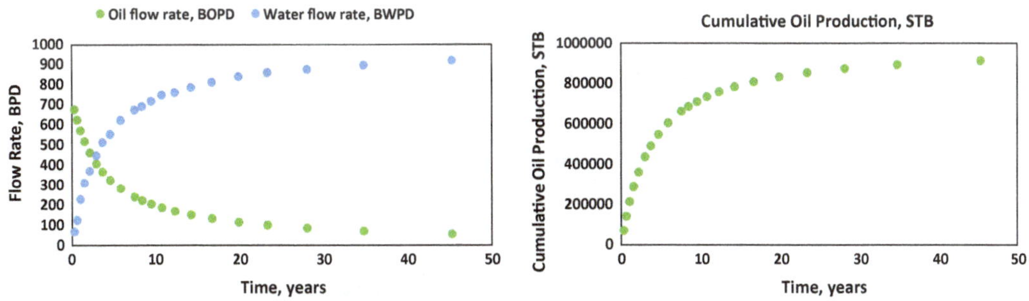

Fig. 12.7—Stiles prediction method. The results assume no crossflow and, therefore, should trend on the conservative side; (left) oil and water flow rates vs. time, (right) cumulative oil production as a function of time.

Convert total production at Year 8.5 (oil + gas + water reported at standard conditions) to total withdrawal (underground withdrawal at reservoir conditions):

- Oil produced (q_oB_o) = (76 STB/D)(1.373 RB/STB) = 104.34 RB/D.
- Free gas produced $\left(q_g - q_oR_s\right)B_g$ = [28.9 Mscf/D – (76 STB/D × 350 scf/STB)]0.00129 RB/scf = 2.967 RB/D.
- Water produced (q_wB_w) = (6 STB/D)(1.0 RBW/STB) = 6 RBW/D.
- Total withdrawal = 104.34 + 2.967 + 6 = 113.307 RBW/D.

Therefore, the minimum water-injection requirements are 113 RBW/D, and because B_w = 1.0 RB/STB, to maintain current reservoir pressure and the current production rate of 76 STB/D, the estimated water-injection rate can be calculated:

$$\frac{76}{113} = \frac{600}{x}, x = 892 \text{ STBW/D to produce 600 STB/D of oil.}$$

12.15 Task 18—Estimate the Production Response to Water Injection

The production rate increases with reservoir pressure. Apply the results of a production test (Table 11.11 in Chapter 11) to estimate the effect of overpressuring:

$$q = 0.043\left(5{,}500^2 - p_{wf}^2\right)^{0.72} \text{ at } p_{bp}\text{; the equation has already been adjusted for } p_{bp}.$$

Adjust the equation to the new $\bar{p}$ value of 3,000 psia. Assume p_{wf} = 2,000 psia to calculate changes for n and J' when $\bar{p}$ = 3,000 psia:

$$n = 0.72\left[1 + 0.06\left(1 - \frac{3{,}000}{5{,}500}\right) - 0.25\left(1 - \frac{3{,}000}{5{,}500}\right)^2 + 0.50\left(1 - \frac{3{,}000}{5{,}500}\right)^3\right] = 0.736,$$

$$J' = 0.043\left[1 - 3.57\left(1 - \frac{3{,}000}{5{,}500}\right) + 4.80\left(1 - \frac{3{,}000}{5{,}500}\right)^2 - 2.31\left(1 - \frac{3{,}000}{5{,}500}\right)^3\right] = 0.0065.$$

The adjusted deliverability equation when $\bar{p}$ = 3,000 psia is $q = 0.0065(3{,}000^2 - p^2_{wf})^{0.736}$. Calculate well deliverability when = 3,000 psia and p_{wf} = 2,000 psia:

$q = 0.0065(3{,}000^2 - 2{,}000^2)^{0.736} = 554$ STB/D.

Calculate changes for n and J' when current = 2,380 psia:

$$n = 0.72\left[1 + 0.06\left(1 - \frac{2{,}380}{5{,}500}\right) - 0.25\left(1 - \frac{2{,}380}{5{,}500}\right)^2 + 0.50\left(1 - \frac{2{,}380}{5{,}500}\right)^3\right] = 0.752,$$

$$J' = 0.043\left[1 - 3.57\left(1 - \frac{2{,}380}{5{,}500}\right) + 4.80\left(1 - \frac{2{,}380}{5{,}500}\right)^2 - 2.31\left(1 - \frac{2{,}380}{5{,}500}\right)^3\right] = 0.0042.$$

The adjusted deliverability equation when $\overline{p}$ = 2,380 psia is $q = 0.0042\left(2{,}380^2 - p_{wf}^2\right)^{0.752}$; when p_{wf} = 2,000 psia and $\overline{p}$ = 2,380 psia, $q = 0.0042\left(2{,}380^2 - 2{,}000^2\right)^{0.752} = 201$ STB/D.

Table 12.2 shows the results of the calculations and the benefit of injecting water to repressurize the reservoir.

The well production rate could increase as much as 2.5 times if the reservoir pressure were increased to 3,000 psia.

Decision point interpretation: We have now reached the point where we know the values of OOIP and primary and secondary reserves, well conditions, and projected flow rates with a fair degree of certainty. The low well primary recovery, with an expected production increase if the reservoir is waterflooded, seems to indicate a viable secondary candidate. We should next look at the surface setup.

Average Reservoir Pressure (psia)	Rate (STB/D)
2,380	201
3,000	554

Table 12.2—Estimating oil flow rates from production tests and comparing the effect of pressure/water injection on the flow rate.

12.16 Task 19—Analyze the Status of the Wells

Table 12.3 summarizes the information from the wells located in the proposed waterflood project. Well testing, workovers, and new drilling should be assigned to the work schedule. These tasks are preparation for project planning and an economic analysis.

Table 12.3 also provides answers to some of the questions that were asked in Section 11.5.21 of Chapter 11:

- Are additional wells required if a waterflood project were installed? If so, where?
- Are well recompletions needed? Which ones? Is any special completion equipment required?
- How many wells should be used?

The answers to the remaining questions in Section 11.5.21 are as follows:

- The recommended water-injection rate was determined in Task 17: 892 STBW/D to produce 600 STB/D of oil.

Well No.	Status	Remarks
1	Shut in because of HGOR	Inspect to ensure wellbore remains open.
2	TA—collapsed casing	Assign for workover or new drill. Gravel pack or special slotted liner should be considered in the completion design.
3	Producing	Inspect well completion. Run a pressure-buildup well test and determine reservoir parameters (k, skin, $\overline{p}$).
4	TA—shut in—bad casing	Assign for workover or new drill. Gravel pack or special slotted liner should be considered in the completion design.
5	TA—sand and water	Assign for workover. Gravel pack or special slotted liner should be considered in the completion design. Analyze water production causes.
6	Shut in from deeper completion	Plug back and convert to a water-injection well. Test after perforating.
7	New drill for water-injection well	Drill and test for oil rate; if good, produce until water out. After water out, convert to water-injection well.

Table 12.3—Proposed waterflood project well status. TA = temporarily abandoned.

- Adequate surface facilities for treating the water before disposal should be included in the project equipment.
- The life of the project was established in Task 15, and the estimated time is approximately 23 years.

The summary indicates that at least one new water-injection well should be drilled in the southwest corner of the reservoir, Wells 2, 4, and 5 should be worked over, and the deeper Well 6 should be recompleted to water injection to promote sweep efficiency.

12.17 Task 20—Calculate Expected Hydrostatic Pressures

Calculate the formation fracture pressure and the initial injection pressure to determine the expected hydrostatic pressures.

Estimate the fracture pressure (at current reservoir pressure and at a given depth *D*) using Eaton's formula:

$$p_{\text{frac}} = \{NOB[\nu/(1-\nu)] + p_{PV}\} \times D,$$

where p_{frac} is the fracture pressure (psi), *NOB* is the net overburden pressure gradient (psi/ft) (overburden gradient – pore pressure gradient), p_{PV} is the pore pressure gradient (psi/ft), is Poisson's ratio, and *D* is the injection mid-interval depth (ft).

Poisson's ratio $= \nu = -\varepsilon_\rho / \varepsilon_\alpha$:

- Poisson's ratio is a measure of the lateral expansion to the longitudinal contraction.
- Elastic properties can be measured using sonic and density logs or by testing the core in the laboratory.
- Typical values are
  - ☐ 0.27 for limestone
  - ☐ 0.33 for sandstone (our case study is a sandstone)

Fig. 12.8 shows stresses applied in a hypothetical test of a rock sample.

Fracture pressure (p_{frac}) = 0.604 psi/ft × 10,800 ft = 6,527 psi.

Therefore, the surface-injection pressure required to initiate a bottomhole fracture would be 6,527 psi – (10,800 ft × 0.465 psi/ft) = 1,505 psi.

12.18 Task 21—Calculate Pressure Rating for Pump Pressure Gradient

Hydrostatic head $= (0.465 \text{ psi/ft}) \times (10{,}800 \text{ ft}) = 5{,}022$ psia, and 5,022 psia < 6,527 psia.

12.18.1 Initial Injection Pressure. The well will not need to have a surface pump and will take fluid in a vacuum. The injection pressure should be less than the fracture pressure.

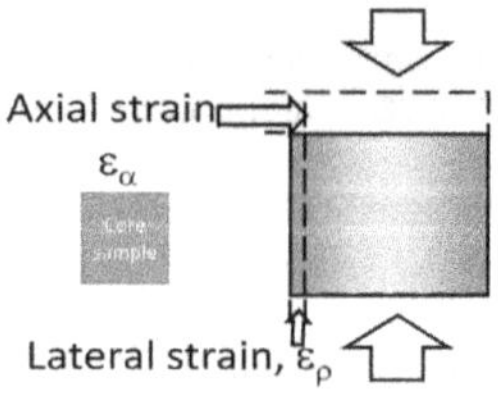

Fig. 12.8—Illustration of Poisson's ratio.

12.19 Task 22—Calculate the Expected Pressure Response Resulting from an Expansion of the Flood Front

The estimated water-injection rate calculated in Task 17 is 892 STBW/D to produce 600 STB/D of oil. The expected pressure response is calculated assuming an injection rate in the lower sand body (intervals C, D, and E in Fig. 11.4 in Chapter 11) that is slightly greater than 892 STBW/D and approximately 1,000 STBW/D. Assume a value for the drainage radius of 2,000 ft.

$$p_{wf} = \frac{i_w \mu_w \ln\left(\frac{r_e}{r_w}\right)}{0.00708hk_w} + p_e = \frac{1{,}000 \times 1.0 \times \ln\left(\frac{2{,}000}{0.25}\right)}{0.00708 \times 17 \times 195.5} + 2{,}380 = 382 + 2{,}380 = 2{,}762 \text{ psia}.$$

The following conclusions can be drawn:

- Pressure does not materially increase above the reservoir pressure.
- Production and facilities engineers should possibly be brought into the discussion regarding the evaluation and ordering of the proper pump rating. The buildup of formation damage affects the injection rate, and higher pumping pressures will be required if a stimulation job is not performed.

12.20 Task 23—Determine the Effects of Formation Damage on the Water-Injection Rate

Assume formation damage extends 3 ft. Calculate skin s as a function of a permeability loss of 10%.

12.20.1 Example Calculation. When k = 195.5 md and is reduced to 90% of the original permeability, k_s = 195.5 × 0.9 = 175.95 md, which extends 3 ft from the wellbore into the reservoir. Skin is then calculated as

$$s = \left(\frac{k - k_s}{k_s}\right)\ln\frac{r_s}{r_w};\ s = \left(\frac{195.5 - 175.95}{175.95}\right)\ln\frac{3}{0.25} = 0.28;\ s > 0 \text{ implies formation damage.}$$

The average permeability in the reservoir is

$$\bar{k} = \frac{\ln(r_e / r_w)}{\frac{\ln(r_d / r_w)}{k_d} + \frac{\ln(r_e / r_d)}{k_{\text{res}}}} = \frac{\ln(2{,}000 / 0.25)}{\frac{\ln(3 / 0.25)}{175.95} + \frac{\ln(2{,}000 / 3)}{195.5}} = \frac{8.98}{0.01271 + 0.03326} = 189.67 \text{ md},$$

or approximately 190 md.

Calculate the required bottomhole pressure to maintain the minimum 1,000-STBW/D water-injection rate:

$$p_{wf} = \frac{i_w \mu_w \ln\left(\frac{r_e}{r_w}\right)}{0.00708hk_w} + p_e = \frac{1{,}000 \times 1.0 \times \ln(2{,}000 / 0.25)}{0.00708 \times 17 \times 190} + 2{,}380 = 393 + 2{,}380 = 2{,}773 \text{ psia}.$$

The calculation indicates that formation damage, when s = 0.28, will not affect the injection rate because hydrostatic head supplies 5,022 psia at the perforations.

Fig. 12.9 relates the effect of formation damage or skin on injection pressure.

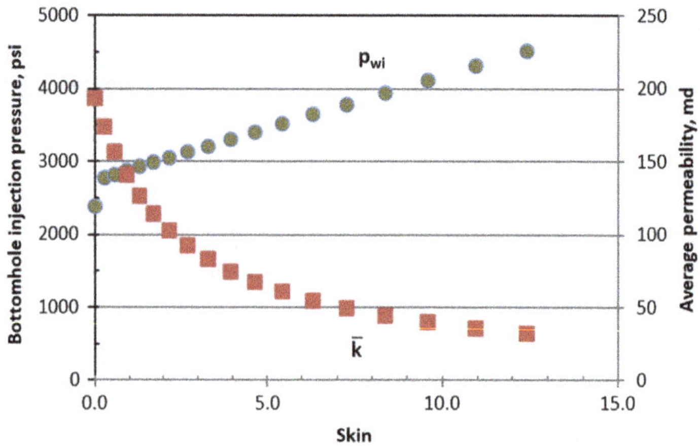

Fig. 12.9—Relating buildup of a 3-ft skin to a loss of average permeability and an increase in injection pressure. Note that the hydrostatic head pressure was not attained when average permeability declines to 32 md.

12.21 Additional Considerations

The evaluation of the results of the analysis indicates that a waterflood project is technically feasible. A basic outline of the development plan includes reviewing the mechanical status of the wells and working over the ones that were temporarily abandoned. Drilling additional wells for injecting water and inspecting available surface facilities for proper water treatment are also needed. Samples of produced water must be tested in the laboratory to provide guidelines.

Estimating capital expenditures, operating expenses, taxes, and oil and gas prices and performing cash-flow future forecasting and calculating the economic aspects of the project will establish its economic life. In this analysis phase, a review of the information available on the exploitation of the reservoir and the technical options and needs for its development have been presented. Additional aspects related to the safety of operations, health and hygienic conditions, the environment, and legal regulations should be considered. Sensitivities should be carried out and risk and uncertainty analyses should be performed to study the options for obtaining the investment capital necessary to recover and optimize production. In particular, a review of the calculations and estimates of the reserves and productivity of the area given the uncertainty of the current state of wells, reservoir pressure, infrastructure availability, and operational safety will improve the basic definition of the project design given in this study.

12.22 Summary

By the end of this discussion, the reader should be aware of the basic principles for applying a waterflood process and should have applied these principles to the study of an example reservoir. In many cases, injection and production behavior is found to differ from the predicted characteristics because of improper assumptions included in the reservoir model. Constant monitoring as a prerequisite for operating a successful waterflood is emphasized.

Simulation studies provide a more-quantitative reservoir character analysis and predictive method for estimating displacement efficiencies for reservoir development options. However, any type of simulation study requires an extensive database and data analysis. One should always be aware that conclusions drawn from the results of simulation studies could result in erroneous predictions because of unforeseen changes in reservoir character. An in-depth study should always be conducted before attempting a reservoir simulation study.

12.23 Nomenclature

A_n = area enclosed by last contour interval, acres
A_{n-1} = all areas lying between first (A_0) and last (A_n) layers, acres
A_0 = area enclosed by the 0-ft thickness contour, acres
B_g = gas formation volume factor, RB/scf
B_o = oil formation volume factor, RB/STB
B_{oi} = initial oil formation volume factor, RB/STB
B_{obp} = oil formation volume factor at bubblepoint conditions, RB/STB
B_w = water formation volume factor, RB/STB
D = Arps' continuous decline rate, %/yr
D = injection mid-interval depth (Eaton's formula), ft
E_A = areal sweep efficiency, fraction or percent
E_D = volumetric sweep efficiency or displacement efficiency, fraction or percent
E_V = vertical sweep efficiency, fraction or percent
h = height or thickness, ft
$h_{(D+E)}$ = sum of thicknesses of D and E intervals, ft
h_t = total thickness in a layered system, ft
i_w = water-injection rate, STBW/D
J' = productivity index intercept, BOPD/psi
k = permeability, md (or darcies when using API correlations)
$\overline{k}$ = average permeability, md
k_d or k_s = permeability of damaged region, md
k_{res} = reservoir permeability in non-altered zone (virgin zone), md
k_s or k_d = permeability of damaged region, md
k_w = effective permeability to water, md
n = turbulence, also defined in the Fetkovich method as an exponent term, dimensionless
N = original oil-in-place, STBO
N_p = production rate or cumulative oil production, bbl, BO, STB, or STB/acre-ft
$\overline{p}$ = average pressure, psia
p_a = abandonment pressure (API correlations), psig
p_{bp} = bubblepoint pressure, psia or psig
p_e = pressure at drainage radius, psia or psig
p_{frac} = fracture pressure, psi
p_i = initial pressure, psia
p_{last} = final pressure, psia
p_{PV} = pore pressure gradient, psi/ft
p_{wf} = bottomhole flowing pressure, psia
p_{wi} = water-injection pressure, psi
q = flow rate, STB/D, BOPD, or B/D
q_g = gas flow rate, Mscf/D
q_o = oil flow rate, BOPD or STB/D
q_w = water flow rate, BWPD
q_1 and q_2 = flow rate at a certain time, RB/D or STB/D
r_d = radius of damaged zone, ft
r_e = drainage radius, ft
r_s = radius of altered zone (damaged zone)
r_w = wellbore radius, ft
RE_{SGD} = solution-gas-drive reservoir recovery, %
RE_{WD} = waterdrive reservoir recovery, %
R_s = solution gas/oil ratio, scf/STB
R_{si} = initial solution gas/oil ratio, scf/STB
s = skin, dimensionless
S_{oi} = initial oil saturation, fraction or %
S_{or} = residual oil saturation, fraction or %
S_{wi} = initial water saturation, fraction or %
$\overline{S}_{wi}$ = average initial water saturation, fraction or %
t = time, years
V = volume, acre-ft
ΔS_o = change in oil saturation, fraction or percent
ε_α = longitudinal contraction (Poisson's ratio)
ε_ρ = lateral expansion (Poisson's ratio)
μ_{obp} = oil bubblepoint viscosity, cp
μ_{oi} = initial oil viscosity, cp
μ_w = water viscosity, cp
ν = Poisson's ratio
ϕ = porosity, fraction or %
$\overline{\phi}$ = average porosity, fraction or %

Chapter 13

Summary of Results

13.1 Reservoir Configuration

- Radiating faults defined by geological and pressure analyses.
- Geological studies indicate the two layers of the C1 Sand are probably noncommunicating.
- A cross section shows that the upper, lower-permeability sand member shales out and trends updip.
- A static-pressure analysis indicates that Well 8 is not located in the AA Unit.
- The well locations and geological settings indicate that a peripheral water-injection project would result in the greatest recovery.

13.2 Production—Pressure History

- Updip Well 1 gassed out while the downdip wells generally produce water and sand, reflecting pressure-depletion conditions.
- Initial production tests indicated that the wells produced 300 to 700 STB/D.
- Production-decline analysis, correlation of well histories, static-pressure histories, and material-balance studies all indicate the existence of pressure-depletion conditions.

13.3 Original Oil-in-Place

- Volumetric studies indicated 6.951 million STB.
- Material-balance studies indicated 6.2 million STB.
- Average value = 6.6 million STB.

13.4 Reserves

- Displacement efficiency = 30% (areal = 80%, vertical = 37%).
- Recovery to date = 1.253 million STB.
- Primary recovery, American Petroleum Institute (API) Subcommittee on Recovery Efficiency (1967) correlation, RE_{SGD} = 18.5%.
- Primary recovery, API correlation = 1.221 million STB.
- Remaining oil = 6.6 million STB – 1.253 million STB = 5.347 million STB.
- Waterflood addition = 484,581 STB (approximately 500,000 STB).

13.4.1 Cautionary Notes.

- Oil in the upper-sand section could be bypassed—possible dual completion?
- Reserves estimates should be viewed as conservative.

13.5 Well Requirements

- Drill one new water-injection well.
- Work over (plug back) and convert Well 6 to a water-injection well.
- Work over Wells 2 and 4 or drill new, and work over Well 5.
- Producing wells should be gravel packed to control sand production.

13.6 Water-Injection Requirements

- Calculated water-injection rate = 892 STBW/D.
- Recommended water-injection rate for project = 1,000 STBW/D.
- Use produced water and consider surface facilities for maximum filtration and conditioning for required future disposal.
- Theoretically, the well will take water in a vacuum.
- Prepare for the effect of formation damage.
- The calculations show fracture pressure = 6,527 psia.
- Drill one new water-injection well and convert Well 6 to a water-injection well.

13.7 Future Performance

- Performance will increase to approximately 200 STB/D after initiation of the water injection, and after achieving a repressuring effect of 3,000 psia, the estimated oil flow rate will be approximately 550 STB/D.
- A majority of reserves will be produced within 5 to 10 years.
- The oil-saturated upper, low-permeability sand should remain at essentially initial conditions after flood out of the lower, high-permeability sand.
- Although the water/oil ratio will rapidly increase, the oil rate declines, so the net throughput will not dramatically increase. Additional water-processing facilities will not be required.

Chapter 14

Waterflood Review

14.1 Introduction

For a successful waterflooding project, one must establish a thorough understanding of the reservoir processes occurring in the field. This knowledge forms the foundation for future performance predictions. However, these predictions often do not materialize. Unexpected changes in well production—i.e., in the flowing tubinghead pressure (FTHP), oil rate, water/oil ratio (WOR), and gas/oil ratio (GOR)—can signal plant, well, or general system problems. Consistent monitoring of the production history and recognizing the onset of these variations can materially affect ultimate recovery. The study of well performance curves provides insight into changing operating conditions. Identifying these changes can readily be accomplished with computer-activated programs. Established systems capture, store, and display production data effortlessly. A predesigned format allows automatic updating and even plots the variables at designated well locations on a field map. Predetermined definitions covering oil, water, and gas production and phase-volume comparisons can be programmed to automatically update a field history. The discussion that follows reviews maps that can be drawn to illustrate a wide variety of field characteristics. Color coding can partially interpret flood-front advance and fluid boundaries.

14.2 Waterflood Event

Data gathering is important because it allows one to perform an in-depth evaluation of factors affecting the water/oil displacement efficiency and water-source selection and handling. Best practices are essential to assist with specific conclusions and recommendations for controlling and modifying performance to guarantee the success of a waterflood project. A flow chart for a summary review of the waterflood event is shown in **Fig. 14.1.**

Waterflood projects have widespread applicability because of water-source availability, relatively low associated costs, and their efficiency in oil recovery. Proper design, planning, execution, monitoring, and surveillance are essential to project success and depend on the application of the best operational practices and an integrated management approach.

Fig. 14.2 shows the water cycle for the management of water injected to enhance oil recovery. Water sources include water wells, streams and rivers, and treated and reused fresh water and seawater. Environmental aspects must be considered, including the depletion of reservoirs, impacts on aquatic life, equipment footprint, and shallow-aquifer contamination. Transportation system pipelines and water storage in above-ground tanks also should be considered. Regulations and disposal criteria should be followed, and one should obtain the proper permits for water sourcing and disposal.

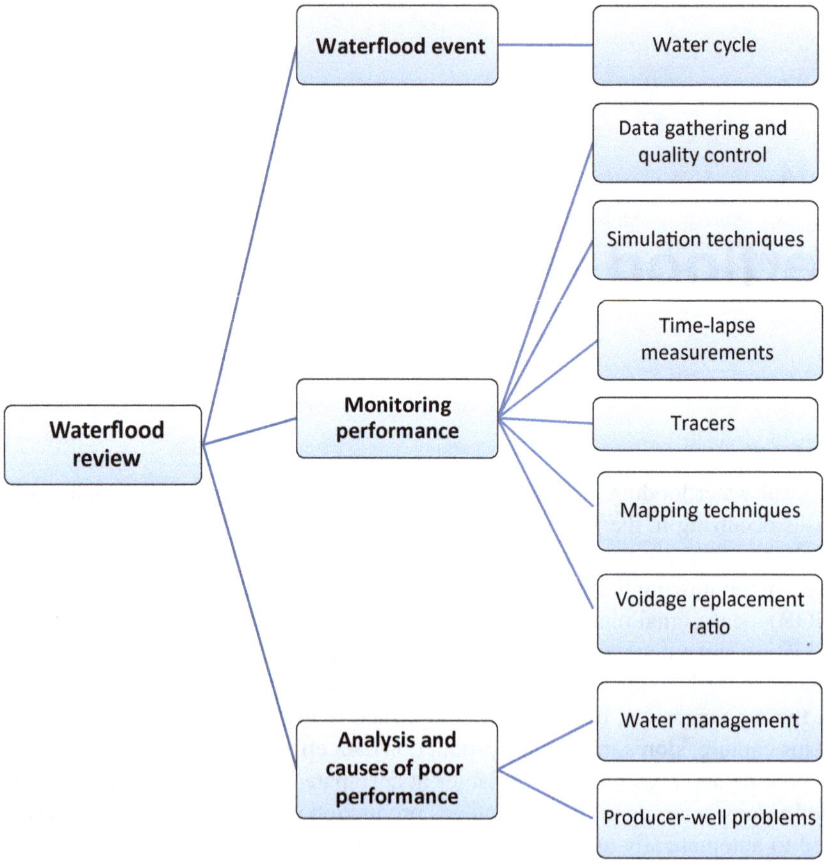

Fig. 14.1—Flow chart of analysis procedure.

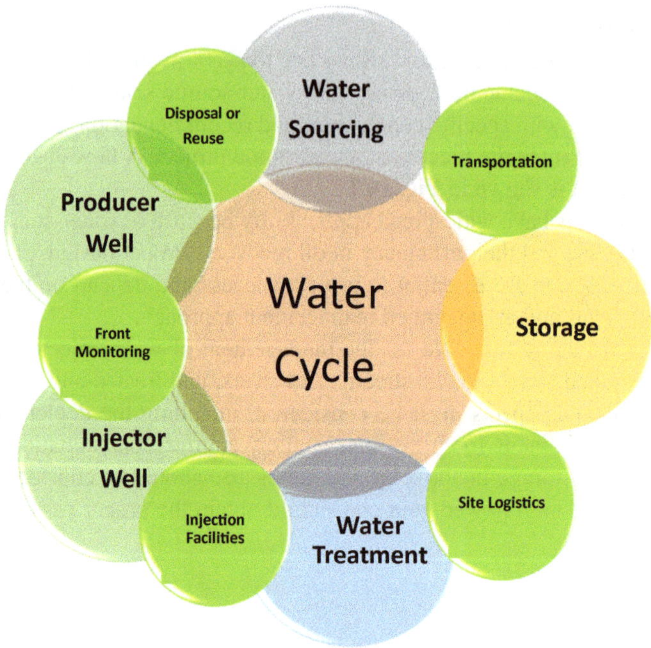

Fig. 14.2—The water cycle—water management—for injecting water into a reservoir to recover oil.

The evaluation and benchmarking of existing production facilities and operations against the design criteria form a first step in the field implementation of a waterflood. Identifying bottlenecks and opportunities, and recommending upgrades and best operational practices, will increase productivity and efficiency.

The surface treatment of water for injection, including water cycling and fluids management, is essential for the optimization of water-injection systems, reducing the environmental impact, and improving project economics.

14.3 Monitoring Performance

Real-time reservoir management is an available technology that aids the decision-making process. Technical and economic impacts, along with health and environmental issues, are analyzed through proper monitoring. This is a continuous process used to improve asset management. **Fig. 14.3** shows the suggested workflow for waterflood project management.

Data gathering and the building of a database for the project constitute the first step. The input requirements for a performance analysis are summarized as follows:

- Production/injection history
  - Water injection
  - Oil production
  - Water production
  - Gas production
  - Wellhead injection pressure
  - Reservoir pressure
- Reservoir properties (field/block/pattern level)
  - Pore volume
  - Original oil-in-place (OOIP) at flood start
  - Displaceable oil-in-place
  - Pressure/volume/temperature data

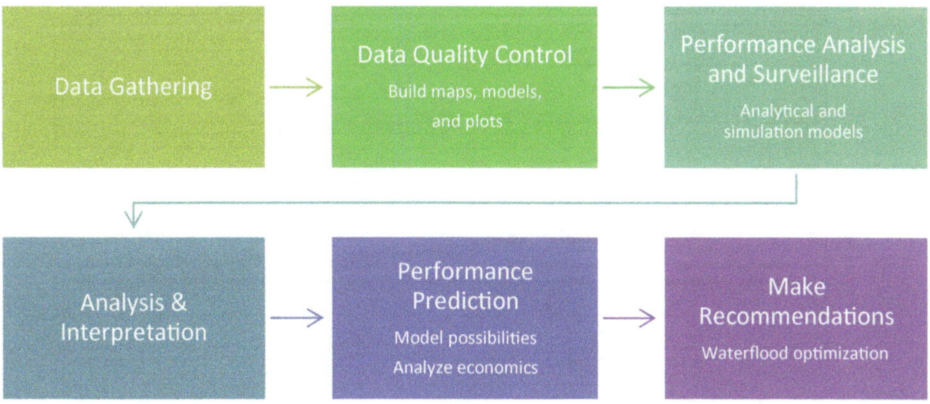

Fig. 14.3—Workflow for waterflood project management to optimize water-injection performance and to aid in making better decisions.

Different analysis techniques can be applied. The use of these methods depends on the number of wells and the history of production under waterflooding. Establishing data quality control at the beginning of all projects is the second step. Erroneous production values can be identified when the production value of one, or sometimes two, months is too high or too low, when compared to that of other months. **Fig. 14.4** shows a production performance graph in which bad data points are identified. Most errant values are at least one log cycle too high.

The use of test data, rather than production data, is recommended where possible. Both unallocated data and allocated data are recommended for comparison and validation purposes. Production logging tests should be used to allocate rate data when possible and to understand pressure support. Multiphase metering technology can be used for permanent installations and for mobile-testing applications. Real-time data acquisition with multiphase meters is an excellent solution for project-management monitoring.

Statistical methods and advanced mapping techniques are commonly applied to understand production performance and water movement. Finite-difference reservoir simulation and streamline simulation techniques are also widely applied for waterflood management and the evaluation of the efficiency of injectors and producers. Streamline simulation is based on saturation mapping along streamlines between wells—the water saturation is calculated along each line. Streamlines change when some of the producer wells are shut in. The pressure variation resulting from production and injection is calculated using finite-difference techniques. Streamline simulation typically is a much-faster process than numerical simulation for cases in which pressure changes are not very significant. **Fig. 14.5** shows streamline trajectories, and **Fig. 14.6** shows finite-difference simulation results.

Time-lapse saturation monitoring, time-lapse pressure monitoring, and time-lapse seismic and production logging are tools used for monitoring fluid displacement and assist with the identification of infill drilling opportunities using tracers. Saturation-log, production-log, and integrity-log interpretation; decline-curve analysis; and nodal analysis performed for individual wells are typically the most-used engineering tools for reservoir management. Saturation monitoring is best performed by comparing the results from one timestep with those from another—this is called time-lapse measurement. Environmental effects can be different from one timestep to the next (e.g., as a result of

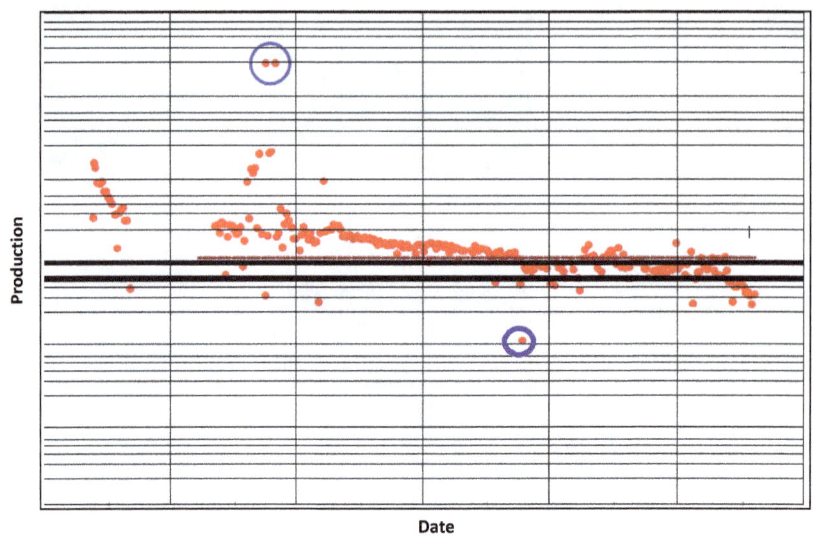

Fig. 14.4—This production graph shows data points that are a log cycle too high or too low. Look for anomalously large maximum month values relative to the majority of the data points.

the type of fluid in hole); therefore, it is always best to make comparisons under the same environmental conditions.

Some saturation monitoring techniques, such as sigma and casedhole resistivity, depend on formation-water salinity or resistivity. These characteristics, in turn, can vary during the producing life of the field, as injection projects get under way and if the alteration of aquifer water will also definitely occur. Time-lapse seismic surveys are also used to determine sweet spots and waterflooded areas such as those shown in **Fig. 14.7.**

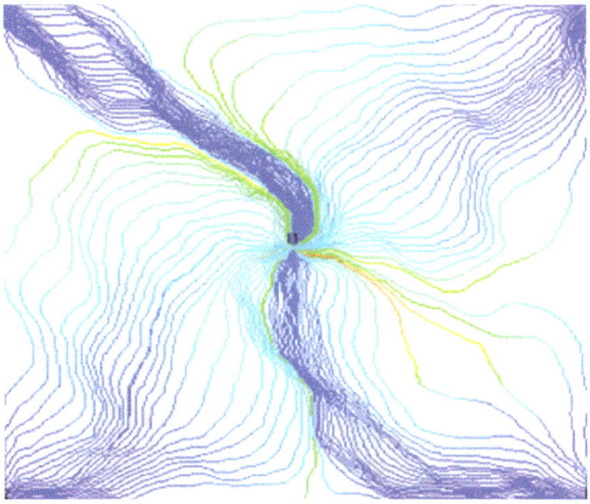

Fig. 14.5—Streamline trajectories for a heterogeneous five-spot pattern (after Datta-Gupta and King 2007).

Tracers are frequently injected to tag injected fluids and to diagnose reservoir problems. The presence of fractures, poor sweep efficiency, uneven permeability distribution, low oil saturation, and similar problems causing inefficient or uneconomic oil production can be identified. The time required for a tracer to travel from an injector to a producer is called the tracer transit time. It is an indication of the magnitude of the channel and can be used to estimate channel volume, which is approximately equal to the volume of water injected during the transit time. Rapid water-injection transit times often confirm that fractures have caused a water breakthrough. Tracers can help determine the injected volume of gel needed to cure adverse water movements.

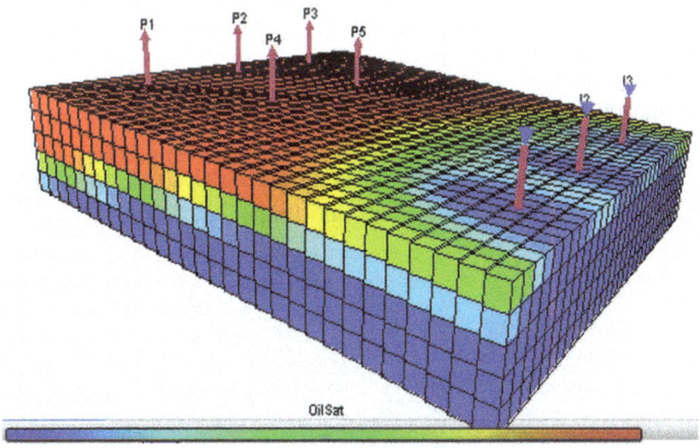

Fig. 14.6—Waterflood surveillance techniques. Visualization using finite-difference models was produced with state-of-the-art reservoir simulation software.

Tracers—such as chemical, radioactive, natural, water-soluble and -insoluble, and oil-soluble tracers—are commercially available. Water-soluble tracers follow the water phase, while oil-soluble tracers partition into the oil phase upon contact. In general, tracers are inexpensive, easily detected at low concentrations in the produced fluid, safe to handle, and soluble in the fluid to be traced and should not be absorbed into the formation. Tracer irruption (breakthrough) times are an estimation of the flow trend. The primary use of tracers is to understand fluid movements and identify reservoir compartments. High-conductivity zones and interwell communication through oriented fractures, which could damage waterflood performance, can be treated with a profile-control agent. Water-control chemical technologies, such as gels and relative permeability modifier treatments to close off those fractures, are sometimes recommended. An example of representations of tracer-irruption

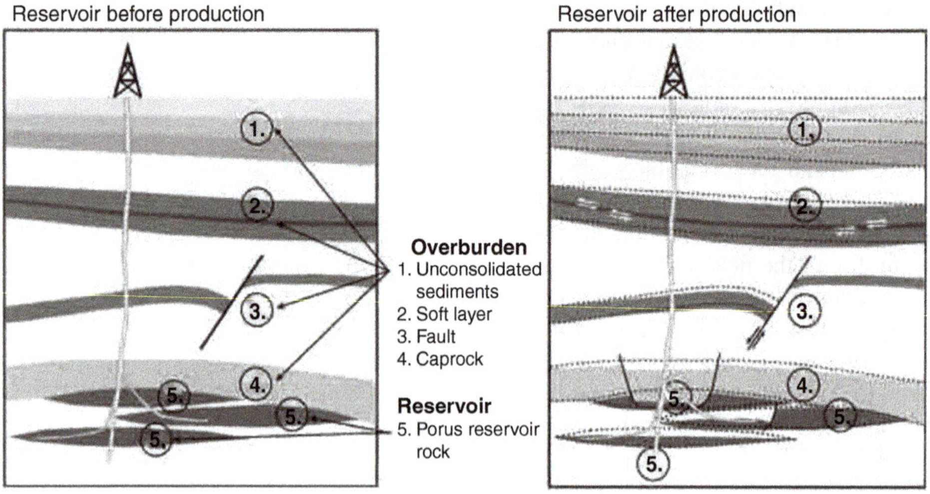

Fig. 14.7—4D (or time-lapse) seismic is now often capable of determining the water saturation in a reservoir between the wells. This is very valuable to understand how the water is flooding the reservoir (after Schlumberger 2021).

times that are based on the tracer data collected in the injector and the associated producers is shown in **Fig. 14.8.** The reservoir zones enclosed in the circles showed the shortest irruption times observed in the producers. These reservoir areas would be considered for chemical treatments for water control.

Statistical techniques based on comparing wells in a particular domain with other wells are used to find anomalies in pressure, production, production ratios, or calculated variables. The identification

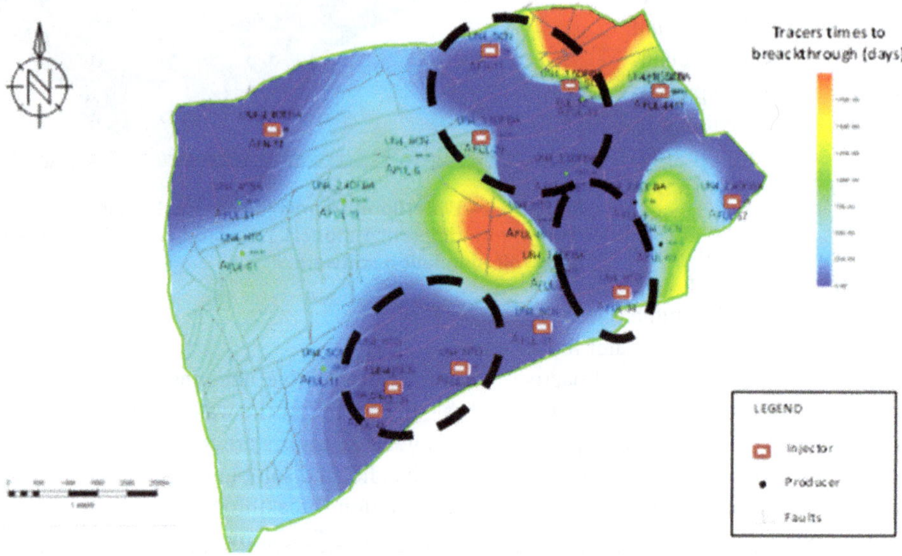

Fig. 14.8—Tracer breakthrough time, El Furrial Field, East Venezuela Basin, east area of SJN FUL 2 Reservoir (after Pérez Amaro et al. 2020). The injectors are located at the flanks of the anticlinal structure, while producers are completed in the uppermost structure. The preferential flow direction is from northeast to southwest along the crest and from east to west at the flanks.

of areas of poor performance or areas needing additional water injection is achieved using raw data, calculated variables, or normalized variables. A properly maintained database allows one to generate calculated variables and plot the results as maps. These maps can be generated as a function of time to show the variation of a particular variable.

Bubble and grid maps are built showing production rates, cumulative volumes, and pressures. **Fig. 14.9** shows the pressure maps for two waterflooded reservoirs.

Calculated variables from a database that are useful for visualization and diagnosis include:

- Time to breakthrough
- Production rates after shut-in
- Production after breakthrough
- Productivity index
- Reserves/production ratio for a pattern
- Reservoir voidage plots
- Production ratios
- Movable hydrocarbon saturation or reserves
- Remaining reserves in a pattern or area or layer
- Water-front movement
- Injection rates
- Cumulative injection per well, pattern, or area

Plotting these variables will assist in understanding the major water flow mechanisms in each area of the field and aid in finding the best candidate wells for workover to improve water flow.

Performing material-balance studies and calculating the voidage replacement ratio (VRR) will assist in determining the amount of water to be injected to maintain reservoir pressure, and a value of VRR = 1 means that a balance should exist between injection and production volumes. In mature waterfloods, values of VRR that are less than unity are common. One typically observes declining oil production rates, increasing water cuts, and increasing GORs in combination with declining reservoir pressures. VRRs greater than unity are necessary before fill-up, and most waterfloods target to operate at approximately this level.

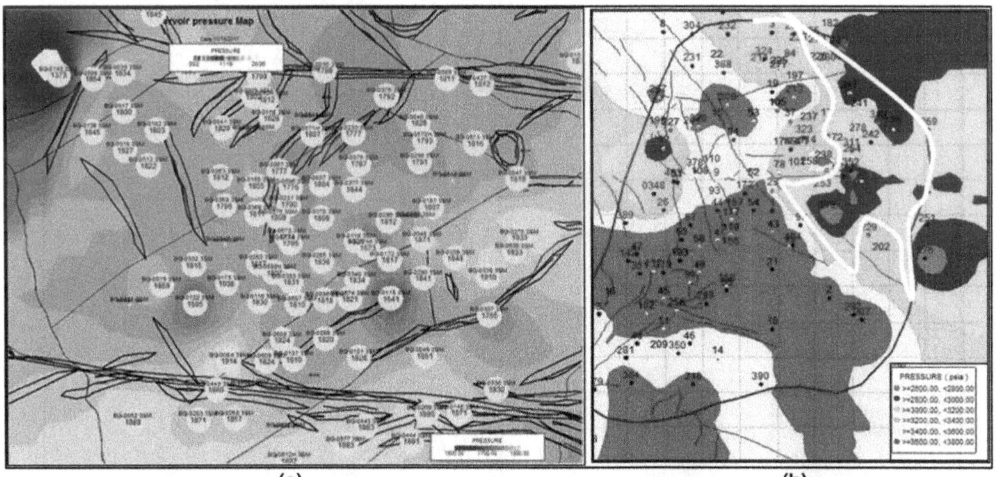

(a) (b)

Fig. 14.9—Pressure maps for two waterflooded reservoirs (after Schlumberger NExT 2010); (a) pressure is uniform throughout the entire reservoir, and (b) pressure shows ranges of variation, with zones in dark indicating low-pressure areas. The northeast part of the field shows relatively lower pressure levels.

Voidage replacement refers to the replacement of the volumes of oil, gas, and water produced from the reservoir by means of injected fluids. VRR is the ratio of reservoir barrels of injected fluid to reservoir barrels of produced fluid. Mathematically (for water injection), it can be expressed as

$$\text{VRR} = \frac{\text{injected reservoir volumes}}{\text{produced reservoir volumes}}, \text{ RB/RB}, \quad \text{(14.1a)}$$

$$\text{VRR} = \frac{B_w\left(\frac{\text{RB}}{\text{STB}}\right)\times i_w\ (\text{STB})}{\left[B_o\left(\frac{\text{RB}}{\text{STB}}\right)\times q_o\ (\text{STB})\right]+\left[B_w\left(\frac{\text{RB}}{\text{STB}}\right)\times q_w(\text{STB})\right]+\left[B_g\left(\frac{\text{RB}}{\text{Mscf}}\right)\times(\text{GOR}-R_s)\left(\frac{\text{Mscf}}{\text{STB}}\right)\times q_o\ (\text{STB})\right]}$$

$$\text{RB/RB}. \quad \text{(14.1b)}$$

In Eq. 14.1b, B_w, B_o, and B_g are the formation volume factors for water, oil, and gas, respectively; q_w and q_o are the production rates for water and oil, respectively; and i_w is the water-injection rate. GOR is the producing gas/oil ratio, and R_s is the dissolved GOR. The third term in the denominator accounts for the free gas produced that is in excess of the gas in the reservoir that is dissolved in the oil. The VRR can be calculated instantaneously by use of injected and produced fluids over any specific time period (typically daily or monthly) or on a cumulative basis by use of the cumulative injected and produced fluids. **Fig. 14.10** shows a composite displaying two maps revealing certain trends and showing prospective areas for infill drilling. Fig. 14.10a shows the cumulative VRR map. In Fig. 14.10b, pressure shows uniformity at approximately 2,800 psi.

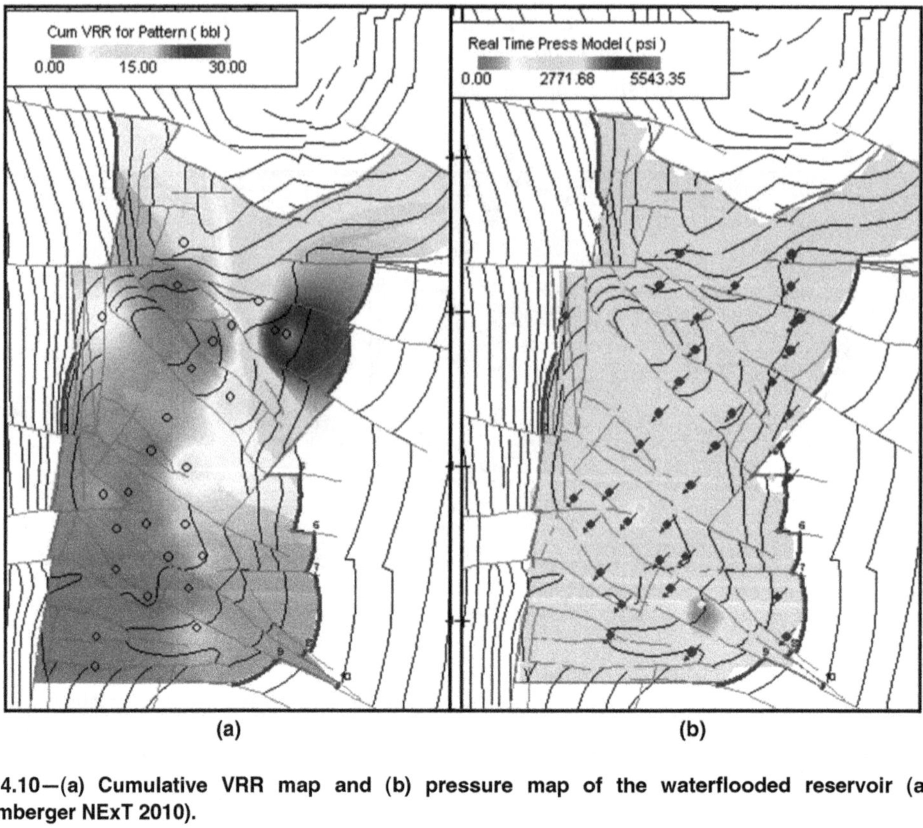

(a) (b)

Fig. 14.10—(a) Cumulative VRR map and (b) pressure map of the waterflooded reservoir (after Schlumberger NExT 2010).

Reserves are estimated by extrapolating the water cut vs. cumulative oil production curve using a straight line until it reaches maximum water cut, whose value is obtained from economic analysis and/or field experience. **Fig. 14.11** provides a map showing remaining recoverable hydrocarbons from a water-cut analysis.

Mapping tools used to evaluate geologic features through log visualization/analysis and calculate the volumetric OOIP using grid-mapping techniques are essential for analyzing the secondary recovery of the reservoir, to optimize the expansion of waterflood patterns. Techniques such as pattern analysis and injection flood-front radius estimations are used to optimize water-injection performance to make improved secondary recovery decisions.

From **Fig. 14.12,** trends of water sweep can be visualized and the quality of petrophysical properties observed. **Fig. 14.13a** is an OOIP map with bubble sizes related to cumulative oil production per well. In Fig. 14.13b, the drainage radius/pressure chart applied throughout the project life is an excellent tool for pattern analysis or for analysis of the entire reservoir. Total fluids cumulative production is expressed by the circles' relative sizes around the well in the

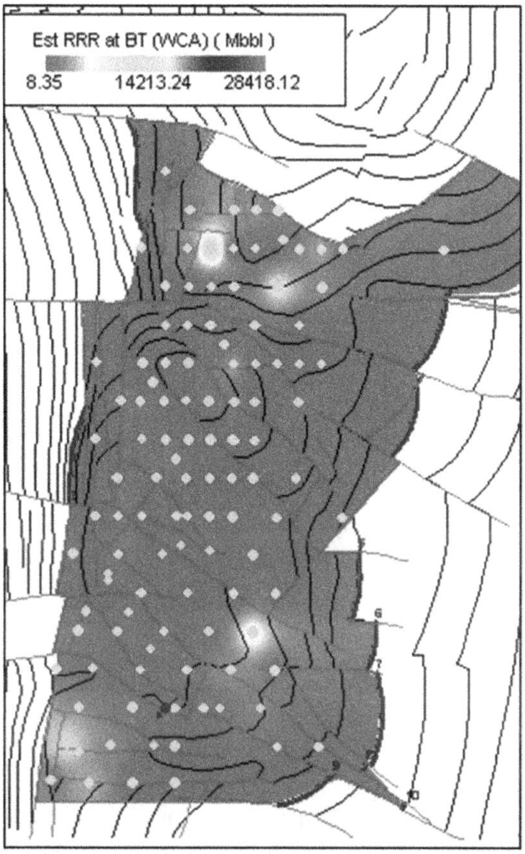

Fig. 14.11—Remaining recoverable hydrocarbon from a water-cut analysis of the waterflooded reservoir (after Schlumberger NExT 2010).

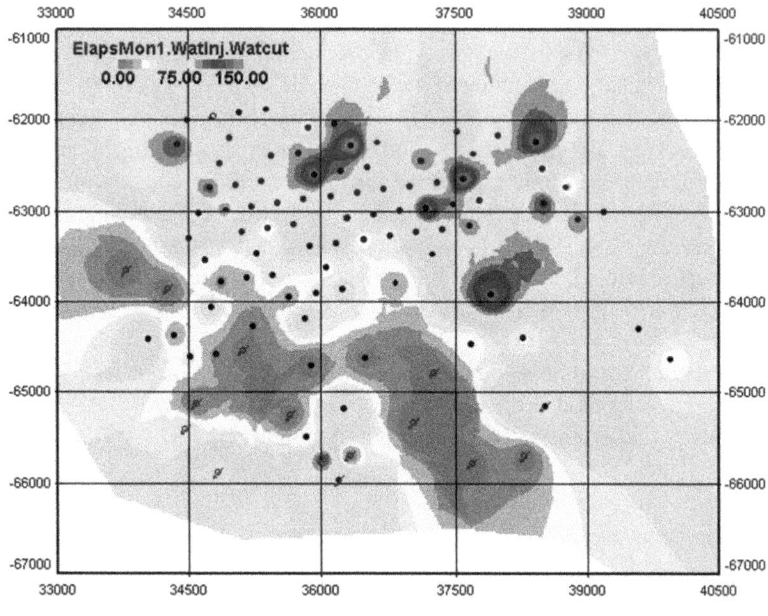

Fig. 14.12—Water breakthrough time (after Schlumberger NExT 2010).

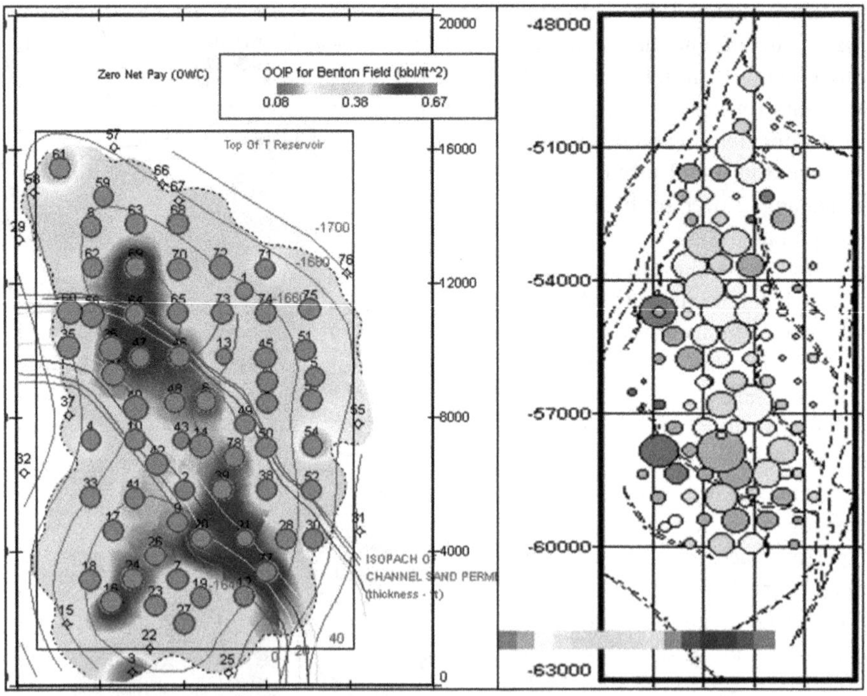

(a) (b)

Fig. 14.13—(a) OOIP map with bubble sizes relating cumulative oil production per well and (b) pressure map of the waterflooded reservoir where bubble size is related to total fluids production on a per-well basis (after Schlumberger NExT 2010).

same map units. This implies that if the bubbles of two wells intersect, there most likely will be interference between those wells. The color of the bubble is related to the pressure value at the well location, which in turn allows simultaneous analysis of two related variables.

14.4 Analysis and Causes of Poor Performance—Producing Wells

Restricted oil production problems can occur in producing wells. Effective water management requires a timely diagnosis of and solution to the problem.

Data management and quality control of the available information on well completions, reservoir pressure, reservoir geology, and petrophysics, and a detailed analysis of historical production data, will enable the performance of a comprehensive diagnosis and the selection of an adequate treatment or technology to apply. Close relationships with all stakeholders, including the manufacturer or developer of the technology, and proper field application will help with the success of the treatment.

Water management requires that there be integrated multidisciplinary teams using a comprehensive approach that requires substantial information, including

- Geological studies and reservoir characterization
- Dynamic and static models
- Geochemistry and extensive analysis of production waters
- Records
- Petrophysics

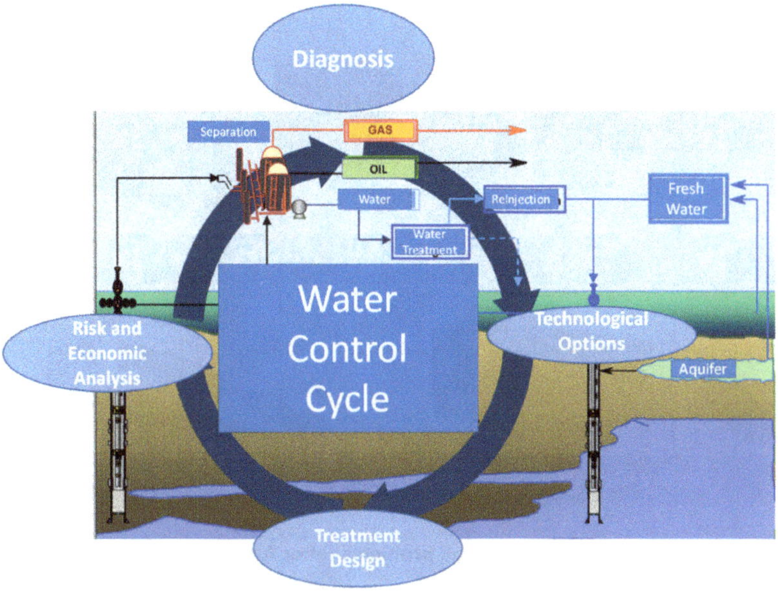

Fig. 14.14—Water control cycle steps to optimize the water management process using an integrated system multidisciplinary approach (after Arnold et al. 2004).

Fig. 14.14 shows the water cycle, where the appropriate diagnosis leads to the analysis of different technological options and a design of the best treatment to optimize economic indicators and reduce the associated risks that apply to the system by analyzing the causes of poor performance.

Producer-well-problem types are shown in **Fig. 14.15.** Additional problem types that are also common in injector wells were presented in Fig. 10.10 from the Waterflood Components chapter.

Fig. 14.15 shows possible causes of water problems, such as poor continuity, in which zones appeared continuous but in actuality were found to be discontinuous and, therefore, were not affected by the water-injection process. Another problem involves flooded zones not completed in the producing well (low areal sweep), which may not be obvious at the start of a project. Other problems are associated with well integrity, such as thief zones (i.e., typically a hole in tubing or a packer leak and wellbore impairment), buildup of scale, movement of formation fines toward the wellbore, asphaltene deposition, or formation fill in the well.

Two important causes of excessive water production are coning and channeling. Water coning occurs when the oil/water contact locally rises toward the completed interval in a well that partially penetrates the pay zone. The growth of the water cone must be halted to avoid this problem and to maximize the production and recovery factor. This is done by perforating the water zone and producing the water at the same pressure as the oil. Increasing the water cut in a well often leads to poor well performance, requiring gas lift or an electrical submersible pump (ESP) to be reoptimized.

Gel treatments have been applied to many wells to reduce water cut. Chemical treatments for water control in producers aim to selectively plug water channeling because reduced pressure gradients help to improve areal sweep efficiency. The treatments in injectors with low vertical efficiency allow the displacement of water toward the producers to be improved, with direct influence on the increase of production.

Waterflood optimization can be obtained by performing changes on the production methods from the surface without intervention into the wellbore to control some water problems. For example, if

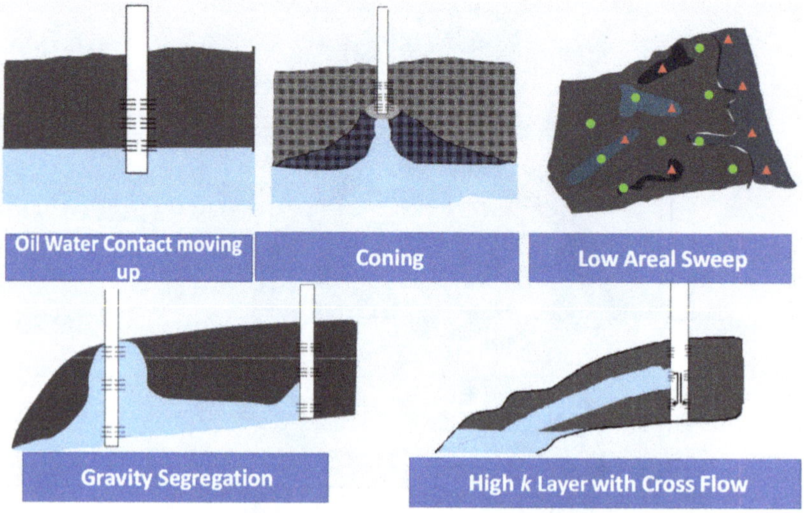

Fig. 14.15—Producer-well problems (after Arnold et al. 2004).

the production manifold pressure is too high for a well to flow, a booster pump could be installed near the wellhead to reduce wellhead pressure and allow for a greater flow rate, even at a high water cut. In the case of coning, production cycling (i.e., alternately shutting in and flowing the well) could be considered.

Poor reservoir areal sweep can often be improved by simply changing the production and injection rates of wells to improve the sweep. This can be done with help from streamline simulators to rebalance producers [i.e., injectors converting producer(s) to injector(s)].

When water cannot be shut off, it is necessary to flow the water into the well. Downhole water separators use a hydrocyclone and an ESP (electric submersible pump) to separate out some of the water and reinject it into disposal layers so that it will not be necessary to flow all of the water to the surface. Plugging can occur in the injection zone, causing the system to fail if oil and fines are present in the water stream. Drilling infill wells or horizontal laterals from existing vertical wells can be useful to minimize coning, access bypassed oil, and change sweep patterns. Other solutions—such as mechanical shutoff tools, patch or bridge plugs set on wireline, or inflated packers set on coiled tubing—are used so that a rig is not required to remove the completion.

14.5 The Waterflooding Process—Summary

Oil, gas, and water production induces a pressure drop between the higher pressured reservoir (p_{res}) and the surface. Oil production rates are a direct function of the FTHP (p_{FTHP}) at the surface. The loss of reservoir pressure resulting from production or an increase in hydrostatic head reduces the flow rate. Water or gas injection is often applied to artificially augment reservoir pressure when natural expansive processes are insufficient to maintain reservoir pressure.

The water-injection process is more commonly applied than gas injection because of the availability of water and its moderate cost. Water injection can be classed as pressure maintenance or as a liquid displacement project. Pressure maintenance involves the injection of sufficient amounts of water to maintain reservoir pressure; hence, the reservoir offtake rate remains stable. Normal water injection attempts to push otherwise-immovable oil toward a producing well.

Waterflooding represents a dramatically different displacement process than the one observed during primary depletion. Primary depletion induces fluid flow through interconnected pores to the producing well. All flow channels connected to the completion zone contribute to the depletion process. Water injection, on the other hand, uses an external energy source to displace injected water and oil liquids toward a producing well. Established flow channels must be connected between the input source and producer. The displacement process does not occur for oil located in discontinuous sands. Knowledge of reservoir architecture and producing character is of paramount importance when contemplating a waterflood.

The applications of fundamental reservoir analysis concepts coupled with commonly available geological and production information have been discussed in the previous chapters. Numerous field examples were included to illustrate the use of these basic concepts. Reservoir simulator studies designed to study multiwell situations require additional technical expertise. However, these preliminary studies should serve as a precursor and should always be conducted before attempting to simulate a reservoir.

14.6 Nomenclature

B_g = gas formation volume factor, RB/Mscf
B_o = oil formation volume factor, RB/STB
B_w = water formation volume factor, RB/STB
ESP = Electric submersible pump
FTHP = Flowing tubing head pressure
i_w = water-injection rate, STB
p_{FTHP} = flowing tubinghead pressure, psia
p_{res} = reservoir pressure, psia
q_o = oil production rate, STB
q_w = water production rate, STB
R_s = dissolved gas/oil ratio, Mscf/STB or scf/STB

References

Abou-Kassem, J.H., Mattar, L. and Dranchuk, P.M. "Computer Calculations of Compressibility of Natural Gas," (Sept.–Oct. 1990) *Jour. Can. Pet. Tech.,* V 29, No. 5, 105–107.

Ahmed, T.: *Reservoir Engineering Handbook*, 4th Ed. Gulf Publishing Company, ELSEVIER, 2010.

Akhil Datta-Gupta and Michael J. King., Streamline Simulation: Theory and Practice, SPE Textbook Series Vol. 11, Society of Petroleum Engineers, Dallas, TX., USA, ISBN: 978-1-55563-111-6

Amyx, J.W., Bass, D.M., and Whiting, R.L.: *Petroleum Reservoir Engineering*, McGrow-Hill Book Company New York, 1960.

API Publication 852–27045:(1980) "Volume Correction Factors," 374.

Arps, J.J.: "Estimation of Primary Oil and Gas Reserves," *Petroleum Production Handbook*, J.J. Frick *ed.* (1962) McGraw Hill Book Co., New York, NY, p. 665.

Bakhmeteff, B.A., and Feodoroff, N.V.: "Flow Through Granular Media, *Trans.* ASME (1937) 59 A97.

Baker, Richard (1997): *Reservoir Management for Waterfloods*," Jour. Canadian Pet. Tech. Ap. 1997 v.36 No. 34.

Beggs, H. D. and Vasquez, M. "Correlations for Fluid Physical Property Prediction," *Jour. Pet. Tech.* (June 1980) 968–970.

Berg, R.R. and Avery, A.H.: "Sealing Properties of Tertiary Crowth Faults, Texas Gulf Coast," AAPG *Bull.* (March 1995) V. 79, No. 3, 375–393.

Berg, R.R. (1986): *Reservoir Sandstones*, Prentice-Hall, Inc., Englewood Cliffs, NJ. p. 479.

Bennion, Brant, "Liquid Hydrocarbon Bearing Reservoirs-Clastic Formations," SPE paper Canada, April 27, 2010.

Bradley, E.F., et.al. (2006) "*Depositional Modeling Champion Field: Assessing the Impact of Reservoir Architecture on Secondary Recovery*" SPE MS 101003.

Bradley, H.B. (1987) "Petroleum Engineering Handbook" Society of Petroleum Engineers, Richardson, TX.

Brown, K.E.: (1973) "Reservoir Mechanics - Chapter 4," *Gas Lift Theory and Practice*, The Petroleum Publishing Co., Tulsa, OK. 924.

Buell, R.S., H. Kazemi, and F.H. Poettmann. Analyzing Injectivity of Polymer Solutions With the Hall Plot SPE 16963. in SPE Annual Technical Conference and Exhibition. 1987. Dallas, TX: SPE

Bruning Aeragraph Charts. Copyright applied for (1957) J. Lessinger.

Camacho-V., R.G. and Raghavan, R.: (May 1989) "Inflow Performance Relation-ships for Solution Gas-Drive Reservoirs," *Jour. Pet. Tech.* 541–550.

Carter, R. D. and Tracy, G. W., "An Improved Method for Calculating Water Influx", AIME *Trans.* 219 (1960) 415–417.

Chang, C.K. (September 1985)," Water Quality Considerations in Malaysia First Waterflood", SPE 12307 JPT,

Chen, H.Y. and Vu, T.: Personal communication at Texas A&M University.

Civan, F, (1999) "Interpretation and Correlations of Clay Swelling Measurements," SPE 52134, SPE Mid-continent Operations Symposium, Oklahoma City, OK, March 29–31, 1999.

Core Laboratories, Inc. "A course in the fundamentals of Core analysis, 1982.

Cosse, R.: *Basics of Reservoir Engineering* (1993) Gulf Publishing, Houston, TX 342.

Craft, B.C. and Hawkins, M.F. (1959) "Applied Reservoir Engineering" Prentiss Hall, New York, NY.

Craig, Jr., F. F., *The Reservoir Engineering Aspects of Waterflooding,* Society of Petroleum Engineers of AIME, Dallas, TX., USA, 1971.

Craze and Buckley (October 1967) "A Statistical Study of Recovery Efficiency", *API Bull.*, D-14.

Dake, L.P., (1978) "Fundamentals of Reservoir Engineering" Elsevier Press, New York

Dandekar, A. Y.: *Petroleum Reservoir Rock and Fluid Properties*, CRC Press, Taylor & Francis Group, Boca Raton, FL., USA. 2006.

Datta-Gupta and King (2007).

Dykstra, H., and Parsons, R. L., "The Prediction of Oil Recovery by Water Flood," *In Secondary Recovery of Oil in the United States,* 2nd ed., pp. 160–174. API, 1950.

Elphick. J, (2010),

Evinger, H.H. and Muskat, M. (1942).: "Calculation of the Theoretical Productivity Factor, *Trans.* AIME

Fetkovich, M.J., "A Simplified Approach to Water Influx Calculations - Finite Aquifer Systems," *Jour. Pet. Tech.* (July 1971).

Fetkovich, M.J.: (1973) "The Isochronal Testing of Oil Wells," paper SPE 4529, presented at the 1973 Annual Fall Meeting, Houston, TX.

Fetkovich, M. J., Vienot, M.E., Bradley, M.D. and Kiesow, U.G.: "Decline Curve Analysis Using Type Curves: Case Histories," SPE 4529, Presented at the 48th Annual Fall Meeting, Las Vegas, NV, Sept 30-Oct 3, 1983. SPE Formation Evaluation (Dec. 1987) 637–656.

Fetkovich, M. J.: "Decline Curve Analysis Using Type Curves," Jour. Pet. Tech. (June 1980), 1065–1077

Gilbert, W.E.: (1954) "Flowing and Gas-Lift Well Performance," A.P.I. Drilling and Production Practices

Gross, S.J. (1985), A Numerical Simulation of Sandstone Reservoir Models" M. Sci., Texas A&M University.

Guerini, A.F.: (December 1991) "Criteria to Determine Gas Reserves Using Short Term Build-Up Tests," MS Graduate Thesis, Texas A&M University.

Guthrie, R.K. and Greenberger, H.M (1955): "The Use of Multiple-Correlation Analysis for Interpreting Petroleum Engineering Data," *Drilling. and Prod. Practices*, API 130–137.

Hadley, H.F., et.al., (2006) "Deposition Modeling of Champion Field, Bruni. Assessing the Impact of Reservoir Architecture on Secondary Recovery," SPE 101033, Presented at the Area Pacific Oil & Gas Conference, Adelaide, Australia, 11–13, September, 2006

Hall, H.N.: "How to Analyze Waterflood Injection Well Performance", *World Oil* (Oct. 1963) 128–130.

Hall, H.N.: (1953) "Compressibility of Reservoir Rocks," *Trans.* AIME 198, 309.

Hall, H: (October 1963) "The Hall Plot" World Oil, 128–130.

Havlena, D. and Odeh, A.S.: "The Material Balance as an Equation of a Straight Line,"*Jour. Pet. Tech.* (August 1963) 896–900.

Hawkins, M.F., Jr.: "A Note on the Skin Effect," *Trans.* AIME (1956) 207 356–357.

Holmgren, R. and Morse, R.A., Junior Members AIME, Stanolind Oil and Gas Co., Tulsa, Okla.
Petroleum Transactions, AIME, Vol. 192, 1951.

Poston, S.W., Laprea-Bigott, M. and Poe, B.D. (2019). *Analysis of Oil and Gas Production Performance*. Society of Petroleum Engineers, Richardson, TX.

Horner, D.R.: "Pressure Build-up in Wells," *Proc.* Third World Petroleum Congress. (1967)

http://www.eureka.slb.com/Files/Rooms/Wettability_White_Paper_Final_1120072.pdf IHS 2016 "Creating a Hall Plot Analysis," copyright 2017 HIS v3.

Jenkins, R.E.: "Typical Core Analysis of Different Formations - Chapter 27," *Petroleum Engineering Handbook* H.R. Bradley, ed. (1987) Society of Petroleum Engineers, Richardson, TX.

Katz, D.I., et.al., (1959) *Handbook of Natural Gas Engineering*." McGraw – Hill Publishing Co., New York, NY

Klins, M.A., and Majcher, M.W. (1989): "Inflow Performance Relationships for Damaged or Improved Wells Producing Under Solution-Gas Drive," paper SPE 19852 presented at the 1989 Fall Meeting, San Antonio, TX.

Klinkenberg, L.J.: "The Permeability of Porous Media to Liquids and Gases," *API Drill. Prod. Practice*, (1941) 200

Krueger, W.C. Jr.: (1968) "Depositional Environments of Sandstones as Interpreted from Electrical Measurements—An Introduction," *Trans.* Gulf Coast Assn. of Geol. Soc. XVII, 226–41.

Lang, W.M., Jr., "Correlation with Multiple Logs," The Log Analyst, Jan.–Feb. 1986

Latil, Marcel: *Enhanced Oil Recovery*, Gulf Publishing Co., Houston, TX (1980) p. 232.

LeBlanc, R.J. (1972): "Geometry of Sandstone Reservoir Bodies," in *Underground Waste Management and Environmental Implications,* (1972) Amer. Assoc. Pet. Geologists Memoir 18, 133–189.

Lee, W.J. (1982), "Well Tersting". SPE Textbook Series Vol. 1, Dallas, TX

Leverett, M.C., and Lewis, W.B.: "Steady Flow of Gas-Oil-Water Mixtures Through Unconsolidated Sand," AIME Trans. Pub. 1206 (1940). Mathews, C.S. and Russell, D.G."Pressure Buildup and Flow Tests in Wells" (1967) Society of Petrolem Engineers, Dallas, TX.

McCain, W.D. Jr.: *The Properties of Petroleum Fluids*, 2nd Ed. PennWell (1990).

McEwen, C.R., "Material Balance Calculations with Water Influx in the Presence of Uncertainty in Pressures," *Jour. Pet. Tech.* (June 1962).

Mihcakan, AltiNay, Ibrahim; "Hall Plot Analysis of a Water Injection Test Affected by Geothermal Reservoir Response", Proceedings World Geothermal Congress 2005.

Miller, H.C.: "Oil-Reservoir Behavior Based Upon Pressure-Production Data", U.S. Bureau of Mines Report of Investigations 3634, April 1942.

Mistrot, G.A.: "Reserves Classifications," *Jour. Pet. Tech.* (July 1987) 798–802.Mitchel, R.W. and Finch, "Water Quality Aspects of North Sea Injection Water", Jour. Pet. Tech (June 1981) 1141–1456.

Moses, P.L. "Engineering Applications of Phase Behavior of Crude Oil and Condensate Systems," Jour. Pet. Tech. (July 1986) 715–723.

Moses, P.L. and Donohoe, C.W. "Gas-Condensate Reservoirs - Chapter 39," in H.B. Bradley, ed. *Petroleum Engineering Handbook* (1987) Society of Petroleum Engineers, Richardson, TX.

Newman, G.H.: "Pore Volume Compressibility of Consolidated, Friable and Unconsolidated Reservoir Rocks Under Hydrostatic Loading," *Jour. Pet. Tech.* (February 1973) 129–134.

OGJ EOR Survey (2004)

Optimización de la Inyección de Agua Mediante Tecnologías Químicas para Mejoramiento del

Perfil de Producción e Inyección en el Área Este del Yacimiento SJN FUL 2 Campo El Furrial, Trabajo de Grado presentado por la Ingeniera de Petróleo Nellys Chek-Yan Chelhod y la Ingeniera Química Iris Margarita Pérez Amaro ante la Universidad Venezolana de los Hidrocarburos, 2017.

Piper, A.M., 1944, A graphic procedure in the geochemical interpretation of water analyses: American Geophysical Union Transactions, v. 25, p. 914–923.

Piper, L.D., McCain, W.D. and Corredor, J.H. "Compressibility Factors for Naturally Occurring Petroleum Gases," paper SPE 26668 presented at the Annual Technical Conference and Exhibition, Houston, TX, October 1993.

Pirson, S.J. (1965) "Handbook of Well Log Analysis" Penwell Publishing. Tulsa, Ok. P. 332

Poston, S.W. and Gross, S.J.: "Numerical Simulation of Sandstone Reservoir Models," SPE Reservoir Engineering (July 1986) 423–429.

Poston, S.W. and Berg, R.R. (1997) 'Overpressure Gas Reservoirs" Society of Petroleum Engineers, Richarson, TX.

Poston, S.W., Laprea-Bigott, M., Poe, B.L. (2019) "Analysis of Petroleum Production Decline Curves" Society of Petroleum Engineers,", Richardson, TX.

Poettman, F.H. and Beck, R.L.: (March 1963) "New Charts Developed to Predict Gas-Liquid Flow Through Tubing," *World Oil* 95–101.

Protos, N.E.: "Application of the Isochronal, Transient p/z Plotting Method for Determination of Original Gas In Place, to Low Permeability Reservoirs," (August 1991) M. Sc. Thesis, Texas A&M University.

Ramon, J.C. and Cross, J. (1997) "Characterization and Prediction of Reservoir Architecture and Petrophysical Properties in Fluvial, Channel Sandstones, Middle Magdalena Basic Colombia" (1997) Ciencia, Technologia & Futuro, V.1, Num. 3, Dec. 1997.

Ramey, J.J., Jr., and Cobb, W.M.: "A General Buildup Theory for a Well in a Closed Drainage Area," *Jour. Pet. Tech.* (Dec. 1971) 1493–1505.

Rankin, C.H.: "Estimating Ultimate Recovery", *The Petroleum Engineer*, (Nov. 1943) 72–74.

Ros, N.C.J.: (October 1961) "Simultaneous Flow of Gas and Liquid as Encountered in Well Tubing," *Jour. Pet. Tech.* (October 1961) 1037–1049.

Satter, A and Thakur, G: *Integrated Petroleum Reservoir Management – A Team Approach*, PennWell Books, Tulsa, OK. USA, 1994.

Schilthuis, R. J., "Active Oil and Reservoir Energy," *Trans.* AIME (1936).

Schlumberger, Dowell, A Classification of Water Problem Types, Jon Elphick.

Schlumberger, Middle East & Asia Reservoir Review (Number 5, 2004)

Schlumberger - Oil Field Review September, 2007 issue of Oilfield Review, Fundamentals of Wettability, p44–61.

Schlumberger –NExT - Network of Excellence Courses Notes – (2002, 2004, 2010)

Shams, M, El-Banbi, Ahmed H., and Khairy, M.: "Effect of Capillary Pressure on Reservoir Performance of Conventional and Naturally Fractured Reservoirs," paper SPE 164658 to be presented at the SPE North Africa Technical Conference and Exhibition, 15–17 April, 2013.

Slider, H.C.: *Practical Petroleum Reservoir Engineering Methods*, Petroleum Pub. Co., Tulsa, OK (1976) 70.

SPE Monograph Volume 11—The Design Engineering Aspects of Waterflooding by S.C. Rose, J.F. Buckwalter, and R.J. Woodhall (Rose et al. 1989)

SPE Reprint Series 2—Water Flooding (1959)—27 reprinted technical papers and 92 papers listed in the bibliography

SPE Reprint Series 2a—Waterflooding (1973)—35 reprinted technical papers and 271 papers listed in the bibliography

SPE Reprint Series 56—Waterflooding (2003)—25 reprinted technical papers and 961 papers listed in the bibliography

SPE-199121-MS, (2020).

SPE 23312 Produced Water Management

SPE 25394 Oil Recovery from Bottom Water reservoir

SPE 38798 Production Water Management

SPE 53254 Produced Water Management - Reservoir and Facilities Engineering Aspects

SPE 61180 Application of Reedbed Technology in Production Water Management

SPE 65162 Water Management in a Closed Loop - Problems and Solutions at Brage Field

SPE 66543 Produced Water Management Strategy with the Aid of Decision Analysis

SPE 70068 Conformance Water-Management Team Development

SPE 73853 Water to Value - Produced Water Management for Sustainable Field Development of Mature and Green Fields.

SPE 77963 and SPE 79012 New Insight into Water Control - State of the Art I

Standings, M.B. "A Pressure-Volume-Temperature Correlation for Mixtures of California Oils and Gases," Drill. and Prod. Prac., API (1947).

Standings, M.B. and Katz, D.L. "Density of Natural Gases," *Trans.*, AIME (1942) 146, 140–44.

Stewart, P.R,: "Evaluation of Individual Gas Well Reserves," *The Petroleum Engineer* (May 1960) 85–89.

Stiff, H.D., (1951) "The Interpretation of Natural Waters, Surface and Ground Water Environments," JPT, v.3 no. 10, p. 15–16.

Stiff, H.A., Jr., 1951, The interpretation of chemical water analysis by means of patterns: Journal of Petroleum Technology, v. 3, no. 10, p. 15–17.

Sullivan, S.A., Poston, S.W. and Piper, L.A.: (1988) Using Short-Term Pressure-Buildup Tests for Reserves Estimation in Tight-Gas Reservoirs," paper SPE 17707 presented at the 1988 SPE Gas Technology Symposium, Dallas, 13–15 June.

Sutton, R.P. (1985) "Pseudo-Critical Properties of Natural Gases," paper SPE 14265 presented at the 1985 Annual Technical Conference, Las Vegas, NV.

Thakur, G and Satter, A: *Integrated Waterflood Asset Management*, PennWell Books, Tulsa, OK. USA, 1998.

Tearpock, D.J. and Bischke, R.E. (1991) "Applied Subsurface Geological Mapping" Prentice Hall, Englewood Cliffs, New Jersey.

The Reservoir Engineering Aspects of Waterflooding, Second Edition

Tiab, D. and Donaldson, E.C. *Petrophysics: Theory and Practice of Measuring Reservoir Rock and Fluid Transport Properties*, (1996) Gulf Publishing Co, Houston, TX p. 706.

Uhri, D.C. and Blount, E.M.: (May 1982) "Pivot Method Quickly Predicts Well Performance," World Oil 53–64.

van der Waals, J.D. *Proc.*, Acad. Sci. Amsterdam (1901) 3. 515.

Vik, B., Djurhuus, K., Spildo, K., and Skauge, Characterization of Vuggy Carbonates" (2007) SPE 111434, (2007). Presented at Abu Dhabi, UAE, Oct. 28–31 - 2007.

Visher, F.S. (1984) *Exploration Stratigraphy* (1984) Pennwell Publishing Co.,Tulsa, OK p. 332.

Vogel, J. V.: (January 1968) "Inflow Performance Relationships for Solution Gas Drive Wells," *Jour. Pet. Tech.* 83–92.

Vu, T.: "Using Short Term Pressure Buildup Tests to Calculate Gas Reserves," (May 1990) M. Sci. Thesis, Texas A&M University.

Vu, T., Chen, H.Y. and Poston, S.W.: "Using Short Term Pressure Tests To Calculate Gas Reserves," SPE paper 20111 presented at 1990 Permian Basin Oil and Gas Recovery Conference, Midland, TX.

Van Everdingen, A.F. and Hurst, W., "The Application of the LaPlace Transformation to Flow Problems in Reservoirs," *Trans.* AIME 186 (1949). Whiting (1963) Class notes, Texas A&M University and Miller, H.C.: "Oil Reservoir Behavior Based Upon Pressure-Production Data," RI 3634 (April 1942) US Bureau of Mines. 36.

Wichert, E., and Aziz, D. "Compressibility Factor for Sour Natural Gases," *Cdn. J. Chem. Eng.* (1972) 49, 269–275.

Willhite, G. P., *Waterflooding,* Society of Petroleum Engineers, Dallas, TX., USA, 1986.

Zeito, G.A. (Oct. 1965): "Interbedding of Shale Breaks and Reservoir Heterogeneities," *Jour. Pet. Tech.* 77–83.

INDEX

www.ingramcontent.com/pod-product-compliance
Lightning Source LLC
Chambersburg PA
CBHW061157240926
51846CB00006B/1035
* 9 7 8 1 9 5 9 0 2 5 0 0 9 *